LECTURE NOTES ON
HUMAN PHYSIOLOGY

CHRIS NUTT   1st MB   (MBChB)
FACULTY OF MEDICINE
1992 — 1993 (1st YEAR)

UNIVERSITY OF DUNDEE.

CHRIS NUTT   2nd MB (MBChB)
FACULTY OF MEDICINE
1993-1994 (2nd YEAR)
UNIVERSITY OF DUNDES.

LECTURE NOTES ON
HUMAN PHYSIOLOGY

EDITED BY

John J. Bray PhD

Patricia A. Cragg PhD

Anthony D. C. Macknight PhD, MD

Roland G. Mills PhD

Douglass W. Taylor MD

all of the
Department of Physiology
University of Otago Medical School
Dunedin, New Zealand

SECOND EDITION

OXFORD

BLACKWELL SCIENTIFIC PUBLICATIONS

LONDON EDINBURGH BOSTON
MELBOURNE PARIS BERLIN VIENNA

© 1986, 1989 by
Blackwell Scientific Publications
Editorial Offices:
Osney Mead, Oxford OX2 0EL
25 John Street, London WC1N 2BL
23 Ainslie Place, Edinburgh EH3 6AJ
3 Cambridge Center, Cambridge
  Massachusetts 02142, USA
54 University Street, Carlton
  Victoria 3053, Australia

Other Editorial Offices:
Librairie Arnette SA
2, rue Casimir-Delavigne
75006 Paris
France

Blackwell Wissenschafts-Verlag
Meinekestrasse 4
D–1000 Berlin 15
Germany

Blackwell MZV
Feldgasse 13
A–1238 Wien
Austria

First published 1986
Reprinted 1987
Second edition 1989
Reprinted 1989, 1990, 1992

Typeset by William Clowes Limited,
Beccles and London
Printed and bound in Great Britain by
Mackays of Chatham PLC, Chatham, Kent

DISTRIBUTORS

Marston Book Services Ltd
PO Box 87
Oxford OX2 0DT
(*Orders:* Tel: 0865 791155
  Fax: 0865 791927
  Telex: 837515)

USA
Blackwell Scientific Publications, Inc.
3 Cambridge Center
Cambridge, MA 02142
(*Orders:* Tel: 800 759–6102
       617 225–0401)

Canada
Times Mirror Professional Publishing, Ltd
5240 Finch Avenue East
Scarborough, Ontario M1S 5A2
(*Orders:* Tel: 800 268–4178
       416 298–1588)

Australia
Blackwell Scientific Publications
(Australia) Pty Ltd
54 University Street
Carlton, Victoria 3053
(*Orders:* Tel: 03 347–0300)

British Library
Cataloguing in Publication Data

Lecture notes on human physiology.—2nd ed.
1. Man. Physiology
I. Bray, John J.
612

ISBN 0-632-02311-2

Contents

List of Contributors, vi

Preface to the Second Edition, vii

Preface to the First Edition, viii

1 Introduction to Physiology, 1

2 Nervous System, 34

3 Muscle, 95

4 Sensory Systems, 122

5 Special Senses, 146

6 Motor System, 189

7 Higher Nervous Functions, 223

8 Endocrine System, 250

9 Reproduction, 293

10 Blood, 314

11 Cardiovascular System, 371

12 Respiration, 465

13 Kidney, Water and Electrolytes, 512

14 Digestive System, 565

15 Energy Metabolism and Temperature Regulation, 610

16 Exercise, 626

Appendix: Units of Measurement, 635

Index, 637

List of Contributors

This book was written by members of the Department of Physiology, University of Otago Medical School, Dunedin, New Zealand.

Dr J. V. Allison BMedSc, MB, ChB, PhD
Chapters 8, 9, 10
Dr D.P.G. Bolton MA, MSc, BM, BCh, PhD, MRCP
Chapter 16
The *late* Mr K. Bradley MSc
Chapter 5
Dr J.J. Bray BAgrSc, PhD
Chapters 2, 8, 9
Dr P.A. Cragg BSc, PhD
Chapters 11, 12
Dr T.A. Day BSc, DipEd, PhD
Chapter 5
Dr E.R. Fawcett BSc, MB, ChB, FFARCS, FFARACS, DIC
Chapter 15
Associate Professor A.J. Harris BSc, PhD
Chapters 2, 3, 5
Dr W.G. Hopkins BSc, BA, MSc, PhD
Chapter 16
Professor J.I. Hubbard BMedSc, MA, DM, PhD, FRACP, FRSNZ
Chapters 2, 4, 6, 7
Dr E.M. Irvine BSc, MB, ChB, DA
Chapter 10
Associate Professor J.P. Leader BA, PhD
Chapter 13
Professor A.D.C. Macknight BMedSc, PhD, MD, FRSNZ
Chapters 1, 13, 14
Dr R.G. Mills BSc, PhD
Chapters 2, 3, 14
Professor J.R. Robinson MA, MD, PhD, ScD, FRACP, FRSNZ, FNZIC
Chapters 1, 13
Professor D.W. Taylor MB, ChB, MD, FRCPE
Chapters 7, 12
Dr P.M.E. Waite* BSc, PhD, MB, ChB
Chapters 4, 5
Illustrators: Mr R. W. McPhee, Mr N.E. Still

*Present address: School of Anatomy, The University of New South Wales, Kensington, NSW, Australia.

Preface to the Second Edition

We are flattered that a second edition of *Lecture Notes on Human Physiology* should be called for within two years of the appearance of the first.

Each chapter has been thoroughly revised. We started with the resolve that any additions should be accompanied by equivalent deletions but, as if by natural law, the former somewhat outweigh the latter. In particular the important topic of 'Blood' has been dealt with at considerably greater length. The subsection 'The immune system' has been written by Professor B. Heslop, Department of Surgery, and we are much indebted to her. We would also like to thank Professor R.D.H. Stewart, Department of Medicine, for help in revising Chapters 8 and 9, and Dr C.P. Bolter for assistance with Chapter 16.

The purpose of the book as an adjunct to the large textbooks remains unchanged, although we have to admit that for students revising for examinations we have in mind a somewhat longer period of time than the few nights that seemed to be envisaged by one reviewer!

Once again it has been a great pleasure to work with Blackwell Scientific Publications.

John J. Bray, Patricia A. Cragg, Anthony D.C. Macknight
Roland G. Mills, Douglass W. Taylor

Preface to the First Edition

Writers of textbooks, like Dr Johnson's lexicographer, are unwise if they aspire to praise and lucky indeed if they escape reproach. This is especially likely to be the case if the textbook attempts to cover in a rather short space, as befits the Lecture Notes series, a very big subject like physiology.

It has not been our aim to replace any of the large standard texts, but rather to produce a book that can be read through reasonably rapidly during, or in revising at the end of, the sort of course in human physiology that is commonly taught to a variety of students in pure science, medicine, dentistry, pharmacy and the like. To this end we have largely omitted detailed discussion of experimental evidence and have refrained from giving references to the original literature. We have also deliberately devoted but little space to applied physiology and to clinical illustration in the belief that these are commonly taught later in the course to students that require them.

The contents of the book largely represent what is taught in the Physiology Department of the University of Otago to students from five different faculties, most of whom study physiology for one year in their second year at university. While the individual chapters have naturally been written by specialists, all have been carefully read and fairly stringently edited by a small panel in an attempt to give a greater uniformity of style than is sometimes found in volumes of multiple authorship. The editors must therefore take the responsibility for the final text.

We are indebted to Professor J.R. Robinson, who began this book and continued to nurture it during his retirement. We also gratefully acknowledge the contributions of the other members of the Department of Physiology, University of Otago Medical School, and of our two illustrators. We wish to thank Mrs S.M. Keast for her patience and care in typing, proof-reading and indexing the manuscript. Mrs L.J. Donaldson for typing assistance and Drs J.G. Blackman and R.J. Harvey for their helpful comments. We thank those authors and publishers who generously gave us permission to reproduce figures or data, the sources of which are acknowledged in the legends. Finally, we are indebted to Mr Per Saugman and his colleagues at Blackwell Scientific Publications. Mr Saugman saw the need for such a book and gently nursed it to fruition.

John J. Bray, Patricia A. Cragg, Anthony D.C. Macknight
Roland G. Mills, Douglass W. Taylor

1 · Introduction to Physiology

1.1 Cells

1.2 Water in the body

1.3 Homeostatic mechanisms

1.4 Passive exchange of water and solutes
Diffusion
Exchange across capillary walls
Passive movement across
    plasma membranes

1.5 Energy-dependent transport across plasma membranes
Primary active transport
Nernst equation
Mechanism of active transport
Secondary active transport
Importance of active transport
Regulation of intracellular Ca^{2+} and pH
Endocytosis, exocytosis and transcytosis

1.6 Resting membrane potential

1.7 Cell volume regulation
Osmosis
Effective osmotic pressure
Net flow of water across membranes
Cell volume
Tonicity

1.8 Osmosis and ultrafiltration across capillary walls
Plasma colloid osmotic pressure
Starling equilibrium
Lymph

1.9 Movement of water and solutes across epithelia
Absorptive epithelia
Secretory epithelia

1.10 Appendix
Patch clamping
Gibbs–Donnan distribution

Physiology is the science of the functions and phenomena of living things. This book deals mainly with the physiology of mammals including man. Their bodies contain vast numbers of cells, which are the fundamental units of all living organisms.

1.1 Cells

Mammalian cells are very small, of the order of 10^{-5} m in diameter. This is about midway between men (of the order of 1 m) and atoms (10^{-10} m). A man has about as many cells—about 10^{14}—as a cell has molecules.

The detailed structure of cells was not revealed by the light microscope, with its limit of resolution of $0.2\ \mu$m. The electron microscope, however, with resolution down towards 1 nm (10^{-9} m) has revealed a complex ultrastructure. In life this is not static; it is maintained by continually assimilating matter from the environment. The cells, even of terrestrial animals, are tiny aquatic organisms—open systems that, like flames and waterfalls (and unlike crystals), need continuous supplies of matter and energy to maintain their form. The details of metabolism are dealt with by biochemistry. Physiology deals rather with getting the supplies to the cells at suitable rates and removing waste products. All the substances concerned travel and react in solution in water, which is the main bulk constituent of all living systems.

1.2 Water in the body

The biological importance of water depends upon a number of peculiar properties.

1 It is a liquid at 'ordinary' temperatures.
2 It has a large heat of fusion.
3 It has a large heat capacity.
4 It has a large heat of vaporization.

The first of these is explained by hydrogen bonding; the others by the energy required to break bonds during melting, warming and vaporization.

5 It has a maximum density at 4°C.
6 It has a large dielectric constant, which reduces electrostatic forces 80 times and makes water a superb solvent for ionic compounds. Moreover, water dipoles are strongly attracted to dissolved ions and to charged surfaces, coating these with relatively immobile layers of water molecules which greatly modify the properties of ions in solution.

About 60%, or two-thirds, of body weight is due to water. The 'average' 70 kg man has about 42 1. This is shared between the cells and the extracellular fluids which surround and supply them. Difficulties in measuring the extracellular fluid volume have resulted in a variety of published values for cellular and extracellular water contents. Older figures, based on inadequate equilibration of markers throughout the extracellular volume, are intracellular water 28 1 and extracellular water 14 1. A better approximation, which allows for the extracellular water in dense connective tissue and bone, is intracellular water 23 1 and extracellular water 19 1. Three litres of the extracellular water is in the *blood plasma*, 15 1 (*interstitial fluid* including lymph) provides an aquatic habitat for the cells and the remaining litre represents the total of the *transcellular fluids* (cerebrospinal fluid (CSF), ocular, pleural, peritoneal and synovial fluids).

The intracellular and the extracellular water have different solutes dissolved in them and so constitute two distinct kinds of fluid, the *intracellular fluid* (ICF) and the *extracellular fluid* (ECF). Muscle cell fluid and plasma are typical of ICF and ECF, respectively, although the composition of ECF varies to some extent in different parts of the body. Their approximate compositions in mmol per kg of water are shown in Table 1.1. Note that the ICF has mainly *potassium* with organic anions; ECF has mainly *sodium* and *chloride*, rather like diluted sea water. Note too that plasma constituents are commonly

Table 1.1. Muscle cell fluid and blood plasma as typical intracellular and extracellular fluids.

Muscle (ICF) mmol kg^{-1}H$_2$O				Plasma (ECF) mmol kg^{-1}H$_2$O			
K$^+$	150	Org P$^-$	130	Na$^+$	150	Cl$^-$	110
Na$^+$	10	HCO^{-3}	10	K$^+$	5	HCO$_3^-$	27
Mg^{2+}	10	Cl$^-$	5	Ca^{2+}	2	Org$^-$	5
Ca^{2+}	10^{-4}	Prot^{17-}	2	Mg^{2+}	1	H$_2$PO$_4^-$	2
						Prot^{17-}	1

pH 7·1
Osmolality: 285 mosmol kg^{-1} H$_2$O

pH 7·4
Osmolality: 285 mosmol kg^{-1} H$_2$O

expressed per litre of plasma rather than per kg of water and, because protein occupies a finite volume, the values are somewhat lower than the above values.

The extracellular fluid is a middle-man fluid, a medium for all exchanges between a cell and any other cell or the external environment. The famous French physiologist Claude Bernard (1813–1878) called the ECF an *internal environment* in which the cells live a secluded life protected from the vicissitudes of a sometimes hostile external environment. The properties and composition of the internal environment have to be kept constant. This constancy, necessary for the well-being of the cells, is maintained by *homeostatic mechanisms* controlling temperature, osmotic pressure, pH and composition of the ECF. Much of physiology is the story of these homeostatic mechanisms.

1.3 Homeostatic mechanisms

A homeostatic mechanism is a regulating mechanism triggered by alteration in some physiological property or quantity, which acts to produce a compensating change in the opposite direction.

Minimum requirements for homeostatic mechanisms are as follows.

1 *Detectors*, often specialized to respond to particular variables.
2 *Effectors*, e.g. muscles and glands. The cardiovascular, respiratory, renal and alimentary systems can be considered as effector systems subserving homeostasis.
3 *Coordinating and integrating mechanisms*, linking **1** to **2**. These are nervous and hormonal.

Nervous system. In essence this consists of afferent nerve fibres linking receptors and efferent nerve fibres linking effectors to coordinating systems in the brain and spinal cord. There are two major subdivisions: *the somatic nervous system* uses striated, skeletal muscles as effectors for purposive behaviour and reflex actions and adjustments; *the autonomic nervous system* sends its efferents to glands, the heart, and smooth muscle in hollow organs and blood vessels.

Endocrine system. This comprises a group of glands which have no ducts, but secrete into the blood *hormones* which affect the function of cells all over the body. The actions are generally slower and less sharply localized than those of the nervous system; they are, however, under the control of the nervous system through hypothalamic and pituitary hormones which influence the other endocrine glands.

1.4 Passive exchange of water and solutes

The heart and blood vessels comprise a great transport system which carries water and solutes including hormones and gases (O_2 and CO_2) throughout the body, and helps to keep the ECF evenly mixed. Circulating blood does not actually reach cells. In tissues, *capillaries* bring blood within 5–10 μm of most cells. To get from blood to the interior of a cell, solutes like glucose and oxygen must:

1 cross the capillary wall;
2 cross a layer of interstitial fluid between capillary and cell; and
3 cross the plasma membrane, which separates the ICF from the ECF.

The second is the simplest. Its mechanism is known as *diffusion*.

Diffusion

This occurs as the spontaneous result of random thermal motions which tend to disperse accumulations of molecules and make concentration uniform. For example, water floating on copper sulphate solution shows *two* processes of diffusion at once—water diffusing into the solution, and copper sulphate diffusing into the water. Movements of molecules through the solution are called *fluxes*, not flow, because the liquid as a whole does not move. *Net flux* of each species is down its own gradient from higher concentration to lower. The rate of net flux for uncharged molecules, that is, the amount moving per unit time, is given by *Fick's law*:

$$\frac{dn}{dt} = -DA\frac{dc}{dx}.$$

Here dn/dt is the number of molecules or mol per unit time diffusing in the x direction across an area A at right angles to the x axis and dc/dx is the concentration gradient (the rate at which concentration increases with distance along the x axis). D, the diffusion coefficient for each substance, depends upon its relative molecular mass (M_r), the temperature and the viscosity of the solvent.

The minus sign is needed because net transport is down the gradient, from higher to lower concentration and is therefore *'downhill' transport*, also called *passive transport* because no energy is needed beyond that inherent in the gradient. The driving force for diffusion is the gradient of concentration of the diffusing substance, and Fick's law states that the rate of transport is proportional to that driving force. This leads to automatic adjustment of rate of transport to demand when, for example, a cell uses O_2 or glucose faster and so increases the gradient from capillary to cell.

Rate of diffusion can also mean distance covered in unit time. The theory was developed by Einstein in 1905 for Brownian movement considered as a 'random walk'. Distance is proportional to the square root of the number of steps, hence to the square root of time. This means it takes four times as long to go twice as far, but half the distance is covered in a quarter of the time. Diffusion is therefore *very rapid over short distances*. Jacobs in 1935 calculated the times taken for 99% equilibration at various distances from a plane where a constant concentration was maintained in water at room temperature (Table 1.2). *Note great speeds over distances of the order of cellular dimensions.*

Table 1.2. Time taken to reach 99% equilibrium by diffusion of solute from a plane boundary in water at room temperature (Jacobs 1935).*

Distance from boundary	Time
10 cm	53 days
1 cm	12·8 h
1 mm	7·6 min
100 μm	4·56 s
10 μm	0·0456 s
1 μm	0·000456 s
0·1 μm	0·00000456 s

*(From Robinson, J. R. (1981) *A Prelude to Physiology*, p. 26. Blackwell Scientific Publications, Oxford.)

Exchange across capillary walls

Capillaries are minute vessels (diameter about 10 μm) specially adapted for the rapid exchange of water and solutes. Their walls are composed of a single layer of endothelial cells about 1 μm in thickness. In contrast to the plasma membrane, in most tissues the capillary wall is a very leaky membrane for it lets substances through whose M_r are less than about 70 000. These include practically all the solutes in the plasma except the plasma proteins. But capillary permeability does differ in different tissues, and in liver, for example, proteins pass through to a greater extent than elsewhere, while, in brain, movement of water-soluble solutes is markedly retarded. These differences reflect the organization of the capillary wall. In liver, obvious gaps between

Chapter 1

adjacent endothelial cells are seen in electron micrographs, whereas in most brain capillaries adjacent cells are held firmly together by *tight* junctions. In some specialized capillaries in the gut and kidney (renal glomerulus), exchange of solutes is facilitated by *fenestrations*—areas within the endothelial cells where little or no cytoplasm separates the plasma membranes on the two surfaces of the cell. In most vascular beds, adjacent endothelial cells are attached to each other at their margins though the attachment offers relatively little resistance to solute exchange. In addition, a continuous *basement membrane* encircles the periluminal surface of all capillaries providing additional support to the endothelial cells. It is now realized that, like those of epithelial cells (p. 24), the plasma membranes of endothelial cells can contain a variety of specific pathways through which ions and other water-soluble solutes can be transported between blood and interstitial fluid. These pathways have been studied particularly in brain capillaries where, because of the tightness of the junctions holding adjacent cells together, the cellular pathway provides the dominant route for transendothelial movements of water-soluble solutes. In other capillary beds, however, much of the exchange of these solutes and of water occurs by diffusion between the cells (the paracellular pathway). Proteins and other macromolecules may also be transported across endothelial cells by a process of transcytosis (p. 15). In contrast, lipid-soluble substances including O_2 and CO_2 diffuse directly across the plasma membranes of endothelial cells.

Passive movement across plasma membranes

Cell membranes (about 7·5 nm in thickness) are thought to be composed of a mosaic of globular proteins embedded in a lipid matrix (Fig. 1.1). The lipids are arranged as a bilayer with their hydrophilic ends oriented towards the outside of the membrane. It is envisaged that both the protein and lipid molecules are free to move in the lateral plane of the membrane—hence the name *'fluid mosaic model'*. Because of the nature of the lipid bilayer it is not surprising that the plasma membrane presents a formidable barrier to diffusing solutes. Jacobs in 1952 calculated the times required for some diffusing solutes to reach 90% equilibrium in a region of water the same size as a red blood cell and compared these with actual times for red blood cells (Table 1.3). Permeability of erythrocytes' membranes is therefore both *low* and extremely *selective* even for solutes of low M_r.

In order to compare substances and cells, Fick's law of diffusion can be written for finite thickness and finite difference of concentration as:

$$\frac{\mathrm{d}n}{\mathrm{d}t} = -D'A\frac{\Delta c}{\Delta x}.$$

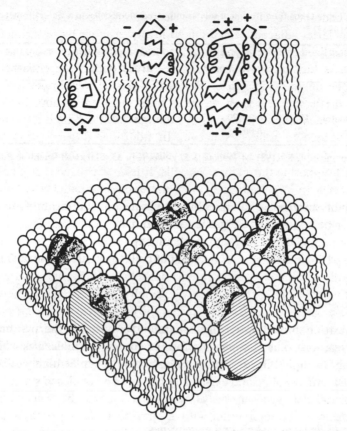

Fig. 1.1. The fluid mosaic model of the cell membrane in cross-section (top) and in three dimensions (below). The globular proteins embedded in the lipid bilayer may be situated on one side of the membrane or extend through it. (From Singer, S. J. & Nicolson, G. L. (1972) *Science* **175**, 720–731.)

Here D' is a restricted diffusion coefficient for diffusion within the membrane. In practice D', A and Δx are often unknown; they can, however, be incorporated into an empirical permeability coefficient $P' = D'A/\Delta x$, so that for diffusion into the cell:

$$\frac{\mathrm{d}n}{\mathrm{d}t} = -P'\Delta c$$

$$= P'(c_{\text{outside}} - c_{\text{inside}}).$$

Generally, though there are many exceptions, most plasma membranes are more permeable to water than to solutes, and, among solutes, far more permeable to gases, organic compounds and small anions than to cations. The orders of magnitude of permeabilities are related as follows:

Table 1.3. Times taken for diffusion of solutes into a red blood cell and a water droplet of similar size (Jacobs 1952).*

Substance	M_r	Calculated time for water droplet	Actual time for red blood cell	Slowing by membrane
Urea	60	0·00035 s	0·5 s	10^3
Glycerol	92	0·00049 s	67 s	10^5
Mannitol	182	0·00070 s	>2 days	10^8
Sucrose	342	0·00095 s	∞	∞

*(From Robinson, J. R. (1981) *A Prelude to Physiology*, p. 53. Blackwell Scientific Publications, Oxford.)

Permeability to water	:	Permeability to gases, organic compounds and small anions	:	Permeability to cations
1	:	0·1 to 0·0001	:	0·0001 to 0·00000001

This wide range suggests that there is more than one mechanism of permeation. Three possible mechanisms may be mentioned.

1 Permeation through water-filled channels (or pores) in the membrane, e.g. water, urea, ions. The channels are formed by protein molecules which span the membrane lipid bilayer. An individual channel can be highly selective for a particular ion or solute and exists in either an open or closed state. Channels can open and close spontaneously (*leak channels*) or can be opened and closed (*gated channels*) by interaction with a specific molecule or by changes in membrane potential. Thus we speak of ligand-gated and voltage-gated channels which are discussed more fully in Chapter 2 (p. 34). When open, the rate of movement of a solute can be some 10^6–10^9 ions s^{-1}, a transport rate much greater than the highest known catalytic rates of enzymes or carriers, and comparable to the rates for free diffusion of ions in aqueous media. One way in which the permeability of the membrane may be regulated is by the insertion or removal of channels. The properties of channels are rapidly becoming known through a variety of powerful new techniques, including patch-clamping (Appendix, p. 29), the use of vesicles formed from cell membranes and the tools of molecular biology. The most studied and best characterized channel is the acetylcholine receptor channel found in skeletal neuromuscular junctions (Fig. 1.2).

2 Permeation by dissolving and diffusing in the membrane lipid, e.g. O_2, CO_2, steroid hormones. In so far as cell membranes offer a fairly complete layer of lipid between the intracellular and extracellular fluids, this mechanism should favour lipid-soluble substances of low M_r. For such substances there is a strong correlation between permeability and oil–water partition coefficients which measure the tendency of a solute to move out of an aqueous into a lipid

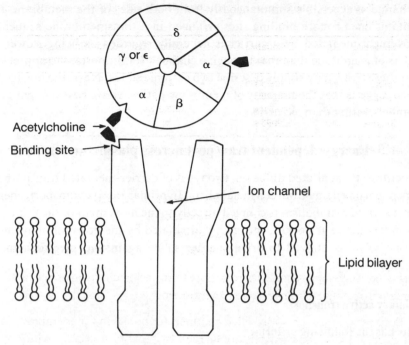

Fig. 1.2. Diagram of the acetylcholine receptor of skeletal muscle showing the subunits (top view), the binding sites and the ion channel. (Modified from Changeaux, J.-P., Devillers-Thiery, A. & Chemouilli, P. (1984) *Science* **225**, 1335–1345.)

environment. The driving force for permeation through the lipid bilayer should therefore be increased roughly in proportion to the membrane–water partition coefficient.

3 Permeation by temporary combination with a membrane component. Traditionally, this has been referred to as *carrier-mediated transport*. Carrier-mediated transport shows (a) specificity, (b) saturation kinetics, (c) competition between similar molecular species, (d) inhibition and (e) a high temperature coefficient. These are like properties of reactions catalysed by enzymes. Carrier-mediated transport can be subdivided into *facilitated diffusion*, where the net movement is passive and driven by the electrochemical potential gradient (e.g. uncoupled entry of Na^+ into cells), and *primary* or *secondary active transport* (discussed below).

In many cases of carrier-mediated transport, it is now realized that movement occurs through channels or pores rather than on carriers. Both channels and carriers are protein molecules which span the membrane lipid bilayer. Membrane carriers can be distinguished from channels in that a carrier mechanism implies a binding site that is exposed sequentially to one side or the other of the membrane (but never to both simultaneously), whereas a channel is accessible simultaneously from both sides of the membrane and contains one or more binding sites arranged in a transmembrane sequence. Movement mediated by a carrier is generally regarded as being slower by orders of magnitude than that through a channel. It is important to appreciate that these terms may simply describe two extreme states. Nevertheless, recent work suggests that the majority of membrane pathways are best thought of as channels, rather than as carriers.

1.5 Energy-dependent transport across plasma membranes

In contrast to facilitated diffusion, two types of carrier-mediated transport are dependent on energy from cell metabolism. In primary active transport, energy derived from metabolism is coupled directly to the movement of the solute. In secondary active transport, the energy dissipated by the downhill movement of one solute is utilized to drive another solute against its electrochemical gradient.

Primary active transport

This has the following criteria:

1 it is coupled directly to a continuous supply of energy;
2 it is independent of the downhill movement of any other solute or of water;
3 it is uphill, which needs to be defined more precisely.

For unchanged molecules, uphill is *from lower to higher concentration*, i.e. against a chemical concentration gradient. For ions, uphill is *from lower to higher electrochemical potential* ($\bar{\mu}$) since there is an electrical potential difference ($\Delta\psi$) across the cell membranes as well as a chemical concentration gradient. The difference in electrochemical potential ($\Delta\bar{\mu}$) in joules per mol for an ion of valency z between solutions dilute enough for us to use concentrations instead of activities is:

$$\Delta\bar{\mu} = \bar{\mu}_2 - \bar{\mu}_1$$

$$= zF\Delta\psi + RT\ln\frac{c_2}{c_1}$$

i.e. electrical work per mol + chemical work per mol, where F is Faraday's constant (96 500 coulombs mol^{-1}), R is the gas constant (8·314 joules K^{-1} mol^{-1}), T is the absolute temperature (K), $\Delta\psi = \psi_2 - \psi_1$ in volts and c_1 and c_2 are the concentrations in mol per litre in solutions 1 and 2, respectively. Dividing by zF gives a useful measure in volts:

$$\frac{\Delta\bar{\mu}}{zF} = \Delta\psi + \frac{RT}{zF}\ln\frac{c_2}{c_1}.$$

The right-hand side of the equation is [the potential difference between 2 and 1] plus [the potential difference required to balance the concentration ratio].

Nernst equation

At equilibrium there is no electrochemical potential gradient across the membrane and $\Delta\bar{\mu}=0$. Therefore the electrical potential difference ($\Delta\psi$) across the membrane at equilibrium, i.e. the equilibrium potential (E), is given by:

$$E = \frac{RT}{zF}\ln\frac{c_1}{c_2}.$$

This is the *Nernst equation* introduced by the German theoretical chemist W. Nernst (1864–1941). In effect it states either the maximum electromotive force that can be generated by a given ratio of concentrations of an ion, or the maximum ratio of concentration that can be sustained by a potential difference imposed from an external source. For use it is often convenient to put in actual values for the constants and to convert the natural logarithms to $\log_{10}$. This gives for monovalent ions at 37°C:

$$E = \pm 61 \log\frac{c_1}{c_2}$$

where E has units of mV and the sign is positive for cations and negative for anions.

Mechanism of active transport

At the molecular level primary active transport involves proteins which probably have their affinity altered by reacting with adenosine triphosphate (ATP). As well as the Na^+–K^+ pump there is evidence for plasma membrane Ca^{2+} and H^+ pumps which involve a Ca^{2+}-ATPase and a H^+–K^+-ATPase, respectively.

An example of active transport is provided by the Na^+–K^+ pump. Cells accumulate K^+ and keep their intracellular concentration of Na^+ low although they live in an ECF which is rich in Na^+ and poor in K^+. The membranes are permeable to both ions, and tracers show continuous exchange of both ions between ICF and ECF. To maintain their composition in a steady state, cells need energy from metabolism to expel Na^+ which diffuses in and to take up K^+ which diffuses out. Poorly metabolizing cells gain Na^+ and lose K^+ down their concentration gradients. If the cells recover, they take up K^+ and expel Na^+, both *against* gradients. These movements of Na^+ and K^+ are coupled directly to the consumption of metabolic energy in the form of ATP. The protein which moves the ions across the membrane is itself an ATPase. Its enzymatic function, which requires Mg^{2+}, is activated by cell Na^+ and extracellular K^+. Because it requires both ions for normal activation, it is often called the Na^+–K^+-ATPase. It is inhibited by cardiac glycosides, of which ouabain is much used in experimental studies on Na^+ and K^+ transport. Though the protein has been isolated and partially purified, there is still much controversy about its precise properties, the sequence of its reactions and the stoichiometry of the transport process, though it is generally assumed that 3 Na^+ are extruded from the cell and 2 K^+ taken up with each cycle.

Secondary active transport

Inasmuch as the Na^+–K^+ pump maintains the low cellular Na^+ concentration and the membrane potential is negative on the inside, there is a favourable electrochemical potential gradient for Na^+ entry to the cells. The energy in this gradient can be used to drive the net movements of other solutes against their electrochemical potential gradient by coupling their movement across the membrane to that of the passive downhill movement of Na^+. If both the Na^+ and the other solute move in the same direction this is termed *co-transport*. Examples include the coupled entry of glucose with Na^+ (Fig. 1.3) and amino acids with Na^+ into small intestinal epithelial cells from the gut lumen, the movement of glucose and amino acids being against their own electrochemical potential gradients. If the Na^+ and the other solute move in opposite directions across the membrane this is termed *counter-transport*. An example is provided by the coupling of Ca^{2+} extrusion from cells with Na^+ entry to cells (Fig. 1.3). Co- and counter-transport are sometimes referred to as symport and antiport, respectively. Note that metabolic energy is not coupled *directly* to these movements. Instead, some of the energy inherent in the Na^+ gradient across the plasma membrane generated by the Na^+–K^+ pump is dissipated by the coupled flow of Na^+ with accompanying solute across the membrane.

Importance of active transport

1 In *all cells* active transport is important for maintaining normal ionic concentrations, in particular of K^+ and Ca^{2+}, which are essential for the regulation of many intracellular activities.
2 Active transport is important for moving ions, water and other substances that accompany ions by sharing common carriers (i.e. co- and counter-transport) across plasma membranes, particularly in kidney, stomach and intestine.
3 Active transport maintains the gradients of ionic concentration which are

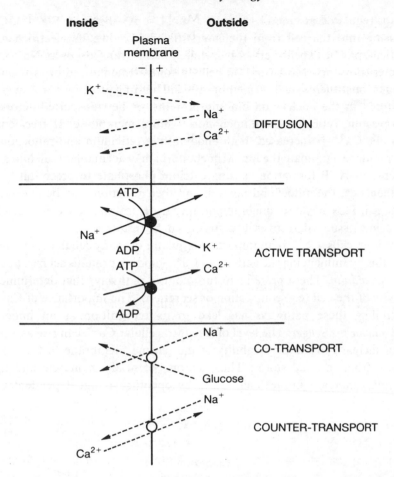

Fig. 1.3. Mechanisms of movement across plasma membranes. The closed circles represent active transport mechanisms; the open circles represent co- and counter-transport mechanisms.

the basis of *resting membrane potentials* and *action potentials* used for signalling in nerves and for the activation of muscles.

4 The active extrusion of Na^+ is important for regulating cellular volume.

Regulation of intracellular Ca^{2+} and pH

Though present in the cytosol at free concentrations (or, more correctly, activities) in the nmolar range, Ca^{2+} and H^+ play important roles in the regulation of a variety of cell functions and so their concentrations need to be controlled precisely.

Intracellular Ca^{2+}. Cellular membranes generally have very low Ca^{2+} permeabilities and the concentration of free Ca^{2+} in the cytoplasm of inactive cells is regulated at around 10^{-7}–10^{-8} mol l^{-1}. As shown in Fig. 1.4, Ca^{2+} diffusing into the cell from the interstitial fluid down its electrochemical gradient may be expelled either by primary active transport or by Na^+–Ca^{2+} counter-transport. Most Ca^{2+} in cells is compartmentalized or is bound to cellular constituents, such as membranes and binding proteins. Ca^{2+} may also be stored in the nucleus on binding proteins, or be transported across the endoplasmic reticulum or mitochondrial inner membrane. If free ionized cytosolic Ca^{2+} is increased, both endoplasmic reticulum and mitochondria will rapidly accumulate the ion. Mitochondria may accumulate calcium at the expense of ATP formation causing calcium phosphate to precipitate. As a consequence, the mitochondria swell and their function may be irreversibly damaged. This may be of major importance in determining the extent to which ischaemic tissue will recover if perfusion can be restored.

Within cells, Ca^{2+} is an important regulator of many cellular functions. It does this by influencing the activity of Ca^{2+}-binding proteins such as *troponin* and *calmodulin*. These proteins in turn control pathways that determine the activity of the cell (e.g. contraction or secretion). The importance of Ca^{2+} in controlling these pathways has lead to its recognition as an important *intracellular messenger*. The level of free intracellular Ca^{2+} can rise as a result of an increase in the permeability of the plasma membrane to Ca^{2+} or its release from internal stores. Thus, action potentials in muscle and nerve terminals increase intracellular Ca^{2+} by opening voltage-dependent Ca^{2+}

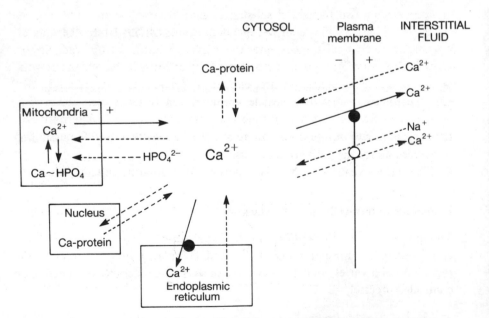

Fig. 1.4. Control of intracellular Ca^{2+}. The filled circles represent active transport; the open circle represents counter-transport.

channels. Many neurotransmitters and hormones also act by opening ligand-dependent Ca^{2+} channels (either directly or indirectly via cyclic AMP) or by releasing Ca^{2+} from intracellular stores, such as the endoplasmic reticulum (via inositol trisphosphate). (See p. 256 for further discussion.)

Intracellular pH. Many cell functions (e.g. enzyme activities, growth) are also influenced by cell pH, and changes in H^+ activity can influence both the activity of cytosolic Ca^{2+} and the levels of cyclic AMP. These in turn influence a variety of specific cellular functions. Since H^+ ions are formed continuously as a consequence of cell metabolism, cells must be able to buffer H^+ and remove them to the interstitial fluid. In general, mammalian cell pH is approximately 7·0–7·2, i.e. slightly more acid than the plasma pH of 7·4. This is considerably more alkaline than the pH of about 6·5 predicted from the Nernst equation if H^+ ions were distributed across the plasma membrane in chemical equilibrium. A major contribution to the removal of H^+ from the cell is Na^+–H^+ counter-transport. In addition, some specialized epithelial cells (e.g. the intercalated cells in the collecting ducts of the kidney) also contain a H^+-ATPase. Of equal importance in cell pH regulation is HCO_3^- transport for which a variety of pathways have been identified (e.g. Cl^-–HCO_3^- counter-transport and Na^+–HCO_3^- co-transport).

Endocytosis, exocytosis and transcytosis

Another mechanism by which substances may be transported into cells is *endocytosis* (Fig. 1.5). This energy-dependent process differs from other modes of transport in that substances enter the cell by inclusion within vacuoles or vesicles. Such membrane-bound structures arise by invagination of the plasma membrane. This invagination first encloses extracellular material and then seals, forming a vacuole or vesicle within the cell. These may fuse with lysosomes where they, including their contents, are degraded. Alternatively, the vesicular membrane may be recycled. Endocytosis is of particular

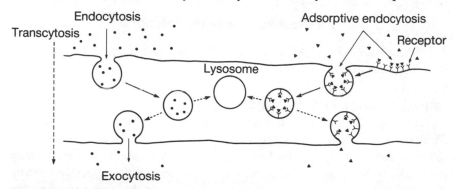

Fig. 1.5. Illustration of endocytosis, exocytosis and transcytosis.

importance for the uptake into cells of substances of large M_r, such as proteins, which do not otherwise cross plasma membranes. For example, sensory and sympathetic neurones depend for their survival and development on the protein nerve growth factor, secreted by their target tissues and taken up by endocytosis at their axon terminals. This uptake is a high-affinity mechanism involving specific membrane receptors and is known as *adsorptive endocytosis*. There also exists a low-affinity non-specific mechanism of endocytosis at nerve terminals that permits uptake of exogenous proteins, e.g. horseradish peroxidase.

Pinocytosis and phagocytosis are forms of endocytosis. *Pinocytosis* ('cell drinking') refers to the endocytotic uptake of soluble materials into cells, whereas *phagocytosis* ('cell eating') refers to the engulfing of particulate matter, e.g. bacteria and viruses, by neutrophils and macrophages.

Exocytosis is the reverse of endocytosis. Here substances are released from cells into the extracellular fluid by fusion of intracellular vesicles with the plasma membrane. This results in an increase in the area of the plasma membrane which is then reduced by endocytosis giving rise to the so-called 'coated' vesicles in the cytoplasm. Important examples of exocytosis include the release of peptide hormones from endocrine glands, of enzyme precursors from exocrine glands in the gut, and of neurotransmitters from nerve terminals. These processes appear to involve contractile proteins associated with the plasma membrane and are triggered by an influx of Ca^{2+} ions in response to specific stimuli.

Transcytosis is a combination of endocytosis occurring at one membrane of a cell (endothelial or epithelial) and exocytosis occurring at the other membrane (Fig. 1.5). Thus macromolecules can be transferred from lumen to interstitial fluid through a cellular pathway without degradation. A variety of plasma proteins (e.g. albumin, low-density lipoprotein, transferrin) can cross endothelial cells in this way.

1.6 Resting membrane potential

A difference of electrical potential is found across plasma membranes between ICF and ECF. This difference, called the *resting membrane potential* (RMP), has a magnitude of up to -90 mV with the internal electrode being negative with respect to that in the bathing fluid.

The RMP is considered to be predominantly a *diffusion potential* arising from the ionic concentration gradients across the membrane which has a selective permeability to ions. Generally plasma membranes are much more permeable to K^+ than to Na^+. Hence K^+ ions tend to diffuse out down their concentration gradient faster than Na^+ ions diffuse in. K^+ ions carry positive charge out with them and leave the inside of the fibre negatively charged. If a simple membrane were permeable only to K^+ this would go on until a sufficient potential difference was built up for the rate of re-entry due to the electrical gradient to balance the rate of loss down the concentration gradient. There would then be a dynamic equilibrium. The potential difference would be the *equilibrium potential* for K^+ (E_K), with the magnitude given by the Nernst equation. The equilibrium potential in mV for monovalent ions at 37°C is given by:

$$E \text{ (inside)} = \pm 61 \log \frac{c_{outside}}{c_{inside}}$$

where the sign is positive for cations and negative for anions. Therefore E_K (inside) is $61 \log 5/150 = 61 \times -1.48 = -90$ mV. However, this is somewhat larger than the measured RMP of most cells, namely -70 to -80 mV. The reason for this is that the plasma membrane is not totally impermeable to Na^+, and Na^+ ions diffusing in down their gradient carry some positive charge into the cell and thus reduce the actual RMP below E_K.

The contribution of the diffusion of other ions to the RMP is expressed in the Goldman equation:

$$\Delta\psi = \frac{RT}{F} \ln \frac{P_K K_o + P_{Na} Na_o + P_{Cl} Cl_i}{P_K K_i + P_{Na} Na_i + P_{Cl} Cl_o}$$

where $\Delta\psi$ is the membrane potential, P is the membrane permeability to the ion denoted by its subscript, o indicates the outside concentration and i the inside concentration. Note that because Cl^- is an anion the Cl^- concentrations are reversed.

A simpler version of the above equation, which assumes no net Cl^- diffusion, is the Hodgkin–Katz equation:

$$\Delta\psi = \frac{RT}{F} \ln \frac{K_o + b Na_o}{K_i + b Na_i}$$

where b is the ratio P_{Na}/P_K and is taken to be 0.01 for the resting axon. The equation predicts $\Delta\psi = 61 \log(5 + 1.5)/(150 + 0.1) = 61 \times -1.36 = -83$ mV. Good agreement of this predicted value with the measured RMP supports the idea that the RMP is largely a *potassium diffusion potential* modified by the membrane's small permeability to Na^+ ions. This is confirmed by the fact that the measured RMP in isolated fibres varies as the Hodgkin–Katz equation predicts if internal or external K^+ concentration is changed experimentally. Note that if we calculate E_{Cl} using the Nernst equation it is -82 mV, which is not significantly different from the measured RMP. Since Cl^- ions are apparently in electrochemical equilibrium, the omission of Cl^- in the Hodgkin–Katz equation is justified.

Rather than writing these equations in terms of ionic permeabilities, it is often convenient to express the membrane potential in terms of conductances (*G*) because the electrical measurements required are easier to perform experimentally. The membrane conductance (ohm^{-1} or siemens) to an ion differs from the membrane permeability (cm s^{-1}) but, as an approximation, it is often used to represent ionic permeability (p. 89). It can be shown that if the total current across the membrane is zero as it must be in the steady state, the membrane potential is given by:

$$\Delta\psi = \frac{E_K G_K + E_{Na} G_{Na} + E_{Cl} G_{Cl}}{G_K + G_{Na} + G_{Cl}}$$

where *E* represents the equilibrium potential and *G* the conductance of the membrane to ions indicated by the subscripts.

A pump which moved one K$^+$ in for each Na$^+$ out would carry no net charge and, therefore, no current across the membrane. However, it is now realized that the pump is probably not electrically neutral but transfers three Na$^+$ out for every two K$^+$ carried in. Thus current is generated by the pump which is therefore termed *rheogenic* (a less satisfactory term is electrogenic). However, provided that the membrane is much more permeable to K$^+$ than to Na$^+$, this rheogenic ion transport contributes relatively little to the resting membrane potential which is dominated by the potassium diffusion potential. To take account of this rheogenic transport, the Hodgkin–Katz equation can be modified by introducing the term *n*, where *n* is the coupling ratio for the pump (i.e. number of K$^+$ transported to Na$^+$ transported in each pump cycle):

$$\Delta\psi = \frac{RT}{F}\ln\frac{K_o + nbNa_o}{K_i + nbNa_i}$$

1.7 Cell volume regulation

Cell volume reflects cell water content. Since plasma membranes are relatively permeable to water (p. 7), water will be distributed between cells and interstitial fluid to maintain constant the *activity* of water in each compartment. Note that water activity in a solution is decreased as the total concentration of solute particles is increased and vice versa. A difference in water activity between compartments will result in a net movement of water across the membrane down its activity gradient—a process termed *osmosis*.

Osmosis

An *ideal semi-permeable membrane* prevents diffusion of solutes but not solvent. Net movement of water across such a membrane (osmosis) will occur down the gradient of water activity and, therefore, from the more dilute to the more concentrated solution. It can be stopped by applying hydrostatic pressure to the more concentrated solution to raise its water activity. The hydrostatic pressure which just stops osmosis is the *osmotic pressure*. In other words it is the pressure required to equalize water activity in the two compartments. A greater hydrostatic pressure will cause water to leave the more concentrated solution and this is called *ultrafiltration*.

The osmotic pressure ($\Delta\pi$) across an ideal semi-permeable membrane is calculated by the *van't Hoff equation*:

$$\Delta\pi = RT\Sigma(c_1 - c_2)$$

where R is the gas constant, T is the absolute temperature (K) and $\Sigma(c_1 - c_2)$ is the sum of the difference in molal concentrations (mol kg^{-1} H$_2$O) of all osmotically active solutes on sides 1 and 2 of the membrane. Traditionally, $\Delta\pi$ has been expressed in atmospheres, in which case $R = 0.082$ kg atm K^{-1} mol^{-1}. The van't Hoff equation predicts that an ideal 1 molal solution of a non-electrolyte should have an osmotic pressure of 22.4 atm at 0°C. Osmotic pressures of actual solutions usually deviate from those expected from their molal concentrations because solute molecules interact with each other and with the solvent. Therefore, strictly speaking, in the van't Hoff equation c refers to the *osmotic activities* of the solutes rather than to their concentrations but in dilute solutions these are approximately the same. Also in dilute aqueous solutions molal concentrations (mol kg^{-1} H$_2$O) closely approximate molar concentrations (mol l^{-1} solution).

The osmolality (osmol kg^{-1} H$_2$O) of a solution can be estimated from its osmotic pressure in atmospheres divided by RT. An electrolyte like NaCl is dissociated in solution into two osmotically active ions, so its osmolal concentration will be approximately twice the molal concentration. The osmolality of a solution can be measured in a commercial osmometer which relies on the fact that the osmotic pressure is related to the depression of the freezing point. Mammalian fluids have an osmolality of about 0.3 osmol kg^{-1} H$_2$O and thus exert an osmotic pressure across an ideal semi-permeable membrane of about $RT \times 0.3 = 7.6$ atm or 5800 mmHg or 770 kPa at body temperature.

Effective osmotic pressure

Biological membranes are not ideal but in varying degree 'leaky'. The osmotic pressure difference across such a membrane ($\Delta\pi$) is less than the ideal and diminishes with time as some solutes diffuse through and eliminate their difference in concentration. The *effective osmotic pressure* will therefore be less than the pressure calculated above and will depend on the degree to which these solutes cross the membrane. This is expressed by including the *reflexion coefficient*, σ, in the van't Hoff equation:

$$\Delta\pi = \sigma RT\Sigma(c_1 - c_2).$$

If none of the solute molecules colliding with the membrane cross it, $\sigma = 1$, i.e. all molecules are reflected. If all the solute molecules pass through the membrane $\sigma = 0$, none are reflected and there is no osmotic pressure. Thus the effective osmotic pressure of the cell fluid, often called its *colloid osmotic pressure*, is due to the large, reflected, non-penetrating solutes (e.g. proteins, organic phosphates). Some of these large solutes are negatively charged and attract small diffusible cations and repel small diffusible anions across the plasma membrane. This leads to an uneven distribution of small ions with a small excess inside the cell called the *Donnan excess*, which also contributes to the effective osmotic pressure of the cell fluid (Appendix, p. 31).

Net flow of water across membranes

The net flow of water across a membrane per unit time (J_v) is given by

$$J_v = AL_p(\Delta P - \Delta\pi)$$

where A is the area of the membrane available for flow, L_p is the hydraulic conductivity which is a measure of the ease with which water flows through the membrane, $(\Delta P - \Delta\pi)$ is the driving force which is dependent on the differences in hydrostatic pressure (ΔP) and effective osmotic pressure ($\Delta\pi$) across the membrane.

Cell volume

In the steady state, cell volume remains constant and $J_v = 0$. This implies either that ΔP and $\Delta\pi$ are both zero or that the effective (colloid) osmotic pressure of the cell is offset by a greater hydrostatic pressure within the cell, i.e. $\Delta\pi = -\Delta P$. The latter is not believed to be true and the determinants of cell volume are considered to be:

1 the total number of osmotically active particles within the cell; and
2 the effective osmolality of the interstitial fluid.

Although there is much less protein in the interstitial fluid and the cell has a considerable excess of colloid osmotic pressure, the cell does not swell. It is believed that the $Na^+ - K^+$ pump, by holding Na^+ extracellularly and preventing its accumulation in the cell, effectively offsets the colloid osmotic swelling force of cell macromolecules. If the Na^+ pump ceases to expel Na^+, for example when cells are chilled, Na^+ enters passively down its concentration gradient bringing Cl^- with it to preserve electroneutrality; H_2O then distributes to equalize water activities across the membrane and the cells swell.

Tonicity

The behaviour of cells in artificial bathing solutions cannot be predicted from the osmotic pressure or osmolality of these solutions as measured with an ideal semi-permeable membrane or by the depression of the freezing point. It depends also on the permeability of the membrane to the solutes. For example, when cells are placed in 300 mmol l^{-1} urea, which is approximately isosmotic with mammalian fluids, the cells will swell because urea enters and H_2O follows. A new term is required to define the strength of a solution as it affects the volume of cells. This is *tonicity* and it can be defined operationally. Thus,

1   if cells *shrink* in a solution, the solution is *hypertonic*,
2   if cells *swell* in a solution, the solution is *hypotonic*, and
3   if the cell volume is *unchanged* the solution is *isotonic*.

Tonicity may also depend upon the functional activity and metabolism of the cells. Most cells swell grossly if their metabolism is inhibited, even in isosmotic NaCl solutions, because the cells gain NaCl. Consequently, isolated tissues need substrates and oxygen as well as the appropriate inorganic ions in bathing solutions designed to be isosmolal with normal ECF.

1.8 Osmosis and ultrafiltration across capillary walls

The structure of the capillary wall allows free permeability to water and solutes of M_r less than 70 000 (p. 5). Small molecules diffuse through the capillary wall and therefore should achieve equal concentrations on both sides and make no contribution to the effective osmotic pressure. Plasma proteins with M_r greater than 70 000 cannot cross the capillary wall (i.e. proteins with $\sigma = 1$ are reflected) and thus cause an effective osmotic pressure. Since plasma proteins are large enough to be classed as colloids, this effective osmotic pressure is known as the *plasma colloid osmotic pressure*.

Plasma colloid osmotic pressure

Plasma proteins amount to approximately 1 mmol l^{-1} of plasma and, as a simplification, there are almost no proteins in the interstitial fluid. This concentration difference for protein would cause an effective osmotic pressure for plasma of 16 mmHg (calculated using the van't Hoff equation). However, proteins are negatively charged and therefore attract diffusible cations and repel diffusible anions. This leads to an uneven distribution of small ions with a small excess (about 0.5 mmol l^{-1}) inside the capillary. This uneven distribution is referred to as the *Gibbs–Donnan membrane distribution* and the small excess of diffusible ions in the plasma is known as the *Donnan excess* (Appendix, p. 32). This excess contributes a further 9 mmHg to the effective osmotic pressure of plasma. Hence the total plasma colloid osmotic pressure is 25 mmHg, first measured by E. H. Starling (1866–1927). This pressure is very small compared with the total osmotic pressure of all the solutes in plasma (5800 mmHg) measured when plasma is separated from pure water by an ideal semi-permeable membrane.

Starling equilibrium

Starling realized that the plasma colloid osmotic pressure of 25 mmHg is very important because its value lies between that of the blood pressure in arterioles and venules. The blood pressure (hydrostatic pressure) at the arteriolar end of a capillary is about 32 mmHg and at the venular end about 15 mmHg. Hydrostatic pressure in the interstitial fluid varies from place to place but on average it is a few mmHg below atmospheric, say about -2 mmHg. The net hydrostatic pressure between blood and interstitial fluid forces fluid out of the capillaries by *ultrafiltration* and this force is greater at arteriolar than venular ends (Fig. 1.6). This ultrafiltration is opposed by *osmosis* returning fluid into the capillaries, the osmotic force being the difference in colloid osmotic pressure between plasma and interstitial fluid. Note that this osmotic force remains constant along the length of the capillary. At the arteriolar end, the net balance of hydrostatic and osmotic forces causes ultrafiltration with fluid leaving the capillary. At the venular end, the net balance causes reabsorption with fluid returning to the capillary. This balancing of ultrafiltration and osmosis is referred to as the *Starling equilibrium*. The balance is not complete as there is slightly more ultrafiltration than osmosis (Fig. 1.6). The Starling equilibrium provides an automatic control of circulating blood volume because alterations in capillary blood pressure change the rate of ultrafiltration leading to a redistribution of fluid between blood and interstitial fluid (p. 433).

In some capillary beds, blood perfusion is intermittent and ultrafiltration when blood pressure is high is balanced by osmosis when blood pressure is low. In particular organs, special relations exist in capillary beds. For example, in the glomeruli of the kidneys blood pressure is high and there is only ultrafiltration. In the alveolar capillaries in the lungs, blood pressure is below the colloid osmotic pressure, hence there is only osmosis and liquid does not accumulate in the alveoli. In the liver, the interstitial fluid outside the capillaries is rich in protein and a very small capillary hydrostatic pressure is balanced by an equally small difference in colloid osmotic pressure.

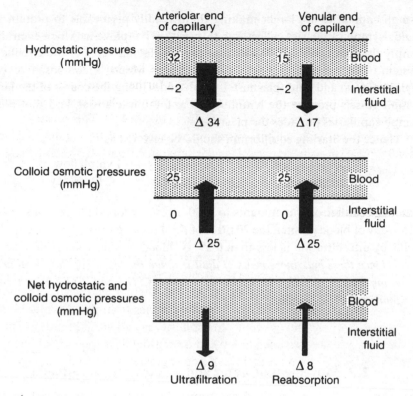

Fig. 1.6. The Starling equilibrium across the capillary wall.

Net flow of fluid per unit time (J_v) across the capillary walls of an organ will not only be altered by changes in hydrostatic pressures and colloid osmotic pressures but, since $J_v = AL_p(\Delta P - \Delta \pi)$, it will be affected by the number of capillary channels that are open and the hydraulic conductivity of the capillary walls. For example, during increases in metabolism more capillary channels are open and J_v will increase because of the increased area available for flow across the capillary walls.

Lymph

In fact, some protein escapes from most capillaries. If it accumulated in the interstitial fluid it would impede the return of fluid by osmosis. This problem is solved by the *lymphatic system* which carries away this protein and any excess fluid resulting from an imbalance between ultrafiltration and osmosis. The lymphatic system consists of a network of blind-ended lymph capillaries lying in the interstitial fluid close to blood capillaries. The walls of lymph and blood capillaries are similar except that the gaps between endothelial cells in lymph capillaries are larger making them readily permeable to protein and fluid. Lymph capillaries collect into thin-walled lymph veins which eventually empty through two main vessels into the subclavian veins of the circulatory system. The lymphatic system carries *lymph* which is composed of cells (lymphocytes) and fluid (plasma). Lymph nodes along the course of the larger lymph vessels produce the lymphocytes and the interstitial fluid that enters lymph capillaries provides the plasma.

Hence the Starling equilibrium should be restated as:

$$\text{Ultrafiltration} = \text{Osmosis} + \text{Lymph flow}$$
$$\text{20 litres/day} \quad \text{16–18 litres/day} \quad \text{4–2 litres/day}$$

Since the cardiac output amounts to 8000 litres/day for adult man, or say 4000 litres/day of blood plasma, the 20 litres of fluid passing through blood capillary walls by ultrafiltration is less than 0·5% of blood plasma volume flowing per day. *Hence these bulk movements of fluid through the blood capillary wall are of little importance for the exchange of nutrients, O_2 and CO_2, which move predominantly by diffusion.*

1.9 Movement of water and solutes across epithelia

Epithelia separate the internal from the external environment and regulate movement of solutes and water to and from the body. Examples include the surface layers of the skin, lungs, wall of the alimentary canal and lining of the urogenital systems. The epithelial cells, which may be one or more layers thick, are separated by their basement membranes from a variable amount of supporting connective tissue containing blood vessels, nerves and smooth muscle fibres. It is believed that the transport functions reside only in the epithelial cell layer.

All epithelial cells are separated from their neighbours by a space—the *lateral intercellular space*—whose size may vary (Fig. 1.7). They are held to each other at their luminal edges by junctions—*'tight' junctions*. Tight junctions were originally thought to be impermeable, but it is now appreciated that to a variable extent solutes and water may diffuse across them. There are thus two pathways by which solutes and water can cross epithelia, either through cells or between cells. Substances passing between cells must cross tight junctions and pass through the lateral intercellular spaces—the so-called *paracellular* or *shunt pathway*. The amount passing through this shunt pathway differs between epithelia, depending on the difference in permeability characteristics of the tight junctions. Epithelia in which the shunt pathway makes a significant contribution to total movements (e.g. proximal renal tubule, small intestine)

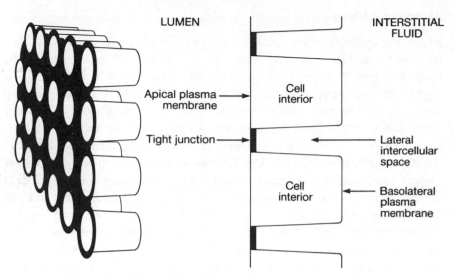

LUMEN

INTERSTITIAL
FLUID

Apical plasma
membrane

Cell
interior

Tight junction

Lateral
intercellular
space

Cell
interior

Basolateral
plasma
membrane

Fig. 1.7. Transporting epithelia.

are referred to as *'leaky'* whereas those in which the cellular pathway predominates are referred to as *'tight'* (e.g. renal collecting duct, salivary gland ducts, amphibian skin and urinary bladder). 'Tight' epithelia can generate and maintain, often under hormonal control, steep salt gradients between lumen and interstitial fluid and thus cause some dissociation between ion and water movements. In contrast, 'leaky' epithelia cannot generate salt gradients and allow passage of isosmotic fluid.

Net flow of water across an epithelium per unit time (J_v) is determined by $AL_p(\Delta P - \Delta \pi)$ (p. 20) while net movements of solutes are determined by the following considerations. There are specific cellular transport pathways for a variety of solutes which allow unidirectional movement between lumen and interstitial fluid. In addition, solutes may diffuse in both directions through the paracellular pathway. The net solute flux (amount moved/unit time · unit area) is, therefore, the difference between two unidirectional fluxes. *Net movement* from lumen to interstitial fluid is termed *absorption*; net movement from interstitial fluid to lumen is called *secretion*. The net movement of any solute across an epithelium is determined by the following.

1 The available surface area (since amount moved/unit time = area × flux).
2 The time in contact with the available area.
3 The electrochemical potential gradient, which will be influenced by blood and lymph flow.
4 The properties of the epithelium, i.e. 'tight' or 'leaky', together with the availability of specific transport mechanisms.

Absorptive epithelia

In general, epithelia which actively transport ions primarily move Na^+. This transepithelial Na^+ transport generates a potential difference across the epithelial layer. The transepithelial potential difference in 'tight' epithelia is high (>20 mV) whereas in 'leaky' epithelia it is low (0–5 mV). Since electrical neutrality must be preserved, Cl^- follows passively, as does water when water permeability of epithelia is sufficient. The detailed mechanisms by which Na^+ crosses epithelia remain poorly understood but are summarized for a typical 'tight' epithelium in Fig. 1.8. Unlike muscle, nerve and blood cells whose plasma membranes are symmetric, the plasma membranes of epithelial cells are asymmetric. The apical portion, which faces the lumen, has unusual permeability and transport characteristics. It has a high Na^+ and low K^+ permeability and no Na^+–K^+-ATPase. In contrast, the basolateral portion consisting of the membrane beneath the 'tight' junctions and adjacent to the capillaries is similar in its permeability and transport properties to other plasma membranes. In general, in 'tight' epithelia only Na^+ readily crosses the apical membrane, probably passively via aqueous channels.

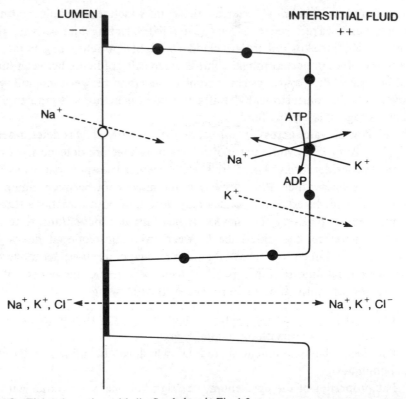

Fig. 1.8. 'Tight' absorptive epithelia. Symbols as in Fig. 1.3.

In 'leaky' epithelia (Fig. 1.9), not only does Na$^+$ cross the apical membrane through specific Na$^+$ channels as in 'tight' epithelia, but entry of glucose, some amino acids and phosphate may be coupled to Na$^+$ entry. The electrochemical potential gradient for Na$^+$ entry is used to drive the accompanying solute into the cells from the lumen (i.e. co-transport). The active transport of Na$^+$ from cell to interstitial fluid is thought to involve the same Na$^+$-K$^+$ pump as already described (p. 11). The coupling of net water movement to net solute movement across 'leaky' epithelia may involve the generation within the lateral intercellular spaces of a local region of Na$^+$ concentration, somewhat higher than that in interstitial fluid, reflecting the distribution and activity of the Na$^+$-K$^+$ pump. This provides an osmotic gradient moving water, either through the cells or tight junctions, from the lumen to the lateral intercellular space. This movement in turn creates a local hydrostatic pressure gradient so that fluid flows from the lateral intercellular spaces to the interstitium. Note,

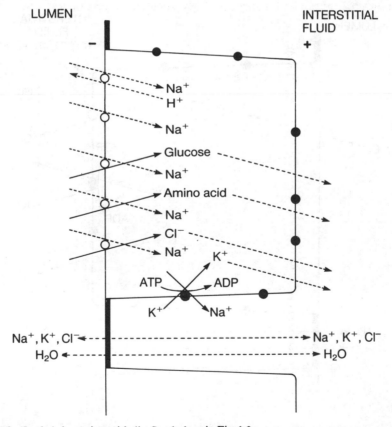

Fig. 1.9. 'Leaky' absorptive epithelia. Symbols as in Fig. 1.3.

however, that the magnitude of any local osmotic gradient in the lateral intercellular spaces is now thought to be only a few mmol kg^{-1} H_2O.

Counter-transport of Na^+ for cell H^+ is also often found. In addition, though often represented as coupled co-transport with Na^+ (Fig. 1.9), movement of Cl^- across the apical plasma membrane may be by counter-transport with cell HCO_3^- (or OH^-). In a cell with both Na^+–H^+ and Cl^-–HCO_3^- exchangers, the net effect will be Na^+ and Cl^- entry to the cell with H_2O and CO_2 formed from the H^+ and HCO_3^- ions extruded from the cell.

Secretory epithelia

A variety of inorganic and organic ions are secreted by epithelia and specific mechanisms are discussed in the appropriate chapters. However, of particular interest because of its contribution to the production of many secretions (sweat, tears, saliva, pancreatic, biliary and intestinal fluids) is the mechanism of isosmotic NaCl secretion, summarized in Fig. 1.10. The secretory cells

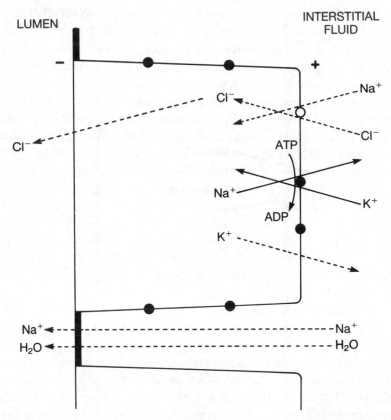

Fig. 1.10. Secretory epithelia. Symbols as in Fig. 1.3.

comprise a leaky epithelium. A Na^+–Cl^- (or Na^+, K^+, $2\,Cl^-$) co-transporter in the basolateral membrane is responsible for driving Cl^- into the cells against its electrochemical gradient. The accompanying Na^+ is recycled back to the interstitial fluid. In the quiescent state, the apical plasma membrane is impermeable to Cl^-. When secretion is stimulated, however, a Cl^- conductance in this membrane is activated. The anion, therefore, moves from the cell to the lumen down its electrochemical gradient. As a consequence, Na^+ passes from the interstitial fluid to lumen through the 'leaky' paracellular pathway to maintain electroneutrality and water moves through the same pathway down the osmotic gradient. Thus a primary isosmotic NaCl secretion is produced.

1.10 Appendix

Patch clamping

It has already been said (p. 8) that passive ion movement across membranes is through water-filled pores or channels. These channels may always be open, or they may fluctuate between open and closed states. If channels fluctuate between open and closed states, the number of ions crossing an area of membrane in a given time will also fluctuate. In macroscopic measurements of ionic conductance we are not aware of this since the high density of channels smoothes out these oscillations to give an average or steady value. There are, however, two ways in which the behaviour of individual channels may be studied. If ionic currents passing across a given area of membrane are subjected to very high amplification, small variations in current can be recorded, particularly if the density of channels is low or if the number open at any given time is reduced by experimental means. These variations in current are referred to as membrane 'noise' and their study as 'fluctuation analysis' or 'noise analysis'. In essence, this involves the separation of 'noise' due to ion flow through the membrane from other 'noise' sources in the system and the investigation of the properties of the former by statistical analysis. By these means, the average density of the channels and their individual properties can be determined.

A more elegant technique involves isolating a small piece of membrane containing only a few channels in which current flow through individual channels can be recorded directly. This technique, developed by Neher and his colleagues in the mid-1970s and known as 'patch-clamping', requires that specially prepared micro-electrodes be brought up to a membrane surface and a tight seal formed. Isolation of a small piece of membrane affords the opportunity of studying the channels uninfluenced by other regions of the cell, either by manipulating the external environment (the outside-out patch-clamp) or the inside environment (the inside-out patch-clamp) (Fig. 1.11). Experiments of this type have shown that the primary mechanism by which many specific external stimuli affect cellular activity is through combination with an external receptor on the cellular surface and opening or closing of ionic channels. Some unexpected results of this research include the surprisingly large number of different channels that co-exist within the surface membrane and that many channels, even those activated by ligands, spontaneously open and close.

Experiments of this type have shown that a large number of different ionic channels co-exist within the surface membrane in either an open or closed configuration. The opening and closing (i.e. gating) of these channels may be voltage dependent, ligand dependent or spontaneous. For example, the

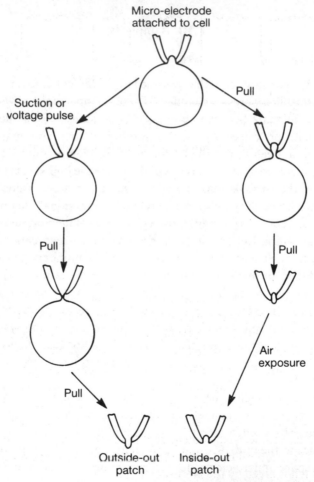

Fig. 1.11. An illustration of different membrane configurations used in patch-clamp analysis. (After Hamill, O.P., Marty, A., Neher, E., Sakmann, B. & Sigworth, F. (1981) *Pflügers Arch.* **391**, 85–100.)

acetylcholine receptor–ionophore complex at skeletal neuromuscular junctions is a 'ligand-gated' channel, the ligand being the neurotransmitter acetylcholine. The currents generated across a patch of membrane containing only a few acetylcholine channels exposed to a low concentration of acetylcholine is shown in Fig. 1.12. It can be seen that the mean channel open time is about 1 ms and it can be calculated that approximately 4×10^7 ions pass through the channels each second.

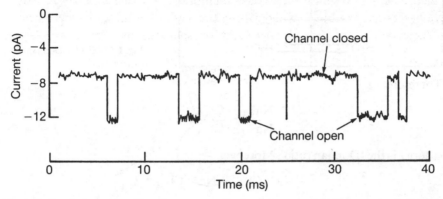

Fig. 1.12. Patch-clamp recording of current passing through an acetylcholine receptor channel of skeletal muscle.

Gibbs–Donnan distribution

Consider two compartments containing aqueous solutions of fixed volume separated by a membrane permeable to water and small M_r solutes but impermeable to protein (Fig. 1.13). Suppose that the solution in side 1 contains protein (Pr) with a net negative charge $(z-)$ and that both solutions contain Na^+ and Cl^- ions. Because of the negative charge on the protein, diffusible cations, i.e. Na^+, will be attracted to side 1 and diffusible anions, i.e. Cl^-, repelled to side 2. As a consequence of the redistribution of ions, side 1 containing the protein will be slightly electrically negative with reference to side 2, that is, an electrical potential difference $(\Delta\psi)$ will exist across the membrane. At equilibrium it will be equal to the equilibrium potential (E) for each ion which is given by the Nernst equation:

$$E = \frac{RT}{F}\ln\frac{Na^+_1}{Na^+_2} = \frac{RT}{F}\ln\frac{Cl^-_2}{Cl^-_1}$$

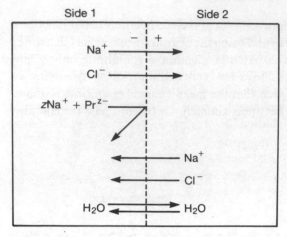

Fig. 1.13. Gibbs–Donnan distribution.

Therefore

$$\frac{Na^+{}_1}{Na^+{}_2} = \frac{Cl^-{}_2}{Cl^-{}_1} = r$$

where r is the 'Donnan ratio'. Moreover

$$Na^+{}_1 Cl^-{}_1 = Na^+{}_2 Cl^-{}_2.$$

The condition of bulk electroneutrality demands that the sum of all positive charges shall equal the sum of all negative charges in any solution, i.e.

$$Na^+{}_1 = Cl^-{}_1 + Pr^{z-}{}_1$$

and

$$Na^+{}_2 = Cl^-{}_2.$$

Therefore

$$Na^+{}_1 Cl^-{}_1 = (Na^+{}_2)^2 = (Cl^-{}_2)^2.$$

Now $Na^+{}_1 Cl^-{}_1$ is not a square and since the sum of the sides of a rectangle is greater than the sum of the sides of a square of equal area,

$$Na^+{}_1 + Cl^-{}_1 > Na^+{}_2 + Cl^-{}_2$$

or

$$Na^+{}_1 + Cl^-{}_1 = Na^+{}_2 + Cl^-{}_2 + e$$

where e is called the 'Donnan excess'. This excess concentration of diffusible ions, e, contributes to the difference in effective osmotic pressure between the two compartments so that:

$$\Delta\pi = RT(Pr^{z-} + e).$$

Note that the system can only reach a steady state when the volumes of the two compartments are fixed, for then $\Delta\pi$ is offset by an increased hydrostatic pressure in compartment 1, i.e. $\Delta\pi = -\Delta P$. Without this restriction an osmotic flow of water and a diffusion of permeant solutes from compartment 2 to 1 will continue until all the solution is found in compartment 1.

Note too that the resting membrane potential of cells is *not* a consequence of the Gibbs–Donnan distribution as the intracellular and extracellular concentrations of ions do not comply with the Donnan ratio. In the case of the capillary wall where a Gibbs–Donnan distribution does exist, it can be calculated that the potential difference across the wall is only a few millivolts.

2 · Nervous System

2.1 Cells of the nervous system
Neurones
Neuroglia

2.2 Electrical signalling in the nervous system
Membrane potentials
Local (electrotonic) potentials
Action potentials
  Ionic basis of the action potential
  Propagation of the action potential
  Extracellular recording of nervous
    activity
  Spontaneous action potentials
Synaptic transmission
  Ionic basis of synaptic potentials
  Presynaptic inhibition

2.3 Growth, development and regeneration of the nervous system
Axonal transport
Axonal growth
Aspects of neural development
  Nerve growth factors
  Cell death
  Role of activity
Neural regeneration
  Regeneration in the peripheral nervous
    system
  Regeneration in the central nervous
    system

2.4 Divisions of the peripheral nervous system
Somatic nervous system
  Neuromuscular transmission

Autonomic nervous system
  Sympathetic system
  Parasympathetic system
  Enteric system
  Activity in autonomic neurones
  Synaptic transmission in
    autonomic ganglia
  Neuroeffector innervation
  Neuroeffector transmission
  Opposing actions of the
    parasympathetic and sympathetic
    systems
  Circulating catecholamines

2.5 Basic design of the central nervous system
Spinal cord
Brain
  Brainstem
  Cerebellum
  Diencephalon and cerebrum
Meninges, ventricles and cerebrospinal
  fluid
  Blood–brain barrier
  Cerebrospinal fluid

2.6 Appendix
Basic electrical properties of biological
  membranes.
Current, voltage, conductance and
  capacitance
Membrane space and time constants
Number of ions crossing the membrane
Voltage clamping
Reversal potential

The nervous system is a network of cells specialized for the rapid transfer and integration of information. It comprises the brain and spinal cord (the *central nervous system*, CNS), and sensory and motor nerve fibres that enter and leave the CNS or are wholly outside the CNS (the *peripheral nervous system*, PNS). The fundamental unit of the nervous system is the neurone which with its processes transmits electrical signals. The cell bodies of neurones tend to be segregated into compact groups (nuclei, ganglia) or into sheets (laminae) that lie within the grey matter of the CNS or are located in specialized ganglia in the PNS. Their nerve fibres course in the white matter of the CNS or along peripheral nerves. Groups of nerve fibres running in a common direction usually form a compact bundle (nerve, tract, peduncle, brachium, pathway).

Many of these nerve fibres are surrounded by sheaths of lipid material called myelin which gives rise to the characteristic appearance of the white matter. In addition to neurones there are glial cells which play a supporting role. There are about five times more glial cells than neurones and they occupy approximately half the volume of the brain.

2.1 Cells of the nervous system

Neurones

Neurones are more diverse in size and shape (Fig. 2.1) than cells in any other tissue of the body. Nevertheless they have certain features in common: they usually possess dendrites, a cell body, an axon and synaptic terminals (Fig. 2.2).

Dendrites are branches which leave the cell body and share with it the function of receiving information from synaptic connections with adjoining neurones. They are usually tapered and possess various protuberances, such as dendritic spines which are sites of contact with other cells.

The *cell body*, also known as the soma or perikaryon, contains the nucleus. In the cytoplasm surrounding the nucleus is an extensive array of rough endoplasmic reticulum, the basophilic-staining so-called Nissl substance which synthesizes proteins for the cell body and its processes. Large neurones replace as much as one-third of their protein content every day.

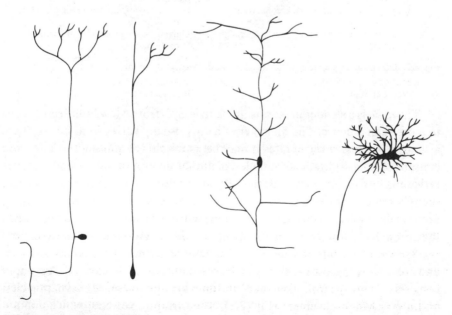

Fig. 2.1. Various types of neurones.

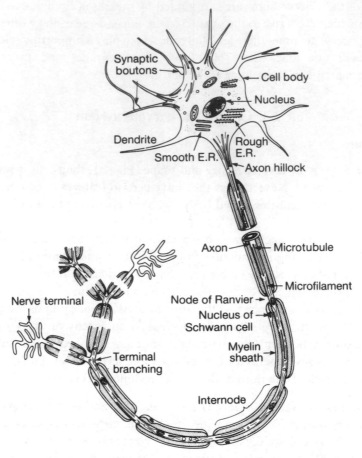

Fig. 2.2. Diagram of a motor neurone. (E.R. = endoplasmic reticulum.)

The *axon* is a slender process ranging in length from a few hundred microns to more than a metre. The diameter of axons usually varies from 0·1 to 20 μm although in some invertebrates it may be as much as a millimetre. The axon contains mitochondria, microtubules, neurofilaments and smooth endoplasmic reticulum. At its junction with the cell body, also known as the *axon hillock*, there is some rough endoplasmic reticulum but beyond the axon hillock there are no ribosomes. Axons are usually longer than dendrites and at their ends they branch to form terminals which make contact with other cells at *synapses*.

Synaptic terminals are the sites of release of chemicals (*neurotransmitters*) and are usually in close proximity to another neurone or effector cell. A synapse consists of a presynaptic element containing synaptic vesicles, a synaptic cleft and a postsynaptic element (Fig. 2.3). The synaptic vesicles, which contain the neurotransmitter, vary in size in different cells from 25 to 250 nm. The

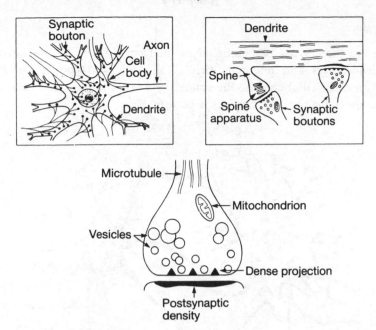

Fig. 2.3. A typical synapse found in the central nervous system. *Left inset:* numerous synaptic terminals on a cell body and dendrites of a neurone. *Right inset:* typical synapses on a dendrite.

synaptic cleft between neurones is 20–30 nm wide and at skeletal neuromuscular junctions it is 50–100 nm wide. In electron micrographs, electron-dense areas on the cytoplasmic faces of pre- and postsynaptic membranes indicate sites of specialization for exocytosis of vesicle contents and for localization of postsynaptic receptors, respectively. While most neurotransmitters are released close to effector cells, some neurones release substances which act at some distance from their site of release and thus are more like endocrine cells. Thus neurones are specialized:

1   to receive information from the internal and external environment;
2   to transmit signals to other neurones and to effector organs;
3   to process information (integration); and
4   to determine or modulate the differentiation of sensory receptor cells and effector organs (trophic functions).

Rapid processing of information by nerve cells relies on their electrical activity, coupled with neurosecretion at synapses; trophic functions may be the result of electrical activity but may also rely on axonal transport of substances to and from the cell body.

Neuroglia

There are three types of neuroglial or glial cells in the CNS—astrocytes, oligodendrocytes and microglia (Fig. 2.4)—and one type in the PNS—Schwann cells. Glial cells fill the spaces between neurones and accordingly have a role in maintaining nervous structure.

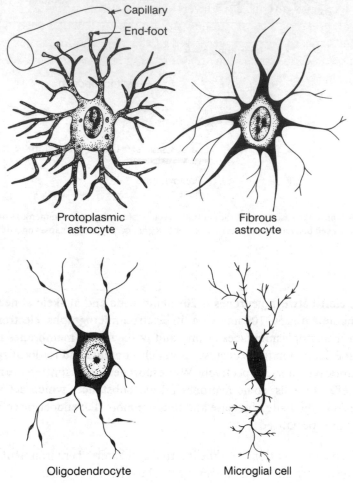

Fig. 2.4. Neuroglial cells.

Astrocytes are star-shaped cells with processes extending into the surrounding tangled network of unmyelinated nerve fibres (neuropil). These processes contain bundles of fibrils and usually run at right-angles to the direction of the nerve fibres. Some processes expand into end-feet which are applied to the surfaces of blood vessels. They are also in contact with one another through gap junctions and have a high permeability to K^+ (and resting membrane potentials close to E_K), which help to dampen increases in extracellular K^+ concentration that might arise in local areas of the brain where there are high levels of activity.

Oligodendrocytes can be distinguished from astrocytes by having fewer and thinner processes; they do not possess gap junctions. They form myelin sheaths around axons in the CNS, as do *Schwann cells* in peripheral nerves, by enveloping them with concentric layers of plasma membrane (Fig. 2.5). *Myelin* forms an insulating sheath around an axon with small areas of axonal membrane exposed between successive myelin segments known as the *nodes of Ranvier*. Schwann cells also encircle unmyelinated axons in the PNS without forming concentric layers of myelin.

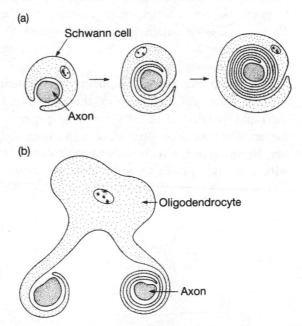

Fig. 2.5. Myelination of axons. (a) A Schwann cell in the peripheral nervous system forms myelin around a portion of a single axon. (b) An oligodendrocyte in the central nervous system can form myelin around several axons.

Microglia are smaller glial-type cells of uncertain origin that appear to act as scavenger cells. Injury to the brain activates microglia which undergo proliferation and migrate to the point of injury where they turn into macrophages and remove the debris. Astrocytes then proliferate and wall off the damaged area forming a glial scar.

In summary, the functions of glial cells are:

1 to maintain the structure of the nervous system;
2 to form myelin sheaths around axons;
3 to repair damage to nervous tissue; and
4 to help maintain the surrounding ionic environment, particularly with respect to K^+ ions.

Chapter 2

2.2 Electrical signalling in the nervous system

Neurones receive, conduct and transmit information and have regional specializations for these functions. For example, a sensory neurone has terminals at the periphery for processing information, an axon for conduction and terminals in the CNS for transmission of signals to other neurones. Electrical signals are derived from changes in the resting membrane potential of the neurones and can be divided into two types—local potentials and action potentials. In general, *local potentials* occur at points of discontinuity (e.g. environment–sensory receptor, CNS synapses and nerve–muscle junctions) and they have their maximum amplitude at their point of origin; 0·2–2 mm away from this the amplitude falls rapidly. In contrast, *action potentials* are transient changes in membrane potential that are conducted over longer distances, in some instances a metre or more, with little change in amplitude. The conducted action potentials can be likened to the transmission of morse code along telegraph wires.

The role of these signals in the nervous system can be best illustrated by outlining a simple common reflex, the *withdrawal reflex*. This reflex results in the rapid withdrawal (latency about 30 ms) of a limb when a part of it is subjected to a noxious stimulus. For example, foot placed on a sharp object may be withdrawn initially without conscious direction. Underlying this reflex withdrawal is a specific sequence of neural events. In brief, the pathway (Fig. 2.6) comprises a sensory (afferent) neurone, an interneurone, a motor (efferent)

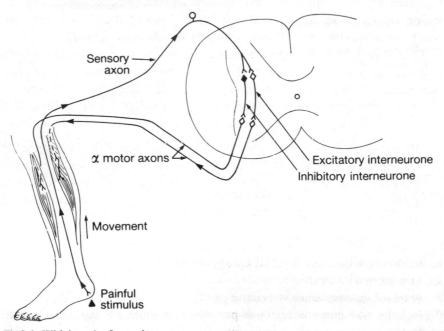

Fig 2.6. Withdrawal reflex pathway.

neurone and skeletal muscle fibres. The noxious stimulus causes a local change of potential (receptor potential) in the sensory nerve terminal (Section 4.1) resulting in the generation of action potentials conducted along the sensory fibre to the CNS. Their arrival at the central terminal of the sensory fibre is followed by release of a neurotransmitter, which produces local potentials in the dendrites and cell body of the interneurone whereupon action potentials are again initiated. These action potentials are conducted along the axon of the interneurone to its point of synaptic contact with the motor neurone. Again, a transmitter substance is released causing a local excitatory potential followed by action potentials in the motor neurone. These action potentials are then conducted down the peripheral axon to the skeletal muscle fibres. At the nerve–muscle junctions, a similar sequence of events occurs, i.e. transmitter release, generation of a local excitatory potential in the muscle fibre followed by action potentials, conduction of the action potentials along the muscle fibre, and finally contraction. It should be noted that this is a very simple model and that withdrawal reflexes require the coordinated activity of many muscle groups, some of them excited, some of them inhibited.

Membrane potentials

Neurones, like other cells, are electrically polarized so that their interior is negatively charged with respect to the outside of the cell (Section 1.6). Typical resting values for neurones are about -70 mV, for glial cells -90 mV, for skeletal muscle cells -80 mV and for smooth muscle cells -70 mV.

The value of the membrane potential can be estimated from the equilibrium potentials of the various ions and their conductances across the plasma membrane (p. 18). Ions move through channels which have the special property of being selectively permeable to particular ion species. At rest, the conductance through K^+ specific channels (G_K) is much greater than that through Na^+-specific channels (G_{Na}) and so the membrane potential approaches the K^+ equilibrium potential ($E_K \sim -90$ mV). When a tissue is stimulated and channels are activated, the current flow through them makes a greater contribution to the membrane potential.

The major types of *gated* channels in excitable membranes include:

1 *voltage-gated channels*, sensitive to the voltage across the membrane in which they are inserted, e.g. the voltage-sensitive Na^+ channel involved in action potential generation;

2 *chemically gated channels*, sensitive to a chemical signal such as the binding of a ligand to a receptor which forms part of the channel, e.g. the acetylcholine receptor channel involved in neuromuscular transmission; and

3 *mechanically gated channels*, sensitive to mechanical deformation of the membrane, e.g. touch receptors on sensory nerve terminals.

The properties of membrane channels have been studied by the patch-clamp technique, which can measure the current flow through a single channel in a very small patch of membrane (Appendix, Chapter 1, p. 29). All membrane channels so far described are either fully open or closed, in an all-or-none fashion. 'Gating' agents affect the probability of a channel being open or closed. Single-channel conductances range between 3 and 200 pS and members of the same class of channel have similar conductances. A well-studied example is the voltage-sensitive Na^+ channel involved in action potential generation. This channel is specifically located in excitable membranes of nerve and muscle. At the normal resting potential, the channels have a very low probability of being open. If the membrane is depolarized, the probability of channel opening increases. Once a channel is opened, it automatically closes again within a millisecond or so and cannot be re-opened (i.e. it becomes *refractory*) until the membrane has been repolarized back to the normal resting potential.

Local (electrotonic) potentials

A net movement of charge across the membrane changes the potential difference. A decrease in the magnitude of the potential difference with the inside becoming more positive is called a *depolarization*; an increase in magnitude is called a *hyperpolarization*. A depolarization is caused by the movement of positive charge into (or negative charge out of) the cell and a hyperpolarization is caused by movement of charge in the opposite direction. If the movement of charge (current) responsible for the change in potential is limited to a particular site on the membrane of an elongated cell, the change is greatest at that point and falls off exponentially with distance. As a result of the new potential gradients established along the cell and across the membrane, currents will flow between adjoining regions and through the membrane. As the potential gradients inside the cell and across the membrane decline with distance (Fig. 2.7), so do the currents. Such local potentials, sometimes referred to as *electrotonic potentials*, are typically restricted to within 1 or 2 mm of their point of origin. In addition to their being thus *decremental* such local potentials are *graded*, i.e. they increase in proportion to the increase in current. Because the membrane behaves as a capacitor (i.e. it stores electrical charge), the changes in potential are *distorted*, i.e. the rate of change in potential is slowed and lasts longer than the stimulus (Fig. 2.7). Local potentials also *summate*.

These local changes in potential are important for the integration of electrical activity in dendrites and nerve cells and as receptor potentials in

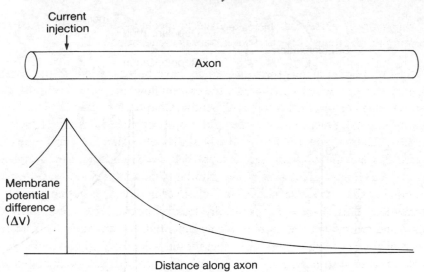

Fig. 2.7. Electrotonic spread of potential in an axon.

sensory receptors, and are the basis for the initiation and propagation of action potentials.

Action potentials

Nerve and muscle plasma membranes have the capacity to generate action potentials. Though small currents produce the local potentials discussed above, large inward currents (such as the influx of Na^+ at a sensory nerve terminal) depolarize the membrane to a point, termed *'threshold'*, beyond which an all-or-nothing response, termed an action potential, ensues. Studies using intracellular micro-electrodes show that during an action potential the polarity of the membrane potential temporarily reverses (i.e. the cell interior becomes positive with respect to the interstitial fluid) before returning to its original value. For example, the membrane potential of a neurone may move from a resting potential of -70 mV to a peak of $+40$ mV and back again in less than 2 ms (Fig. 2.8). The magnitude of the action potential is not dependent on the magnitude of the signal initiating it. Moreover its duration is fixed and does not vary with the duration of the stimulus.

Immediately after each potential there is a period called the *absolute refractory period* during which a second action potential cannot be initiated. Hence, no matter what the stimulus, summation of action potentials is impossible. The absolute refractory period is followed by the *relative refractory period* during which the cell membrane can be brought to threshold only with a larger than normal stimulus. The durations of the absolute refractory periods of fibres of different sizes are shown in Table 2.1.

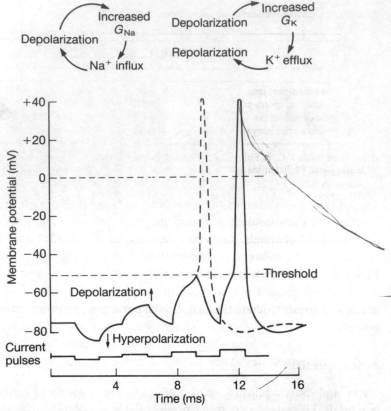

Fig. 2.8. Generation of an action potential. *Top:* diagrams showing the effects of changing the membrane potential on Na⁺ and K⁺ conductances during the depolarizing and repolarizing phases respectively of the action potential. *Bottom:* the responses of a neurone to localized current pulses beginning with a hyperpolarizing current and followed by a series of depolarizing currents of increasing strength. At threshold half of the stimuli will initiate all-or-nothing action potentials.

Ionic basis of the action potential

This was first described by Hodgkin and Huxley in the early 1950s on the basis of voltage-clamp experiments (Appendix, p. 93). When an excitable cell is depolarized slightly (but is still below threshold), Na⁺ conductance (G_{Na}) is increased by the opening of a small number of rapidly activated, voltage-dependent Na⁺ channels. Each opens for only a short time (milliseconds) but the total increase in Na⁺ influx results in Na⁺ making a greater contribution to the membrane potential and the cell is further depolarized (Section 1.6). At the same time, there is a net movement of K⁺ out and Cl⁻ into the cell because of the shift of the membrane potential away from the equilibrium potentials of these ions ($E_K \approx -90$ mV; E_{Cl} = resting membrane potential). As long as the depolarizing influence remains, the movement of K⁺ and Cl⁻ tends to *desensitised!*

Table 2.1. Properties of mammalian nerve fibres.

	Nerve fibre type		
	A	B	C
Fibre diameter, μm	1–20	<4	0·3–1·5
Conduction speed, m s^{-1}	5–120	3–15	0·6–2·5
Spike duration, ms	0·3–0·5	1·2	2·0
Absolute refractory period, ms	0·4–1·0	1·2	2·0

(Data from Ruch T.C. & Fulton J.F. (1960). *Medical Physiology and Biophysics*, 18th ed., p. 77. W.B. Saunders, Philadelphia.)

counter the increased contribution of Na$^+$ and limits the change in membrane potential. If the stimulus increases, the resulting depolarization opens more and more Na$^+$ channels until with a depolarization of about 15 mV the current generated by the influx of Na$^+$ is balanced by the movement of K$^+$ and Cl$^-$. This potential is the threshold for the initiation of action potentials. Beyond this point the process becomes self-regenerating, the increase in Na$^+$ influx leading to greater depolarization and the opening of many more channels which then allow Na$^+$ to enter. This explosive increase in G_{Na} continues until the membrane potential reaches a value that approaches E_{Na} ($\approx +70$ mV) and the potential would stay at this value but for the automatic closure of the Na$^+$ channels.

This automatic shutting of the voltage-sensitive Na$^+$ channels is known as *sodium inactivation* and inactivated Na$^+$ channels cannot be reactivated until the membrane potential has returned to a value that is near the resting membrane potential. The self-regenerating nature of the increase in G_{Na} and its inactivation ensures that the action potential is all-or-none and transient and that it is followed by a brief period of inexcitability—*the absolute refractory period*.

Repolarization following an action potential also depends on the outward movement of K$^+$ ions to bring the cell back to a potential near E_K. This is brought about by the opening of voltage-dependent K$^+$ channels in the membrane that react more slowly than Na$^+$ channels to depolarization. As a consequence K$^+$ conductance (G_K) reaches a maximum some time after the peak in G_{Na} (Fig. 2.9). The increased G_K remains until the potential has returned to resting membrane potential. This process, called *delayed rectification*, is responsible for the rapid repolarization of the membrane during termination of the action potential and together with residual Na$^+$ inactivation is the cause of the relative refractory period.

During each action potential the cell gains some Na$^+$ ions and loses some K$^+$ ions. In the Appendix (p. 92) it is calculated that only about 1 in 3000 of the K$^+$ ions in a nerve fibre of 20 μm diameter is actually exchanged for Na$^+$

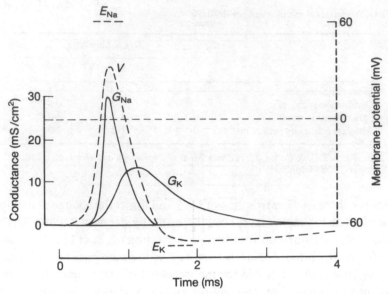

Fig. 2.9. The time-course of the computed ionic conductances during an action potential in a squid giant axon. (Adapted from Hodgkin, A.L. & Huxley, A.F. (1952) *J. Physiol.* **117**, 500–544.)

ions during the course of an action potential and this is insignificant. However, after several hundred impulses have passed, the total amount of K$^+$ ions lost becomes significant (particularly in small diameter fibres) and the ionic gradients must be re-established by the Na$^+$–K$^+$ pump using metabolic energy.

The existence of specific membrane channels is supported by the action of specific pharmacological blocking agents. For example, *tetrodotoxin* blocks voltage-dependent Na$^+$ channels and hence the increase in G_{Na}; *scorpion venom* prevents Na$^+$ inactivation and so, after initiation of an action potential, the cell remains in a depolarized condition; and *tetraethylammonium* blocks K$^+$ channels and leads to a delayed recovery of the action potential. *Local anaesthetics* prevent the generation of the action potential by inhibiting the voltage-dependent opening of the Na$^+$ channels.

Propagation of the action potential

A propagated action potential can be considered as a wave of excitation passing down an axon. At any one time only a portion of the axon is depolarized. This local depolarization gives rise to a passive spread of current up and down the axon (Fig. 2.10) and this current in its turn depolarizes adjacent regions ahead of the action potential, bringing them to threshold and so generating an action potential in a new region. As regions which have just been excited are refractory, propagation of the action potential is *unidirectional*.

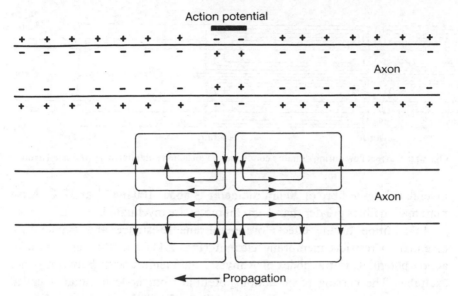

Fig. 2.10. Flow of current associated with the propagation of an action potential in an unmyelinated axon.

Because the action potential mechanism is relatively slow, the further local currents can spread the faster the action potential can travel. The *conduction velocity* can, in principle, be enhanced by increasing the *membrane space constant*, because current spreads further along the fibre, and by minimizing the *membrane time constant*, because the time taken to reach threshold decreases (Appendix, p. 90). The intracellular resistance is dependent on fibre diameter because in larger fibres more ions are available to carry the current, whereas the extracellular resistance is not normally a limiting factor because of the large volume of the extracellular fluid. In considering biological adaptations that can increase conduction velocity, one solution is to increase fibre diameter (thus decreasing intracellular resistance and increasing the space constant), as for example in squid whose giant nerve fibres have diameters of up to 1 mm. The conduction velocity of such a squid axon is about 25 m s^{-1} compared with about 1 m s^{-1} for small (about $1 \mu\text{m}$) unmyelinated fibres. An alternative development has been to increase the membrane resistance and to decrease the membrane capacitance by enveloping axons with myelin sheaths. Since resistances in series sum and capacitances in series sum inversely (Appendix, p. 89), by surrounding an axon with a spiral layer of myelin, the total membrane resistance is increased and the total membrane capacitance decreased. This greatly increases the space constant without increasing the time constant (Appendix, p. 90) and the conduction velocity of small nerve fibres is enhanced. The conduction velocity of a

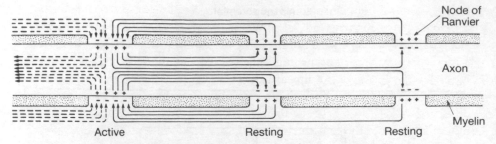

Fig. 2.11. Current flow during saltatory conduction of an action potential in a myelinated axon.

myelinated nerve fibre of 20 μm diameter is about 100 m s^{-1} at 37°C. Most mammalian fibres greater than 1 μm diameter are myelinated.

In addition to the effects on membrane resistance and capacitance myelination restricts membrane current (Fig. 2.11) and the generation of action potentials to the nodes of Ranvier; the membrane in between is not excitable. The passing of an action potential from node to node is called *saltatory conduction*. However, it should be realized that at any instant an action potential travelling along a myelinated axon will occupy a large number of nodes and not simply a single node. For example an action potential lasting 0·5 ms travelling at 100 m s^{-1} will occupy a 50 mm length of the axon. In summary the benefits of myelination are:

1 higher conduction velocities for rapid signalling;
2 smaller diameters for conserving space; and
3 higher metabolic efficiency because of the reduced flux of ions and hence the reduced expenditure of energy required to restore ionic concentrations.

Extracellular recording of nervous activity

Peripheral nerves contain both sensory and motor axons ranging from large-diameter, myelinated fibres with high conduction velocities to small, unmyelinated, slowly conducting fibres. During the conduction of an action potential along these axons, current is caused to flow in the tissue or medium surrounding the nerve. As the extracellular resistance is low (compared with the axonal membrane resistance and the intracellular longitudinal resistance), the field potentials generated in the surrounding medium will be small. However, if the whole nerve or an isolated axon is immersed in mineral oil, the increase in extracellular resistance by minimizing current dissipation results in the recording of greater potentials when two electrodes are placed on the surface of the nerve or axon.

Action potentials recorded by *extracellular electrodes* and a differential voltage amplifier are seen to be waves of *negativity* moving along the axon.

Before an action potential reaches the recording electrodes there is no potential difference between them (Fig. 2.12). When an action potential reaches the first electrode, the underlying membrane reverses its polarity and becomes negative with respect to the membrane under the second electrode. By convention this is shown graphically as an upward deflection. When the action potential is affecting the membrane under both electrodes simultaneously, the potential difference falls to zero between the electrodes. Later as the electrical activity moves away from the first electrode to affect predominantly the area of the membrane under the second electrode, the potential difference is reversed. As this recording shows both negative and positive components it is referred to as a *di-* or *biphasic* action potential. This recording can be simplified if the action potential fails to reach the second electrode. This can be accomplished

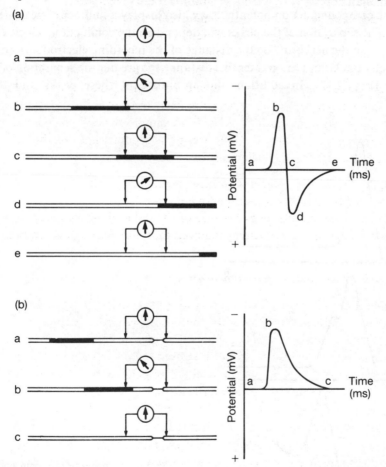

Fig. 2.12. Extracellular recording of an action potential conducted along an axon. (a) Biphasic record. (b) Monophasic record after crushing a portion of the axon between the recording electrodes. The potentials corresponding to the positions of the conducted wave are indicated by a, b, c, d and e.

if a small segment of the axon between the electrodes is crushed or a local anaesthetic applied. Then the action potential is recorded as a single negative deflection—a *monophasic* action potential (Fig. 2.12b).

These mono- or biphasic action potentials, when recorded from a single axon, show the all-or-none and refractory characteristics of the intracellularly recorded action potential. In contrast, when the activity in a whole nerve is recorded, the record is now the summed activity of action potentials being conducted along all the fibres. This is called the *compound action potential*. It is not all-or-none because of the different thresholds of various fibres. It increases in amplitude as the stimulus is increased until all fibres are excited. Its absolute refractory period is the absolute refractory period of the largest axons (Table 2.1).

The compound action potential may also display a number of peaks (Fig. 2.13). The separation of the peaks will depend on the conduction velocities of the fibres in the nerve and on the distance of the recording electrodes from the point of stimulation; the greater this distance the greater the separation of the peaks (Fig. 2.13). In addition, the amplitude of these peaks and their

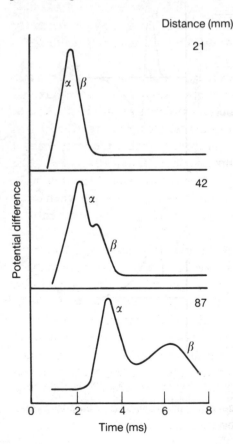

Fig. 2.13. Components of a compound action potential. (After Erlanger, J. & Gasser, H.S. (1937) *Electrical Signs of Nervous Activity*, p. 13. University of Pennsylvania Press, Philadelphia.)

relationship to each other depend on the fibre composition of the nerve. As the contribution of each fibre to the compound action potential is proportional to its diameter, the large fast-conducting fibres, which also have the lowest threshold, dominate the record even though the small fibres are more numerous.

For convenience, Gasser and his colleagues in the 1930s divided the axons in peripheral nerves into three broad groups (A, B and C) according to their conduction velocities. The A fibres included all the peripheral myelinated fibres from 1 to 20 μm in diameter; this group was further subdivided into groups of decreasing size and conduction velocity, the α, β, γ and δ fibres. The B group comprised the small myelinated fibres in visceral nerves (preganglionic autonomic) and the C group all small unmyelinated, afferent and efferent fibres. The conduction velocities and other properties of the fibres in the various groups can be seen in Table 2.1. A different grouping (I, II, III and IV) is often used for sensory fibres (p. 128). As mentioned earlier, the conduction velocity of a myelinated fibre is closely related to its fibre diameter since this determines the cross-sectional area and the intracellular resistance (p. 47). However the relationship between diameter and velocity is not linear; in the case of large fibres the conduction velocity (in m s^{-1}) is approximately the diameter (in μm) multiplied by 6, for small myelinated fibres the factor is 4·5 and for unmyelinated fibres it is 1·7.

Conduction along fibres may be blocked by cold, anoxia, compression and drugs. The conduction velocity decreases by approximately 3% of the maximal velocity for each 1°C fall. This is important in limbs or superficial tissues where the temperature may be well below the core temperature. Conduction block occurs in the large myelinated fibres at about 7°C and in the unmyelinated fibres at about 3°C. Compression and anoxia preferentially block large myelinated fibres which have a higher metabolic rate. Thus painful stimuli, which are transmitted in small unmyelinated fibres, can still be felt when other modes of sensation are lost. In contrast, local anaesthetics tend preferentially to block small fibres, the surface area to volume ratio of which is high.

As changes in the conduction velocity of peripheral nerves can also occur as a result of accident or disease, studies of the compound action potential can be useful clinically. Thus neural degeneration and subsequent regrowth lead to the formation of thinner fibres with lower conduction velocities. Similarly, demyelinating diseases, e.g. *multiple sclerosis* and *diphtheria*, may result in a reduction in thickness or a localized loss of myelin. Upon remyelination there is also a reduction in the distance between the nodes of Ranvier (internodal length). Under these conditions there is a decrease in the length constant and an increase in the membrane capacitance, both of which diminish the longitudinal spread of current and hence the conduction velocity. With localized loss of myelin there may also be a conduction block.

Spontaneous action potentials

Some neurones spontaneously undergo fluctuations in their resting membrane potentials. These may generate action potentials even in the absence of external excitation. Such fluctuations in membrane potential have recently been shown to be generated by specific changes in membrane conductances which may involve Na^+, Ca^{2+}, and Ca^{2+}-dependent K^+ conductance channels. Cells with these properties are important in the generation of rhythmic activity reflected in respiration and the electroencephalogram (EEG). They have also been shown to underlie the expression of some behavioural patterns in invertebrates.

Synaptic transmission

As already mentioned, neurones communicate with one another at synapses (Fig. 2.3). Transmission across the cleft separating the pre- and postsynaptic faces of opposing neurones is accomplished by *chemical* means. Rarely in the mammalian CNS but more commonly in invertebrates there are *electrical* synapses where the adjacent cells are joined by gap junctions allowing current to pass directly from one neurone to another.

A typical neurone in the central nervous system may receive inputs from many other neurones (*convergence*) and make synaptic contact with many other neurones (*divergence*) (Fig. 2.14). For example, a motor neurone in the ventral horn of the spinal cord may receive some 20 000–50 000 synaptic contacts. The so-called *synaptic boutons* (Fig. 2.3) are located mainly on the dendrites (axo–dendritic synapses) and cell body (axo–somatic synapses) of the neurone but in some cases they are found on the axon (axo–axonal synapses). Synaptic boutons release the contents of their synaptic vesicles into the intercellular space (*neurosecretion*). They contain mitochondria and the enzymes necessary to synthesize the neurotransmitter.

When an action potential invades a nerve terminal it causes an influx of Ca^{2+} ions into the terminal which triggers the release by exocytosis of neurotransmitter. This neurotransmitter diffuses across the cleft and binds to receptor molecules linked to ion-selective channels (*ionophores*) in the

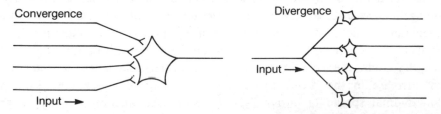

Fig. 2.14. Convergence and divergence of neural inputs.

postsynaptic membrane. This usually produces a permeability change the nature of which is characteristic of the receptor–ionophore complex, not of the transmitter. The *synaptic delay*, which is the time between excitation of the nerve terminal and the permeability change in the postsynaptic membrane, is less than 0·5 ms. Diffusion across the narrow cleft separating the pre- and postsynaptic membranes is very rapid and contributes only a small fraction to this delay; most of the delay can be attributed to the time for presynaptic release of the transmitter. The action of the transmitter is terminated by diffusion away from the cleft, by enzymatic breakdown or by re-uptake into the terminal or surrounding cells.

The change in permeability may produce a small local change in membrane potential (*synaptic potential*) which is either a depolarization (*excitatory postsynaptic potential*, EPSP) or a hyperpolarization (*inhibitory postsynaptic potential*, IPSP) (Fig. 2.15a & b). These small synaptic potentials last longer than an action potential and when they succeed each other by a sufficiently short interval they summate (*temporal summation*). Synaptic potentials elicited

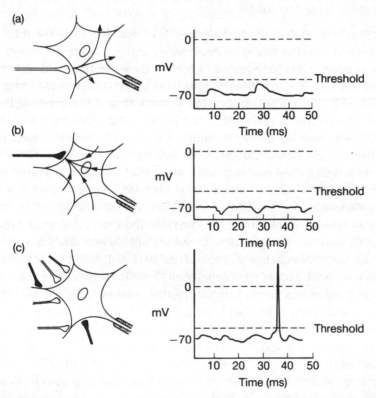

Fig. 2.15. Synaptic transmission in the central nervous system. (a) Excitatory postsynaptic potentials. (b) Inhibitory postsynaptic potentials. (c) Summation to threshold and generation of an action potential.

simultaneously at different sites on a neurone can also summate (*spatial summation*). If the net result is a depolarization of sufficient magnitude an action potential is triggered (Fig. 2.15c). Inhibitory synaptic potentials tend to prevent the cell from reaching threshold.

One particular part of the neurone, the so-called *initial segment*, which is situated close to the axon hillock, has a lower threshold than that of the dendrites and cell body, and so the action potential is usually initiated in this region. Action potentials are not normally fired in the dendrites or the cell body. While synapses located close to the initial segment may be more effective than those synapses located further away, there is evidence to suggest that some of the more distal synapses of motor neurones may compensate by generating larger synaptic potentials. The co-existence of excitatory and inhibitory synaptic inputs and their capacity for temporal and spatial summation allows the neurone to *integrate* signals from a variety of other neurones.

Ionic basis of synaptic potentials

By altering the relative conductances to Na^+ and K^+ ions, the membrane potential can be set to any value between E_K and E_{Na}. Synaptic transmission is mainly effected by changing one or both of these conductances, the relative contributions of which can be estimated from the *reversal potential* (Appendix, p. 94). An EPSP is a depolarization of a few mV (Fig. 2.15a) resulting from an increased conductance to both Na^+ and K^+ ions. Na^+ ions move into the cell and K^+ ions move out but, as the movement of Na^+ ions predominates, the net effect is slight depolarization of the neurone. This brings the membrane closer to threshold and makes it more likely that an action potential will be triggered. Some inhibitory transmitters activate only G_K, so that a slight hyperpolarization, i.e. an IPSP, occurs (Fig. 2.15b) moving the membrane further away from threshold. More commonly, there is a simultaneous increase in both G_K and G_{Cl}. As E_{Cl} is close to the membrane potential, an increase in G_{Cl} alone would not change the membrane potential, but it means that more excitatory current is required to depolarize the cell as the membrane resistance is lowered and in addition any depolarization is opposed by Cl^- moving into the cell. The combined effect of increasing G_K and G_{Cl} is to move the membrane potential to a value between E_K and E_{Cl}. Cells already depolarized by excitatory activity will produce larger IPSPs but cells in a hyperpolarized state will generate smaller or no IPSPs while still being effectively inhibited.

Synaptic potentials are local potentials that spread passively along the membrane. Both EPSPs and IPSPs can arise in the same neurone because of the co-existence at separate sites of different permeability mechanisms in the same cell. These permeability mechanisms in the membrane are ion-selective channels coupled to receptors which are activated by specific transmitters. In some cases, the interaction of a transmitter with its receptor may not be coupled directly to the opening of an ion channel but may instead result in an increase in the concentration of a specific intracellular messenger (e.g. cyclic AMP, Ca^{2+} and derivatives of inositol lipids) (see p. 254). Such messengers then start a cascade of events that modulate the activity of specific ion channels. The activation of these channels can be studied by the *patch-clamp* technique (Appendix, Chapter 1, p. 29).

Many transmitter substances have been discovered in brain and peripheral tissues, including acetylcholine, noradrenaline, dopamine, serotonin, glutamic acid, aspartic acid, λ-aminobutyric acid, glycine and also many peptides, such as substance P and opioid peptides. Some are excitatory, some are inhibitory, and some have both actions, their effects being determined not by their receptors but rather by the particular ion-selective channels to which their receptors are linked. Receptors may be clustered in the postsynaptic membrane or in some cases scattered over the surface of the cell. Indeed many substances appear to exert slow modulatory influences on the excitability of groups of nerve cells. Such an action is more aptly called *neuromodulation*. A list of established and putative neurotransmitters and neuromodulators is given in Table 2.2.

Table 2.2. Some of the substances that act, or are thought to act, as neurotransmitters or neuromodulators in the brain, spinal cord and peripheral nervous system.

Acetylcholine
Monoamines: adrenaline, noradrenaline, dopamine, histamine, serotonin (5-hydroxytryptamine)
Amino acids: aspartic acid, glutamic acid, glycine, γ-aminobutyric acid
Purine derivatives: adenosine, adenosine triphosphate
Peptides: adrenocorticotrophic hormone, angiotensin II, antidiuretic hormone, calcitonin gene-related peptide, cholecystokinin, β-endorphin, leu-enkephalin, met-enkephalin, gastrin, glucagon, gonadotrophin-releasing hormone, neuropeptide Y, neurotensin, oxytocin, somatostatin, substance P, thyrotrophin-releasing hormone, vasoactive intestinal polypeptide

---

Presynaptic inhibition

This is a different type of inhibition in which transmitter release is suppressed. It occurs where the terminal of one neurone makes synaptic contact with the synaptic ending of a second neurone which in turn excites a third neurone (Fig. 2.16). The first neurone controls the amount of transmitter released by the second neurone by depolarizing its terminal and causing partial inactivation of the voltage-dependent Na^+ mechanism. This in turn reduces the amplitude of any action potential that arrives at the axon terminal and presumably the influx of Ca^{2+} ions which determine the amount of transmitter released.

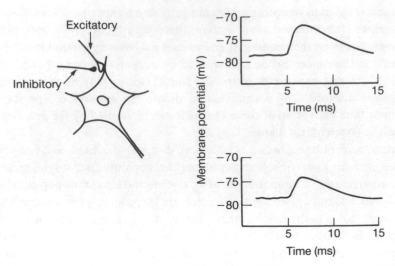

Fig. 2.16. An excitatory postsynaptic potential (a) before and (b) during presynaptic inhibition.

2.3 Growth, development and regeneration of the nervous system

Axonal transport

The transport of substances along the axon from the cell body of the neurone to its terminals (i.e. anterograde or centrifugal transport) or in the opposite direction (i.e. retrograde or centripetal transport) is called axonal or axoplasmic transport. Since there are no ribosomes in axons and nerve terminals, proteins required for the functions of nerve cell processes must be synthesized in the cell body and transported to the cell periphery. Rates of axonal transport have been determined by measuring the rate of accumulation of enzymes and transmitters in occluded nerves or, more precisely, by tracing the movement of radioactively labelled proteins after injection of labelled amino acids into regions containing cell bodies. This technique has revealed that there are several distinct rates of transport in axons—a fast rate of up to 400 mm per day, and two slower rates of 5–10 mm per day and 1–2 mm per day.

The slowest rate of transport appears to be that of the microtubules and neurofilaments. Other structural elements, notably actin, move somewhat faster. Fast axonal transport involves the movement of organelles such as vesicles along the axon. It is a local energy-dependent mechanism and is independent of the cell body and electrical impulses. Recent studies of the movement of vesicles in extruded axoplasm show that they travel along the microtubules.

A most elegant demonstration of retrograde transport comes from studies in which horseradish peroxidase is injected into the terminal regions. The intracellular localization of this enzyme can be demonstrated by histochemical techniques. It is taken up into the nerve terminal by endocytosis and subsequently appears in the cell body. This technique has proved to be of considerable value for tracing neural pathways in the brain. Rates of such retrograde transport approach that of fast anterograde transport. Substances of physiological and pathological importance that are taken up at the nerve terminal and transported retrogradely include nerve growth factor, tetanus toxin and poliomyelitis virus.

Axonal growth

The rate of growth of regenerating axons corresponds to the rate of slow axonal transport, i.e. about 1–2 mm per day. Structural components synthesized in the cell body and carried by axonal transport can be incorporated at the growing tip of the axon called the *growth cone*. The growth cone is motile and makes and breaks contact with its environment, apparently 'tasting' its surroundings to make sure its direction of growth is appropriate. For example, sympathetic and sensory fibres grow towards a source of nerve growth factor.

Aspects of neural development

Nerve growth factors

Substances derived from glial cells and target tissues can promote the development and survival of neurones. The best characterized of such substances is *nerve growth factor* (NGF), a protein of M_r 135 000. Sympathetic and sensory neurones have an absolute requirement for NGF during the early stages of their development and disappear following the administration of antibodies to NGF at this time. Sympathetic neurones also require NGF for optimal maintenance at later stages. NGF is produced by target tissues and is specifically taken up into the nerve terminals of sympathetic and sensory neurones by adsorptive endocytosis (p. 15) and then transported to their cell bodies.

Cell death

Many more nerve cells differentiate than actually survive into adult life. The amount of cell death is greatly increased if the targets for that particular group of cells are removed. Cells may also die if deprived of their afferent input. Once the initial development of the nervous system is complete, mammalian neurones do not divide.

Role of activity

There are certain critical periods in the development of young animals during which particular groups of neurones must be used if their synaptic connections are to be functionally maintained. For example, if one eye of a young kitten is covered during the second to the fourth month of age the animal remains functionally blind in that eye for the rest of its life and there is a permanent deficit in the number of cortical synaptic connections relaying information from the deprived eye to the visual cortex. Strabismus (squint or 'lazy eye') in children has a similar result; the critical period here is from 2 to 5 years of age.

Neural regeneration

Regeneration in the peripheral nervous system

If a peripheral nerve fibre is severed, the distal portion of the axon dies. Thereupon, Schwann cells surrounding the axon dedifferentiate and undergo mitosis, filling the nerve sheath distal to the cut. Cell bodies of cut neurones undergo chromatolysis, a process involving dispersion of Nissl substance and synthesis of new RNA. Thus RNA coding for proteins involved in synaptic functions is lost and replaced by newly synthesized RNA, which codes for proteins necessary for regrowth of the axon. The proximal ends of cut axons sprout and, provided that they can find the way to the distal portion of the nerve guided by Schwann cells, they grow within the old nerve sheath. This applies to both motor and sensory neurones. If, however, the more central portion (dorsal root) of a sensory nerve is damaged, regeneration into the spinal cord usually does not occur. There is little specificity in the regrowth of the peripheral axons, so that normal relations between particular motor neurones and muscles, and particular sensory neurones and their peripheral fields of innervation, are not accurately restored. The success of re-innervation of the original target area depends on the re-establishment of the original pathway. Hence, in attempting to restore function, it is helpful to suture the cut ends of the corresponding bundles (fascicles) together again.

Regeneration in the central nervous system

Damaged neurones in the CNS of higher vertebrates have a very limited capacity for regeneration. Axotomy often results in the death of the injured neurone, possibly because cells must be nourished by their connections in order to be maintained. Cells surviving axotomy do produce axonal sprouts, but these do not grow into their proper pathway and form a tangled mass near the point of section. If cells in the CNS are denervated, intact synapses in their vicinity may sprout collaterals which form inappropriate synapses on the denervated cells.

2.4 Divisions of the peripheral nervous system

The peripheral nervous system is that portion of the nervous system that lies outside the spinal cord and brain; it comprises both the somatic and the autonomic divisions.

Somatic nervous system

This division contains all the peripheral pathways responsible for direct communication with the environment. It includes all sensory (*primary afferent*) fibres from tissues such as eyes, ears, skin, joints and skeletal muscles and it includes the motor (*efferent*) fibres to skeletal muscles. The cell bodies of the sensory fibres are in the dorsal root ganglia or brain; the cell bodies of the motor nerves are in the spinal cord or brain. The sensory fibres and motor fibres are collected together to form *compound nerves* which are covered by a connective tissue sheath (*epineurium*) (Fig. 2.17). Within the nerve the fibres are arranged into bundles (*fascicles*); each fascicle is surrounded by a *perineurium* and each fibre by an *endoneurium*. The somatic sensory fibres range in size from small unmyelinated to large myelinated axons; the motor nerves are all myelinated. The most distal portions of the somatic nerves are usually branched. The terminal regions of any one sensory fibre go to adjacent receptors of the same type and the intramuscular portions of motor nerves branch to innervate a number of identical muscle fibres. Somatic motor neurones have only an excitatory effect on skeletal muscle; there are no peripheral inhibitory actions exerted on skeletal muscle.

Neuromuscular transmission

Motor fibres give off unmyelinated terminal branches which form synapses (*neuromuscular junctions*) with skeletal muscle fibres. In adult mammalian skeletal muscles there is only one neuromuscular junction in the middle of each skeletal muscle fibre. At the neuromuscular junction the terminal branch of the axon lies in a shallow depression of the muscle fibre surface. The pre- and postjunctional membranes are separated by the junctional cleft which is about 100 nm wide and contains a basement membrane (*basal lamina*). The postjunctional membrane is thrown into a series of folds, which together with the prejunctional membrane and junctional cleft, is called the *endplate* (Fig. 2.18).

(a)

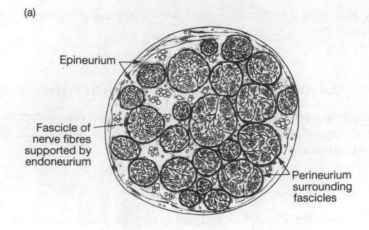

(b)

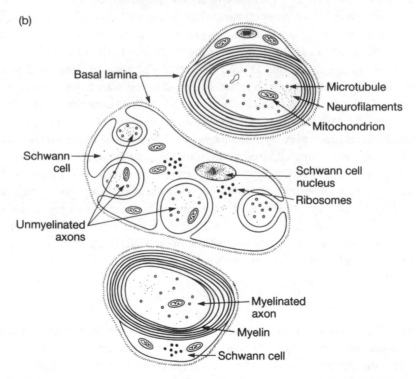

Fig. 2.17. (a) Cross-section of a peripheral nerve. (b) Enlargement of part of a fascicle showing myelinated and unmyelinated axons. Note that a number of unmyelinated axons are enclosed by a Schwann cell while each myelinated axon is enveloped by layers of myelin from a single Schwann cell.

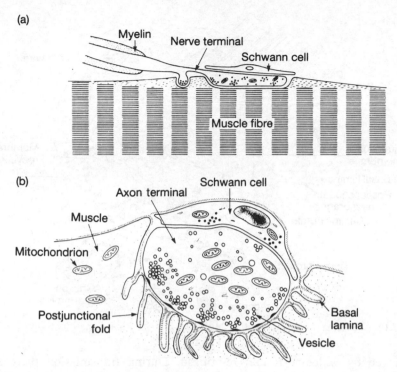

Fig. 2.18. Fine structure of a mammalian neuromuscular junction, (a) in longitudinal section and (b) in cross-section.

The neurotransmitter at the neuromuscular junction is *acetylcholine* (ACh). It is synthesized in the cytoplasm in the reaction:

$$CH_3COCoA \qquad + \quad HOCH_2CH_2N^+(CH_3)_3$$
acetyl coenzyme A   choline

$$\rightarrow \; CH_3COOCH_2CH_2N^+(CH_3)_3 \quad + HCoA$$
acetylcholine   coenzyme A

which is catalysed by the enzyme *choline acetyltransferase*. ACh is then packaged, together with ATP, into vesicles of about 50 nm diameter.

The mechanism of release of ACh (Fig. 2.19) is similar to that of other transmitters (p. 52). Briefly, when a nerve fibre is stimulated an action potential invades the nerve terminal causing the influx of Ca^{2+}. As a consequence ACh is released into the junctional cleft by exocytosis (Fig. 2.19), together with ATP and, probably, soluble proteins contained within the vesicles. This step in the release of ACh is blocked by *botulinus toxin* produced by the bacterium *Clostridium botulinum* and by high Mg^{2+} (*in vitro*). The vesicular membrane which fused with the prejunctional membrane during exocytosis is subsequently

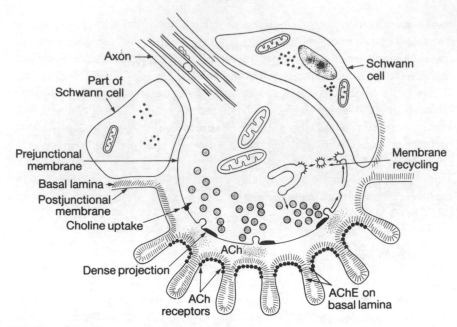

Fig. 2.19. Chemical transmission at the endplate.

removed by endocytosis and re-utilized. During transmission there is a synaptic delay of 0·2–0·3 ms. Much of this delay is attributed to the release of ACh, rather than to its subsequent diffusion across the cleft and the conductance changes that it provokes. Several hundred quanta of ACh, each quantum containing approximately 10 000 molecules, are released by each nerve impulse. A quantum represents the number of ACh molecules contained within each vesicle. More than sufficient numbers of ACh molecules are released from the nerve terminals during each nerve impulse to ensure the generation of an action potential in the skeletal muscle fibre that it innervates. For this reason neuromuscular transmission is said to have a high safety margin.

The ACh diffuses across the cleft and binds to receptor sites on channels (Fig. 1.2) that are localized on the crests of the postjunctional folds (Fig. 2.19). The receptors are referred to as *nicotinic* because nicotine mimics the effects of ACh. Interaction of ACh with its receptors is blocked reversibly by *tubocurarine* (curare) and almost irreversibly by α-*neurotoxins*. The neurotoxins, e.g. α-bungarotoxin, α-cobrotoxin, are small proteins (M_r 8000) obtained from snake venoms and, when labelled with radioactive iodine, have been used to estimate the number of ACh receptors (10^7) at the neuromuscular junction.

The binding of ACh to the two receptor sites on a closed channel induces a conformational change that opens the channel and increases G_{Na} and G_K. Recently it has been possible to estimate the opening time (about 1 ms) of individual channels by patch-clamp analysis (Appendix, Chapter 1, p. 29). The conductance change that results from the opening of a large number of these channels results in a net influx of positive ions and a localized depolarization of the endplate called the *endplate potential* (EPP) (Fig. 2.20a). It has been calculated from the reversal potential that opening of these channels by ACh increases G_{Na} and G_K in the ratio of $1\cdot3:1$ (Appendix, p. 94). Under normal circumstances the EPP exceeds threshold and an action potential is generated but in the presence of curare sub-threshold EPPs can be observed. In the absence of nerve impulses spontaneous release of ACh occurs. This is either *non-quantal*, presumably the result of continuous leakage from the nerve terminal, or *quantal*, giving rise to *miniature endplate potentials* (MEPPs) (Fig. 2.20b).

The ACh released by each impulse is rapidly hydrolysed (within a few ms) to choline and acetate by the enzyme *acetylcholinesterase* (AChE) which is located in the basal lamina of the postjunctional membrane. This step is blocked by organophosphates, such as diisofluorophosphate, which bind to the active site of the AChE. Any that diffuses away from the endplate into the bloodstream is destroyed by pseudocholinesterase. The choline is taken up by a specific transport mechanism in the nerve terminal and re-utilized in the synthesis of ACh.

Myasthenia gravis. This disease affects about 0·01% of the population and is characterized by a failure in neuromuscular transmission with repetitive

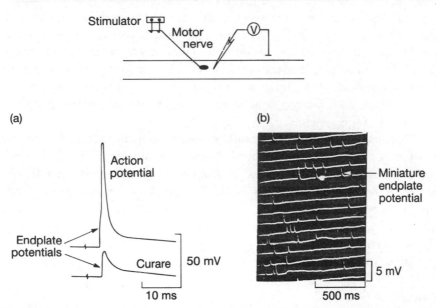

Fig. 2.20. Potentials recorded with a micro-electrode inserted into a skeletal muscle fibre adjacent to the endplate. (a) Evoked endplate potentials in the absence and presence of curare. (b) Spontaneous miniature endplate potentials.

nerve stimulation. The symptoms are weakness and fatigue, particularly of extraocular muscles, difficulty in swallowing and speech, and, in advanced stages, respiratory failure. Hyperplasia of the thymus gland is frequently associated with this disease.

Biopsy samples of intercostal muscles of myasthenic patients show a reduction in the amplitude of the EPPs and MEPPs. Binding studies with ^{125}I-labelled α-neurotoxin have shown that there may be a 90% decrease in endplate ACh receptors—a finding consistent with neurophysiological studies. The symptoms of this disease therefore appear to reflect a deficiency in the number of ACh receptors at the neuromuscular junction.

Removal of the thymus gland often has a beneficial effect and it has long been suspected that myasthenia gravis is an autoimmune disease. This has recently been confirmed by the demonstration of antibodies to ACh receptors in these patients.

Autonomic nervous system

The actions of the peripheral autonomic nervous system are normally involuntary and are directed to the control of individual organ function and to homeostasis. Classically the autonomic nervous system is regarded as being *solely motor* in function, its fibres going to cardiac muscle, to smooth muscle and to glands. However the terms 'autonomic afferents' and 'central autonomic processes' are also used. The peripheral autonomic nervous system is usually divided anatomically into the *sympathetic* and *parasympathetic systems*, and many tissues are innervated by both systems. When this occurs the two systems usually have opposing effects. In addition to these nerves there is a network of nerves that can act independently of the CNS. This network is often considered as another division of the autonomic nervous system and is referred to as the *enteric system*. Both the sympathetic and parasympathetic systems modulate the activity of this system.

The organization of the autonomic nervous system differs from the somatic division as all final motor neurones lie completely outside the CNS. The cell bodies of the peripheral neurones are grouped together to form *ganglia*. The efferent fibres passing from the CNS to the ganglia, the *preganglionic* fibres, are slow-conducting (class B and C) fibres which release the neurotransmitter *ACh*. The final motor neurones from the ganglia to the tissues, the *postganglionic* fibres, are largely slow-conducting, unmyelinated class C fibres. These fibres and fibres from visceral sensory receptors run together as *visceral nerves*.

Sympathetic system

All preganglionic sympathetic nerves have their cell bodies in the thoracic and upper lumbar segments (T_1–L_3). The axons, along with somatic motor fibres, pass out of the spinal cord in the ventral roots (Fig. 2.21). On leaving the spinal column, the preganglionic fibres separate from the somatic nerves to form the *white rami communicantes* and join a distinct group of sympathetic ganglia, the *vertebral* (*paravertebral* or *chain*) ganglia. These ganglia are segmentally arranged and lie along each side of the spinal column. The preganglionic fibres either:

1 synapse with postganglionic neurones in one or more of the vertebral ganglia; or

2 leave the vertebral ganglia in visceral nerves and pass to *prevertebral* ganglia in the abdomen or to the *adrenal medullae*.

Many of the postganglionic fibres from vertebral ganglia leave the ganglia as the *grey rami communicantes* and join the spinal nerves that pass to peripheral

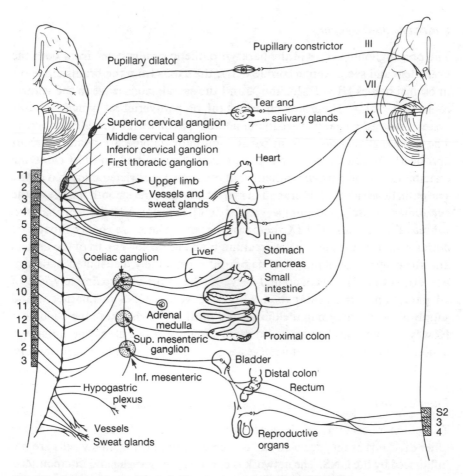

Fig. 2.21. Outflow of sympathetic (left) and parasympathetic (right) fibres.

tissue. Others, and also postganglionic fibres from prevertebral ganglia, join visceral nerves to particular organs. Most of these postganglionic fibres release the neurotransmitter *noradrenaline* (norepinephrine) and are referred to as adrenergic fibres. The adrenal medullae can be thought of as modified ganglia that release the hormone *adrenaline* (epinephrine), and also some noradrenaline, into the bloodstream rather than directly onto effector cells. Some sympathetic nerves that innervate the blood vessels of muscles, sweat glands and hair follicles in the skin release acetylcholine instead of noradrenaline. The distribution of the postganglionic sympathetic nerves is not necessarily the same as the somatic motor nerves from the same segment and there is a lot of overlap between adjacent outflows. In brief, and very approximately, the T_1 outflow passes to the head, T_2 to the neck, T_3–T_6 to the thorax, T_7–T_{11} to the abdomen and T_{12}–L_3 to the legs.

Parasympathetic system

The preganglionic nerves of the parasympathetic system come from both the brainstem and sacral spinal cord (S_2–S_4); the axons from the brainstem leave in cranial nerves III, VII, IX and X and the sacral axons leave in the ventral roots (Fig. 2.21). There are no vertebral or paravertebral ganglia in this system; instead all ganglia are found adjacent to or within the effector organ. The postganglionic fibres from parasympathetic ganglia are relatively short and nearly all release *acetylcholine* as a neurotransmitter. In most cases the distribution of the parasympathetic outflow is more restricted than that of the sympathetic system. Thus cranial nerve III supplies the smooth muscle of the eye (ciliary muscle and pupillary constrictor muscle), VII the lacrimal and submaxillary glands, and IX the parotid gland. The sacral parasympathetic outflow supplies the lower colon, rectum, bladder, the lower part of the ureters and the external genitalia. The major and most widely distributed parasympathetic outflow travels in cranial nerve X, the vagus. Approximately 70% of all parasympathetic preganglionic fibres leave the CNS in this nerve and supply all of the viscera in the thorax and most of the viscera in the abdomen. However, note that the vagus also carries many sensory fibres and a number of non-parasympathetic motor nerves, e.g. to laryngeal and pharyngeal muscle.

Enteric system

The autonomic nerves in the gastrointestinal tract differ from those in the other divisions as they form an extensive network in which many cells are not influenced by the CNS. The network is composed of ganglia and interconnecting bundles that lie in the wall of the intestinal tract (from the oesophagus to the rectum) and form the myenteric and submucosal plexuses (see Fig. 14.3). A number of different neuronal types have been identified in the plexuses but a detailed knowledge of the relationships between ganglia and between cells is lacking. However, it is recognized that this system contains sensory neurones, interneurones (integrative) and excitatory and inhibitory motor neurones. Many of the excitatory neurones (interneurones and motor neurones) release acetylcholine as a neurotransmitter, but others, particularly interneurones, may release 5-hydroxytryptamine (5HT, serotonin) or other compounds. The inhibitory neurones to smooth muscle do not release acetylcholine or noradrenaline as a transmitter and it has been suggested that they may be purinergic and release ATP, but more recent evidence suggests that a peptide (possibly vasoactive intestinal polypeptide, VIP) may be involved.

It should be noted that there is good evidence indicating that non-adrenergic, non-cholinergic nerves are also active in the regulation of other tissues, such as the smooth muscle of the respiratory tract and urinary system. The identity of the transmitter released by these fibres is not proven but it may be VIP or substance P.

Activity in autonomic neurones

The pattern of activity in the preganglionic autonomic nerves is controlled by the *hypothalamus*, and by centres in the *brainstem* and *spinal cord* (Section 2.5). The output to some tissues is phasic; to others it is continuous (tonic) and of low frequency (e.g. 1–2 Hz to blood vessels). The activity in all these pathways is either initiated or modified by sensory input from visceral or somatic receptors or by changes in emotional state. When the change in activity occurs as a result of visceral input it is often referred to as an *autonomic reflex*.

The activity in many postganglionic fibres is wholly dependent on preganglionic activity, and the ganglia act mainly as distribution centres for this activity. The extent of this distribution can be estimated from the ratio of preganglionic to postganglionic axons. It varies enormously but a ratio of 1:190 has been found in the human superior cervical ganglion and 1:30 in sympathetic-chain ganglia. However, distribution is not the only function of autonomic ganglia. They can also show integrative activity, namely:

1 postganglionic summation of preganglionic activity;
2 complex, excitatory–inhibitory synaptic potentials; and
3 inhibitory input from visceral receptors.

The integrative capacity of ganglia appears to be most highly developed in the enteric system as normal patterns of contractions can be maintained even when all connections to the CNS are severed. As local anaesthetics block this activity it is thought to be due entirely to peripheral reflex activity.

Synaptic transmission in autonomic ganglia

Within autonomic ganglia there are both divergence of the presynaptic fibres and convergence on to postsynaptic neurones. In some cases synaptic transmission from individual fibres is always successful and the ganglion cells act as relays; in others the summed activity of converging neurones is required for successful transmission.

In each case the pre- to postganglionic transmission is *cholinergic*, acetylcholine being released from the preganglionic terminal to act on postganglionic *nicotinic* acetylcholine receptors. The resultant changes in permeability lead to a fast EPSP (Fig. 2.22); if it is large enough or if several smaller EPSPs sum to reach threshold, an action potential will be initiated and propagated to the effector tissue. Although this aspect of transmission is similar to that in skeletal muscle (p. 59), ganglionic transmission does differ in that, first, the pharmacology of the nicotinic cholinoceptors is somewhat different and, secondly, the acetylcholinesterase in the synapses does not appear to be important in the termination of transmitter action; rather the transmitter diffuses away.

In addition to the fast excitation, preganglionic activity can lead to slow IPSPs (Fig. 2.22) due to small catecholamine- or peptide-containing 'interneurones' in the ganglia, and to slow EPSPs due to a late cholinergic effect on *muscarinic* cholinoceptors (p. 72).

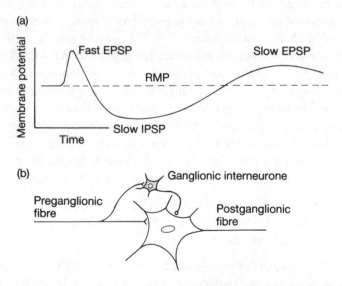

Fig. 2.22. Synaptic transmission in autonomic ganglia. (a) Synaptic potentials in a postganglionic neurone. (b) The neural pathways thought to be responsible for the fast EPSP and slow IPSP.

Neuroeffector innervation

Of the sympathetic and parasympathetic systems, the former has been studied more extensively. As described earlier the axons of these postganglionic adrenergic neurones are long and unmyelinated, but the terminal regions branch extensively. Within effector tissues the extent and arrangement of the terminal branches vary. In some tissues they remain in bundles with only the occasional single terminal branch evident, whilst in others single terminals predominate (Fig. 2.23). The neuroeffector junctions formed by the terminals also vary from narrow (25 nm) to wide (100 nm) junctional clefts. Irrespective of their structure the transmitting regions in these terminals are small swellings (approximately 1 μm in cross-section) called *varicosities*. Each neurone may have thousands of varicosities and each is responsible for the synthesis, storage and release of transmitter.

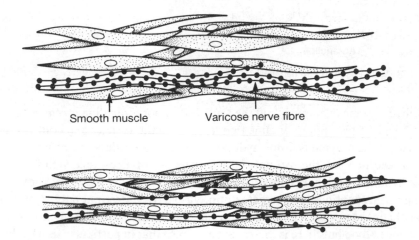

Smooth muscle　　　　Varicose nerve fibre

Fig. 2.23. Terminal branches of autonomic neurones. In some effector tissues they remain grouped together in bundles (top) but in others single fibres branch off from the bundle and the terminal branches make close contact with the effector cells (bottom).

Neuroeffector transmission

In parasympathetic nerves acetylcholine is synthesized as it is in somatic motor neurones (p. 61). In adrenergic neurones noradrenaline is synthesized from tyrosine in reactions catalysed by the enzymes outlined in Fig. 2.24. All the reactions but the last take place in the cytoplasm. The final step, from dopamine to noradrenaline, occurs in the storage vesicles. In addition to containing the transmitter and the enzyme dopamine β-oxidase, vesicles also contain chromogranins (proteins) and ATP, the functions of which are unknown. The rate-limiting step in this pathway is the conversion of tyrosine

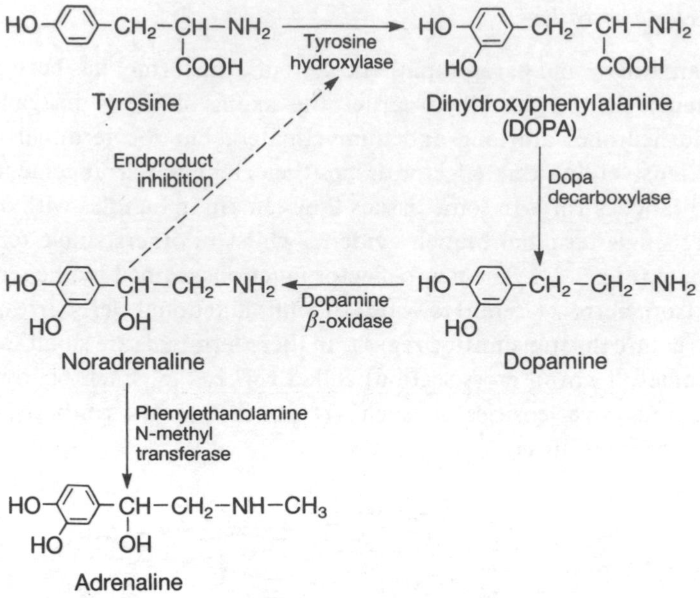

Fig. 2.24. Synthesis of noradrenaline.

to dihydroxyphenylalanine, and the synthesis of noradrenaline is regulated by a small quantity of noradrenaline that leaks from the vesicles and controls the activity of the enzyme tyrosine hydroxylase (i.e. end-product inhibition).

In addition to the primary transmitter substances, acetylcholine or noradrenaline, both parasympathetic and sympathetic neurones are known to release neuropeptides (e.g. somatostatin, vasoactive intestinal polypeptide or enkephalin). At present the physiological significance of these agents is not fully understood but there is increasing evidence that in particular tissues they may have a role as *cotransmitters* while in other tissues they may act as neuromodulators.

Like chemical transmission at other sites, the release of transmitter from the *prejunctional* terminal follows invasion of the terminal by an action potential and an influx of Ca^{2+}. The contents of the granules are then released by exocytosis from the varicosities into the junctional regions where they are free to interact with receptors (Fig. 2.25a). Unlike skeletal muscle where receptors are limited to the neuromuscular junctions, the postjunctional receptors are not all localized beneath the nerve and many are spread over the surface of the effector cell. In addition to these receptors on the effector cell there are receptors on the prejunctional membrane which are thought to control the release of transmitter.

The noradrenaline released from sympathetic nerves (and adrenaline from the adrenal medullae) produce their effects by interacting with α- or β-

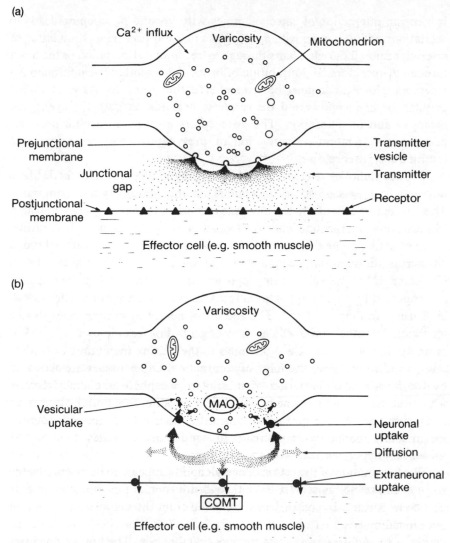

Fig. 2.25. (a) Chemical transmission at the sympathetic neuroeffector junction. (b) Inactivation of noradrenaline at the sympathetic neuroeffector junction.

adrenoceptors as described by Ahlquist in 1948. These receptors are classified according to the effect of natural and synthetic catecholamines and pharmacological blocking compounds. The catecholamines have the following orders of efficacy on adrenoceptors:

1   α_1-receptors, adrenaline > noradrenaline > isoprenaline;
2   α_2-receptors, adrenaline > noradrenaline (isoprenaline is ineffective);
3   β_1-receptors, isoprenaline > adrenaline = noradrenaline;
4   β_2-receptors, isoprenaline > adrenaline > noradrenaline.

In general, interaction of catecholamines with α_1- and β_1-receptors leads to excitation and interaction with β_2-receptors to inhibition. Thus stimulation of arteriolar smooth muscle occurs through α_1-receptors, stimulation of the heart through β_1-receptors, and inhibition of bronchial smooth muscle through β_2-receptors. However it must be remembered that some tissues have mixed populations of receptors and the response depends on both the agent, the receptors and their location. The α_2-receptors are found on both pre- and postsynaptic membranes; activation of presynaptic receptors appears to inhibit transmitter release.

The acetylcholine released from the postganglionic parasympathetic fibres interacts with *muscarinic* cholinoceptors on the postjunctional membrane. These receptors differ from the nicotinic receptors, in that they are stimulated by muscarine, not nicotine, and are blocked by atropine and not tubocurarine.

In most tissues the excitatory effects of transmitters appear to arise through an increase in membrane conductance, mainly to Na^+ but in some also to Ca^{2+} and K^+. The subsequent depolarization of the effector cell usually accompanied by an action potential initiates the response of the effector cell by causing an influx of Ca^{2+}. The actions of neurotransmitters may also be mediated by other intracellular messengers. For example, cyclic AMP promotes the opening of Ca^{2+} channels in the plasma membrane of cardiac muscle and so increases the force of contraction. Other tissues are activated by the intracellular formation of inositol triphosphate and diacylglycerol which release intracellular stores of Ca^{2+} and stimulate protein kinase C, respectively (p. 156). Inhibitory actions of the transmitters can be accounted for in many tissues by an increase in conductance, mainly to K^+, and subsequent hyperpolarization of the membrane.

The termination of the action of acetylcholine appears to be essentially the same as in skeletal muscle (p. 63) although diffusion out of the junction may also be important. In contrast, once released from the neurone the action of noradrenaline is *not* terminated by degradation but it is *inactivated* either by *uptake* or by *diffusion* away from the junctional region. The uptake processes transport the intact amine across plasma membranes either into the neurone (in which case it is called re-uptake, or *neuronal uptake*) or into the effector cell (*extraneuronal uptake*) (Fig. 2.25). Estimates of the relative activities of these two processes vary greatly but it would appear that in densely innervated tissues some 70% of the amine released is removed by neuronal uptake and 25% by extraneuronal uptake. Extraneuronal uptake and diffusion may be more important in diffusely innervated tissues. The major portion of the amine accumulated by the neurone is transported into vesicles for re-use as a transmitter, with obvious benefit to the neurone.

That portion of the transmitter taken up by the nerve and not recycled is degraded by *monoamine oxidase* (MAO) whilst the amine taken up extraneuronally is degraded largely by *catechol-O-methyltransferase* (COMT). As the effects of catecholamines are not potentiated immediately by inhibition of either or both of these enzymes it is believed that they are not involved in the termination of transmitter action. Rather, enhanced responses may be seen in the presence of inhibitors of neuronal uptake (e.g. cocaine and desipramine) or of inhibitors of extraneuronal uptake (e.g. corticosteroids).

Opposing actions of the parasympathetic and sympathetic systems

Many tissues receive a dual autonomic innervation and stimulation of one component usually results in effects opposite to those produced by stimulation of the other (Table 2.3). Further, it is clear that the effect of stimulating any one component may vary from tissue to tissue, being excitatory in some tissues and inhibitory in others. However, a closer consideration of Table 2.3 reveals that the *parasympathetic* effects are largely directed towards *maintenance* and

Table 2.3. Effects of sympathetic and parasympathetic stimulation.

Effector	Parasympathetic stimulation	Sympathetic stimulation	Adreno-ceptors
A. Muscle			
Gastrointestinal:			
Longitudinal and circular	Increase motility	Decrease motility	α, β
Sphincters	Relax	Contract	α
Urinary bladder:			
Detrusor	Contract	Relax	β
Trigone (internal sphincter)		Contract	α
Cardiac:	Decrease rate	Increase rate	β_1
	Decrease force (atria only)	Increase force	β_1
Blood vessels:			
Arterioles			
skin and mucosa		Vasoconstrict	α
abdominal, mesenteric		Vasoconstrict	α
skeletal		Vasoconstrict	α
		Vasodilate (circulating adrenaline)	β_2
		Vasodilate (cholinergic)	
coronary	Vasodilate	Vasodilate	β_2
		Vasoconstrict	α
pulmonary	Vasodilate	Vasoconstrict	α
brain	Vasodilate	Vasoconstrict	α
penis, clitoris	Vasodilate	?Vasoconstrict	α
Veins		Vasoconstrict	α
Splenic capsule		Contract	α
Tracheal–bronchial	Contract	Relax	β_2

Table 2.3. contd.

Effector	Parasympathetic stimulation	Sympathetic stimulation	Adreno-ceptors
Eye:			
Dilator of pupil		Contract	α
Sphincter of pupil	Contract		
Ciliary	Contract	Slightly relaxed	β_2
Genital organs:			
Seminal vesicle		Contract	α
Vas deferens		Contract	α
Uterus		Relax (depends upon hormonal status)	β_2
Pilo-erector		Contract	α
B. Glands			
Salivary	Voluminous serous secretion	Decrease mucous secretion (submaxillary)	α
Pancreas and liver	Secrete	Decrease secretion or no effect	α
Nasopharyngeal	Secrete	Decrease secretion or no effect	
Bronchial	Secrete	Decrease secretion or no effect	
Sweat		Secrete (cholinergic)	
Tear	Secrete		
C. Metabolism			
Liver		Glycogenolysis Gluconeogenesis	β_2
Adipose cells		Lipolysis	β_1
Insulin		Reduced	α

conservation of bodily function. Thus, responses to parasympathetic stimulation include slowing of the heart, constriction of the pupils, contraction of the bladder (detrusor muscle) and increased secretion and motility in the digestive tract.

In contrast, the effects of *sympathetic* stimulation are directed toward coping with stress and these tissue responses together comprise the '*fight or flight*' response described by Cannon in 1939. These changes include increased heart rate and contractility, bronchodilatation, pupillary dilatation, inhibition of intestinal motility, constriction of the splanchnic vascular bed, decreased muscle fatigue, and elevated blood glucose and free fatty acids. However, this is not to say that the sympathetic system always acts *en masse* or that it is active only in stress. Many sympathetic effects are associated with normal activity and are localized (e.g. pupillary dilatation and regional changes in blood flow), and these can occur without substantial changes in other tissues.

The activity of tissues having a dual innervation depends on the balance between parasympathetic and sympathetic discharge. Often one of the systems is dominant, for example, the diameter of the pupils and the resting heart rate are largely determined by the level of activity (tone) in their parasympathetic nerve supplies. Changes in effector activity are usually the result of reciprocal changes in both parasympathetic and sympathetic activity. Mutual antagonism in the periphery can occur:

1 as a result of dual reciprocal innervation of effector cells, such as occurs in heart, bronchial smooth muscle and detrusor smooth muscle; and
2 from suppression of synaptic transmission in ganglia.

The latter has been found in parasympathetic ganglia supplying detrusor muscle and in the ganglia that form the myenteric plexus of the intestinal tract. In fact most of the sympathetic supply to the intestinal muscle layer, excluding that accompanying blood vessels, ends in the myenteric plexus and appears to act by presynaptic inhibition of transmitter release.

Circulating catecholamines

The activity of some autonomically innervated tissues is influenced by both noradrenaline released from nerves and by catecholamines released into the bloodstream from the adrenal medullae. In man, adrenaline comprises some 80% of the catecholamines released from the gland, the remainder being noradrenaline. The adrenaline and noradrenaline are synthesized and stored in different cells. An additional step in the synthetic chain (Fig. 2.24) produces adrenaline by *N*-methylation of noradrenaline. The activity of the enzyme *phenylethanolamine*-N-*methyltransferase* responsible for this step is increased by steroids from the adrenal cortex.

The free plasma concentrations of these amines are low at rest, being 1–2 nmol 1^{-1} for noradrenaline and 0·2–0·8 nmol 1^{-1} for adrenaline. All of the adrenaline comes from the adrenal medullae and the noradrenaline from the adrenal medullae and sympathetic nerves. Low levels are maintained at rest because the rate of release from the adrenal medullae is normally low and the half-life of the circulatory catecholamines is relatively short (less than 1 min in experimental animals). During periods of stress, physical or emotional, the rate of release of adrenaline can increase eight- to tenfold.

Adrenaline has many actions similar to noradrenaline but, because it is a more potent β-agonist, it has more pronounced metabolic actions. These include elevation of blood glucose and free fatty acids and an increase in metabolic rate. Adrenaline also produces vasodilatation in those vascular beds in which the β- to α-receptor ratio is high, e.g. in skeletal muscle and heart muscle.

2.5 Basic design of the central nervous system

The CNS comprises the *brain* lying within the skull and the *spinal cord* lying within the vertebral column. During embryonic development the CNS initially develops as an infolding of the ectoderm, forming a hollow tube called the neural tube. On either side of this tube is found neural crest tissue which gives rise to dorsal root ganglion cells, autonomic ganglion cells and a wide variety of non-neural cells including Schwann cells and endocrine cells. The inner surface of the neural tube is the principal site of generation of nerve cells which migrate in an orderly way to regions where they differentiate. Migrating nerve cells are closely associated with specialized glial cells which apparently act as guides for this migration.

As the neural tube grows it forms three longitudinal swellings at its anterior end. These swellings develop into the *fore-*, *mid-* and *hindbrain*. They retain their cavities (*ventricles*) which remain in communication with the canal in the *spinal cord* which is formed from the remainder of the tube. The forebrain is divided into two regions, rostrally the *telencephalon* and caudally the *diencephalon*. The telencephalon is formed by a midportion and two lateral outpouchings, the primitive *cerebral hemispheres* (*cerebrum*), which grow first forward and then backward to cover most of the mid- and hindbrain (Fig. 2.26a, b). The diencephalon gives rise to the *thalamus* and *hypothalamus*. The hindbrain gives rise to the *pons* and *medulla*, which together with the midbrain from the *brainstem*. Further smaller raised swellings develop in the hindbrain to form the *cerebellum*. The cerebellum and the cerebral hemispheres have two unique features:

1   their surfaces are extensively folded, forming depressions called fissures or *sulci* and raised portions called *gyri*; and
2   unlike the brainstem and spinal cord they have their grey matter on the outside and white matter on the inside.

Spinal cord

The spinal cord has a segmental structure with pairs of nerve roots, one pair on each side, arising at more or less regular intervals (Fig. 2.27a). Each pair comprises the *dorsal root* carrying information into the spinal cord from peripheral receptors and the *ventral root* carrying signals out which control affector organs such as muscles. In cross-section, each segment of the spinal cord shows a central butterfly-shaped area of grey matter and a peripheral zone of white matter (Fig. 2.27b). The grey matter contains cell bodies of neurones. In the ventral horn of the grey matter lie the cell bodies of motor neurones whose axons leave in the ventral roots to innervate muscles; in the dorsal horn of the grey matter lie cell bodies of interneurones concerned with the processing of signals entering in the dorsal root axons.

(a)

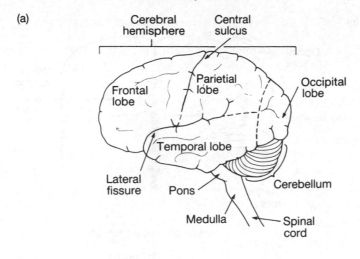

(b)

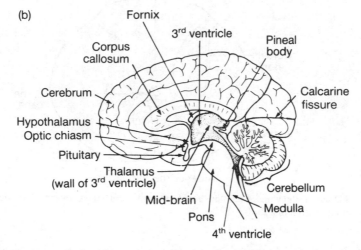

Fig. 2.26. The human brain: (a) lateral view; (b) view of a midsagittal section.

The white matter consists of the axons of cells in the brain descending to spinal cord cells and the axons of cells in the spinal cord ascending to synapse with cells in the brain. As like axons travel together, well-formed bundles (*tracts*) can be distinguished. Thus, in the dorsal quadrant of the white matter we can distinguish the dorsal columns composed of axons ascending to the brainstem with information from skin, joint and muscles; in the lateral quadrant there are both ascending tracts to the cerebellum and brainstem and descending tracts from the cerebral cortex and brainstem. The spinal cord segments perform the initial processing of afferent information and contain the neural circuits for many reflexes, some of which are the basis of movement and posture.

(a)

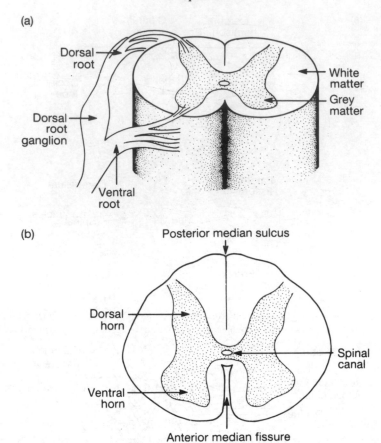

(b)

Fig. 2.27. (a) Diagram showing the relationship of the dorsal and ventral roots to the spinal cord. (b) Cross-section of the spinal cord.

Brain

The brain consists of three major subdivisions—the *brainstem*, the *cerebellum*, which is attached to the brainstem, and the *diencephalon and cerebrum* (Figs 2.26 & 2.28).

Brainstem

This is an ancient part of the brain in an evolutionary sense. Although much smaller than the cerebellum, the brainstem is vital to life while the cerebellum may, with considerable residual but not life-threatening disability, be dispensed with. The brainstem links spinal cord and cerebrum and is composed of three regions—the medulla oblongata, pons and midbrain (Fig. 2.28).

The *medulla* is continuous with the spinal cord and contains the same fibre tracts. The grey matter, however, is not organized in a continuous column but

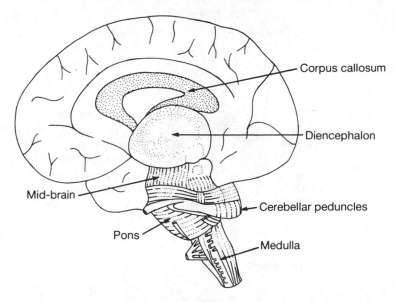

Fig. 2.28. A lateral view of the brainstem after removal of the left cerebral hemisphere and the cerebellum. (Redrawn from Curtis, B.A., Jacobsen, S. & Marcus, E.M. (1972) *An Introduction to the Neurosciences*, p. 13. W.B. Saunders Co., Philadelphia.)

is broken into discrete nuclei, including motor and sensory nuclei for the throat, mouth and neck, and nuclei involved in the control of the respiratory and cardiovascular systems and of movement and posture.

The *pons* may be recognized on its ventral surface (Fig. 2.29a) by the bulge of the brainstem formed by axons descending from the cerebrum and turning up into the cerebellum which, in the intact brain, conceals the dorsal surface of the pons. The pons is continuous with the medulla and contains the same ascending and descending tracts in the reticular grey matter (*reticular formation*) that are involved in the control of the respiratory and cardiovascular systems. The pons also contains the nuclei of nerves for the motor and sensory functions of the face.

The *midbrain*, the smallest part of the brainstem, is continuous with the pons below and the diencephalon above. The ventral surface is characterized by the large *cerebral peduncles* (Fig. 2.29a) lying laterally and carrying axons from the cerebrum to the brainstem and to the cerebellum. The dorsal surface may be recognized by two pairs of protuberances, the *superior* and *inferior colliculi* (Fig. 2.29b). The superior collicular neurones are concerned with the processing of visual information and the inferior collicular neurones with auditory signals. The midbrain also contains nuclei concerned with the state of wakefulness of the brain and ascending tracts from the spinal cord on their way to the diencephalon.

(a)

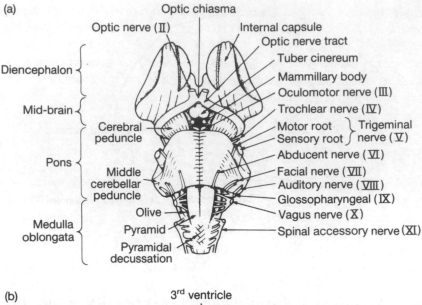

Optic chiasma

Optic nerve (II)   Internal capsule
                                    Optic nerve tract
Diencephalon   Tuber cinereum
                                    Mammillary body
Mid-brain   Oculomotor nerve (III)
                                    Trochlear nerve (IV)
            Cerebral   Motor root   ⎫ Trigeminal
            peduncle   Sensory root ⎬ nerve (V)
Pons   Abducent nerve (VI)
            Middle   Facial nerve (VII)
            cerebellar   Auditory nerve (VIII)
            peduncle   Glossopharyngeal (IX)
            Olive   Vagus nerve (X)
Medulla   Pyramid   Spinal accessory nerve (XI)
oblongata
            Pyramidal
            decussation

(b)

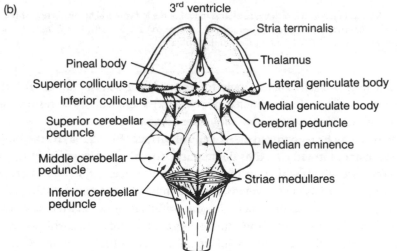

3rd ventricle

                                    Stria terminalis
Pineal body   Thalamus
Superior colliculus   Lateral geniculate body
Inferior colliculus   Medial geniculate body
Superior cerebellar   Cerebral peduncle
peduncle   Median eminence
Middle cerebellar
peduncle   Striae medullares
Inferior cerebellar
peduncle

Fig. 2.29. The brainstem: (a) ventral view; (b) dorsal view. (Redrawn from Truex, R.C. & Carpenter, M.B. (1969) *Human Neuroanatomy*, 6th edn, p. 31. Williams & Wilkins Co., Baltimore.)

From the brainstem at different levels the various *cranial nerves* have their superficial origins (Fig. 2.29a). There are twelve pairs of these and they innervate the skin and muscles of the face, structures in the head and neck and, in the case of the vagus (X) nerves, the thoracic and many of the abdominal viscera. Strictly speaking, the olfactory (I) nerves are connected to the forebrain, and not the brainstem. The optic (II) nerves are joined to the diencephalon via the optic chiasma. Cranial nerve pairs III and IV emerge from the midbrain, V, VI, VII and VIII from the pons and the remaining four pairs from the medulla.

The nuclei from which these cranial nerves have their origins, in some instances, extend through a considerable length of the brainstem. Note that many of the nerves contain a mixture of fibres, motor and sensory, somatic and autonomic.

Cerebellum

This is a large midline organ overhanging the dorsal aspect of the brainstem to which it is attached by three large bundles of white matter on each side— the superior, middle and inferior *cerebellar peduncles* which carry neural signals in and out of the cerebellum (Fig. 2.30a). Three cerebellar lobes can be recognized. On the superior aspect (Fig. 2.30b) the anterior lobe is separated from the larger middle lobe by the primary fissure. The anterior and middle lobes are further divided into a narrow median strip—the *vermis*—and two lateral *cerebellar hemispheres*. Rostrally, and seen only on the inferior aspect (Fig. 2.30c), are the *flocculus* and *nodule* together forming the *posterior lobe*. The anterior and middle lobes are transversely folded, the folds being termed *folia*. This folding vastly increases the surface area of the cerebellum.

All parts of the cerebellum have the same structure—a thin superficial layer of grey matter, the *cerebellar cortex*, covering a much larger expanse of white matter formed by axons entering and leaving the cortex. Deep in the white matter, close to the brainstem, lie the *intracerebellar* nuclei. A section at right angles to the long axis of a folium shows that it is much wider laterally than medially and that subsidiary foldings of the folium, each with cortical grey matter and white matter, give the section the appearance of a tree, hence the name *arbor vitae* (Fig. 2.30a). The cerebellum has its major role in the control of the rate, range and direction of movement. The small, phylogenetically old, posterior lobe is connected to the organ of balance (labyrinth) and is concerned with the reflexes which ensure an upright posture.

Diencephalon and cerebrum

The *diencephalon*, nearest to the cerebrum, is covered by it in the intact brain and can be seen only if a brain is sectioned transversely or if one hemisphere is removed (Fig. 2.28). Developmentally it represents that part of the forebrain from which the rudimentary cerebral hemispheres have budded off. Its walls have been thickened by the growth on either side of the mass of grey matter that constitutes the *thalamus*, and ventrally by the structures that make up the

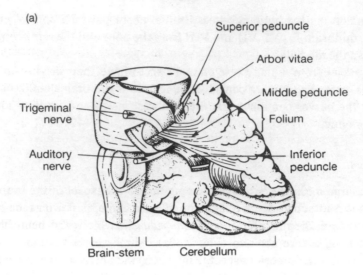

(a)

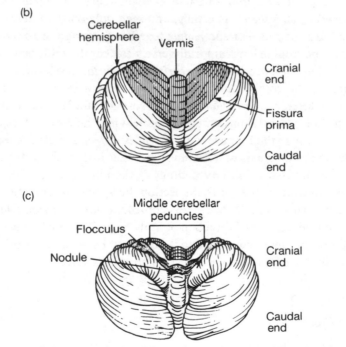

(b)

(c)

Fig. 2.30. The cerebellum: (a) lateral aspect; (b) superior aspect; (c) inferior aspect. (Redrawn from *Gray's Anatomy* (1949), eds T.B. Johnson & J. Willis, 30th edn, pp. 949 & 954. Longmans, Green & Co., London.)

hypothalamus, so that the original cavity has become the narrow cleft of the *third ventricle* (Fig. 2.26b). The *thalamic nuclei* function in at least three ways. One group relays signals concerned with all types of afferent information, except olfaction, to the cerebral cortex on the same side. Another group relays signals to the motor cortex on the same side and receives information from the cerebellum and basal ganglia. A third group is concerned with sleep and wakefulness. The *hypothalamic nuclei* serve as regulating centres for autonomic functions (e.g. body temperature, heart rate, blood pressure). They control the release of hormones by the pituitary gland, and are involved in the expression of the emotions and the regulation of food and water intake.

The *cerebrum* consists of the right and left *cerebral hemispheres*. Connecting them in the midline is a large thick band of white matter known as the *corpus callosum* (Fig. 2.26b) made up of axons passing between the hemispheres. Each hemisphere comprises an outer layer, about 1 cm thick, of grey matter which is the *cerebral cortex*, covering a dense thick inner layer of white matter. The grey matter has a rich blood supply; the white matter is less well endowed. The white matter is composed of axons connecting different regions of the cortex with each other and connecting the cortex with the rest of the brain. Within the white matter of each hemisphere is a collection of neuronal cell bodies lying between the thalamus and the cerebral cortex. These nuclei, known as the *basal ganglia*, are concerned with the initiation and control of movement.

The cerebral cortex is folded into gyri and sulci, thereby increasing the surface area. The deepest sulci are called *fissures*. On the lateral surface of each hemisphere (Fig. 2.26b) the *lateral fissure* partially separates off the temporal lobe from the rest. This lobe contains the *primary auditory cortex* which receives signals from the auditory receptors in the inner ear.

The *central sulcus* runs from the medial surface to the lateral fissure (Fig. 2.26a). It is not as prominent as the lateral fissure but there are usually well-formed and continuous gyri on either side of the sulcus. The central sulcus forms the posterior boundary of the frontal lobe. The gyrus forming this posterior boundary and the anterior wall of the central sulcus is the *precentral gyrus*. This is the *primary motor cortex* containing neurones whose axons run down through the brainstem and spinal cord and synapse with motor neurones.

The *parietal lobe* lies behind the central sulcus. The posterior border of the sulcus is the postcentral gyrus of the parietal lobe—the *primary sensory cortex* which receives information from receptors in skin, joint and muscle.

The *occipital lobe* is most posterior and little of it can be seen on the lateral surface. Examination of the medial surface (Fig. 2.26b), however, shows the *calcarine fissure* of the occipital lobe where the *primary visual cortex* is located. The occipital lobe is entirely concerned with the processing of visual information.

The *corpus callosum* connects the two hemispheres and is surrounded by the *cingulate gyrus* which is involved in the control of emotional behaviour.

Meninges, ventricles and cerebrospinal fluid

The CNS is covered by three membranes or *meninges* (Fig. 2.31) comprising a tough outer layer, the *dura mater*, a middle layer, the *arachnoid mater*, in which lie the blood vessels, and an inner layer, the *pia mater*. Between the arachnoid mater and the pia mater is the *subarachnoid space* which is in communication with the ventricles and contains *cerebrospinal fluid* (CSF), a specialized extracellular fluid. At the base of the brain, the subarachnoid space becomes enlarged to form cisterns, the largest being the *cisterna magna* (Fig. 2.32a).

The ventricles comprise the *lateral ventricles* in the cerebral hemispheres, the *third ventricle* in the diencephalon and the *fourth ventricle* in the hindbrain. The third and fourth ventricles are connected in the midbrain by a canal called the *cerebral aqueduct* (Fig. 2.32a).

Blood–brain barrier

Slow diffusion of many substances of low M_r between the blood and CSF and between the blood and brain suggests the existence of a *blood–CSF barrier* and a *blood–brain barrier*. These barriers are permeable to respiratory gases, to glucose and to fat-soluble drugs like volatile anaesthetics. The endothelial cells of the capillaries within the CNS are held together by 'tight' junctions and it is these which limit diffusional exchanges of water and water-soluble solutes. However, these cells are involved in transport of solutes between blood and brain interstitial fluid. Note that the blood–brain barrier is absent in certain structures called *circumventricular organs* which abut on the third and fourth

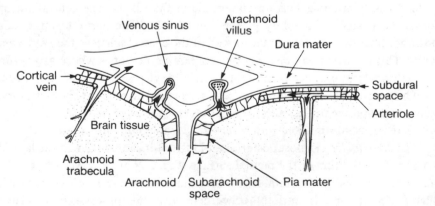

Fig. 2.31. Meninges of the brain.

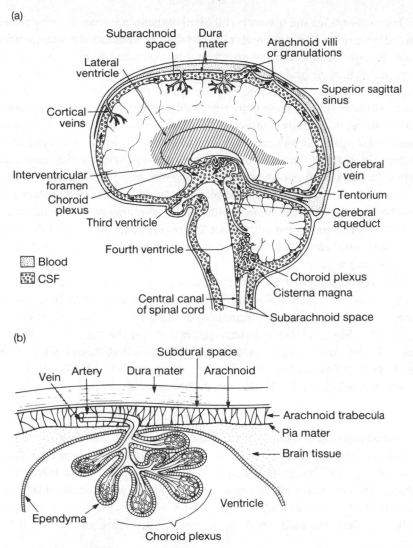

(a)

Subarachnoid space
Dura mater
Arachnoid villi or granulations
Lateral ventricle
Superior sagittal sinus
Cortical veins
Cerebral vein
Interventricular foramen
Tentorium
Choroid plexus
Cerebral aqueduct
Third ventricle
Fourth ventricle
Blood
CSF
Choroid plexus
Cisterna magna
Central canal of spinal cord
Subarachnoid space

(b)

Subdural space
Vein
Artery
Dura mater
Arachnoid
Arachnoid trabecula
Pia mater
Brain tissue
Ventricle
Ependyma
Choroid plexus

Fig. 2.32. (a) Flow of cerebrospinal fluid. (After Rasmussen, A.T. (1937) *The Principal Nervous Pathways*, p. 4. Macmillan Co., New York.) (b) Choroid plexus.

ventricles, e.g. the subfornical and pineal organs, the area postrema and the median eminence.

Cerebrospinal fluid

The CSF is formed predominantly by the *choroid plexuses*—rich networks of blood vessels covered with epithelial cells (*ependyma*) projecting into the ventricles (Fig. 2.32b). Fluid formed in the lateral ventricles passes to the third and fourth ventricles and to the central canal of the spinal cord. This movement is aided by the cilia on ependymal cells. The fluid escapes into the subarachnoid space through foramina in the ependymal lining of the fourth ventricle and circulates around the brain and spinal cord (Fig. 2.32a). Finally it is reabsorbed through the arachnoid villi into the sinuses of the venous system. In most regions of the brain, substances are free to diffuse between the ependymal cells and so there is a ready exchange of solutes between the CSF and the extracellular spaces of brain and spinal cord. The local environment of the neurones is further controlled by the activity of glial cells which can adjust both K^+ and H^+ ion concentrations. Of the 700 ml of CSF formed per day about 70% is derived from choroid plexuses and the other 30% comes from endothelial cells lining the brain capillaries. The volume of CSF in adult man is about 140 ml (compared with about 250 ml for brain interstitial fluid) and in the horizontal position its pressure is about 10 mm Hg, i.e. a little less than local venous pressure.

The composition of CSF differs from what one would expect if it were simply an ultrafiltrate of plasma (Table 2.4), indicating that it is actively secreted. One model for the formation of CSF by the choroid plexuses is illustrated in Fig. 2.33. Note the unusual feature that the Na^+–K^+-ATPase is located in the apical plasma membrane. Note also that though CO_2 crosses the blood–brain barrier readily, HCO_3^- does not. The HCO_3^- in CSF is synthesized within the epithelial cells, in the reaction

$$CO_2 + H_2O \rightleftarrows H_2CO_3 \rightleftarrows H^+ + HCO_3^-.$$

The initial reaction is catalysed by carbonic anhydrase which is present in the epithelial cells. The primary secretion has more HCO_3^- than plasma (about 45 mmol l^{-1}) and it buffers H^+ ions produced by neuronal and glial metabolism, reducing the concentration to about 23 mmol l^{-1}. This decrease in $[HCO_3^-]$ may also reflect the mixing of choroidal CSF with fluid secreted by the capillary endothelial cells which is thought to have a higher Cl^- to HCO_3^- ratio.

Table 2.4. Composition of cerebrospinal fluid compared with protein-free plasma.*

	CSF/plasma water
Na^+	1·0
K^+	0·7
Ca^{2+}	0·5
Cl^-	1·1
HCO_3^-	0·9
glucose	0·6

*Note that the protein content of CSF is 0·3 g l^{-1} compared with 70 g l^{-1} for plasma.

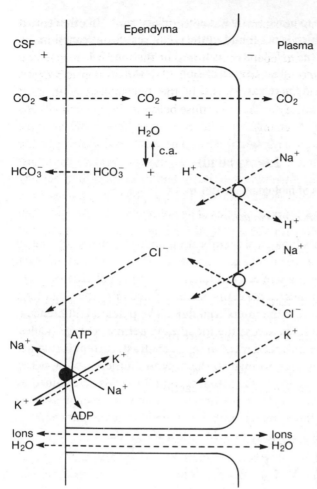

Fig. 2.33. Model for the formation of CSF by the choroid plexuses (c.a. is an abbreviation for carbonic anhydrase).

The functions of the CSF are as follows.

1 *Conferring of buoyancy.* The brain has little mechanical strength or rigidity. It weighs 50 g in CSF compared with 1500 g in air and so flotation in CSF protects it against deformation and damage from the innumerable accelerations imposed by movements of the head. Counter-pressure of CSF surrounding the blood vessels within the cranium and the spinal cord compensates for gravitational effects of alterations in posture (p. 453) or external acceleration.

2 *Maintenance of a constant ionic environment in brain.* This is important because the activity of neurones is highly sensitive to ionic changes. The CSF also provides another route for removal of substances which have limited lipid solubility or are too large to move easily across capillary walls. In other tissues the lymphatic system, which is not found in the CNS, serves this function.

3 *Control of respiration and pH.* This is discussed in Section 12.4.

4 *Control of water balance.* The concentration of NaCl in CSF may be a factor mediating thirst and the release of antidiuretic hormone (p. 266).

2.6 Appendix

Basic electrical properties of biological membranes

Current, voltage, conductance and capacitance

Small ions can flow across biological membranes, and this current can result in the separation of electrical charge. Some of the properties of such membranes can be expressed in electrical terms. The potential difference (V) across biological membranes is normally some 60–100 mV, the cytoplasm being negative with respect to the interstitial fluid. The potential difference is a measure of energy separated across the membrane per unit charge (joules/coulomb) and is therefore independent of area. In contrast, current (coulomb/second), which is directly proportional to the potential difference, increases with the available area. For any V, the actual current flow (I) is determined by the electrical conductance (G) of the membrane. Conductance expresses the ease with which current flows. Its reciprocal, the membrane resistance (R), is more familiar from elementary physics. Ohm's law states that:

$$I = \frac{V}{R} \quad \text{or} \quad I = GV.$$

Conductance (siemens) and resistance (ohms) are also determined by the area involved. When area is taken into account, the resistance is often referred to as the *specific membrane resistance* and has the units ohms × area (since V/I = ohms, V/I area^{-1} = ohms × area).

In practice, it is sometimes possible to determine the resistance of a plasma membrane by inserting micro-electrodes into a cell, passing a known current across the membrane, and measuring the change in potential difference which results. (This is permissible since the cytoplasm offers comparatively little resistance to current flow.) In some cells, membrane resistance so measured is found to be influenced by the direction of the induced change in potential, i.e. there is not a linear relationship between current and voltage—Ohm's law being obeyed strictly over only a narrow range. Thus the resistance calculated from the change in potential following a hyperpolarizing current (one which makes the interior of the cell more negative) may differ from that calculated when potential is decreased by the same amount following a depolarizing current. This asymmetrical behaviour is referred to as *rectification*. It indicates that current flows more easily across the membrane in one direction than in the other.

The conductance (reciprocal of the resistance) of a membrane can be thought of as reflecting the number and selectivity of water-filled ion-conduction pathways spanning the lipid bilayer. For example, K^+ ions generally pass more readily across plasma membranes through these pathways than do Na^+ ions. Therefore, replacement of Na^+ by K^+ in the medium produces a large increase in measured membrane conductance, though the number of available pathways may be unaltered. Conductance, measured by electrical techniques, is not to be confused with *permeability*, measured by isotope exchange or other chemically based methods. For example, the electrically 'neutral' one-to-one coupled movement of Na^+ and Cl^- across a membrane (co-transport) can be detected using Na^+ isotopes. In such a case the membrane is permeable to Na^+ but the co-transport mechanism does not have an electrically measurable conductance. (However, if an ion moves only by simple passive diffusion across a membrane, then it is possible to relate measured conductances and measured ion permeabilities by using appropriate conversion factors and transference numbers which express the fraction of total current carried across the membrane by the ionic species of interest.)

Plasma membranes do not respond to current passed across them with an instantaneous change to a new steady potential difference. Initially, some of the current passed changes the amount of charge (Q) stored on the membrane itself. The membrane, in separating charge across it, acts like a capacitor and the charge separated per unit driving force (coulombs/volt) is the *capacitance*, C, equal to Q/V, and its unit is the farad (F). The capacitance per unit area or *specific membrane capacitance* is similar for most biological membranes, about $1\ \mu F\ cm^{-2}$.

Because use is made of electrical analogues in understanding ion movements across single membranes and epithelia and because such concepts as resistance and capacitance are employed in discussing circulatory and respiratory physiology as well, the following (Fig. 2.34) should be noted:

1 the total resistance (R_T) in a circuit containing resistors in series is $R_1 + R_2 + R_3$. Since $G = 1/R$, the total conductance (G_T) for this circuit is given by $1/G_T = 1/G_1 + 1/G_2 + 1/G_3$ or $G_T = G_1 G_2 G_3/(G_1 + G_2 + G_3)$;

2 the total resistance (R_T) in a circuit containing resistors in parallel is given by $1/R_T = 1/R_1 + 1/R_2 + 1/R_3$ or $R_T = R_1 R_2 R_3/(R_1 + R_2 + R_3)$. Here the total conductance is given by $G_T = G_1 + G_2 + G_3$;

3 the total capacitance (C_T) in a circuit with capacitors in series is given by $1/C_T = 1/C_1 + 1/C_2 + 1/C_3$ and with capacitors in parallel it is $C_1 + C_2 + C_3$.

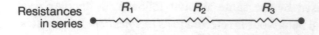

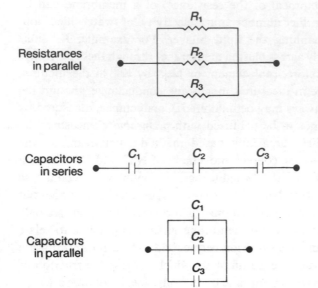

Fig. 2.34. Resistors and capacitors in series and in parallel.

Membrane space and time constants

Nerve and muscle fibres are elongated, like a submarine cable, and are endowed with so-called 'cable' electrical properties. Because the membrane is not a perfect insulator, when current is injected into a fibre it flows longitudinally in both directions and progressively leaks out across the membrane. By inserting glass micro-electrodes into large fibres at points of increasing distance from the site of current injection (Fig. 2.35a), it is possible to record electrotonic potentials along the fibres. A simple electrical model for the passive properties of an axon is shown in Fig. 2.35b. In such experiments, it can be seen that the amplitude of the electrotonic potential declines exponentially with distance from the current electrode (Fig. 2.35c). This exponential process is expressed by the *membrane space constant* (λ) which is defined as the *distance* in which a voltage falls to $1/e$ (about 37%) of its original value. It is independent of the size of the event. It can be shown that

$$\lambda = \tfrac{1}{2}\sqrt{\frac{DR_\mathrm{m}}{R_\mathrm{i}}},$$

where D is the diameter (cm) of the fibre, R_m is the specific membrane resistance (ohm cm^2) and R_i is the specific intracellular resistance (ohm cm).

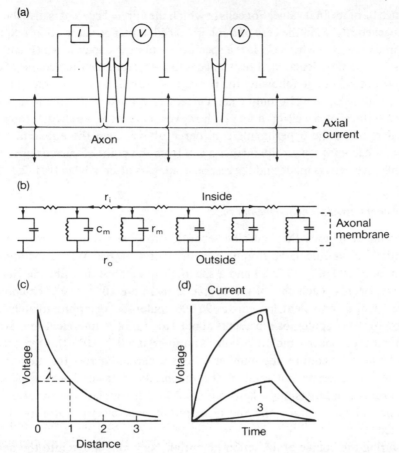

Fig. 2.35. Electrotonic potentials generated by the brief injection of current. (a) The injection of current (*I*) and the recording of potential (*V*). (b) A circuit diagram representing the cytoplasmic resistance and the membrane resistance and capacitance. (c) The attenuation of potential away from the injection site. (d) The voltage changes produced by a current pulse at 0, 1 and 3 space constants from the site of injection.

Note that R_m refers to the specific transverse resistance of 1 cm^2 of membrane, i.e. $R_m = r_m \times \pi D$ where r_m is the membrane resistance of 1 cm length of fibre (ohm cm), and R_i refers to the specific resistance through 1 cm^3 of axoplasm, i.e. $R_i = r_i \times \pi \dfrac{D^2}{4}$ where r_i is the intracellular resistance per 1 cm length of fibre (ohm cm^{-1}).

Moreover, when current is passed across the membrane of a cell the membrane capacitance will slow the time-course of the consequent voltage change. Therefore, the electrotonic potential will rise exponentially before it reaches a plateau (Fig. 2.35d). This exponential rise is expressed in the *membrane time constant* (τ) which is defined as the *time* required for the voltage to reach $1/e$ of its final value. For cells in which the applied current is distributed homogeneously, as in the case of a spherical cell, it can be shown that τ equals the product $R_m C_m$ where C_m is the specific membrane capacitance (F cm^{-2}). Thus, C_m can be calculated, if resistance is known, from the time-course of the change in the voltage following the passage of current across the membrane. This relationship is valid only when current is applied uniformly across the whole surface of a cell or fibre. If, however, current is applied focally by injection, as in the experiments considered above, then the apparent time constant becomes longer the further away from the point of current injection the measurement is made and the response appears after a delay (Fig. 2.35d).

Number of ions crossing the membrane

The number of ions that must be separated to give rise to the membrane potential can be calculated from the relationship $C = Q/V$. With a membrane potential of 100 mV (10^{-1} V) and a specific capacitance of 1 μF, the charge separated across each cm^2 of membrane is therefore $10^{-1} \times 10^{-6}$ coulombs. Since 1 mol of univalent ion carries 96 500 coulombs, or approximately 10^5 coulombs, 10^{-7} coulombs represents about 10^{-12} mol of univalent ion. Since the number of ions in a mol (Avogadro's number) is $6 \cdot 02 \times 10^{23}$, this represents 6×10^{11} ions. Therefore, the number of ions separated across the membrane to give a membrane potential of the magnitude measured across plasma membranes is infinitesimal when compared with the molar concentrations of ions in biological fluids, and the principle of bulk electroneutrality is not violated.

During the course of an action potential, Na$^+$ ions move into the nerve fibre during the rising phase and K$^+$ ions move out to repolarize the membrane during the falling phase. The net gain of Na$^+$ ions and net loss of K$^+$ ions have been determined in squid giant axons using radioactive isotopes of Na$^+$ and K$^+$ as tracers. These experiments show that 3–4×10^{-12} mol of each ion actually crosses each cm^2 of membrane during the passage of one impulse.

The number of K$^+$ ions crossing the membrane can be calculated in relation to their intracellular concentration by equating a nerve fibre to a hollow cylinder. Its surface area is given by $2\pi r L$ and its volume by $\pi r^2 L$, where r is the radius and L the length. Therefore the volume enclosed by a cylinder of 1 cm^2 surface area is:

$$\pi r^2 \, 1/2\pi r = r/2.$$

In the case of a 20 μm axon this equals 5×10^{-4} cm^3. Since the axon contains about 140 mmol l^{-1} K$^+$, there are 7×10^{-8} mol of K$^+$ ions in each segment enclosed by 1 cm^2 of membrane. Therefore, in comparison, the 3–4×10^{-12} mol of K$^+$ ions per cm^2 of membrane that actually leave the interior of the nerve fibre during the passage of one impulse represent only about 1/3000 of the internal K^+ pool and in the short term this is insignificant. In the smallest axons, however, it may become significant and such fibres can conduct only a few action potentials before their gradients must be recharged by the $Na^+–K^+$ pump.

Voltage clamping

The ionic basis of the action potential which has been discussed in Section 2.2 was described by Hodgkin and Huxley in the early 1950s. Their interpretation was based on voltage-clamp experiments on squid giant axons. In these experiments a feedback apparatus was used to deliver a current across the membrane in order to clamp membrane potential at any selected voltage. The current passed in order to clamp this voltage was monitored.

If the command voltage was set to the resting potential, no current flowed. When the potential was changed rapidly to more positive values and an action potential invoked, there was a brief inward current followed by a strong prolonged outward current (Fig. 2.36). Replacing external Na^+ with the impermeant cation choline abolished the inward current, indicating that it was carried by Na^+ ions. This was confirmed by clamping the cell to a voltage more positive than E_{Na}. This reversed the inward current, so there were now two outward currents, with different time-courses, the first carried by Na^+, the second by K^+. From such experiments it was possible to derive the ionic conductance changes underlying the action potential as shown in Fig. 2.9.

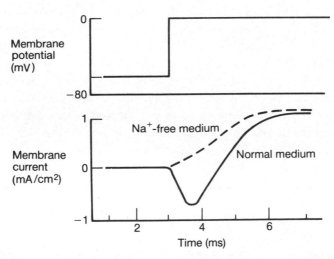

Fig. 2.36. Voltage clamping of a squid giant axon showing the membrane potential set by the command voltage and the current flow in Na^+-containing and Na^+-free media. (After Hodgkin, A.L. & Huxley, A.F. (1952) *J. Physiol.* **166**, 449–472.)

Reversal potential

Synaptic potentials result from current flow through ion-selective channels. This flow tends to change the membrane potential towards some particular value. For example, at the neuromuscular junction, activation of acetylcholine receptors increases Na^+ and K^+ conductances in the ratio 1·3:1. Assuming $E_{Na} = +60$ mV and $E_K = -90$ mV, the potential value the cell tends towards is (see equation on p. 18, ignoring G_{ce}):

$$= \frac{1·3}{1·3+1} \times 60 + \frac{1}{1·3+1} \times -90$$

$$= -5 \text{ mV}.$$

As the current flow is local and is opposed by current flow drawn from neighbouring regions of the cell, this value is not actually reached. If, however, an experimenter injects current into the cell and changes the value of the resting potential so that the cell is made positive inside, the synaptic potential will still tend towards −5 mV. Thus, instead of being a depolarizing potential, it will now be a hyperpolarizing one. This technique is used experimentally to determine the reversal of the synaptic potential and so to measure the relative conductances of Na^+ and K^+.

3 · Muscle

3.1 Skeletal muscle
Cellular structure of skeletal muscle
Contractile process
  Excitation–contraction coupling
  Energy balance
Contraction of muscle
  Length–tension relationship
  Force–velocity relationship
  Muscle fibre types
Regulation of contraction
  Motor units
  Gradation of tension
  Recruitment of motor neurones
Development and maintenance of
  skeletal muscles

Development of muscles
Effects of training
Effects of aging
Effects of damage to nerve or muscle

3.2 Smooth muscle
Smooth muscle structure
Contractile activity of smooth muscle
Resting membrane potential and action
  potentials
Types of smooth muscle
  Spontaneously active smooth muscle
  Electrically inexcitable smooth muscle
  Intermediate smooth muscle
Activation of smooth muscle

There are three kinds of muscle in the body, classified according to their structure and function. *Skeletal muscles* are characterized by the presence of thin light and dark bands (striations) that are seen to lie across fibres when viewed through a microscope. These muscles form some 40% of the fat-free body weight, they are under voluntary control, and they are the only tissue through which man can directly influence his environment. In contrast to skeletal muscles, *smooth muscles* (which are also called involuntary or visceral muscles) lack transverse striations and are not under conscious control; smooth muscles are found in viscera and blood vessels. *Cardiac muscle*, like skeletal muscle, is striated but unlike skeletal muscle it is spontaneously active; it generates the pressures required to drive blood around the vascular system and is described in Section 11.2.

3.1 Skeletal muscle

Skeletal muscles have two major functions: to exert force and to produce heat. The force developed by muscles is used to move or resist the movement of limbs, to close sphincters that control the emptying of hollow organs, to move the tongue and to regulate the vocal cords; and to perform other specialized functions. The heat produced by muscles is used to maintain body temperature (p. 622). The level of non-shivering heat production is regulated by hormones whilst shivering and the heat it produces are under direct nervous control.

Muscles vary enormously in their capacity to generate force and in the rate at which this force can be developed. As the maximal force that muscles develop is proportional to their cross-sectional area (up to 40 N cm^{-2}), so the 'strength' of a muscle is dependent on the number of muscle fibres and on their diameters. In general, a pennate arrangement of fibres gives rise to contractions that are more powerful than a simple parallel arrangement (Fig. 3.1). As the contractions of muscles depend on the shortening of a large number of subcellular units (sarcomeres, see below) arranged in series, the speed with which the ends of a muscle shorten depends on the number of units in the series and on the rate of their shortening. Muscles also contain varying amounts of fibrous and connective tissue which contribute to their mechanical properties.

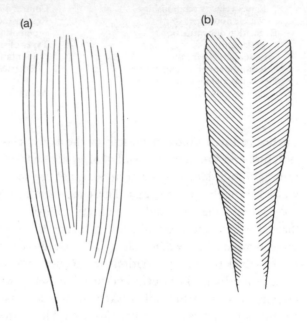

Fig. 3.1. Parallel (a) and bipennate (b) arrangements of muscle fibres.

Cellular structure of skeletal muscle

Muscle fibres are long multinucleate cells 50–60 μm in diameter and ranging in length from a few millimetres to many centimetres. Each fibre usually runs from tendon to tendon and terminates at each end in the connective tissue of the tendons, but in large muscles two or more fibres may be arranged in series. The force developed by the muscle fibres is generated by intracellular contractile proteins which are arranged into *myofilaments* (Fig. 3.2). The myofilaments are in bundles, called *myofibrils*, which run the whole length of the fibre. Each myofibril is surrounded by the *sarcoplasmic reticulum*, and between the lateral cisternae of the sarcoplasmic reticulum are fine *transverse (T) tubules* opening out on the surface membrane (the *sarcolemma*) (Fig. 3.3).

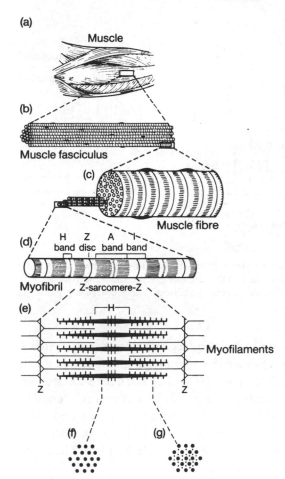

Fig. 3.2. Organization of muscle structure from whole muscle to myofilament (a to c) and transverse sections (f and g) showing the pattern of myofilaments. (Redrawn from Bloom, W. & Fawcett, D.W. (1975) *A Textbook of Histology*, 10th edn, p. 306. W.B. Sanders Co., Philadelphia.)

The complex of a T tubule and two adjacent portions of cisterna is known as a *triad* (Fig. 3.3) and in human muscles these are located at the junction of the A and I bands.

The myofilaments are arranged into *sarcomeres* which are the contractile units of muscle fibres. Each sarcomere is approximately 2 μm in length and its limits are defined by a Z line at each end. From the Z line *thin* (approximately 5×1000 nm) *actin* myofilaments project toward the middle of each sarcomere (Fig. 3.2) and in the central region of each sarcomere the filaments interdigitate with *thick* (approximately 12×1600 nm) *myosin* filaments. Each thick filament is surrounded by an hexagonal array of thin filaments (Fig. 3.2) and from each thick filament helically arranged crossbridges extend towards the thin filaments.

The striated appearance of skeletal (and cardiac) muscle fibres is a result of the serial and parallel repetition of the myofilaments and the differing

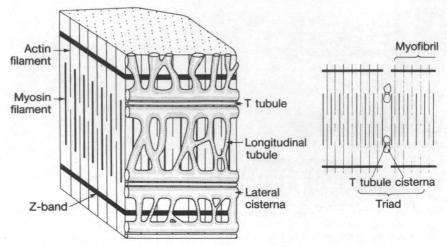

Fig. 3.3. *Left:* diagram illustrating the sarcoplasmic reticulum and T tubules in mammalian skeletal muscle. *Right:* a transverse section through the sarcoplasmic reticulum and T tubules illustrating the relationship between a T tubule and the two adjacent lateral cisternae (a triad).

abilities of the actin- and myosin-containing regions to transmit light. As polarized light is not transmitted through the myosin-containing region (i.e. it is anisotropic) this region is called the A band (Fig. 3.2). But light is transmitted through the actin-containing region (i.e. it is isotropic) and so it is referred to as the I band. In the middle of the A band where the myosin and actin filaments do not overlap, there is a lighter H band which marks the region devoid of crossbridges and in the middle of this a finer dark M line. The Z line lies in the middle of each I band.

Contractile process

If a muscle changes its length the sarcomeres also change in length. However, the thin and thick filaments remain the same length and the change in length is achieved as a result of the filaments sliding over each other—the sliding filament theory developed by Hansen and Huxley in the 1950s. The forces generated during contractile activity arise in the regions where the actin filaments overlap the crossbridges. A rod-like light meromyosin component of myosin molecules forms the backbone of the thick filament while a heavy meromyosin component forms the crossbridges. During contractions the myosin crossbridges attach to adjacent actin filaments and flex toward the centre of the sarcomere (Fig. 3.4). As this occurs at both ends of the myosin filament the opposing actin filaments are drawn in toward the centre, the Z lines are pulled closer and the muscle fibre shortens.

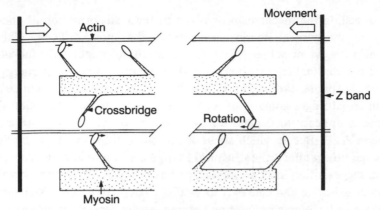

Fig. 3.4. Relative movement of actin and myosin filaments. This is accomplished by the rotation of the crossbridge head which contains the myosin ATPase. A second flexible point appears to exist where the crossbridge joins the backbone of the filament.

This process has at least three requirements:

1 that the actin and myosin must interact;
2 that the myosin crossbridges must flex; and
3 that the system must be able to convert chemical energy to mechanical energy.

The actin filaments are composed mainly of globular actin molecules in a helical arrangement (Fig. 3.5) and it has been shown that purified actin molecules combine with the crossbridge component of myosin molecules. This interaction is inhibited in normal resting cells by the presence of a troponin–tropomyosin complex. Early X-ray diffraction studies of muscles at rest and in *rigor mortis* (see below) showed that the crossbridges could have two stable positions, the resting position and the flexed position. But it is only recently that pulsed irradiation–diffraction studies of muscles have demonstrated movements of the crossbridges during contractions.

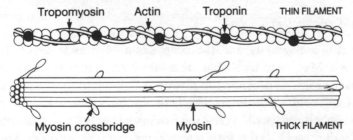

Fig. 3.5. Each actin filament is composed of two chains of actin monomers (5 μm). In the grooves between these chains lie strands of tropomyosin and at regular intervals of about 40 μm are troponin molecules. Each myosin filament is composed of many myosin molecules, each of which comprises a rod-like component and neck and head regions.

Unravelling the mechanism involved in the transformation of chemical energy to mechanical energy has proved the most difficult. There is overwhelming evidence that ATP (adenosine triphosphate) is the fuel used to operate the contractile machinery and that the other sources of energy, e.g. creatine phosphate, carbohydrates and fatty acids, are used to produce ATP. Myosin crossbridges contain an ATPase and the rate of contraction of the sarcomeres appears to depend on the speed with which this enzyme can hydrolyse ATP. But at which stage of the crossbridge cycle the hydrolysis occurs and how it affects the structural change is still problematic. A common schema suggests that after a crossbridge has flexed ATP combines with the myosin crossbridge and causes it to be dissociated from actin, and then the hydrolysis and energy consumption are involved in repositioning (extending) the crossbridge. After this the ATPase is not yet free to hydrolyse further ATP, possibly as a result of continued association of reaction products. When the crossbridge next combines with actin the reaction products dissociate and flexion involves the release of stored energy. When muscles are depleted of ATP the actin and myosin can no longer dissociate and *rigor mortis* occurs, that is, the muscles become stiff and inextensible.

However it should be realized that a single cycle of crossbridge movement would not cause a large contraction as the actin filaments would be moved by only some 1% of the length of a sarcomere. Significant muscle shortening requires that the myosin crossbridges undergo repetitive cycles. In each cycle the crossbridge attaches to the actin, flexes, and then dissociates before returning to its initial configuration and a new binding site on the actin filament. The repetitive asynchronous activity of the crossbridges from surrounding myosin filaments ensures that the force exerted on an actin filament is maintained during a contraction. This activity can result in the actin filament being pulled between the myosin filaments. When the muscle is unable to shorten, the elastic properties of the muscle fibres allow the crossbridge mechanism to operate and force to be generated.

Excitation–contraction coupling

Myosin ATPase activity and the contraction of muscle depend on both cytoplasmic Mg^{2+} and Ca^{2+}, but these ions have completely different roles. Whereas the Mg^{2+} is necessary for myosin ATPase activity and is in adequate supply, Ca^{2+} controls the interaction between the actin and myosin filaments and its supply is regulated. At rest the free cytoplasmic Ca^{2+} concentration is so low ($\sim 10^{-8}$ mol l^{-1}) that little interaction occurs. However during activity the concentration rises sharply, thus the development of tension within a muscle fibre is regulated by the free cytoplasmic Ca^{2+}.

The interaction between actin and myosin (and the ATPase activity) is inhibited by the *troponin–tropomyosin* complex (Fig. 3.5). The troponin component is a complex molecule spaced regularly along the actin filament and has specific tropomyosin-binding (T), calcium-binding (C) and inhibitory (I) subunits. The inhibitory effect of the complex on actin–myosin binding is removed when Ca^{2+} binds to the C subunit. The change is thought to be associated with movement of the tropomyosin strands which lie in the grooves between the strands of actin molecules. Ca^{2+} concentrations of 10^{-5} mol l^{-1} cause maximal inhibition of the troponin–tropomyosin influence; lowering the Ca^{2+} concentration allows the complex again to exert its influence.

In resting muscle, the free cytoplasmic Ca^{2+} concentration is low because the sarcoplasmic reticulum contains a membrane-bound pump that avidly collects Ca^{2+} and then transports it to the lateral cisternae. However, this Ca^{2+} can be released by depolarization of T-tubule membranes that come into close apposition with the lateral cisternae. An action potential propagating along the muscle and down the T tubules causes Ca^{2+} release and an elevation in the free intracellular Ca^{2+} concentration (Fig. 3.6); the propagation of the

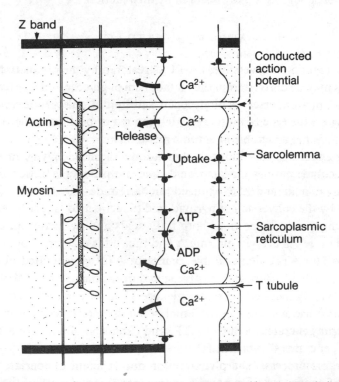

Fig. 3.6. The release of Ca^{2+} from the sarcoplasmic reticulum and its re-accumulation by an active transport process.

action potential down the T tubule ensures that the contractile activity in adjacent myofibrils is synchronized. The ionic basis of the action potential in skeletal muscle is similar to that in axons (p. 44).

Energy balance

As mentioned above, ATP is the immediate source of energy for muscle contraction. Very little ATP is stored in a muscle, and it must constantly be renewed because of its consumption in both the contractile process and the Ca^{2+}-transporting mechanism of the sarcoplasmic reticulum. The ability of muscles to accomplish this restoration is shown by our inability to demonstrate changes in the level of ATP unless its synthesis is blocked. The short-term reserve for replacement is creatine phosphate (PC), which forms a dynamic balance with free ATP, and the enzyme creatine phosphokinase (CPK) ensures that this equilibrium is reached rapidly. Thus the hydrolysis of ATP results in

$$ATP \rightleftharpoons ADP + P_i$$

but the level of ATP is rapidly restored by the reaction

$$ADP + PC \overset{CPK}{\rightleftharpoons} ATP + C.$$

As the last reaction is reversible, the PC is restored by the production of new ATP. This may be derived from the metabolism of glucose and free fatty acids from blood, or from reserves of glycogen and lipid droplets in muscle fibres. Depending on the type of muscle (p. 109) the major portion of the ATP may derive from either anaerobic or aerobic metabolism.

Under anaerobic conditions the breakdown of muscle glycogen proceeds via the glycolytic pathway to lactic acid. The end-product is lactic acid rather than pyruvic acid and the oxidized nicotinamide dinucleotide (NAD^+) generated by the conversion of pyruvic acid to lactic acid is used in an earlier step. During aerobic conditions both fatty acids and pyruvate can enter the citric acid cycle via acetylcoenzyme A and thus a far greater amount of ADP is converted to ATP. For example, 3 mol of ATP is generated per mol of glucose-6-phosphate formed from glycogen during anaerobic metabolism (i.e. 2 mol of ATP is generated per mol of glucose). In contrast, aerobic catabolism through the citric acid cycle of the 2 mol of pyruvate produced per mol of a 6-carbon sugar generates 36 mol of ATP, and oxidation of a mol of a 6-carbon fatty acid generates 44 mol of ATP.

The whole process, that is contraction and relaxation, operates with an *efficiency* of conversion of metabolic energy into external work of the order of 20%; the remainder is dissipated as heat.

Contraction of muscle

Muscles are said to be contracting when the contractile machinery is active and energy is being consumed. This term applies whether the muscle is shortening, remaining at constant length or lengthening. Whilst in everyday use the muscles perform in all these ways, it is convenient to study muscle contraction when either the length of the muscle or its load is constant. When the length remains constant (*isometric* contractions), we measure the force (tension) generated by the contractile machinery. When the load remains constant (*isotonic* contractions), we measure the shortening of the muscle. These forms of contraction are, of course, also used in everyday activities, e.g. isometric contractions are involved in the maintenance of posture and isotonic contractions in the lifting of limbs.

Isometric studies of muscle show that shortly after a muscle is stimulated by a *single* stimulus there is an increase in muscle tension which then decays. The time-course of an action potential and a contraction are shown in Fig. 3.7.

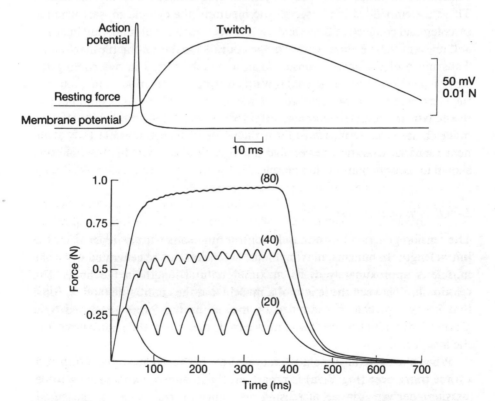

Fig. 3.7. *Above:* the time-course of an action potential recorded intracellularly from a single fibre and the accompanying isometric twitch recorded from many fibres. *Below:* isometric contractions from a rat extensor digitorum longus muscle showing the response to a single stimulus and to bursts of increasing frequencies (Hz) indicated in parentheses.

The time taken for the development of peak tension varies from 10 to 100 ms; its rate of decline also varies and both depend on the type of muscle being studied. A single contraction of this type is called a *twitch*. If the muscle is stimulated a second time, before it has had time to relax completely, the second response may add to the first and a greater peak tension developed. This is referred to as *mechanical summation*. If the stimulation of the muscle continues it fails to relax completely and during the period of stimulation the tension fluctuates (Fig. 3.7). With increasing frequency of stimulation the maximum tension is increased, the oscillations become smaller and, eventually, at *fusion frequency* a smooth *tetanic* contraction is produced. The tension produced in a tetanus may be 2–3 times as great as that produced in a twitch.

The substantial difference between the maximal tensions reached in a twitch and in a tetanus has been attributed to the physical properties of the muscle and to changes in the cytoplasmic Ca^{2+} concentration. First, muscles are not rigid and the forces generated by the contractile machinery are transferred to limbs by elastic structures (the tendons and myofilaments). These are embedded in a visco-elastic medium (the cytoplasm, sarcolemma, sarcolemmal connective tissue and the connective tissue around fibre bundles) and many of these elements are arranged parallel to the contractile machinery. Thus much of the energy consumed in a twitch is used in overcoming the damping action of these elements. With continued activation as in a tetanus, the elastic elements are stretched, the movement of filaments is minimal and the maximum muscle tension is attained. Second, there is evidence from invertebrate striated muscles that a higher level of cytoplasmic Ca^{2+}, and hence muscle activation, is reached during a tetanus, but this has not been shown to occur in mammalian muscle.

Length–tension relationship

The tension generated by a muscle contracting isometrically depends on the initial length. In humans, maximal muscle tension can be generated when the muscle is approximately at its maximal natural length in the body. The relationship between the length of a muscle and the contractile (active) force that it develops can be examined by measuring the forces generated by a muscle at different lengths. Two forces can be measured: the *passive* force and the *total* force.

When a relaxed (unstimulated) muscle held between a movable clamp and a force transducer (Fig. 3.8a) is progressively stretched, an increasing force (tension), derived from an increasing resistance to stretch, can be measured (Fig. 3.8b). As the contractile machinery is not active this force is passive and is due to the resistance exerted by elastic elements in the muscle. It should be noted that the force generated is not directly proportional to the increase in

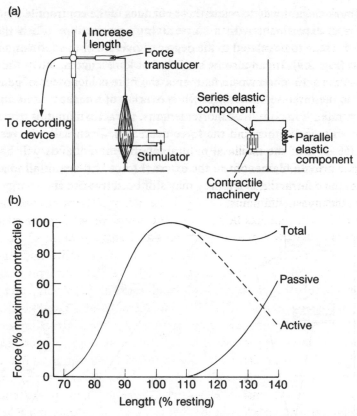

Fig. 3.8. (a) The experimental set-up used to study isometric contraction at different lengths (left) and a model of the contractile and elastic elements in muscle (right). (b) The relation between force and muscle length. Note that the total force generated at each length is the sum of the active force generated by the contractile elements and the passive force due to extension of the elastic elements.

length (the elastic modulus increases with length), and that the elasticity varies greatly amongst muscles. Some muscles, for example the back muscles of man and the hind-leg muscles of kangaroos, powerfully resist extension by purely passive mechanisms.

The curve of *total* tension is constructed by stimulating the muscle at various lengths and measuring the forces generated (Fig. 3.8b). Each of these forces will be the sum of both the passive force and the *active* force developed by the contractile machinery. The amplitude of the active force at the various lengths is obtained by arithmetically subtracting the passive force from the total force. The maximal active force is seen to occur near the natural resting length and to decrease with changes in length.

The most elegant way to relate these changes to the contractile machinery is to do this experiment with a single living muscle fibre. Then the force generated is seen to be related to the degree of overlap of the actin and myosin filaments (Fig. 3.9). It can also be seen that at long lengths, when there is no overlap of the actin and myosin filaments, the fibre is incapable of generating a force; in the intermediate range, when overlap of filaments is optimal, the force generated is maximal; at shorter lengths, the actin filaments overlap and interfere with each other and the force decreases. Eventually, at very short lengths (60–70% of the maximal natural length) the Z bands will be pulled against the myosin filaments and the external force will again fall to zero. At this point the contractile machinery may still be active but the energy is used to distort the myosin filaments.

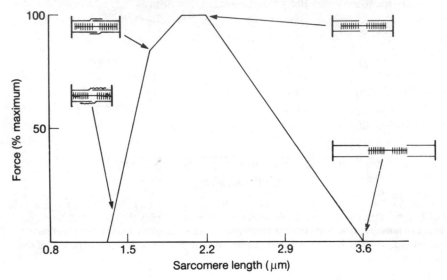

Fig. 3.9. The relationship between the contractile force and sarcomere length in a single muscle fibre. The insets illustrate the degree of overlap of the myofilaments at the sarcomere lengths indicated. (After Gordon, A.M., Huxley, A.F. & Julian, F.J. (1966) *J. Physiol.* **184**, 170–192.)

Force–velocity relationship

The length–tension curve has described the ability of muscles to develop tension when the muscle is held at fixed lengths (isometric contractions). But as mentioned earlier, the movement of limbs may be associated with the shortening of muscles under a constant load (isotonic contractions). It is an everyday experience that the lighter the load the more rapidly it can be lifted. In fact, both the rate and the degree of muscle shortening depend on the load. The relationship between the rate of shortening and the load carried by a muscle is illustrated by the force (load)–velocity curve.

This relationship is determined by measuring the rate of shortening of a muscle as it lifts a variety of loads. The muscle is not initially subject to each load as this would alter the starting length of the muscle. However before it can shorten the muscle must obviously first lift each load. Such an event is called an *after-loaded* contraction.

When stimulated tetanically an after-loaded muscle starts to contract. Initially and until the tension exceeds the load, the contraction is isometric. After this, the muscle shortens isotonically and continues to shorten until it reaches the length at which (according to the length–tension curve) the maximal force it can develop is equal to the load. It is clear that with zero load the time required initially to shorten (the latency) will be minimal and the velocity of the contraction maximal (Fig. 3.10); as the load is increased the latency is increased and the velocity decreases. Finally, when the load is too heavy, the velocity of shortening is zero and the muscle is contracting isometrically. It can be seen from the force–velocity curve that the *power* (force × velocity) which a muscle develops is not constant. The power output of a muscle is in fact optimal when both the load and the velocity are moderate—hence the advantage of ten-speed bicycles.

The reasons for the shape of the force–velocity curve are not known. One suggestion is that the myosin crossbridges move continually as a result of thermal agitation and that there is only a limited space within which a crossbridge and an actin site can interact. If this is correct and the actin filament is moving, the probability of successful union will decrease as the velocity of movement increases. Thus at high velocities few crossbridges will be formed and, as the force generated is dependent on the number of crossbridges, it will be low at high velocities. Accordingly, the velocity of shortening will increase until the force generated by the muscle equals the

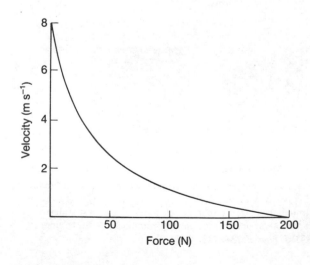

Fig. 3.10. The effect of force (load) on the velocity of shortening of human muscle. (Adapted from Wilkie, D.R. (1950) *J. Physiol.* **110**, 249–280.)

load. This idea is supported by the observation that the velocity of shortening in isotonic contractions is relatively constant.

Muscle fibre types

The position of our body and the complex pattern of movements it performs are the results of the pulling actions of muscles. However, the diversity of these actions requires that the muscles used to perform them have different properties. Thus, some muscles are called upon to maintain a high level of tension for long periods without fatigue while others are required to produce intermittent rapid movements. These two extremes of activity are illustrated by the postural soleus muscle that reaches a peak tension in 80–200 ms and the extraocular eye muscles that develop their peak tension in 7–8 ms (Fig. 3.11a). The soleus muscle contains predominantly slow-contracting muscle fibres and the extraocular muscles predominantly fast-contracting muscle fibres. Muscles that have to perform both endurance and rapid actions have a more even mixture of these fibres types. When the properties of the slow (type I) and fast (type II) muscle fibres are compared, pronounced differences are evident. The slow fibres have a low myosin ATPase activity and a high capacity to produce

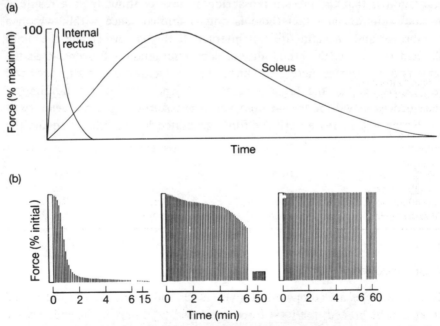

Fig. 3.11. (a) Isometric twitch contractions of cat internal rectus and soleus muscles scaled to the same peak height. (Adapted from Cooper, S. & Eccles, J.C. (1930) *J. Physiol.* **69**, 377.) (b) Fatigue of fast (left), intermediate (middle) and slow (right) muscle fibres that were stimulated through their nerve supply at 40 Hz for 330 ms once each second. (From Burke, R.E., Levine, D.W., Tsairis, P. & Zojac, F.E. (1973) *J. Physiol.* **234**, 723.)

ATP by oxidative phosphorylation which is aided by a well-developed blood capillary network and high levels of intracellular *myoglobin*. The latter is an oxygen-binding protein (like haemoglobin) which both facilitates the diffusion of O_2 into these muscle cells and stores a small quantity of O_2 in the cells. The combined effect of these characteristics is slow, fatigue-resistant contractions as the rate of production of ATP is sufficiently rapid to replace that split by the myosin ATPase. The high concentration of myoglobin and capillary density in these muscles have led to the use of the term 'red muscle'.

There are two distinct groups of fast-contracting fibres. Both have a greater diameter and a higher myosin ATPase activity than the slow fibres but their resistances to fatigue differ (Fig. 3.11b); the resistance to fatigue is correlated with a high oxidative capacity and those fibres with a high resistance are often referred to as *intermediate* fibres. The largest and fastest contracting type II fibres (the so-called *fast* fibres) have a poorly developed oxidative metabolism and depend largely on glycolysis for the production of ATP. Although ATP production by glycolysis is rapid, the high rate of consumption of these fibres during powerful contractions results in their rapid fatigue. A summary of these and other properties of the different fibre types is given in Table 3.1.

Table 3.1. Muscle properties.

	Type I	Type II	
	high oxidative	high oxidative	low oxidative
Rate of contraction	Slow	Fast	Fast
Myosin ATPase activity	Low	High	High
Main pathway for ATP production	Oxidative phosphorylation	Oxidative phosphorylation	Glycolysis
Number of mitochondria	Many	Many	Few
Myoglobin content (muscle colour)	High (red)	High (red)	Low (white)
Capillary density	High	High	Low
Glycogen reserves	Low	Intermediate	High
Rate of fatigue	Slow	Intermediate	Rapid
Fibre diameter	Small	Intermediate	Large

Regulation of contraction

Because muscles are composed of many fibres and the activity of individual fibres can be graded, the total force generated by a muscle depends on the number of active fibres and the level of activity in each fibre. Each motor axon entering a muscle makes contact with a number of muscle fibres; each of these fibres is innervated by a single terminal branch of that axon. Thus groups of muscle fibres are activated synchronously.

Motor units

A motor unit comprises a motor neurone and the group of muscle fibres innervated by the branches of its axon (Fig. 3.12). Motor units vary greatly in size, ranging from one or two muscle fibres in the smallest units of muscles controlling fine movements of fingers or eyes to more than 2000 in the largest units in limb muscles. All the muscle fibres in a motor unit are of the same type, and they tend to be very homogeneous in their properties and so the terms type I and type II are used for both motor units and muscle fibres. In general, the type I units of slow muscles are rather similar in size although they are not particularly large; type II units, in contrast, supply fast muscles and are often large. The larger a motor unit is, the larger the axon and the nerve cell body of the motor neurone supplying it. This probably reflects the need for production by the cell of all the materials needed to keep every one of its nerve terminals functioning.

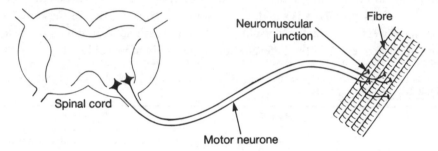

Fig. 3.12. Two motor units, each consisting of a motor neurone and the muscle fibres that it innervates.

Gradation of tension

Increments in tension can result from an increase in the force generated by individual motor units or by bringing into action (*recruitment*) additional units. Extracellular recordings of the electrical activity of muscle fibres (electromyography) have shown that both these events occur, but not at the same rate. Thus, the initial development of muscle tension is thought to be largely due to recruitment of units. As explained below there is, in addition, an increase in firing frequency but the contribution of this to increments in tension is thought to be important mainly in the generation of larger forces.

The recruitment of motor units is not random but occurs in an orderly fashion from small to large. Low tensions are produced and precisely controlled by the selective mobilization of a number of small units. In fact, under most circumstances, a small proportion—paradoxically, the smallest ones—do most of the work. The largest units are activated only when a maximal effort is required and even then their activity is often brief.

Recruitment of motor neurones

The ordered recruitment from the *pool* of neurones supplying a muscle arises because the smallest cells are the most easily excited. The smaller surface area of the small motor neurones results in these cells having a higher input resistance. When similar excitatory synaptic currents are generated in the small and larger motor neurones the small ones reach threshold first. In contrast, inhibitory synaptic currents generated in neurones within an activated pool appear to be more effective on the larger neurones as these are the cells closest to their threshold. As the intensity of excitatory synaptic activity in a motor neuronal pool increases, larger and larger motor units are recruited and at the same time the frequency of discharges increases. However there are also neural mechanisms that limit the discharge frequency of individual motor neurones to a frequency appropriate to the type of muscle fibres they innervate.

It should be noted that the contractions of skeletal muscles that move objects, maintain posture and adapt to changes in load and fatigue are not regulated solely by the motor units. These activities also make use of sensory information including that from the muscles and limbs involved. The role of the muscle receptors (the muscle spindles and Golgi tendon organs) in motor control is discussed in Chapter 6.

Development and maintenance of skeletal muscles

The speed with which muscles can contract and their ability to do work are not constant throughout life, but change as a person grows; their performance is also influenced by exercise. The development, growth and maintenance of muscles are all dependent on the presence of an intact motor nerve supply.

Development of muscles

The number of fibres in skeletal muscles appears to be genetically determined but the expression of the full genetic capacity is dependent on the normal development of the nerve supply to the muscles. If during early development the motor nerves fail to maintain contact, for example, as a result of physical or drug-induced damage, the muscles will be smaller than normal due to a decrease in the number of their fibres.

As well as influencing the number of fibres in a muscle the nerves also appear to determine the type of fibre that will develop. The properties of the muscle fibres within each motor unit are determined by the nerve controlling that unit. This has been demonstrated in a number of ways, but most obviously in that the muscle fibres within the one unit are homogeneous with respect to such things as contraction time, resistance to fatigue, enzymes of anaerobic and aerobic metabolism and myosin ATPase.

These properties are determined early in development but they are not irreversible and changes can be seen in both developing and adult muscles, for example after denervation (see below). The ability of nerves to regulate the properties of muscles is referred to as a *neurotrophic* influence but it is not known precisely how this influence is exerted. There is good evidence that nerve-induced muscle activity at the appropriate frequency (tonic low frequency for slow muscles and phasic high frequency for fast muscles) is important. There is also evidence suggesting that specific messengers, *neurotrophic factors*, are released by motor nerves to influence the muscle fibres they innervate.

Effects of training

Type I fibres make up about 30–40% of the cells in human muscles and they are approximately the same size in men and women (the mean diameter being approximately 60 μm). Type II fibres are larger in men (average diameter 69 μm) than in women (50 μm). Two distinct responses to regularly performed strenuous exercise can be seen in muscle: hypertrophy of the fibres with an increase in strength (e.g. weightlifters) and an increased capacity for aerobic metabolism (e.g. long-distance runners, cross-country skiers, swimmers).

Endurance exercise training gives rise to an increased capacity for oxidation of pyruvate and long-chain fatty acids. This is due to an increase in the absolute amount of enzymes, for example, those of the tricarboxylic acid cycle and those involved in the activation, transport and oxidation of long-chain fatty acids. There is an increase in myoglobin, which speeds the rate of diffusion of O_2 from cell membrane to mitochondria. Trained individuals have increased intramuscular stores of triglyceride and lowered concentrations of serum triglycerides and their muscle can utilize lipids directly from blood.

The consequences of these changes are that during submaximal exercise trained individuals derive more energy from fat and less from carbohydrate than do untrained individuals. Furthermore, in the trained individual, liver and muscle glycogen stores are better maintained during exercise and a greater proportion of oxygen is extracted from the blood supply to muscles. Fatigue is thought to be associated with depletion of muscle glycogen stores, accumulation of a high concentration of lactate and hypoglycaemia (lowered blood glucose) due to depletion of liver glycogen. All these changes are minimized by training, so that endurance is enhanced.

Effects of aging

The total number of fibres in a muscle decreases with age, and it has been suggested that the average number of fibres per motor unit gets larger. It is a generally held opinion that the units lost are the smaller ones and that the progressive loss of muscle fibres is due to a loss of the nerve cells that supplied them. The potency of synaptic transmission also declines with age, possibly as a result of the observed decrease in the synthesizing ability of aged nerve cells.

Effects of damage to nerve or muscle

After the nerve to a muscle is sectioned, there are changes in both the muscle and the axons. The loss of the neurotrophic influence results in pronounced changes in the muscle fibres. The earliest changes, such as the partial decrease in resting membrane potential and the increase in sensitivity to applied acetylcholine (which results from the addition of acetylcholine receptors to all of the sarcolemma), can be seen in a few days. But other changes, such as a pronounced decrease in the ability to develop tension, the change in enzymic composition, and the decrease in fibre diameter (*atrophy*), may take longer to develop. In man, the fibres may shrink down to some 10 μm and remain so for months or until they are re-innervated; re-innervated fibres grow and develop according to the characteristics of the motor neurone. If muscle fibres remain denervated for prolonged periods (months to years), they will gradually be replaced by connective tissue and fat.

When the nerve to a muscle is sectioned, some of the motor neurones die, but others regenerate their axons. However, in higher vertebrates there is little or no specificity in the re-establishment of nerve–muscle connections. Regrowth of the axons is aided and directed by the presence of the old nerve sheaths (hence the accurate suturing together of the cut ends of a nerve is important). Normal muscles have fibres within individual motor units well scattered across the muscle, but, following regeneration of a cut nerve, muscle fibres of a single motor unit now occur as a clump of cells (Fig. 3.13), as if the ingrowing nerve made connections with all the muscle fibres in its immediate vicinity. However re-innervation may not always be successful and when a whole limb is denervated there is very little evidence of orderliness in nerve regeneration to muscles and normal coordination of movement is never fully restored.

Damaged muscle fibres do not possess an intrinsic capacity for regeneration. However, damage is repaired very efficiently by the activation of small mononuclear cells (satellite cells) that normally lie beneath the basal lamina of muscle fibres. These cells undergo mitosis, increase in numbers, and ultimately fuse to repair the damaged fibres or to make new multinucleate muscle cells.

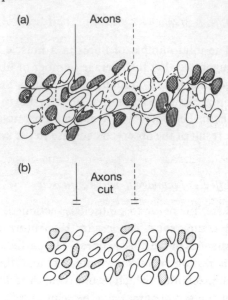

(a) Axons

(b) Axons cut

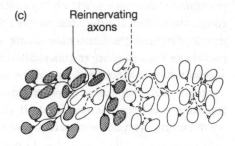

(c) Reinnervating axons

Fig. 3.13. Re-innervation of muscle fibres. (a) Prior to denervation the fibres innervated by each motor neurone are intermixed. (b) The muscle fibres atrophy following section of the nerve. (c) After re-innervation the muscle fibres of each motor unit tend to be grouped together.

In animals, it is actually possible to remove a muscle, mince it, pour the mince back into the appropriate place in the animal, sew up the skin, and produce a new but smaller functional muscle. Regeneration of the muscle in such cases is critically dependent on the presence of the nerve.

3.2 Smooth muscle

Smooth muscle has a wide range of functions and all muscular forces not generated by skeletal or cardiac muscle are generated by smooth muscle. Thus, smooth muscles control the movement of material through most hollow organs; for example, they propel material in the gastrointestinal tract, they restrict flow in blood vessels and bronchi, and they expel material from the bladder and vas deferens. Smooth muscles also control pilo-erection and influence the input of sensory information into receptors as when the dilator and constrictor muscles of the iris affect the amount of light reaching the retina.

Smooth muscles vary in their level of activity, from those that show more or less continuous activity to those that are quiescent for prolonged periods. In some, only localized contractions occur (e.g. intestinal sphincters), whilst in others the whole organ may be involved (e.g. bladder). Activity in smooth muscles may depend on a number of factors including the character of the smooth muscle cells, their environment, neural input and hormones. All neural influences are exerted by the autonomic nervous system; some tissues are innervated by only one division, while others are innervated by both the parasympathetic and sympathetic divisions.

Smooth muscle structure

A connective tissue sheath, the *epimysium*, surrounds the smooth muscle of each organ. Thin septa extend inward from the epimysium to form the *perimysium*, which contains fibroblasts, capillaries, nerves and collagenous elastic fibres; the collagen fibres are synthesized by the muscle cell and form the major component of the extracellular space. The perimysium divides smooth muscle into discrete *bundles* of fibres. These bundles range from 20 to 200 μm in width, and anastomose with one another; these anastomoses can be seen at roughly 1 mm intervals along a given fibre bundle (Fig. 3.14). An exception is found in arteriolar walls which may be only a few cell diameters in thickness and the smooth muscle of which is not organized into bundles.

The individual smooth muscle cells within a bundle are 2–10 μm in diameter, and vary in length from about 50 μm in arterioles to 400 μm in most other organs. The smooth muscle cells within each bundle are fusiform, or

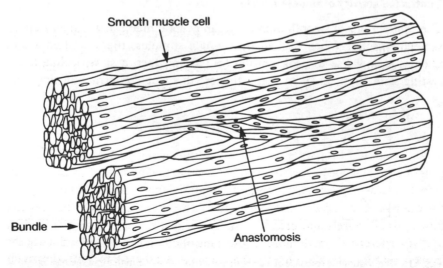

Fig. 3.14. Smooth muscle cells arranged in bundles that are interconnected by an anastomosis.

irregular elongate cells that interweave and overlap with each other (Fig. 3.14) to form a network interlaced with collagen.

Individual smooth muscle cells come into close contact with ten or so neighbouring cells; at these points they may be connected by specialized intercellular junctions of relatively low electrical resistance called *gap junctions* (nexuses). At these junctions the sarcolemma of the cells is separated by 3·5 nm but the gap is bridged by structures which allow small ions to pass from cell to cell. The relatively low electrical resistance of these junctions allows current, which may have either an excitatory or an inhibitory effect, to pass from cell to cell. Where bundles exist the direct coupling of cells within each bundle may result in the bundles being the functional (contractile) unit.

Pronounced differences between the structure of smooth muscles and striated muscles are seen at the ultrastructural level. For instance, smooth muscle cells possess few mitochondria, the sarcoplasmic reticulum is poorly developed and located close to the sarcolemma, there is no postjunctional thickening or specialization at the neuromuscular junction and the myofilaments of actin and myosin are irregularly arranged. The actin filaments appear to be inserted into specialized structures in the sarcolemma, so-called dense bodies, and radiate out in a longitudinal direction from these. A detailed knowledge of the actin and myosin filaments in smooth muscle is lacking but relative to skeletal muscle the actin filaments are very long, there is a much higher ratio of actin to myosin and there is very little troponin. There is instead a high concentration of a specific Ca^{2+}-binding protein, *calmodulin* (p. 256) associated with the contractile proteins.

Contractile activity of smooth muscle

The contractions of smooth muscles are in general much slower than those of skeletal muscles. When excited by a single stimulus, there is often a long latency, a slow rise to peak tension (>1 s) and then a slow decline to the resting state (Fig. 3.15). In many tissues, this single contraction may take

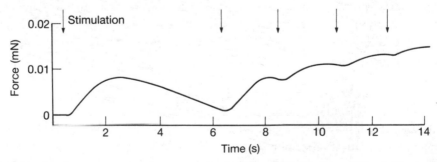

Fig. 3.15. The contractile response of a smooth muscle to a single stimulus and the response to repetitive stimuli.

several seconds; with repetitive stimulation the forces generated by smooth muscles increase and can reach levels similar to those found in skeletal muscles (30–40 N cm^{-2}). However, unlike many skeletal muscles, smooth muscles can maintain their tension at a high level for long periods and over a wide range of muscle lengths. It seems probable that the low activity of the myosin ATPase may account for both the slow development of force and the relatively low oxygen consumption during contractions ($< 1/100$ that of skeletal muscle). The ability to contract over a wide range of lengths (up to four times the resting length) may be a result of the irregular arrangement of the myofilaments.

As with skeletal muscle, the force generated by smooth muscle is controlled by the level of intracellular free Ca^{2+}. In smooth muscle during stimulation, this Ca^{2+} may come from the interstitial fluid as a result of a change in membrane permeability to Ca^{2+} or it may be released internally from bound stores. The incoming Ca^{2+} appears to contribute substantially to the rising phase of the smooth muscle action potential (see below). As a result of these changes the cytoplasmic concentration of Ca^{2+} may rise from a resting level of 10^{-8} mol l^{-1} to 10^{-6} mol l^{-1} or higher. With this rise in concentration, more Ca^{2+} combines with the calmodulin regulatory protein to activate a protein kinase that phosphorylates myosin. As phosphorylation of the myosin is a prerequisite for the activation of the smooth muscle actin–myosin complex, it is the level of free Ca^{2+} that regulates the contractile activity of smooth muscles.

Resting membrane potential and action potentials

The resting membrane potential of many smooth muscles is in the range of -60 to -70 mV and the basis of this membrane potential is similar to that found in other excitable cells. However, the resting membrane potential is some 20 mV below the equilibrium potential for K^+ and this may be the result of a relatively large resting permeability to Na^+.

Not all smooth muscles exhibit action potentials, but in those that do they are usually spike-like, but somewhat slower than in skeletal muscle; plateau-type action potentials are seen in some tissues (e.g. in the ureter) (Fig. 3.16). A depolarization of some 20 mV is required to reach threshold and initiate an action potential. The action potentials reach a peak potential of some $+10$ mV and if the stimulus is maintained repetitive firing may occur, the frequency depending on the degree of depolarization.

The inward current responsible for the action potential in nerves and skeletal muscles is carried by Na^+ ions. This can be shown by the fact that the magnitude of the overshoot of the action potential is directly proportional to E_{Na}, and that action potentials fail in solutions depleted of Na^+. This is not true for some smooth muscles, where removing Na^+ from the bathing fluid

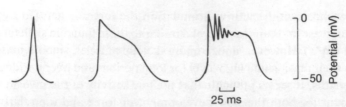

Fig. 3.16. Action potentials of smooth muscle may be spike-like (left), plateau-type (middle), or a mixture of these (right).

actually makes the action potential larger. On the other hand, removing Ca^{2+} ions from the bathing fluid does abolish action potentials, while increasing Ca^{2+} produces larger action potentials. Mn^{2+}, which blocks Ca^{2+}-mediated action potentials in other tissues, blocks the smooth muscle action potential.

Types of smooth muscle

As mentioned earlier, the activity of smooth muscles may be rhythmical and in these tissues dependent on spontaneous myogenic mechanisms; other tissues are quiescent until stimulated by an incoming signal. The former have often been referred to as 'unitary' (cells acting together) and the latter as 'multi-unit' (cells acting independently) smooth muscles, respectively, but these divisions are now of little use as they represent extremes. Smooth muscles are now conveniently divided into three groups according to their membrane properties.

Spontaneously active smooth muscle

Many organs containing smooth muscle contract rhythmically (e.g. stomach, small intestine, ureter, parturient uterus). As this coordinated activity is seen in the presence of neurotoxins or local anaesthetics, it is concluded that it is initiated and coordinated by the smooth muscle cells, i.e. it is *myogenic* (as in the heart). Such activity usually depends on the spontaneous generation of action potentials, and the presence of a conducting system (the gap junctions). Two types of mechanism are responsible for the spontaneous generation of action potentials, *pacemaker* potentials and *slow waves*.

   In some smooth muscles (e.g. uterus and taenia coli), it appears that focal pacemaker regions slowly depolarize a group of cells to threshold; the subsequent action potentials are then conducted through the tissue. The pacemaker regions are not constant in location, and it is thought that all regions within these tissues have the capacity to assume the role of pacemaker.

   The rhythmic activity of the stomach and intestine result from depolarizations called slow waves (Fig. 14.7). These are discrete, plateau-type depolarizations, lasting 2–8 s, that can be recorded from all regions of the external muscle layer. As slow waves can be initiated by depolarizing currents, they are propagated and their frequency is determined by the cells having the highest rate. In the intestine, where they are about 20 mV in amplitude, and, in the antrum of the stomach, where they are about 40 mV in amplitude, the slow waves initiate action potentials or spikes. In these tissues, the amplitude of contractions depends on the number of action potentials. The situation is less certain in the fundus and body of the stomach where the slow waves are larger but no spikes are seen. The ionic mechanism responsible for the generation of the slow waves is not fully understood at present but it has been attributed to changes in membrane conductance and the activity of a rheogenic pump.

The spontaneous contractile activity can be altered by nervous activity which may be either excitatory or inhibitory. For example, in the intestine, acetylcholine, the transmitter released from parasympathetic nerves, causes the smooth muscle to depolarize. As a consequence, the number and frequency of action potentials on each slow wave is increased and the contractions are more forceful. In contrast, the inhibitory action of noradrenaline, the transmitter released from sympathetic neurones to the detrusor muscle, and the inhibitory actions of non-adrenergic, non-cholinergic autonomic neurones to the gut (p. 67) are due to hyperpolarization and movement of the membrane potential away from threshold. This may result in complete cessation of contractile activity while the spontaneous fluctuations in membrane potentials continue at a subthreshold level.

Electrically inexcitable smooth muscle

This term applies to an extreme, but not unimportant, group of smooth muscles that do not generate action potentials (e.g. bronchial, tracheal and specific arterial smooth muscles of some species). In these tissues, the membrane potential remains stable until the tissue is stimulated. Stimulation may be the result of neurotransmitter release or the activity of local or blood-borne agents (e.g histamine, bradykinin). Stimulation is accompanied by depolarization and subsequent contractions. In tissues with a sparse innervation, excitation can spread because of the presence of gap junctions. The physiological advantage of these muscles may reside in their generally slow and sustained response to nerve stimulation.

Intermediate smooth muscle

This category is the most widely distributed (e.g. iris, pilo-erector, blood vessels, vas deferens, seminal vesicles). These have a stable resting membrane potential and when stimulated they exhibit spike-like action potentials. The cells are linked by gap junctions but conduction is decremental and so the contractions fail to spread throughout the tissue. The force of contraction is proportional to the frequency of the action potentials and is usually under neural control.

Activation of smooth muscle

Contraction of all smooth muscles is dependent on changes in the intracellular Ca^{2+} level. This can occur as a result of inherent myogenic mechanisms which regularly depolarize the muscle fibres or from neural or hormonal action. Contractions may also be induced by other means. For instance, some smooth muscles are relatively plastic when slowly stretched but rapid stretching results in a stretch-induced depolarization and contraction. Such behaviour may be important in *autoregulation* of blood vessels. In other tissues, local agents modify the force of contraction (e.g. the actions of O_2 and CO_2 on blood vessels of the lungs, and histamine on bronchial smooth muscle).

In many tissues, the dominant influence is exerted by the nerves. Irrespective of whether they are sympathetic, parasympathetic or enteric, excitation is usually the result of a depolarization (an *excitatory junction potential*, Fig. 3.17) which is caused mainly by an increased conductance to Na^+ and to a lesser extent K^+. In the case of inhibition, the hyperpolarization of the smooth muscle membrane (the *inhibitory junction potential*) (Fig. 3.17) may be due to an increased conductance to K^+ (and possibly Cl^-).

In both cases, the increase in conductance arises as a result of the neurotransmitter interacting with specific surface receptors on the muscle fibres (p. 72). If the neuromuscular junction has a relatively small junctional cleft (20 nm), then the junctional potential is distinct with a fast rate of rise and may last 0·5 s; it seems that wider (400–500 nm) neuromuscular junctions may not exhibit the same rapid junctional potentials and the response to nerve stimulation is relatively slow. In fact in some tissues (e.g. the tunica media of blood vessels) many of the muscle fibres may not be directly innervated; they may, however, be under some neural influence, current spreading from neighbouring innervated regions (through gap junctions).

The contractile activity of many smooth muscles is also influenced by circulatory hormones and parahormones (p. 570) and discussions of these are presented in the relevant chapters. From these it can be seen that the actions of many hormones are limited to particular smooth muscles. Thus, the stimulatory actions of the gastrointestinal hormone, gastrin, on smooth muscle, are largely limited to the stomach and gall-bladder; and the stimulatory actions of oxytocin are limited to the uterus. The specificity of these interactions is demonstrated by the inability of hormones with similar structures (e.g. gastrin and cholecystokinin, and oxytocin and vasopressin) to have the same effects;

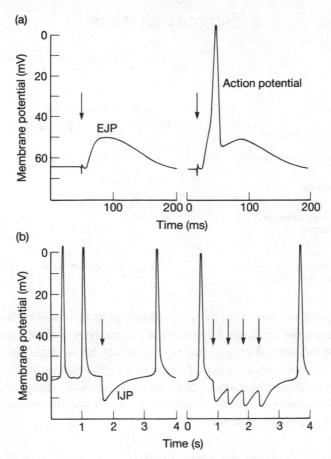

Fig. 3.17. Junction potentials in smooth muscles. (a) Excitatory junction potentials (EJP) recorded intracellularly from the vas deferens following stimulation (arrows) of its sympathetic nerve supply (left, subthreshold EJP and right, suprathreshold EJP leading to an action potential). (b) Inhibitory junction potentials (IJP) recorded intracellularly from guinea-pig taenia coli following stimulation (arrows) of the intramural nerves (left, single stimulus; right, repetitive stimuli).

this presumably reflects the highly specific nature of the receptors on the plasma membrane.

Other hormones have widespread activities, thus adrenaline can change the contractile activity of many tissues (e.g. blood vessels, bronchioles, the intestines). Similarly, the highly potent prostaglandins (PG) and thromboxanes have pronounced effects on a number of smooth muscles. Thus PGF_2 is a potent stimulator of uterine contractility (and is used to induce labour at term), of intestinal smooth muscle and of bronchial smooth muscle. However, the physiological roles of these compounds are uncertain.

4 · Sensory Systems

4.1 Receptors
Specificity
Transduction mechanism
Adaptation
Coding of sensory information

4.2 Sensory neurones

4.3 Somatosensory pathways
Dorsal column pathway
Anterolateral pathway
Function of relay nuclei
Surround inhibition
Descending inhibition

4.4 Somatosensory cortex
Somatotopy
Evoked potentials

4.5 Modalities of sensation
Mechanoreception
Touch
Kinaesthesia
Thermoreception
Pain
Somatic pain
Visceral pain
Referred pain
Phantom limb pain
Opioids

The term 'sensory systems' applies to those parts of the nervous system concerned with the detection, transmission and analysis of information about stimuli from the internal and external environments. Sensation is the conscious perception of such stimuli; much of the sensory information that is received by the body is outside the realm of consciousness. Sensory systems include the *receptors*, the *afferent nerve fibres* of sensory neurones and the *central pathways* activated by appropriate stimuli.

4.1 Receptors

Receptors are structures which convert different forms of stimulus energy into nerve impulses. They may be endings of sensory neurones or separate adjacent cells. The impulses generated then travel along the afferent nerve fibres to the central nervous systems (CNS). Receptors can be grouped as follows according to their location in the body.

1 Somatic receptors:
    superficial receptors —on the body surface and accessible mucous membranes
    deep receptors —in muscles, tendons and joints
2 Visceral receptors —in the walls of blood vessels, gastrointestinal tract, bladder and in membranes and linings of body cavities
3 CNS receptors —in the brain and spinal cord

4 Receptors of special senses (Chapter 5):

visual receptors	—in the eye
auditory receptors	—in the ear
orientation receptors	—in the vestibular apparatus
olfactory receptors	—in the nose
gustatory receptors	—in the tongue and oral mucosa.

Specificity

Receptors are, to a large extent, specific or selective in their response, being sensitive primarily to one particular kind of energy. This energy, or change in energy, forms the *adequate stimulus* and a receptor transforms or *transduces* this particular kind of stimulus into a change in membrane potential. It may also be able to transduce other forms of energy but its *threshold* will then be higher, i.e. the stimulus will need to be of higher intensity. Receptors are usually classified according to the kind of energy for which they are most specific, i.e. as mechanoreceptors, thermoreceptors, chemoreceptors, photoreceptors or nociceptors (Table 4.1). Sometimes specificity can be associated with a particular structure (e.g. Pacinian corpuscle) but often receptors with different specificities have no obvious structural differences (e.g. some cutaneous mechano- and thermoreceptors).

Transduction mechanism

As shown in Fig. 4.1, receptors may be highly specialized and a stimulus may act directly on the *membrane of the nerve ending*, e.g. free nerve endings in the skin, or indirectly via an *accessory structure*, e.g. a capsule or hair shaft, or via a *receptor cell*, e.g. rod and cone cells in the eye, taste buds in the tongue, hair cells in the ear. But even where the stimulus acts directly on the nerve terminal, the relationship of the ending to the surrounding tissue will still affect the transmission of the stimulus to the ending. Where an accessory structure is present, this usually plays an important role in transmission of the stimulus to the receptor.

A stimulus to a receptor cell causes a change in the conductance of the receptor membrane and this in turn causes a local change in the membrane potential, the *receptor potential*. The receptor potential evokes a local *generator potential* in the sensory nerve terminal either directly by electrotonic spread or indirectly by release of a chemical transmitter (Fig. 4.2). The generator potential initiates action potentials which are propagated along the sensory nerve to the CNS. Where the stimulus acts directly on a sensory nerve ending, the receptor potential and the generator potential are one and the same thing. There are, however, advantages in having separate receptor cells. For example,

Table 4.1. Classification of receptors.

Receptor type	Adequate stimulus	Location	Examples of effective stimulus
Mechanoreceptors	Mechanical deformation	Skin	Touch, pressure, vibration
		Muscles and tendons	Changes in muscle length and tension
		Joints	Joint position and movement
		Viscera (e.g. blood vessels, lung, stomach, bladder)	Distension
		Cochlea	Sound vibrations, about 16 Hz to 20 kHz
		Vestibule (e.g. semicircular canals, utricle)	Linear acceleration, angular acceleration
Thermoreceptors	Heat changes	Skin	Warming or cooling
		Hypothalamus	Warming or cooling
Chemoreceptors	Certain chemicals	Carotid and aortic bodies	Changes in plasma P_{O_2}, P_{CO_2}, pH
		Medulla oblongata	Local changes in pH
		Tongue and gut	Acids, salts, sugars
		Nose	Odorous chemicals
		Hypothalamus	Changes in plasma osmolality
Photoreceptors	Electromagnetic radiation of particular wavelengths (400–700 nm)	Eye	Light
Nociceptors	Mechanical, thermal or chemical but only at an intensity which threatens or causes tissue damage	Skin	Pinch, crush, sting, heat above 45–50°C
		Deep structures (muscle, joint and viscera)	Excessive stretch

synthesis of cellular material can be regulated locally by the receptor cell and regeneration of a damaged receptor cell is quicker than regeneration of a sensory fibre.

The *receptor potential* usually arises from a change in membrane conductance to Na^+ ions. The potential change is commonly one of depolarization although there are exceptions. For example, vertebrate photoreceptors are hyperpolarized by light while vestibular hair cells are either depolarized or hyperpolarized according to the direction of hair displacement. The receptor potential is non-propagating, has no refractory period and its amplitude is graded and dependent on the strength of the stimulus. Repeated stimuli can therefore summate. The relationship between stimulus intensity and receptor potential ('transfer function') is as a rule logarithmic.

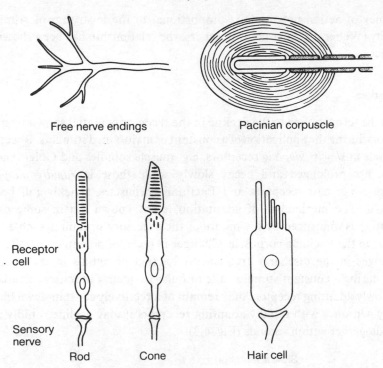

Fig. 4.1. Structure of various sensory receptors.

Where there is a separate receptor cell, the *generator potential* is a synaptic potential probably due to an increase in both G_{Na} and G_K. Its properties are similar to those of the receptor potential and it may provide amplification of the receptor potential signal. Both receptor and generator potentials are more resistant to local anaesthetics than are action potentials. The relationship between generator potential amplitude and the frequency of action potentials in the sensory fibre is approximately linear. The final result of this combination of transfer functions is different in different systems. In some cases, the

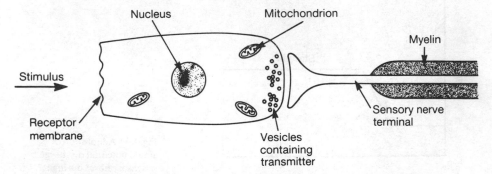

Fig. 4.2. Diagram illustrating synaptic contact between a receptor cell and sensory fibre.

frequency of action potentials is proportional to the logarithm of stimulus intensity (Weber–Fechner law); in others, the relationship has been shown to be linear.

Adaptation

This is the term applied to the decline in the receptor potential shown by most receptors during the application of a constant or maintained stimulus. Receptor potentials in *slowly adapting* receptors, e.g. muscle spindles and Golgi tendon organs, are prolonged and decay slowly while those in *rapidly adapting* receptors, e.g. hair receptors and Pacinian corpuscles, quickly fall below threshold. The mechanism of adaptation is not known but in some cases adaptation is influenced by the nature of the accessory structures such as the lamellae in the Pacinian corpuscle. Changes in receptor potential are reflected as changes in the discharge frequency of action potentials in the afferent fibres; during a constant stimulus, the impulse frequency in sensory neurones with slowly adapting receptors may remain at a relatively constant level but in sensory neurones with rapidly adapting receptors it may decline rapidly and cease altogether within seconds (Fig. 4.3).

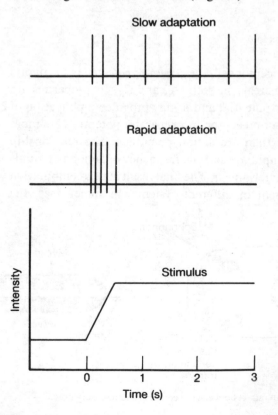

Fig. 4.3. Adaptation of action potential discharge in sensory fibres during a stimulus.

Coding of sensory information

A stimulus is transformed within a *sensory unit* into one or a number of stereotyped action potentials. The term sensory unit applies to the afferent nerve fibre of a single sensory neurone, all its peripheral branches and central terminals, and any non-neural transducer cells associated with it. The impulses so generated provide information to the CNS about the nature, position and intensity of the stimulus.

The nature or *modality* of sensation (i.e. the type of sensation experienced, e.g. vision, touch, etc.) depends on the specificity of the receptor for a certain form of energy. If a receptor is activated by other types of stimuli, necessarily at higher intensities, the brain still interprets the sensation as that of the adequate stimulus. Thus, activity in optic nerve fibres is interpreted as the sensation of vision whether the activity was produced by light or by sharp pressure to the eye.

Information on *stimulus position and shape* is conveyed to specific parts of the brain in a way that depends on the particular sensory unit or groups of sensory units that are activated. For example, each cutaneous sensory unit responds only to appropriate stimuli applied over a localized region of the body surface. This region forms the *receptive field* for that sensory unit. Its size reflects the threshold of the receptors and the degree of branching of the neurone. Certain regions of the body, e.g. fingertips and lips, receive a denser innervation, individual units having smaller receptive fields than in other regions, e.g. trunk and thigh. These densely innervated areas are therefore capable of providing more precise information about the position, size and shape of a stimulus. Receptive fields of individual sensory units overlap so the stimulus is usually effective in exciting a number of sensory units in the skin. In an analogous fashion, visual fibres have effective visual fields and some joint receptors have effective angles of movement.

A suprathreshold stimulus usually produces a burst or train of impulses rather than a single impulse. The response from tonic or phasic receptors increases as the intensity or rate of change of the stimulus increases. The number of neurones responding also increases and this is known as '*recruitment*' of sensory units. *Stimulus intensity* is therefore coded by:

1 the frequency of impulses in individual sensory units;
2 the number of active sensory units.

4.2 Sensory neurones

Leaving aside the special senses, sensory nerve fibres may be classified as somatic or visceral. Their cell bodies are located outside the CNS in various ganglia. *Somatic sensory neurones* have their cell bodies either in the trigeminal or dorsal root ganglia. Their peripheral axons (somatic afferent fibres) travel in peripheral nerves carrying information from the skin, skeletal muscles, tendons, joints and bone. Their central processes travel in trigeminal or dorsal roots to the CNS. *Visceral sensory neurones* have their cell bodies in the various cranial nerve ganglia or in the dorsal root ganglia of the thoracic, lumbar and sacral regions. Their peripheral axons (visceral afferent fibres) travel with sympathetic and parasympathetic nerves and carry information from receptors in blood vessels and internal viscera, particularly the heart, lungs, bladder, rectum and genital organs. Their central axons travel in the cranial VII, IX and X nerves to the nucleus of the tractus solitarius in the brainstem, or, in the thoracic and sacral regions, join the dorsal roots to enter the dorsal horn of the spinal cord with the somatic afferent fibres.

As already discussed (p. 51), peripheral nerve fibres are classified according to their diameters. In general, the greater their diameter the greater their conduction velocities. Unfortunately, there are two overlapping classification systems for afferent fibres. However, it is common to use the alphabetical A, B and C system for cutaneous afferents and the numerical I, II, III and IV system for muscle afferents.

Associated with the differences in diameter, peripheral nerve fibres also show differences in the following functional properties.

1 *Modality*. Although there are no rigid correlations between diameter and modality, some general trends are clear, as shown in Table 4.2.

2 *Conduction velocity*. This increases with increasing fibre diameter and with myelination (p. 47).

3 *Electrical threshold*. When a nerve is stimulated by electrical pulses, the strength of the stimulus required decreases with increasing fibre diameter.

Table 4.2. Cutaneous sensory nerve fibres and their receptors.

Fibre type	Numerical equivalent	Diameter (μm)	Receptors	Modalities represented
Aβ	Smaller fibres of I and all II	4–16	Merkel's discs, Pacinian corpuscles, Ruffini endings, Meissner's corpuscles, larger hairs	Touch, pressure, vibration, position sense
Aδ	III	1–4	Small hairs Specialized nerve endings Specialized nerve endings	Touch, pressure Cold Pricking pain
C	IV	0·5–2	Unknown mechanoreceptors* Unknown thermoreceptors Unknown nociceptors	Touch, pressure Cold, warmth Dull aching pain

* Not found in man.

Consequently, unmyelinated fibres require the greatest stimulation. This is a property of the axons and does not reflect thresholds to physiological stimuli.

4 *Refractory period.* This decreases with increasing diameter; action potentials are of shorter duration and recovery is faster in large fibres. Thus, the frequency of firing in unmyelinated fibres seldom exceeds 10 Hz while short bursts of 500 Hz or more are occasionally seen in large myelinated fibres.

5 *Sensitivity to local anaesthetics.* Local anaesthetics, such as procaine and xylocaine, act on excitable membranes to reduce the voltage-dependent influx of Na^+ associated with the action potential and hence block activity. Small fibres are the most susceptible because of their high surface area to volume ratio. Thus after application of local anaesthetic to a cutaneous nerve, sensations of temperature change and pain are blocked before touch and pressure associated with large fibres.

6 *Sensitivity to ischaemia.* A reduction in blood supply to a nerve, for example, as a result of an embolism, a thrombosis or an occlusion due to a pressure cuff, blocks large fibres first, probably because of their greater volume and hence higher metabolic demands. Thus, in this type of block, vibration sensitivity and position sense are lost before the ability to appreciate pain or changes in skin temperature.

4.3 Somatosensory pathways

Afferent nerve fibres, except for those of the special sense organs, enter the CNS mainly via the spinal dorsal roots, and the trigeminal and vagal roots. The innervation areas of peripheral nerves to the skin show little overlap and hence section of a nerve gives a characteristic region of sensory loss. However, fibres from one peripheral nerve enter the spinal cord over several dorsal roots, so that a particular innervation area is represented over a number of segments. The innervation area of the skin supplied by a single dorsal root is known as a *dermatome*. Dermatomes from adjacent dorsal roots overlap considerably so that section of a single root does not produce a region of complete anaesthesia.

After entering the spinal cord, fibres usually bifurcate giving ascending and descending branches which mostly end in adjacent grey matter. Inputs giving rise to sensations usually follow one of two ascending pathways within the spinal cord—the *dorsal column pathway* and the *anterolateral pathway*.

Fibres destined for the dorsal column pathway are mainly large myelinated (Aβ) fibres which enter the spinal cord via the more medially positioned dorsal rootlets. They bifurcate, giving one branch which enters the *dorsal horn* and a second branch which enters the ipsilateral *dorsal column* (Fig. 4.4).

The branches in the dorsal horn have a variety of destinations. Branches of the largest fibres from the annulospiral endings of the muscle spindles make

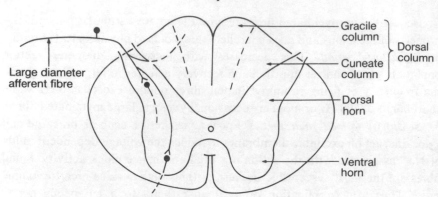

Fig. 4.4. Diagram showing the entry of a large diameter (Aβ) afferent nerve fibre into the spinal cord and the passage of its branches into the dorsal horn, ventral horn and dorsal column.

direct synaptic contact with motor neurones in the ventral horn. Branches from muscle, joint and skin mechanoreceptors synapse in the dorsal horn with the cells of origin of the *spinocerebellar tracts* which supply the cerebellum with information used in the control of posture and movement. Other branches contact dorsal horn cells and are involved in local segmental activity.

Dorsal column pathway

The branches of the primary afferent fibres in the dorsal column ascend without synapsing to the medulla. There they synapse with neurones in the ipsilateral *dorsal column nucleus* (Fig. 4.5). These neurones send their axons to the opposite side of the medulla, forming a well-marked histological feature— the *sensory decussation*—where the fibres from the two sides cross. After crossing, the fibres ascend in a tract known as the *medial lemniscus*, which passes through the midbrain and terminates in the *ventrobasal nuclear complex* of the thalamus. The thalamic neurones send their axons to layer IV of the six layers in the postcentral gyrus of the cerebral cortex (*somatosensory cortex*).

The dorsal column pathway carries information necessary for fine tactile discrimination, vibration sensitivity and position sense. Throughout the pathway the sensory modalities are preserved. Because the dorsal column pathway provides for these rather refined sensibilities, it is usually called the *discriminative* (specific) pathway. Since the medial lemniscus is a major component of this pathway it is also referred to as the *lemniscal system*.

Somatotopy. Receptive fields in this pathway remain fairly small and localized and, throughout the whole pathway from dorsal columns to the somatosensory cortex, adjacent fibres or cells correspond to adjacent areas in the periphery. Thus, the body surface is represented in an orderly, topographic fashion

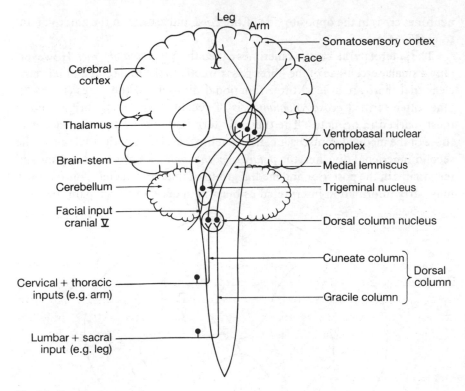

Fig. 4.5. Dorsal column pathway.

known as *somatotopic organization*, which is more fully discussed in Section 4.4. Moreover, certain peripheral areas have higher innervation densities peripherally and therefore disproportionately larger central representations throughout the whole pathway.

Conduction properties. Activity in this pathway is relatively stable, showing little alteration with arousal and attention and a lower susceptibility to blockage by general anaesthetics and hypoxia than that in the anterolateral pathway.

Anterolateral pathway

Fibres contributing to this pathway are the smaller myelinated (Aδ) and unmyelinated (C) fibres which enter the spinal cord via the more lateral rootlets, together with some branches of larger fibres. Many bifurcate, sending branches rostrally or caudally in the dorsolateral tract (tract of Lissauer) before synapsing in the dorsal horn with second-order neurones most commonly found in laminae I and V (Fig. 4.6). The axons of second- or third-order neurones cross to the opposite side of the cord and ascend in the anterolateral tract (Fig. 4.7).

The anterolateral tract is sometimes called the *spinothalamic tract*. However, only a small percentage of the fibres pass directly to the thalamus and it is now clear that there are at least three functional divisions of the fibres (Fig. 4.8). One, often termed *neospinothalamic*, is of recent phylogenetic origin and is most marked in primates. The thalamic terminations are found adjacent to those of the medial lemniscus, i.e. in the ventrobasal part of the thalamus. The second functional division, the *palaeospinothalamic*, is slower conducting and terminates in the posterior and medial group of thalamic nuclei. This pathway may carry information interpreted as burning, poorly localized pain, because

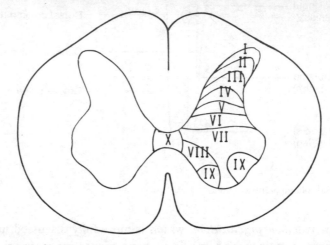

Fig. 4.6. Laminae of the spinal cord—ten anatomically defined layers. (From Rexed, B. (1964) *Prog. Brain Res.* **11**, 58–90.)

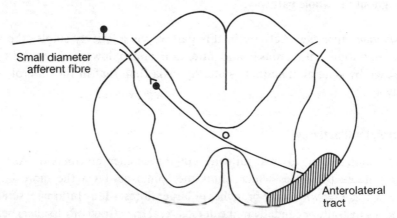

Fig. 4.7. Diagram showing the entry of a small diameter (Aδ or C) afferent nerve fibre into the spinal cord and its first relay in the dorsal horn; the fibre of the second-order neurone then crosses to the anterolateral segment.

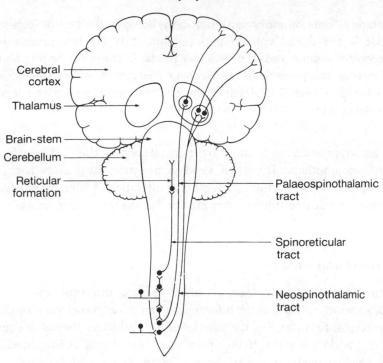

Cerebral cortex

Thalamus

Brain-stem

Cerebellum

Reticular formation

Palaeospinothalamic tract

Spinoreticular tract

Neospinothalamic tract

Fig. 4.8. Anterolateral pathway.

lesions in the intralaminar thalamic nuclei relieve such pain without affecting the appreciation of pricking pain. The third functional division, sometimes called the *spinoreticular tract*, is found in all vertebrates and terminates in the reticular formation of the medulla, pons and midbrain. These cells in turn relay to the thalamus and hypothalamus. The extent to which these pathways project to the somatosensory cortex and the role of this projection are still uncertain.

Fibres entering the anterolateral pathway carry information about touch, pressure, cold, warm and noxious stimuli. However, this pathway shows far more convergence and interaction than the dorsal column pathway. Hence, modality specificity is less precise and cells may respond to several different types of stimuli; at thalamic level; some cells respond to somatic, auditory and visual inputs. This pathway is therefore often referred to as the *non-discriminative* (non-specific) pathway.

Because this is the only pathway carrying impulses resulting from noxious and thermal inputs, damage to the anterolateral tract can give rise to a dissociated sensory loss; touch of a particular skin region can still be felt (via the dorsal column path) but both types of pain (pricking and burning) and temperature changes are not perceived.

Somatotopy. Somatotopic organization in the neospinothalamic tract is similar
to that in the dorsal column pathway but it is far less precise in the
palaeospinothalamic and spinoreticular tracts. Because of the much greater
convergence, receptive fields are frequently larger and may extend over the
whole hand or even limb. Hence, there is far more overlap and less clear
topographic representation.

Conduction properties. In contrast to the dorsal column, this pathway is, except
for the neospinothalamic portion, slowly conducting. Responses, particularly
in the spinoreticular portion, show marked alterations with attention and can
habituate to repeated stimuli. They are also far less resistant to anaesthesia.

Function of relay nuclei

On any sensory pathway, there are always several interruptions at synaptic
regions known as *relay nuclei*; information can never travel via a single fibre
from receptor to cortex. For the somatosensory pathways, the nuclei consist of
cells in the dorsal horns of the spinal cord, the dorsal column nuclei, the
corresponding nuclei in the trigeminal system and the ventrobasal nuclear
complexes of the thalamus. These nuclei are important in the functioning of
the sensory pathways as they provide for interaction and modification of the
input by excitatory and inhibitory mechanisms. These mechanisms are
activated by ascending and descending pathways.

   The mechanisms of excitation and inhibition at synapses have been dealt
with in Chapter 2, p. 52. Inhibition can be either presynaptic or postsynaptic.
One situation in which both pre- and postsynaptic inhibition are thought to
contribute to the functioning of sensory pathways is surround inhibition.

Surround inhibition

This is also called lateral or afferent inhibition. It is important in all sensory
systems, particularly the visual and auditory systems, in the sharpening of
contrasts, in the localization of stimuli and in spatial discriminative ability.

   A sensory unit commonly has an excitatory receptive field surrounded by a
region, the inhibitory surround, which when stimulated reduces the output of
the unit. The inhibitory surround of a unit is located in the excitatory fields of
other adjacent sensory units which, when stimulated, exert an inhibitory effect
on that unit by way of interneurones (Fig. 4.9). These interneurones may act
either by pre- or postsynaptic inhibition. Thus, the unit with the greatest
afferent input imposes the greatest inhibition on its neighbours.

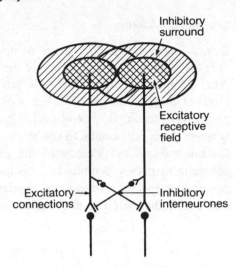

Fig. 4.9. Surround inhibition.

At each central synapse, diverging connections may result in activity becoming widespread. Surround inhibition is important because it reduces the spread of activity. It occurs at synaptic relays at all levels of the sensory system, thus ensuring that the focus of activity is kept sharply localized, with weakly excited surround connections being inhibited by the strongly active centre. An example of how this works in two-point discrimination (p. 140) is shown in Fig. 4.10.

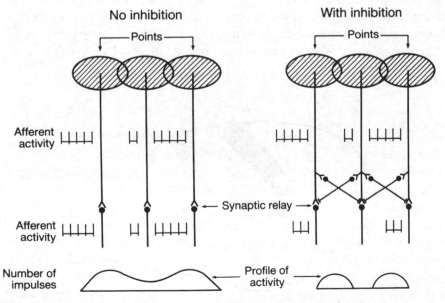

Fig. 4.10. Surround inhibition in two-point discrimination.

Descending inhibition

Descending pathways originating in various regions of the brain exert control over the sensory as well as the motor functions of the spinal cord. They act at every synaptic level to reduce irrelevant activity and to sharpen contrasts to improve discriminate ability. For example, the inhibition of descending pathways by the application of a cold block to the spinal cord alters the activity of dorsal horn cells caudal to the block. Conversely, electrical stimulation in the midbrain in the vicinity of the periaqueductal grey matter produces analgesia in man and animals. This analgesia is produced by nerve fibres descending from the midbrain to the relay nuclei in the dorsal horn where they take part in the inhibition of dorsal horn neurones.

4.4 Somatosensory cortex

The primary or *first* somatosensory cortex (SI) lies along the central sulcus (Fig. 4.11). It is classically considered to lie behind the sulcus in the postcentral gyrus, though many sensory responses occur precentrally. A *second* and smaller somatosensory area (SII) lies along the lip of the lateral fissure. Layer IV of the somatosensory cortex receives inputs from the thalamus. From there, activity spreads to more superficial and deeper layers before passing to other cortical regions. Recordings in this area have shown that neurones in columns perpendicular to the surface have similar receptive fields and respond to inputs from similar receptors. Responses can be obtained from cutaneous as well as deep mechanoreceptors (mainly in muscle). Thus, the somatosensory cortex, like the visual and auditory cortex, is organized functionally into columns.

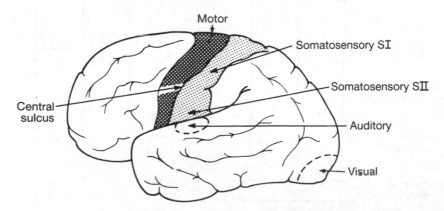

Fig. 4.11. Somatosensory cortex—primary (SI) and secondary (SII) areas.

Somatotopy

Responses in the first somatosensory cortex result from stimulation on the opposite side of the body, except for parts of the face, and are somatotopically organized. For example, responses from the foot and leg are represented medially in the somatosensory cortex, responses from the face laterally and responses from the hand and arm in between (Fig. 4.12). As in the dorsal column pathway, the density of peripheral innervation determines the size of the corresponding cortical area. The particular area which is enlarged varies in different animals and reflects the behavioural use of different peripheral areas in discrimination. For instance, in the pig the snout representation is particularly large; in rats the whiskers, while in man the hands and lips are disproportionately represented. SII also shows somatotopic organization but responses are frequently bilateral and mainly to cutaneous mechanoreceptors.

Lesions to the somatosensory cortex cause severe impairment of fine tactile discrimination in the corresponding peripheral area. Electrical stimulation of the cortex gives rise to a tingling sensation in the corresponding part of the periphery, though normal tactile sensations are not elicited.

In primates, somatosensory neurones which respond to nociceptive stimuli have been found together with neurones that respond with a slowly adapting discharge to pressure on the skin (Table 4.2). These neurones are found in the sulcus between the pre- and postcentral gyri. Two populations that respond to nociceptive stimuli have been described. One has restricted contralateral fields and could be involved in the sensory–discriminatory aspects of pain. The other has large, whole-body receptive fields and is thought to be involved in the

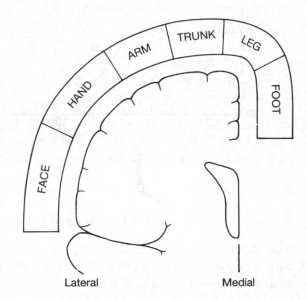

Fig. 4.12. Somatotopic organization. The sensory representations are indicated on a cross-section through the somatosensory cortex. (Adapted from Penfield, W. & Rasmussen, T. (1968) *The cerebral cortex of man: a clinical study of localization of function*, p. 44. Hafner Publishing Co., New York.)

motivational and arousing aspects of pain. Two lines of evidence suggest there is a similar representation in man. Epilepsy generated in this region has as its aura pricking pain, and superficial lesions of the region produce a major impairment of sensitivity to cutaneous pain.

Evoked potentials

On stimulation at the surface of the body, a mass response consisting of the IPSPs and EPSPs of many activated cells can be recorded from the surface of the somatosensory cortex, or even through the scalp if averaging techniques are used to improve the signal to noise ratio. This mass response is known as an *evoked potential* (Fig. 4.13). Evoked potentials have also been studied in the visual and auditory systems. They most commonly consist of a series of alternating positive and negative waves, the first wave being positive.

The *latency* of an evoked potential varies in different sensory systems and in different animals but is very consistent under similar stimulating and recording conditions. This has proved valuable in the early diagnosis of certain diseases like *multiple sclerosis*, in which the latency of the evoked potential is increased because of loss of myelin along the pathway. Evoked potentials can also be used to assess the integrity of a sensory path in infants or comatose patients.

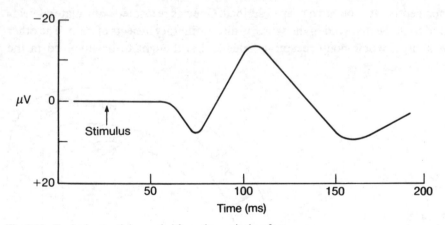

Fig. 4.13. Evoked potential recorded from the cortical surface.

4.5 Modalities of sensation

Now that the nature of the receptors, the sensory neurones and the central pathways involved in sensation have been discussed, we can consider in more detail the various modalities of sensation.

Mechanoreception

The sensations usually associated with gentle mechanical stimulation of the skin are *touch* and *pressure*, although more complex sensations such as *vibration* and *tickle* are also recognized. Figure 4.14 illustrates the kinds of mechanoreceptors to be found in hairy and glabrous skin of mammals. Muscles and joints contain receptors which contribute to *position sense* as well as responding to *pressure* and *stretch*. Viscera have only a few mechanoreceptors.

Mechanoreceptors are specialized to detect particular properties of the stimulus such as its position, its intensity, its duration, its velocity or its acceleration. Receptors that respond when the stimulus is stationary are *position* or *length* detectors. As the stimulus is applied, the frequency of nerve impulses increases with the amplitude of deformation and then remains steady or declines gradually when the displacement is static. Such slowly adapting receptors can also gauge the *intensity* and *duration* of mechanical deformation of the skin or muscle. During a change in position or length they may also provide information about the velocity of deformation. Examples of such receptors are Merkel's discs and Ruffini endings. Both are probably involved in the sensation of pressure on the skin.

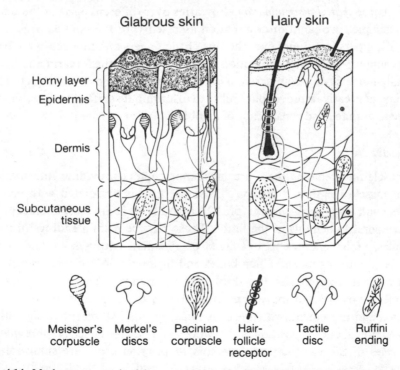

Fig. 4.14. Mechanoreceptors in glabrous and hairy skin. (From Schmidt, R.F. (1981) *Fundamentals of Sensory Physiology*, p. 86. Springer-Verlag, New York.)

Receptors detecting primarily *velocity* respond mainly during movement and cease to fire after movement has stopped. Such rapidly adapting receptors are called '*phasic*' receptors in contrast to '*tonic*' receptors that continue to respond to a constant stimulus. Phasic receptors are involved in sensations such as touch and tickle. Examples of this type of receptor are Meissner's corpuscles in glabrous skin and hair follicle receptors in hairy skin.

A third type of receptor responds to rapid transients in skin displacement, i.e. to *acceleration*. They are very rapidly adapting and may discharge only one impulse for each stimulus. Two types are known—Pacinian corpuscles and certain hair receptors. Rapidly repeated stimulation of these receptors evokes a sensation of vibration. Pacinian corpuscles are extremely sensitive to vibration and detect very small amplitudes of the order of 1 μm, particularly between 150 and 300 Hz.

Touch

Tactile sensations can be produced by indenting the skin on the fingertips by as little as 10 μm. Localized regions or *touch points* occur which are more sensitive to touch or pressure than surrounding areas. In the fingertips the thresholds of the receptors are significantly lower than surrounding areas. The capacity for *spatial discrimination* also varies over different parts of the body. For example, two-point discrimination assessed with the pointed arms of a pair of calipers is 1–3 mm over the tips of the fingers and tongue; in contrast the minimum spatial discrimination is 20–50 mm over the forearm and back. Usually touch involves an active movement of the skin over an object. This movement greatly improves our ability to discriminate surface texture, as well as the size, shape and consistency of an object.

Kinaesthesia

In muscle and tendons there are mechanoreceptors providing information about muscle length and tension. These are usually associated with *muscle spindles* and *Golgi tendon organs* (p. 192). In joints there are mechanoreceptors that respond to bending of the limbs. These may be Ruffini endings in joint capsules, Golgi endings in ligaments or Pacinian corpuscles often found associated with joints and along bones and ligaments. All of these receptors (and mechanoreceptors in the skin) may be involved in the subconscious control of posture and movement, and also in the conscious awareness of position and movement, which is called *kinaesthesia*. All receptors signalling information about the position of the limbs and body are called *proprioceptors*. The roles of the various proprioceptors in position sense are summarized below.

Recordings from joint receptors indicate that they respond mainly at the extremes of movement of a joint and many respond at both extremes or to movements in more than one axis of rotation (e.g. flexion–extension and abduction–adduction). Thus, the majority do not, as was once thought, appear to be able reliably to signal joint position over most of the range of movement.

Muscle spindles and Golgi tendon organs also give information about the state of joints and can provide accurate information about position. Indeed various lines of evidence suggest an important role for these receptors in kinaesthesia. Thus, vibration of muscles (a powerful stimulus for muscle spindles) can give rise to illusions of movement and inaccurate estimation of joint position. These illusions remain even if the joints and skin are anaesthetized. One organ, the tongue, has muscle spindles but no joints or joint receptors. If the tongue is anaesthetized so that information from mucous membranes is lost, its position can still be distinguished. The lack of importance of joint receptors in kinaesthesia is also indicated by the fact that patients with a joint completely replaced by a prosthesis retain position sense for that joint.

The contribution of cutaneous afferents to joint position is still uncertain. Cutaneous afferents can facilitate detection of passive finger movement, but skin anaesthesia often has little effect on kinaesthetic performance.

In conclusion then, it appears that muscle spindles and Golgi tendon organs are the principal receptors for normal position sense. However, joint and skin receptors may play a role in situations where information from muscles is impaired. Joint receptors presumably also have a role at the extremes of movement.

Thermoreception

There are two types of cutaneous receptors associated with thermal sensations—*cold receptors* that respond to a decrease in skin temperature and *warm receptors* that respond to an increase. These receptors are active in detecting localized changes in temperature (e.g. when a cold object is touched) as well as in detecting changes in ambient temperature. These receptors, together with receptors in the hypothalamus and spinal cord, are also involved in the regulation of body temperature (p. 622). The structure of cutaneous thermoreceptors is not well defined but they are probably free nerve endings. They are supplied by small myelinated (Aδ) and unmyelinated (C) fibres.

The afferent fibres associated with cold receptors usually show a tonic discharge at skin temperatures between 15 and 20°C. They have a peak frequency at about 30°C. Although normally silent above 40°C, they may give a transient paradoxical response around 45°C. The afferent fibres of warm receptors show a tonic discharge at skin temperatures about 35°C and give maximum frequencies around 45°C. At temperatures a few degrees above this, their activity ceases. Cold and warm receptors both give a phasic response to a fall or rise in temperature of as little as 0·1°C.

Our ability to detect actual skin temperature and temperature changes depends on factors like the area of skin affected, the initial skin temperature and the rate of temperature change. For steady temperatures, the body has a 'neutral zone' from about 20–36°C (the actual range being variable) in which there is no persisting sensation of body temperature. For temperatures above or below this range, persisting sensations of cold or warmth are experienced. Above about 45°C, pain rather than warmth is usually reported. Changes in temperature of the order of 0·1–0·3°C can be detected if rapid; changes of more than 5°C may be required if they occur slowly.

Pain

Pain can be described as the sensation resulting from stimuli which are intense enough to threaten or to cause tissue injury. Such stimuli may be mechanical (e.g. scratch), chemical (e.g. acid) or thermal (e.g. burn); neurologists commonly test pain sensitivity by a pinprick. However, pain sensations may show no simple correlation with the intensity or extent of tissue injury. Areas of apparently normal skin may show increased sensitivity (hyperpathia) while massive tissue destruction may occur without pain. Thus, the location and type of injury is important. Pain sensitivity also varies in different individuals, races and cultures. Even the same injury in the same individual can cause different degrees of pain depending on the situation. For example, injuries in accident victims are often not remembered as painful at the time of the accident. It is clear then that the sensation of pain is complex and that the body can alter the threshold in different situations.

It is usual to consider pain as having two components, the sensation *per se* and the emotional overtones of suffering and distress which are associated with it. While these two components usually occur together, certain procedures may leave one without the other. Thus, sensation may occur without distress after frontal lobotomy operations, while abnormal activity in certain thalamic areas may lead to a sensation of pain that is particularly distressing and intractable. Certain drugs may also have different effects on the two components; morphine for instance is especially effective on the emotional, reactive component. These considerations make studies on pain mechanisms difficult.

Somatic pain

The receptors in skin, muscle and joint that respond to painful stimuli— nociceptors—are probably free nerve endings. They are supplied by either small myelinated ($A\delta$) fibres or unmyelinated (C) fibres. The endings of $A\delta$ fibres signal high-intensity mechanical stimuli; while the endings of C fibres signal high-intensity mechanical or heat stimuli or are less selective, responding to high-intensity mechanical, thermal and noxious chemical stimuli (polymodal nociceptors). The response is usually a vigorous burst of activity. Sometimes, the two groups of nerve fibres give rise to a 'double' sensation, the pain differing in latency and quality. There is a sharp initial pain signal which is caused by faster conducting A fibres, then later a long-lasting, aching pain due to activity in C fibres.

Many nociceptors show *sensitization*, that is, an increase in activity in response to repeated stimulation. Such sensitization probably underlies the hyperalgesia found in an injured area. Various factors such as K^+, acetylcholine, kinins and prostaglandins released by tissue damage have been implicated as mediators activating nociceptors but no single factor has yet been identified. There is also evidence that factors such as substance P may be released from peripheral nociceptor terminals themselves and may contribute to the inflammatory process. *Cramping pains* may occur with excessive use of skeletal muscle as a result of stimulation of nociceptors by unknown factors. *Itch* is probably a distinct sensation with its own receptors; it also requires the liberation of a chemical substance, probably histamine.

Visceral pain

Responses from visceral receptors, including nociceptors, have been studied in the vagus and other visceral nerves. As in the skin, visceral nociceptors are thought to be free nerve endings and occur in the walls of most hollow viscera, mesenteries and blood vessels. Like somatic nociceptors they are supplied by small myelinated and unmyelinated afferent fibres.

It is unclear to what extent viscera contain specific nociceptors. Many afferents from the heart, lungs, intestine and bladder respond to innocuous mechanical stimuli as well as to damaging stimuli. However, specific responses have been reported from heart, ureter and gall-bladder. Effective stimuli are either excessive stretch or ischaemia. For example, in the heart afferents have been described which respond only to interruption of coronary blood flow, while in the ureter there are afferents responding specifically to overdistension. Nociceptors are not present at all in some tissues, for example, brain, so brain tissue can be cut without giving rise to pain. The dura, on the other hand, is extremely sensitive.

The sensations most commonly detected from viscera are either fullness or pain. Thus, stretch in the rectum or bladder gives rise to the sensation of fullness and triggers emptying reflexes. Excessive stretch or distension of many viscera gives pain which is often colicky or intermittent, for example, biliary colic or ureteric colic. Alternatively, severe visceral pain can also occur with ischaemia as for instance in angina pectoris. Visceral pains are commonly poorly localized and may be referred to other parts of the body.

Referred pain

Damage to an internal organ is commonly associated with pain or tenderness not in the organ but in some skin region sharing the same segmental innervation. A classic example of this is the referral of cardiac pain to the left shoulder and upper arm. The most likely explanation for referred pain is that some central cells receive both cutaneous and visceral inputs. Such cells have been described in the lamina V of the dorsal horn. In the normal situation, lamina V cells are activated by skin inputs. When the visceral organ is injured, it activates the same cell in the dorsal horn and this is interpreted as injury to the skin. The skin, though uninjured, feels tender and local anaesthetics applied to the skin help to relieve pain by reducing the total sensory input to central cells.

Phantom limb pain

Amputation of a limb is often followed by the feeling that the limb is still present. The phantom area gradually shrinks and may completely disappear, but a small percentage (5–10%) of amputees are left with a persistent and severe pain apparently from the absent region. The pain is thought to arise centrally because sectioning of the contralateral anterolateral tracts may give only temporary analgesia, and the pain may return and be felt at the same site as preoperatively. However in some cases the pain may be relieved for long periods by a prolonged or severe stimulus to the stump, although in others the stump may be re-amputated or anaesthetized without any relief.

Opioids

Opiates, such as morphine, heroin and codeine, are amongst the most powerful analgesics known. They act centrally by combining with receptors in many regions, including those associated with pain perception, such as the periaqueductal grey matter of the brainstem, parts of the limbic system and the substantia gelatinosa (lamina II) of the spinal cord.

An increased understanding of the way in which they are active has occurred with the discovery of the endogenous *opioid peptides*, comprising enkephalins, endorphins and dynorphins, which are powerful analgesics when administered intracranially. The *enkephalins* are pentapeptides (tyrosine–glycine–glycine–phenylalanine, then methionine or leucine) produced within the brain in regions like the limbic system and thalamus and in the substantia gelatinosa of the spinal cord. The *endorphins* and *dynorphins* are larger molecules containing a core of amino acids similar to the enkephalins (p. 264). They were first found in the pituitary and later shown to be present in several regions of the brain. In the CNS, opioid peptides are localized in nerve terminals and are thought to act as neurotransmitters or neuromodulators. They have pre- and postsynaptic inhibitory actions when iontophoresed on to neurones.

Although quite different in structure from the opiates, opioid peptides act by binding to the same type of receptors. The opioid peptides are as potent, or even more potent, in relieving pain than opiates when injected into the brain. They are not effective when injected intravenously, however, because they do not cross the blood–brain barrier and are rapidly degraded. It is claimed that their levels are decreased in chronic pain states and increased by electrical stimulation of the periaqueductal region. Such stimulation produces powerful analgesia, presumably by activating descending inhibitory controls. Opioid peptides may play a role in *acupuncture* as some of the effects of acupuncture can be blocked by the opiate antagonist naloxone.

As well as acting at higher levels of the CNS, the opioid peptides may also act at the level of the spinal cord and in peripheral tissue. *Substance P*, a peptide present in the terminals of afferent fibres, has been suggested as the transmitter for nociceptive afferents in the dorsal horn. Opioid peptides may suppress pain by acting presynaptically to block release of substance P from these afferent fibres. Substance P may also be released from peripheral nerve terminals and play a role in vasodilatation and inflammation. There is recent evidence that, besides their central action, opioids may act peripherally to modify this release.

5 · Special Senses

5.1 Vision
Anatomy of the eye
Image formation
Focusing of the image on the retina
Defects in focusing
Light reflex
Eye movements
Receptor mechanisms of the eye
Receptors
Photopigments
Adaptation
Receptor potentials
Electro-oculogram and
electroretinogram
After-images
The visual pathway
Retina
Central pathways
Neural responses in the visual pathway
Binocular vision
Colour sensation

5.2 Taste
Gustatory receptors
Afferent nerves and central pathways in
gustation

5.3 Smell
Olfactory receptors
Afferent nerves and central pathways in
olfaction

5.4 Hearing
Aspects of a sound stimulus
Functional anatomy of the ear
Stimulation of Corti hair cells
Action potential discharges in the
auditory nerve
The central auditory pathways
Deafness

5.5 Vestibular function
Stimulation of vestibular hair cells
The semicircular canals
The utricle and saccule
Functions of the vestibular organs
The central vestibular pathways
Disorders of vestibular function

5.1 Vision

Anatomy of the eye

The eye is a hollow sphere, the periphery of which is composed of three layers—a tough outer fibrous layer (the *sclera* and *cornea*), a middle layer comprising the vascular *choroid*, the muscular *ciliary body* and the *iris*, and an inner neural layer, the *retina* (Fig. 5.1). There are many blood vessels running over the surface of the retina which can be seen through an ophthalmoscope. Such examination is important in the diagnosis of diseases affecting blood vessels. The two inner chambers of the eye contain the *aqueous humour* and *vitreous humour* and suspended between them is the *lens*. Other parts of the eye are as shown in Fig. 5.1. The shape of the eye is maintained by an intraocular pressure of around 15 mmHg.

Image formation

The stimulus for vision is electromagnetic radiation with wavelengths in the range of 400–700 nm, detected by receptors in the retina. Light which bypasses

146

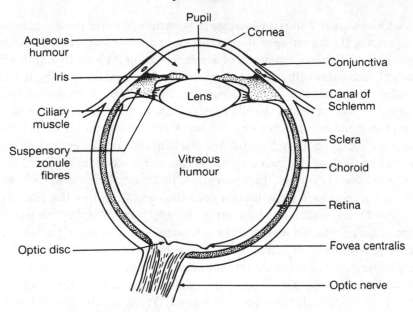

Fig. 5.1. The eye in horizontal section.

the receptors is absorbed by the pigment epithelium on the outer surface of the retina and by the choroid layer. For sharp vision the eye must bend (refract) incoming light rays so that they converge and focus on the retinal receptors and for the most detailed vision onto the region of highest acuity, the *fovea centralis* (Fig. 5.1). This is a depression in the retina at the centre of a small area called the *macula*, which is darker than the rest of the retina (due to greater pigmentation in its epithelium). Blood vessels at the periphery of the macula are sparse and at the centre are absent; such avascularity enhances light transmission.

Focusing of the image on the retina

When focused for infinity the eye has a total converging power of about 60 dioptres (power in dioptres = 1/focal length in m). Most of this converging power is achieved by the refraction of light rays at the curved air–cornea interface (45 dioptres). The lens provides further converging power (15 dioptres at rest) which can be increased for focusing objects closer than infinity.

The changes involved in looking from an object at infinity to one nearby comprise the *near response*, which has three components.

1 *Accommodation.* This is the increase in refractive power of the lens achieved by increasing the curvature of its anterior and posterior surfaces (particularly anterior). In childhood, the power of accommodation is about 10 dioptres but it steadily decreases with age (*presbyopia*). When focused for infinity, the lens is pulled flat by tension in the *zonule* (lens ligament). Accommodation is achieved by contraction of the ciliary muscles which are under the control of parasympathetic nerve fibres (cranial nerve III or oculomotor nerve). This causes the zonule to move forward and slacken, thus releasing the lens from tension and allowing it to round up due to its own elasticity (Fig. 5.2).

2 *Constriction of the pupil.* This provides a better depth of focus, since with constriction light rays pass through only the central part of the lens. This reduces spherical and chromatic aberrations, which are caused by the aspheric shape of the cornea and lens, and the higher refractive index of the lens core compared with its outer layers. Pupil size is controlled by both sympathetic and parasympathetic nerve fibres to the muscles of the iris. Sympathetic fibres (cervical sympathetic nerve) contract radial muscle and thereby *dilate* the pupil; parasympathetic fibres (cranial nerve III) contract circular muscle and thereby *constrict* the pupil.

3 *Convergence of the eyes.* This ensures that the light rays from an object fall on corresponding parts of each retina thus giving a fused image. It is achieved by contraction of the left and right medial recti muscles (see Fig. 5.4) controlled by cranial nerve III.

Defects in focusing

There are three kinds of refraction error that occur in eyes, namely hypermetropia, myopia and astigmatism. Another defect in focusing is presbyopia in which there is a loss of accommodative power.

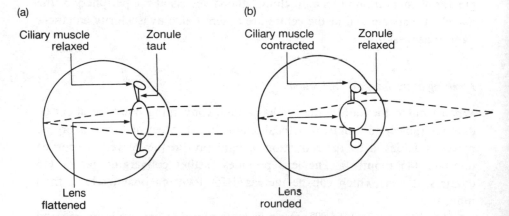

(a)

Ciliary muscle Zonule
relaxed taut

Lens
flattened

(b)

Ciliary muscle Zonule
contracted relaxed

Lens
rounded

Fig. 5.2. Accommodation. In distant vision (a), the ciliary muscles are relaxed; in near vision (b), they are contracted to focus the image on the retina.

1 *Hypermetropia* (long-sightedness). In this condition, the eyeball is too short (or the focusing system too weak) and even with full accommodation objects nearby are focused behind the retina (Fig. 5.3a). This can be corrected by using a *convex* lens.

2 *Myopia* (short-sightedness). The eyeball is too long (or the focusing system too strong) and even with full relaxation objects at infinity are focused in front of the retina (Fig. 5.3b). This can be corrected by using a *concave* lens.

3 *Astigmatism*. Here the curvature of the cornea (or occasionally the lens) is not uniform, thus giving different degrees of refraction in different planes. Therefore, the image is in focus in some planes and blurred in others. This can be corrected by using a *cylindrical* lens which gives additional refraction in the required meridian.

4 *Presbyopia*. The *near point of vision* is the distance between the eye and the closest object which can be brought to a clear focus. It recedes with age, being approximately 10 cm from the eye at 20 years and 80 cm at 60 years. This change is due to a reduction in the elasticity of the lens which decreases its ability to accommodate.

Light reflex

The pupil reflexly constricts when looking at near objects or in response to light. There are two light reflexes—the *direct* and the *consensual*. When light is shone into one eye the pupil constricts and this is called the direct light reflex. At the same time, the other pupil also constricts and this is called the consensual light reflex. The afferent pathway for the light reflex is via the optic nerve to the pretectal region of the midbrain and the efferent pathway is along parasympathetic fibres of cranial nerve III.

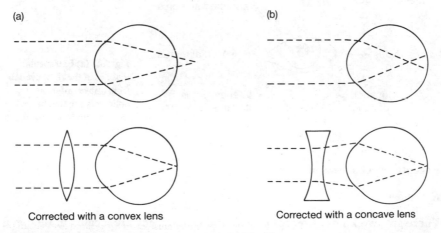

(a) (b)

Corrected with a convex lens   Corrected with a concave lens

Fig. 5.3. Defects in focusing. (a) Hypermetropia. (b) Myopia.

Eye movements

Each eyeball is moved by four rectus muscles (superior, lateral, inferior and medial) and two oblique muscles (superior and inferior) (Fig. 5.4a). The lateral rectus is controlled by cranial nerve VI (abducent nerve), the superior oblique by cranial nerve IV (trochlear nerve) and all the others by cranial nerve III (oculomotor nerve). When looking straight ahead, these muscles have the actions shown in Fig. 5.4b. The two eyes may be moved in the same direction (conjugate movements or versions) or in opposite directions (disjunctive movements or vergences) as occurs for instance during accommodation. Eye movements are also distinguished as *saccadic*, the fast step-wise movement employed when altering fixation, and *smooth pursuit*, the movement employed when the eyes follow a moving object. A *squint* occurs when the visual axis of one eye is not fixed on the object being viewed by the other.

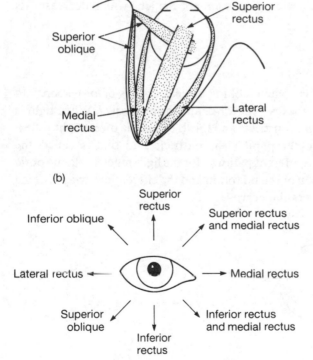

Fig. 5.4. (a) Extraocular muscles of the right eyeball from above (inferior oblique is situated below). (b) Directions of movement of the right eye.

Receptor mechanisms of the eye

Receptors

There are two kinds of receptor, *rods* and *cones*. They are located at the back of the retina, furthest from the entering light rays and next to the pigmented

Table 5.1. Ways in which rods and cones differ.

Rods	Cones
120×10^6 per eye	7×10^6 per eye
Not found in fovea but throughout the rest of the retina	Highly concentrated in fovea, less concentrated peripherally
High sensitivity; used in night vision	Lower sensitivity; used in day vision
Poor acuity	High acuity, especially in fovea where limit is diameter of one cone (2–3 μm)
Do not give colour sensation	Give colour sensation

choroid which absorbs stray light and thus prevents blurring. They are supplied by blood vessels of the choroid rather than by those on the retinal surface, which is why retinal detachment leads to extensive damage of the receptor cells. The two types of receptor differ in structure (see Fig. 5.8) and also in other respects (Table 5.1).

Photopigments

The transduction of light energy into a change in receptor membrane potential begins when it reacts with photopigments located in the receptors. Each photopigment consists of a protein, *opsin*, and a chromophore, *retinal* (the aldehyde of vitamin A). There are four forms of opsin giving four photopigments—the rod photopigment, *rhodopsin*, and the three cone photopigments, *erythrolabe* (red-sensitive), *chlorolabe* (green-sensitive) and *cyanolabe* (blue-sensitive). Their absorption spectra are shown in Fig. 5.5. The ability to detect different colours results basically from different degrees of excitation of the three cone types.

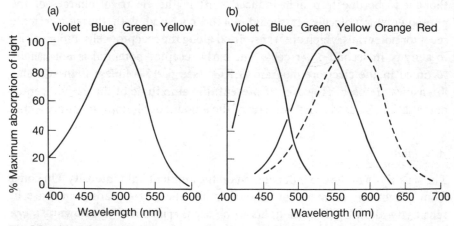

Fig. 5.5. Absorption spectra (a) of the rod photopigment and (b) the three cone photopigments.

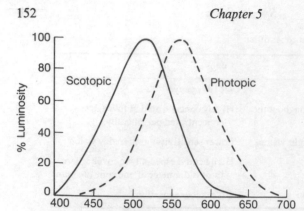

Fig. 5.6. Luminosity functions in eyes adapted for scotopic and photopic vision.

The perceived brightness of light depends on its wavelength and on whether the eye is adapted for night vision (*scotopia*) or day vision (*photopia*). The scotopic spectral luminosity curve (Fig. 5.6) has a peak at 507 nm and is characteristic of the absorption spectrum of rhodopsin, while the photopic spectral luminosity curve (Fig. 5.6) has a peak at 555 nm and is characteristic of the absorption spectra of the three cone pigments combined. Thus, in scotopic vision, the eye is most sensitive to green while in photopic vision it is most sensitive to yellow. The shift from scotopic vision to photopic is called the *Purkinje shift*.

The effect of light on the photopigments is to isomerize the chromophore retinal and hence to start a series of reactions in which a number of intermediate compounds are formed. The reactions lead eventually to the splitting of retinal from opsin, a process known as 'bleaching'. This series of reactions is associated with a potential change across the plasma membrane of the receptor cell known as the *early receptor potential*. This transient electrical response is thought to be due to a light-induced shift in the electrical charges of the photopigment molecule. It should not be confused with the slower ionic receptor potentials which are transmitted along the receptor cells and give rise to activity in second-order cells. The early receptor potential is not usually recorded in the electroretinogram (ERG) (see p. 154) unless high stimulus intensities are used. However, if present, the amplitude of the early receptor potential can be used as an estimate of the amount of photopigment present.

Adaptation

The sensitivity of the eye is dependent on the ambient light intensity. On going from a light environment into a darker one there is a gradual increase in sensitivity allowing dimmer lights to be seen, a mechanism known as *dark adaptation*. This is due to pupil dilatation and to changes in receptor sensitivity.

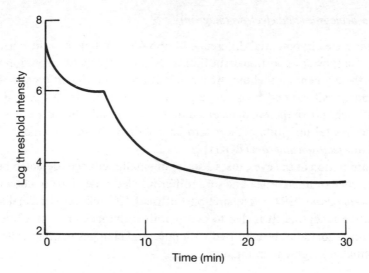

Fig. 5.7. Dark adaptation.

The latter has two components—an initial one involving cones and a later one involving rods (Fig. 5.7). The final change in sensitivity is about 10^5, of which only a small fraction is due to dilatation of the pupil and most to regeneration of photopigment and to neural mechanisms within the retina, occurring over approximately 30 min.

A much faster (3–5 min) period of *light adaptation* occurs when going from a dark environment into a brighter one, which results from pupil constriction and bleaching of photopigments.

Receptor potentials

The bleaching of photopigment which results in the early receptor potential is followed by a change in permeability within the receptor that results in a *receptor potential*. It begins with the photoisomerization of the chromophore retinal which activates another protein in the disc membrane called *transducin*. Transducin is one of the family of G-proteins (p. 256) which when activated binds guanosine triphosphate (GTP) and stimulates the activity of a phosphodiesterase that hydrolyses guanosine-3',5'-monophosphate (cyclic GMP). Cyclic GMP provides the link between the membraneous discs and the plasma membrane in which it opens Na^+ channels. Thus, light decreases cyclic GMP levels, closes Na^+ channels and hyperpolarizes the receptor cells. In the dark, the receptor cells are relatively depolarized and show a steady release of transmitter; in the light, they are hyperpolarized and transmitter release is reduced.

Electro-oculogram and electroretinogram

There is a steady potential difference of about 6 mV between the cornea and
the fundus (posterior portion of the interior of the eye) generated predominantly
across the pigment epithelium. As a result of this potential difference between
the front (positive) and back (negative) of the eye, electrodes placed on the
skin at each side of the eye detect a change in potential when the eye is moved
in a horizontal direction. The record so obtained during eye movements is
called an *electro-oculogram* (EOG).

Illumination of the eye causes a series of smaller changes detected between
an electrode on the cornea and an indifferent electrode. This is known as an
electroretinogram (ERG). The first part of the ERG reflects electrical activity
within the receptors. It is due to permeability changes and associated ionic
fluxes and should not be confused with potential changes in other cells within
the retina and pigment epithelium.

After-images

Visual sensations outlast the stimulus and give rise to both *positive* and *negative*
after-images. When an object is viewed briefly, it continues to be seen as a
positive after-image. However, with more prolonged exposure, it appears as a
negative after-image, i.e. a light source appears dark against a brighter
background. If the light is coloured, the negative after-image is tinged with
the complementary colour. The existence of positive after-images is used in
film and television techniques since, if the picture being viewed changes
frequently enough, the after-image will last from one frame to the next. The
minimum frequency required to give a non-flickering picture is known as the
critical fusion frequency. It is around 30–50 Hz and is higher at higher light
intensities. Modern projectors work well above this (72 Hz) so that no 'flicks'
are experienced.

The visual pathway

Retina

The retina is a complex outgrowth of the CNS and contains, besides the
receptor cells (rods and cones), four other types of neurones—*bipolar* cells,
horizontal cells, *amacrine* cells and *ganglion* cells, with a number of subtypes of
each. Ganglion cells are the output cells of the retina; their axons, of which
there are about 1×10^6, leave the eye at the *optic disc* to form the *optic nerve*.
There are no receptor cells overlying the optic disc and thus light focused onto
it cannot be seen—hence the term the *blind spot*. The retinal cells are all
arranged in orderly layers with the receptors next to the pigment epithelium

and the axons of the ganglion cells running over the surface of the retina (Fig. 5.8). As a consequence, light has to traverse these layers first before reaching the receptor cells.

Ganglion cell axons are unmyelinated and, hence, transparent, until they reach the optic nerve. At the fovea, where visual acuity is greatest, this orderly layering is modified; the fovea consists mainly of receptors (cones), with the other neural elements displaced to the sides and retinal blood vessels absent.

Rods and cones make different connections within the retina. Thus, cones synapse with bipolar cells (*cone–bipolars*) which in turn can connect with ganglion cells. Rods, in contrast, connect with ganglion cells more indirectly via two pathways. Some rods have gap junctions onto cones and hence connect with ganglion cells via cone–bipolars. Other rods synapse with bipolar cells

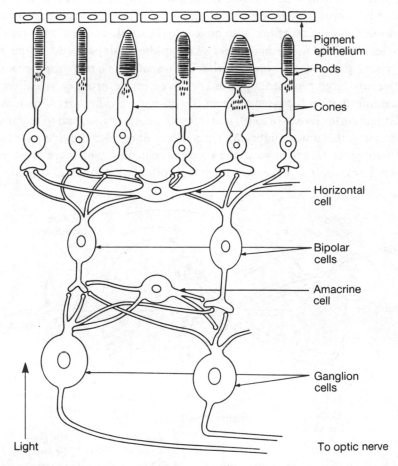

Fig. 5.8. Structure of the retina. (Adapted from Dowling, J. E. & Boycott, B. B. (1966) *Proc. R. Soc. (B)* **166**, 80–111.)

(*rod–bipolars*) which in turn connect to amacrine cells; these amacrine cells have synaptic connections with ganglion cells. There is considerable convergence in the rod–bipolar pathway to ganglion cells (1500:1) but relatively less in the cone–bipolar pathway, especially at the fovea centralis (16:1).

Central pathways

Axons of the ganglion cells travel in the optic nerve and end in that part of the thalamus called the *lateral geniculate nucleus* (Fig. 5.9). Fibres of the geniculate cells travel to the *primary visual cortex* (also called striate cortex and area 17) of the occipital lobe. There is a partial crossover of the optic nerve fibres. Fibres from the nasal halves of the retinae cross while those of the temporal halves remain ipsilateral, so that visual stimuli in the left half of the visual field of both eyes excite the right visual cortex and vice versa. Besides these connections optic nerve fibres also project to the striate cortex via the pulvinar complex of the thalamus, though less is known about this pathway. Throughout the visual pathway the organization of responses is topographic with a particularly large representation of the fovea in the primary visual cortex. Visual information is passed on from the primary to other areas, such as the peristriate cortex (areas 18 to 19) and inferotemporal cortex, and to subcortical regions (e.g. thalamus, midbrain). Optic nerve fibres also provide an input to midbrain pretectal regions and the superior colliculus, which are involved in light reflexes, control of eye movements and orientation to visual stimuli.

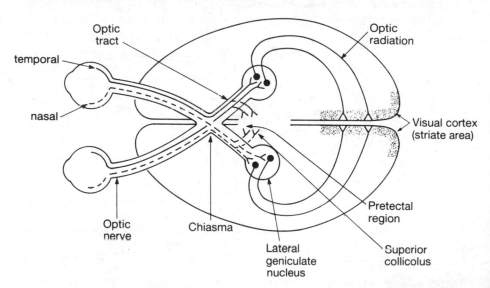

Fig. 5.9. Visual pathway viewed from below.

Neural responses in the visual pathway

More is known about the responses of cells in the visual pathway than about cells in any other sensory system. Most of our understanding comes from experiments on animals in which recordings were made from single cells in the visual cortex and the most effective stimulus for each cell was determined. More recently, new anatomical techniques have allowed individual cells, their processes and connections to be identified and correlated with their functions. Experiments in the 1960s, particularly those performed by D. H. Hubel and T. N. Wiesel, led to the idea of hierarchical processing of visual information but later work has shown that there is also parallel processing. Particular aspects of the visual stimulus such as contrasts are selected. Thus, constant levels of illumination are relatively ineffective as stimuli while changes in intensity with time or position are readily detected.

Responses in the retina. The complex interconnections and convergences which occur within the retina provide for considerable modification of responses to visual inputs between receptors and ganglion cells. Each ganglion cell in the retina receives inputs from receptors over a larger or smaller area of retina, which is called the *receptive field* of that ganglion cell. Many ganglion cells fire spontaneously even in the dark. This background activity can be changed by shining a small light spot within the receptive field. If the firing rate increases, this is called an 'on' response, if it slows, an 'off' response. When a number of ganglion cells are tested in this way, they fall into two groups. In the first, the receptive field is made up of a small more-or-less circular 'on' area surrounded by a larger ring-shaped 'off' zone, and these are called *'on' centre fields* (Fig. 5.10). The second group has the converse arrangement: an 'off' centre field surrounded by an 'on' ring and these are called *'off' centre fields*. Some fields

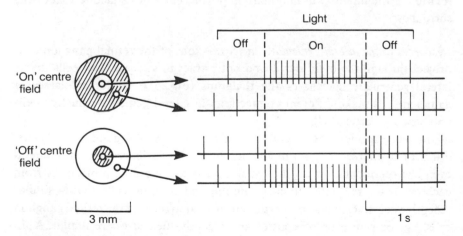

Fig. 5.10. Receptive fields of ganglion cells in the retina.

may have an intermediate zone in which both 'on' and 'off' discharges are recorded. In the primate, but not the cat, these concentric field cells may also be sensitive to spectrally opponent colours, e.g. red centre, green surround and vice versa (see later).

The intensity of the response, that is, the amount of increase or decrease in discharge frequency, depends on how much of a given area is stimulated, thus a process of summation is at work. Ganglion cells have been classified depending on whether this summation within their receptive fields is linear (X *cells*) or non-linear (*Y cells*). For X cells, the response is a simple summation of the excitatory and inhibitory areas stimulated, and spread of the stimulus into the 'surround' inhibitory area will reduce the response. Consequently, uniform illumination of the whole field is relatively ineffective. Besides these concentric field cells, other cells sensitive to movement or edge contrasts do occur but are relatively rare.

All the ganglion-cell fields are the result of complex cross-connections between receptors and other retinal cells. Thus, horizontal cells make both excitatory and inhibitory connections with receptor cells and are probably important in providing the inhibitory surrounds. The amacrine cells, on the other hand, form gap junctions with some bipolar cells and synaptic junctions with ganglion cells. They can give either transient or sustained responses and may be important in movement detection. Amacrine and ganglion cells are the only retinal neurones which generate impulses.

In summary, analysis of the responses of retinal nerve cells shows that they do not simply relay all signals unchanged to the next cell in line, but that due to convergence and lateral inhibition the first stage of *integration* is taking place in the visual system. The receptors of adjacent areas combine their effects so that the ganglion cell discharges in a way that is determined by the relation of the intensity of illumination of the centre of its field to that of the surround.

Responses in the lateral geniculate nucleus. Axons of the retinal ganglion cells project through the optic nerve and optic tract to synapse with cells in the lateral geniculate nucleus of the thalamus (Fig. 5.9). Different classes of ganglion cells make different connections to lateral geniculate nuclear cells and hence to cortical cells.

In most species, the lateral geniculate nucleus is organized in layers, different layers receiving an input from either the ipsilateral or contralateral eye. The representation of the visual field is topographic, projections from each eye being registered in each lamina, and adjacent cells having similar receptive fields. Response properties are similar to those of the retinal ganglion cells, e.g. 'on-centre' or 'off-centre', and receptive field sizes are similar. As in the retina, cells can show linear summation (X cells) or be non-linear (Y cells).

Responses in the visual cortex. Axons from the lateral geniculate nucleus travel to the visual cortex via the optic radiation (Fig. 5.9). This has a representation of the entire contralateral visual field, and a small extension over the vertical midline. Responses are topographic with relatively much more cortex devoted to central vision (e.g. in the macaque 90% of the cortical area is devoted to the central 10°).

Most area-17 cells respond best to linear contrasts between light and dark, such as edges, lines or slits rather than to circles. The majority are sensitive to the orientation of the line or slit, within about 20–30°, and all orientations are represented. Cells are usually classified as 'simple', 'complex' and 'hypercomplex' based on their response properties although many modifications and subdivisions of this classification have been proposed.

Simple cells have clearly defined 'on' and 'off' regions between edges or slits at certain orientations as in Fig. 5.11. The response is a simple summation of the stimuli within the 'on' and 'off' areas of the receptive field. Moving the edge or slit, usually in only one direction, is often a particularly effective stimulus.

Complex cells also require edges or slits at a particular orientation but the

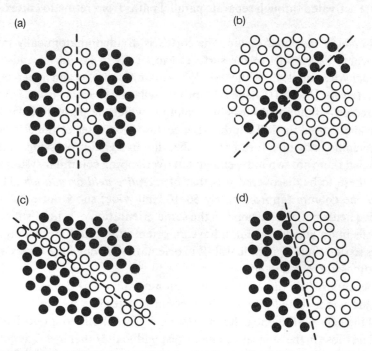

(a) (b) (c) (d)

Fig. 5.11. Receptive fields of 'simple' cells in the visual cortex. (a)–(d) Examples of the responses of four cortical cells in the cat to spots of light shone on the retina of one of the eyes. Areas indicated by open circles produce an 'on' response to retinal stimulation, areas indicated by closed circles produce an 'off' response. The broken lines indicate the axes of orientation of the receptive fields. (Adapted from Hubel, D.H. & Wiesel, T.N. (1962) *J. Physiol.* **160**, 106–154.)

response cannot be explained as simple summation of 'on' and 'off' areas in the receptive field. Thus, a complex cell might respond to an edge or slit of the right orientation anywhere in the field, or it might give an 'on' and 'off' response to a light or dark edge. Many complex cells also respond to movements, often in two directions.

Hypercomplex cells require an edge or a slit of fixed length or one that ends within the receptive field. This 'end-inhibition' requirement in fact may apply to both simple and complex cells. Therefore, hypercomplex cells, which have been subdivided into hypercomplex I (simple cells with end-inhibition) and hypercomplex II (complex cells with end-inhibition), do not really comprise a class of their own.

The original idea that simple cells received their input from the lateral geniculate nucleus and projected to complex cells in a hierarchical scheme of processing visual information has now been challenged. Thus, complex cells can be activated monosynaptically from the lateral geniculate nucleus and some respond to stimuli that do not excite simple cells. It is more likely that simple cells represent inputs from X ganglion cells with linear summation while complex cells receive inputs from non-linear Y cells. It is likely that these are activated through separate parallel paths from retina to cortex.

Organization of the visual cortex. The cortex is divided anatomically into six layers which lie parallel to the surface. Most cells in layers IV and VI are simple or hypercomplex I, while layer V has mainly complex and hypercomplex II cells. Layers II and III have all types of cells. The visual cortex is further organized on a functional basis into columns which lie perpendicular to the surface and extend the full depth of the cortex. Responses of the cells within these columns indicate a further specialization in signal processing. It is now known that there are two independent but overlapping columnar systems. The first of these to be discovered was that of *receptive field orientation*. Thus all cells in one column (approximately 50–100 μm wide) show not only similar receptive fields but also respond to the same orientation of a light slit or dark bar. Cells in an adjacent column have an orientation differing by only a few degrees so that there is a gradual shift in orientation through the cortex and all orientations are represented.

Superimposed on these orientation columns are *ocular dominance columns* which are wider (400–500 μm). These columns alternate with each other so that all the cells within one column respond preferentially to inputs from one eye, while those in the adjacent columns respond to the other eye. The columns extend across the cortex as bands (Fig. 5.12). Normal development of ocular dominance columns requires the normal use of both eyes. Monocular deprivation leads to a reduction in width of the corresponding columns or an associated expansion of those of the functional eye.

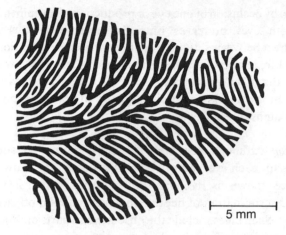

Fig. 5.12. Ocular dominance columns over the exposed part of the striate cortex in the right occipital lobe of the monkey. The dark stripes correspond to one eye and the light stripes to the other. (From Hubel, D.H. & Freeman, D.C. (1977) *Brain Res.* **122**, 336–343).

Thus an area on the retina of each eye is represented by a corresponding block of cortex (approximately 1–2 mm²) which contains an ocular dominance column for each eye, with these further subdivided into orientation columns. An adjacent point on the retina is represented by an adjacent block of cortex. There is now evidence that other cortical regions concerned with responses to colour and movement may also have a columnar organization, so that this may be fundamental to the analysis of visual information.

Binocular vision

The visual fields of the two eyes overlap in the central part to give a region which can be viewed binocularly. The images falling on the two retinae are not identical but neural mechanisms allow them to be interpreted as single images. For this fusion to occur the image of an object must fall on corresponding points of each retina. If such correspondence is prevented in the young child (e.g. by a squint), permanent impairment of fusion can occur.

The neural mechanisms which allow for fusion occur within the cortex. Many cortical cells can be excited by inputs from both eyes. The fields of such binocular cells occur at approximately corresponding points on each eye. However, one difference is in the relative effectiveness of the two eyes. In most cases, one eye is more effective than the other, a condition known as *ocular dominance*.

Cells responsive to both eyes are present at birth. Binocular responsiveness of most cells is gradually lost during early life if normal binocular vision is prevented (e.g. by occlusion of one eye or production of a squint). Interestingly, closure of both eyes preserves binocular responses, thus suggesting a competitive effect between activity from the two eyes. Prevention of binocular vision in later life, in both man and animals, is far less likely to lead to failure of fusion. As discussed previously, the explanation for these observations appears to lie in the development during early life of the appropriate ocular dominance columns for each eye.

Depth perception. Although most binocular cells require very similar stimulus characteristics for each eye, receptive fields of some cells show slight lack of correspondence known as *receptive field disparity*, which is thought to be important in discrimination of depth. However, in man, other clues are also highly relevant. Such clues include the obscuring of parts of objects by others, shadow effects, relative sizes and relative movements.

Colour sensation

Sensations of colour are due primarily to the presence of three types of cones containing different photopigments and maximally sensitive to different wavelengths of light (Fig. 5.5). Thus there are *red*-sensitive cones (peak 565 nm, actually yellow but extends into red), *green*-sensitive cones (peak 540 nm) and *blue*-sensitive (peak 440 nm) cones, and sensations of colour will depend on the extent to which each type is excited. This is termed the *Young–Helmholtz* or *trichromatic theory* of colour vision.

The three cone types do not have separate connections to the central nervous system. Certain horizontal cells and ganglion cells, called *spectrally opponent cells*, show inhibitory interactions between light of different wavelengths. These opponent cells are of four basic types:

1   excited by red light, inhibited by green;
2   excited by green light, inhibited by red;
3   excited by yellow light (i.e. presumably connected to both red and green cones), inhibited by blue; and
4   excited by blue light and inhibited by yellow.

The receptive field organization of spectrally opponent cells may be identical over its whole extent (e.g. excited by red and inhibited by green anywhere in the field) or it may show centre–surround antagonism (e.g. excited by red in the centre and inhibited by green in the periphery). Such interactions explain why red and green or yellow and blue are complementary colours. Responses of the opponent cells are also the basis for the after-image of a coloured stimulus appearing as its complementary colour. In primates, spectrally opponent cells also occur in the lateral geniculate nuclei and in the primary visual cortex. In area 17 of the cortex, spectrally opponent cells often have the properties of simple cells, requiring an edge or slit at the right orientation and spectral range. Other regions of the cortex have been found where there are colour columns, the cells all responding to one particular colour.

Defects in colour vision. Individuals with normal colour vision can match any spectral colour by using a mixture of the three primary colours and are therefore known as *trichromats*. Those who require only two primary colours to match the spectrum are called *dichromats* and those who can match any colour by varying the intensity of only one primary colour are called *monochromats*. Monochromats are totally colour blind and perceive only different shades of grey.

If a trichromat requires more of one primary colour than normal to effect a match, he is said to be anomalous and may be classified as *protanomalous* (weak red sensitivity), *deuteranomalous* (weak green sensitivity) or *tritanomalous* (weak blue sensitivity). Dichromats have a more severe colour defect and lack one of the cone pigments. Their colour blindness is distinguished as *protanopia, deuteranopia* or *tritanopia* which literally mean defects of the first (red), second (green) or third (blue) pigment systems, respectively. In practice, both protanopes and deuteranopes confuse red and green because the spectral absorption curves of their respective green and red pigment systems overlap.

5.2 Taste

The sense of taste (gustation) is important in the selection and enjoyment of food, but full appreciation requires tactile, visual and olfactory inputs as well, as is shown by the loss of taste when the nose is blocked.

Gustatory receptors

These are modified epithelial cells (*taste cells*) which, along with supporting cells, are organized into groups called taste buds. Most taste buds are located on nipple-like protruberances (*papillae*) on the tongue surface (Fig. 5.13), with a few on the epiglottis, soft palate and pharynx. They are 50–70 μm in diameter and open on to the surface of the tongue at the taste pore (Fig. 5.14). In man, there are about 10 000 taste buds, the number decreasing with age. Three to five buds are found on each of the *fungiform papillae* at the sides and tip of the tongue while hundreds are found in the grooves of the larger *vallate papillae* at the back (Fig. 5.13). A third type the *foliate papillae*, rudimentary in man, are arranged in folds at the back edges of the tongue. Taste cells are constantly shed, with a half-life of approximately 10 days, and are renewed by division of

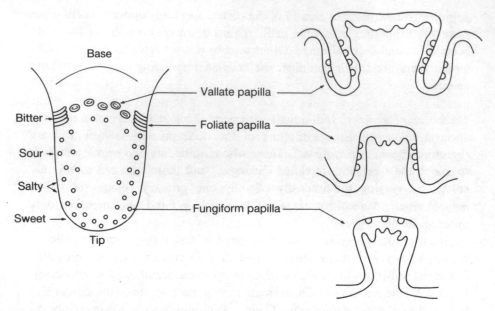

Fig. 5.13. Distribution of taste buds on the tongue and appearances of different papillae.

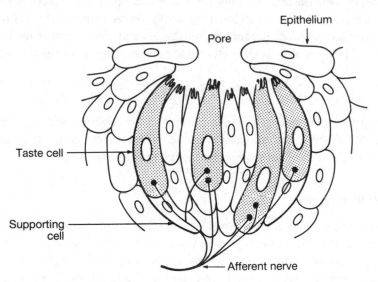

Fig. 5.14. Structure of the taste bud.

underlying basal epithelial cells to form supporting cells which eventually differentiate into taste cells. East taste cell makes synaptic contact at its base with an afferent nerve, upon which it depends for its continual existence; denervated taste buds degenerate and reappear again when the nerve regenerates.

Taste receptors are chemoreceptors. Chemicals enter the pore and react with the microvilli at the apical surfaces of the receptor cells to alter membrane permeability; responses can be depolarizing or hyperpolarizing and may be initiated either directly, or indirectly via inositol trisphosphate or cyclic nucleotides (p. 255). Classically, four basic tastes are recognized: *bitter, sweet, sour* and *salty*. Some taste buds respond to only one of these, the majority to more. The taste experienced varies with position on the tongue; bitter at the back, sweet at the tip, sour at edges and salty in a region overlapping the tip and edges (Fig. 5.13). All complex tastes are thought to be accounted for by combinations of these four basic tastes. Most of what the layman calls 'taste' is a combination of taste and smell.

The sensation of sourness is produced by acids and is related to the H^+ concentration, but not all acids are equally sour at equivalent pH. Salty sensations are elicited mainly by the anions of inorganic salts, particularly chloride, but also by other halides, sulphate and nitrate. Many organic compounds are described as bitter, e.g. quinine, caffeine, nicotine, morphine and strychnine. Most sweet substances are also organic, e.g. sucrose, maltose, lactose, glucose, saccharine, cyclamate, glycerol, but so too are salts of lead and beryllium. Recently, two proteins which are intensely sweet have been discovered.

Afferent nerves and central pathways in gustation

Taste buds on the anterior two-thirds of the tongue are supplied by the chorda tympani branch of the facial nerve (VII), those on the posterior one-third by the lingual branch of the glossopharyngeal nerve (IX). The vagus nerve (X) supplies taste buds on the epiglottis. Each bud receives branches from many fibres. Taste fibres from the chorda tympani, glossopharyngeal and vagus nerves join in the medulla to travel in the tractus solitarius and synapse first in a rostral portion of the nucleus of the tractus solitarius called the 'gustatory nucleus'. Some convergence of inputs may occur here between different taste fibres and between gustatory and thermal inputs, many cells responding to cooling of the tongue as well as to chemical stimuli. The second-order fibres project ipsilaterally to the most medial part of the ventroposterior medial nucleus of the thalamus. The thalamic neurones then project to the somatosensory cortex. Cells of the gustatory nucleus also relay information via the pons to the hypothalamus and the limbic system.

5.3 Smell

In many animals the sense of smell (olfaction) is important for the selection of food, in sexual activities and in the recognition of other animals and of territories.

Olfactory receptors

The receptors lie in the specialized olfactory epithelium lining the roof of the nasal cavity, and cover an area of about $5\ cm^2$ in man but relatively much more in such animals as the dog. The epithelium contains three sorts of cells (Fig. 5.15). The receptors, *olfactory cells*, are, in fact, specialized bipolar neurones. The apical region is a specialized dendrite which ends on the surface in a knob bearing 5–10 long, non-motile cilia, embedded in a layer of mucus. The basal region gives rise to a fine unmyelinated axon. In addition, the epithelium contains supporting cells, which have apical microvilli and secrete mucus, and basal cells which lie on the basement membrane. The olfactory cells are the only nerve cells in adult mammals to show continual degeneration and replacement. They are replaced by division and differentiation of the basal cells.

Several thousand chemicals can be smelled. To be detected they must reach the olfactory epithelium by diffusion, which can be aided by sniffing to increase the air flow. The chemicals are mainly organic, containing 3–20 carbon atoms, but there is no simple relationship between chemical structure and odour. They must be volatile and have some water solubility in order to dissolve in the mucus and also have lipid solubility to interact with the receptor membrane. Threshold concentrations vary, being extremely low for certain substances,

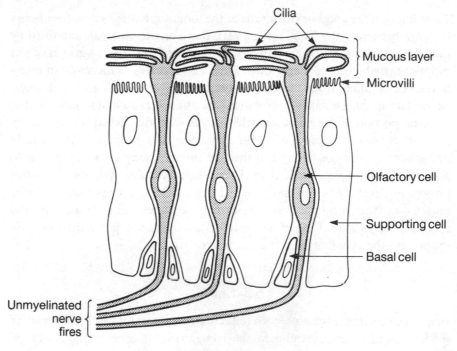

Fig. 5.15. Structure of the olfactory epithelium.

e.g. 0.4 ng l^{-1} air for divinyl sulphide (garlic). Chemical stimuli are thought to combine with receptors on the cilia of the sensory neurones. This leads (either directly, or indirectly via cyclic AMP) to an increased conductance to Na^+ and possibly K^+ and hence to a depolarizing generator potential. The summated generator potentials of the olfactory neurones can be recorded as a mass response or *electro-olfactorogram* (EOG) across the olfactory epithelium. Recordings from single neurones have shown that depolarization and hence impulse activity increases with increasing concentration of odorant, single neurones responding to many odours.

The sense of smell shows marked adaptation and is subject to 'masking' of one smell by another—the basis for 'air fresheners'. Olfactory acuity is also better in women than in men, particularly at the time of ovulation, and has been positively related to oestrogen levels.

Afferent nerves and central pathways in olfaction

The unmyelinated axons of the receptor cells pass in bundles through the cribiform plate and enter the olfactory bulb. The bulb is organized into synaptic clusters, or glomeruli, where the axons make contact with *mitral* and *tufted* cells. In each glomerulus, 10^4–10^5 primary axons synapse with approximately 200 secondary cells so there is marked convergence. Single cells of the olfactory bulbs usually respond to a number of odours and can be excited or inhibited by different stimuli. Olfactory bulb cells project to the opposite bulb via the anterior commissure as well as to various regions of the limbic system (septal nuclei, anterior perforated substance, hypothalamus, amygdala and prepiriform cortex). There is also evidence for connections to the mediodorsal nucleus of the thalamus and hence to the orbitofrontal neocortex. As in the olfactory bulb, cells in the prepiriform cortex and orbitofrontal neocortex respond to a variety of odours.

5.4 Hearing

The auditory system has evolved as a survival system for warning and hunting with the function of bringing slight sounds to attention and locating them with respect to visual space. In man, it also functions as the afferent part of spoken communication. To carry out these functions of *recognition, location* and *communication*, it must detect sounds of particular interest, such as human speech, and ignore irrelevant background sounds. Each ear accepts sound energy from a limited range of directions and then performs a frequency analysis on it so that the auditory nerve carries information on the intensities of the frequency components of a sound. The brain has centres specialized for analysing the relative intensity and phase of individual sounds so as to compute their direction of origin. It also has higher centres dedicated to recognizing the source and nature of sounds and the content of speech.

The sensation of sound detected by the ear is caused by variation in air pressure within a specified range of frequencies and intensities. Outside this range the sensations evoked are variously described as vibration, flutter, tickle and pain.

A source of sound such as a tuning fork causes the surrounding air molecules to oscillate. The disturbance spreads out from the source in sinusoidal waves which consist of areas of compression of air (high pressure) alternating with areas of rarefaction (low pressure). Sound can thus be described, like all waves, in terms of both frequency and amplitude (pressure difference). The *loudness* of the sound is related to the amplitude of the sound wave and the *pitch* to the frequency.

Aspects of a sound stimulus

Frequency. A normal young person can hear sounds ranging in frequency from 20 to 16 000 Hz. Higher frequencies are called ultrasound and lower frequencies infrasound. Frequency discrimination by the human ear is best at about 1 kHz, where changes of as little as 3 Hz can be detected.

Sound of only a single frequency, a *pure tone*, is uncommon in daily experience. Most sounds are composed of numerous, simultaneously occurring frequencies. Fourier (1768–1830) first suggested that all complex waveforms could be analysed by breaking them down into simpler constituent waves of single frequencies. When subjected to Fourier analysis musical sounds are found to be composed of a fundamental frequency plus a number of harmonically related frequencies, i.e. the higher ones (the 'harmonics' of 'upper partials') are integral multiples of the lowest (the 'fundamental'). The number and intensity of each determine the tone or timbre of the sound, whilst the fundamental determines the pitch. Thus the range and strength of the violin's upper partials are greater than for the tuba and give the violin a brighter tone.

Amplitude. This is the objective aspect of a sound stimulus, measurable with instruments, whilst 'loudness' is the subjective, conscious aspect, which is harder to quantify. The amplitude of sound waves is measured using the decibel scale (the unit of sound amplitude, the *bel*, is named after Alexander Graham Bell). This is a logarithmic scale defined as sound pressure level in decibels.

$$\text{Decibel (dB)} = 20 \log_{10} P_t/P_r$$

where P_t is the test pressure and P_r a reference pressure of 2×10^{-4} dynes cm^{-2} (2×10^{-5} N m^{-2}), just loud enough in the range 1–3 kHz to be audible to an average listener. Alternatively, the decibel may be expressed in terms of sound *intensity* (W m^{-2}) in which case it equals 10 times the logarithm of the ratio of the intensity of the test sound to the intensity of the reference sound.

The threshold of hearing varies with the stimulus frequency and is least over the range 1–3 kHz (Fig. 5.16). The resulting displacement of the tympanum is about the diameter of a hydrogen atom. The threshold rises to 60 dB (i.e. 10^3 times this value) at 50 Hz and 30 dB at 10 kHz. Threshold for pain is about 120 dB, although sounds above 100 dB may damage the organ of Corti in the cochlea. Loudness discrimination is about 1 dB under the most favourable experimental conditions (but more usually 2–3 dB), so that at 1–3 kHz there are no more than about 120 steps of loudness from the threshold of hearing to the threshold of pain.

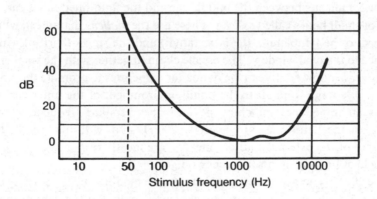

Fig. 5.16. Threshold of audibility curve.

Duration. To establish a pitch requires a minimum duration of 15–20 ms which surprisingly enough hardly varies with frequency.

Direction of sound source. At lower frequencies the direction of origin of a sound can be determined only by sensing the difference in time of arrival of the wavefront, whereas for higher frequencies the shielding effect of the head and pinnae of the ears results in differences in amplitude of the sound between the two ears. As a result, discrimination is best at higher frequencies, where both these mechanisms can be employed. The direction of origin of a pure tone of frequency less than 1 kHz is difficult to determine; real sounds generally contain harmonics at frequencies greater than this, allowing their position to be sensed.

Functional anatomy of the ear

The ear is divided into outer, middle and inner ear compartments (Fig. 5.17).

Outer ear. The *external auditory meatus*, about 27 mm long, conducts air pressure variations from the *auricle* (pinna or ear flap) to the *tympanum* (eardrum). The air column has a resonant frequency of about 3–4 kHz and conducts energy to the eardrum most effectively in this range.

Middle ear. The outer ear and the middle ear are separated by the eardrum or tympanum (Fig. 5.17). The cavity of the middle ear is air-filled and is connected to the pharynx by a narrow passage called the *Eustachian tube* which allows the pressure of the air in the middle ear to equilibrate with the outside pressure. Vibrations of the tympanum are conveyed through the middle ear to the *oval window*, which sits between the middle ear and the fluid-filled inner ear, via a series of small bones called *ossicles*. These are the *malleus* (hammer), which is attached to the tympanum, the *incus* (anvil) and the *stapes* (stirrup) which is attached to the oval window. The ossicle chain, together with the smaller area of the oval window relative to the tympanum, operates to promote the efficient transfer of energy from air to the liquid environment of the inner ear, i.e. it acts as an impedance matching device. The resonant frequencies of the middle-ear mechanism are around 800–1200 Hz and, with the broadly tuned resonance of the external auditory meatus, are mostly responsible for the low threshold of hearing in this frequency range.

The effectiveness of the ossicle chain can be modified by the activity of the middle-ear muscles. The *tensor tympani muscle* pulls the tympanum inwards and this results in the stapes being pushed into the oval window. The *stapedius muscle* pulls the stapes out of the oval window and pushes the tympanum

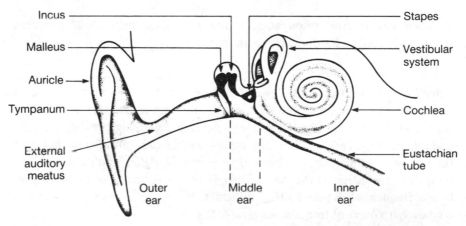

Fig. 5.17. Diagram of the outer, middle and inner ear compartments.

outwards. When they contract together, they stiffen the middle-ear mechanism to reduce the transfer of energy into the inner ear. They do so in response to high-intensity sounds, with a latency of about 10 ms, but soon relax again during constant stimulation.

Inner ear. This has a bony coiled tube of two and a half turns, the *cochlea*, divided by membranes along its length into three parallel canals or scalae, called the *scala vestibuli*, the *scala media* or *cochlear duct* and the *scala tympani* (Fig. 5.18). The scala vestibuli and scala tympani are filled with *perilymph* and join at the *helicotrema*; the cochlear duct is filled with *endolymph*. (Perilymph is very close to extracellular fluid in composition while endolymph, which is continuously secreted by a plexus of blood vessels called the *stria vascularis*, is very close to intracellular fluid in composition.) The cochlear duct is separated from the upper scala vestibuli by *Reissner's membrane* and from the lower scala tympani by the *basilar membrane*. When the stapes is pushed into the oval window, a wave is conducted along the scala vestibuli and through the flexible Reissner's membrane to the cochlear duct. The basilar membrane is depressed and the round window bulges out into the middle ear where the sound energy is dissipated into the air.

The sound receptors of the ear are found in the *organ of Corti* which lies on the basilar membrane within the cochlear duct (Fig. 5.19). It is composed of an epithelium of *hair cells* (Corti cells) and supporting cells. The hair cells are arranged into inner and outer groups. Each hair cell is anchored on the basilar membrane and has a bundle of hairs (*stereocilia*) projecting from its tip and embedded in the shelf-like *tectorial membrane*. The hair cells make contact at their basal surface with afferent cochlear fibres which pass into the bony *modiolus* where their cell bodies form the *spiral ganglion*. About 90% of the spiral ganglion cells in each ear innervate inner hair cells, about 10 axons

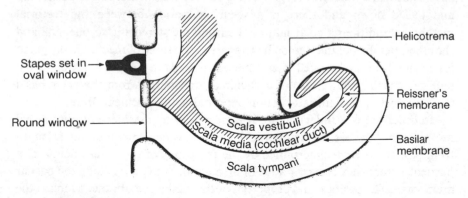

Fig. 5.18. Diagram of the inner ear. (The coiling of the cochlea is reduced for simplicity.)

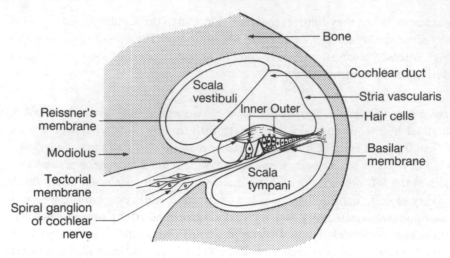

Fig. 5.19. A cross-section through the three canals of the cochlea.

converging on each hair cell; the remainder diverge to innervate many outer hair cells.

In addition to afferent cochlear fibres which run with the vestibular fibres in the auditory nerve (cranial nerve VIII), there is a small tract of efferent fibres ending on or close to the Corti hair cells, the olivocochlear bundle, which has an inhibitory action, reducing the afferent discharge at a given stimulus intensity. Furthermore, outer hair cells may be induced to vibrate and so actively dampen local regions of the basilar membrane.

Stimulation of Corti hair cells

The basilar membrane is a fibrous structure running from the inner core or modiolus to the outer side of the bony tube (Fig. 5.19). When the basilar membrane moves up and down, it tends to move rather rigidly, bending about an axis near the modiolus (Fig. 5.20). The tectorial membrane is pushed up and pulled down and there is a shearing motion between the tectorial membrane and the reticular lamina, bending the stereocilia to one side and the other. Bending of the stereocilia triggers the opening of mechanically gated ion channels in the hair-cell membrane. This leads to a depolarization of the hair cell causing the release of a chemical transmitter from the cell's basal surface which in turn initiates action potentials in the cochlear fibres.

In order to get a fuller picture it is necessary to consider not only radial movements of the basilar membrane, but also how movements along its length from base to apex are determined by the frequency of the sound. Sound of a particular frequency causes a wave of displacement to travel along the basilar membrane. The position and properties of the basilar membrane, which at the window end is narrower and stiffer than at the apex or helicotrema, are such

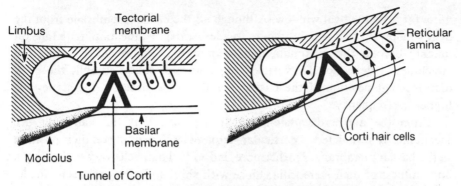

Fig. 5.20. Movement between the tectorial membrane and the reticular lamina causing the hairs of the Corti cells to bend. (From Davis, H. (1957) Initiation of nerve impulses in cochlea and other mechanoreceptors. In *Physiological Triggers and Discontinuous Rate Processes*, ed. Bullock, T.H., pp. 10–71. American Physiological Society, Washington.)

that the region nearest the window resonates best with high-frequency sound while more distant regions resonate best with low-frequency sound. Figure 5.21a shows the displacement caused by a 200 Hz wave instantaneously (solid line) and the envelope of all displacements over a period of many cycles (dashed line). It can be seen that there is a region of maximum displacement

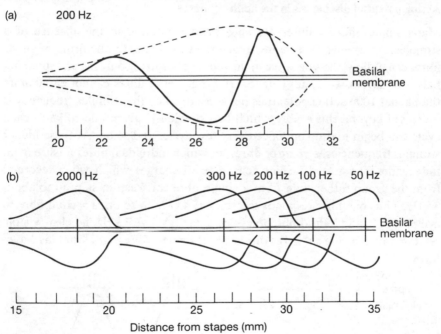

Fig. 5.21. (a) Deflections at one instant (solid line) and an envelope of all deflections over a period (dashed line) caused in a basilar membrane by a 200 Hz sound wave. (b) Envelopes of basilar membrane deflection for various frequencies of sound stimuli. (Adapted from von Bekesy, G. (1974) *J. Acoust. Soc. Am.* **19**, 452–460.)

quite far from the oval window. Although all the basilar membrane from the oval window up to this point moves to a lesser degree, the amplitude falls off quickly beyond the maximum due to damping. Figure 5.21b shows the envelopes of deflection for various frequencies of sound stimuli; the peak displacement position of the basilar membrane is closer to the stapes the higher the frequency.

From the above discussion, it can be seen that the hair cells are mechanically stimulated at particular frequencies depending on their position in the basilar membrane. Furthermore, individual hair cells vary in the length and stiffness of their stereocilia; those with short stiff cilia are mechanically more sensitive to high-frequency stimuli than those with longer and more flexible stereocilia. Finally, at least in some lower vertebrates hair cells are electrically tuned due to spontaneous oscillations in their membrane potential which show a characteristic frequency for each cell. In summary, each hair cell is tuned over a very narrow range by their position along the basilar membrane, by their individual mechanical properties, by active damping of local regions of basilar membrane by outer hair cells, and possibly by their electrical resonant properties.

Action potential discharges in the auditory nerve

Many single afferent fibres are spontaneously active in the absence of a stimulus. At sine-wave stimulus frequencies up to 1 kHz, the firing of single fibres can 'follow' the sine-wave in the sense of occurring only in corresponding halves of successive cycles (Fig. 5.22). Since nerve fibres cannot carry more than about 1000 action potentials per second, when the stimulus frequency is increased beyond this a fibre which had previously given one spike in each cycle will begin to drop out once every few cycles and so on. At these higher stimulus frequencies, a *group* of fibres, in which individual fibres misfire in an independently assorted fashion, can give discharges which when recorded from the group rather than from a single fibre can keep in step up to about 3 kHz. Thus one fibre responds to cycles 1, 4, 7, and so on, a second fibre to cycles 2, 5, 8, and a third fibre to cycles 3, 6, 9. Above 3 kHz, no clear visible relation can be made out in the discharge of a single fibre to the stimulus waveform.

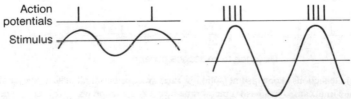

Fig. 5.22. Action potential discharges in a single afferent fibre produced by two different intensities of a fixed frequency.

If the stimulus thresholds are determined for a single fibre at various frequencies and intensities (Fig. 5.23), it is found that they rise rapidly on both sides of one small range of frequencies (the 'best' frequency) but more steeply on the higher frequency side, so that usually some high frequencies cannot provoke discharge, however intense they are. Note that as the stimulus intensity is raised the fibre responds over a wider frequency range.

We are now in a position to describe how the auditory information is encoded in auditory nerve fibre discharges with reference to the four properties of a sound stimulus discussed earlier.

1 *Pitch* discrimination is mainly determined by the area of basilar membrane occupied by firing receptors for a given stimulus frequency (the 'place theory' of hearing). High notes stimulate receptors closer to the base of the basilar membrane and low notes stimulate receptors closer to the apex (Fig. 5.21b).

2 *Intensity* sensation is derived from the total number of impulses per second in the auditory nerve fibres. At a given stimulus frequency only certain fibres will respond to a low-intensity stimulus. As the stimulus intensity is raised the discharge frequency in these firing fibres is increased (Fig. 5.22) and previously silent fibres with higher thresholds at that stimulus frequency are recruited.

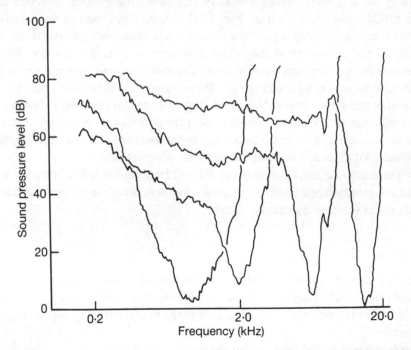

Fig. 5.23. Tuning curves of four units from the cat auditory nerve. (Adapted from Kiang, N.Y.S. (1984) Peripheral neural processing of auditory information. In *Handbook of Physiology*, Section 1, *The Nervous System*, eds. Brookhart, J.M. & Mountcastle, V.B., Vol. III, pp. 639–674. American Physiology Society, Washington.)

3 *Duration of stimulus* is signalled by the total duration of the afferent spike discharge caused by the stimulus.

4 *Direction of the sound source* is signalled by preservation in the auditory pathway of (part of) the initial time difference in receptor activation on the two sides of the head, as well as continuing intensity differences.

This information is now delivered to the cochlear nuclei in the medulla, where all the auditory fibres in cranial nerve VIII terminate. The auditory system as a whole is characterized by orderly segregation of nerve fibres and their cell bodies concerned with particular frequencies (tonotopic organization). A series of relay nuclei between the cochlear nuclei and the auditory cortex processes this information further, allowing, for example, analysis of sound location (superior olivary nucleus) and correlations with visual location of objects (inferior colliculus).

The central auditory pathways

The auditory pathway from the spiral ganglion to the primary auditory cortex has several major relay stations including the cochlear nucleus, the superior olivary nucleus (many axons bypass this nucleus), the inferior colliculus and the medial geniculate nucleus (Fig. 5.24). Axons from the cochlear nucleus ascend to the medial geniculate nuclei in either the ipsilateral or the contralateral lateral lemniscus, often after synapsing in the superior olivary nucleus, the first binaural relay station. The main cortical area is deep in the lateral or Sylvian fissure, emerging at the surface near the middle of the dorsal border of the superior temporal gyrus. Both ears are represented in both right and left cortices. Collateral branches from the pathway travel to spinal levels, the cerebellum and reticular formation of the brainstem and to the superior colliculi. All these are for coordination with other systems.

There are also inhibitory endings derived from higher levels of the CNS at each afferent synaptic level so that some modification of the afferent discharge must occur in these integrating centres.

Deafness

Hearing impairment in the sense of a raised threshold to sound stimuli may be due to impaired sound transmission in the outer or middle ear (conduction deafness) or to damage to the receptors or to the neural pathways (sensorineural deafness).

Conduction deafness may be caused by:

1 narrowing or blockage of the external meatus as by pus, wax, coal-dust etc.;

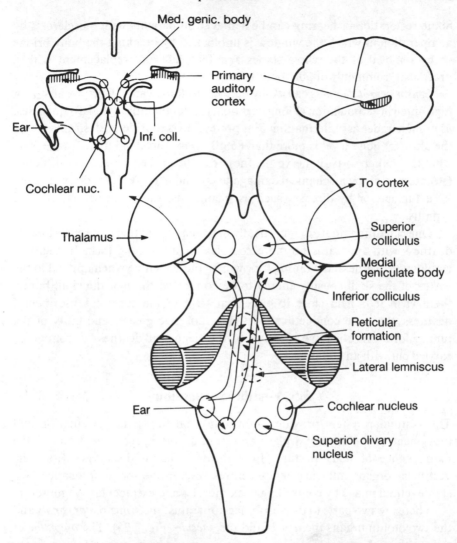

Fig. 5.24. A dorsal view of the brainstem (cerebral cortex and cerebellum removed) showing the central auditory pathways. *Top left inset:* simplified representation of the pathways in a roughly sagittal section of the brain. *Top right inset:* a lateral view of the left cerebral hemisphere showing the location of the primary auditory cortex.

2 thickening of the tympanum following scarring or repeated middle ear infections;

3 exudate in the middle ear (otitis media);

4 dislocation or fixation (ankylosis) of the ossicle chain; and

5 otosclerosis, which is a narrowing of the gap between the footplate of the stapes and the surrounding bone, progressing to fixation of the footplate when the gap is bridged.

Some restoration of hearing can be achieved in patients with otosclerosis by an operation in which the window is unblocked by breaking the bony bridge or by removal of the whole stapes and footplate and replacement with a prosthesis, commonly of polythene tubing.

Sensorineural deafness caused by damage to the organ of Corti results from high-intensity stimulation of long duration, predominantly of high frequencies, as in such trades as boilermaking. Ear protection is considered desirable where the ambient noise level is more than 85 dB above threshold in the range 300–2400 Hz. Other sorts of nerve deafness can be caused by some antibiotics (streptomycin), by mechanical damage to cranial nerve VIII (auditory nerve), by a tumour, or by diseases such as meningitis, rheumatism, malaria and syphilis.

One can differentiate diagnostically between conduction and sensorineural deafness with apparatus no more complex than a tuning fork. In bilateral conduction deafness, the tuning fork will be more audible when applied to the vertex of the skull causing energy to be conducted through the skull bones, than when it is held close to either ear. But in the case of sensorineural deafness, the bone conduction route does not give greater audibility of the tuning fork. An objective measure of the degree of deafness is commonly carried out with an audiometer.

5.5 Vestibular function

The vestibular system provides information on the spatial orientation and movement of the head and plays an essential role in movement and in the maintenance of body posture. In addition, afferent discharges from the vestibular organs influence reflex centres responsible for maintenance of a stable retinal image by controlling neck muscles and extraocular eye muscles.

There are two parts to the vestibular apparatus: the *semicircular canals* and the two otolith organs the *utricle* and the *saccule* (Fig. 5.25). The *ampullae* of the semicircular canals contain sense organs which respond to rotatory acceleration of the head. The otolith organs are sensitive to the direction of the force of gravity and to linear accelerations of the head. All these tissues constitute the *membraneous labyrinth*, a system of tubes filled with endolymph and surrounded by perilymph. This in turn is encapsulated within the *bony labyrinth*, a series of bony tubes containing both the auditory and vestibular sense organs and which lies in cavities in the temporal bone.

Stimulation of vestibular hair cells

Vestibular sensation results from stimulation of hair cells in the ampullae and otolith organs. Bundles of 40–70 *stereocilia* project from the free surface of

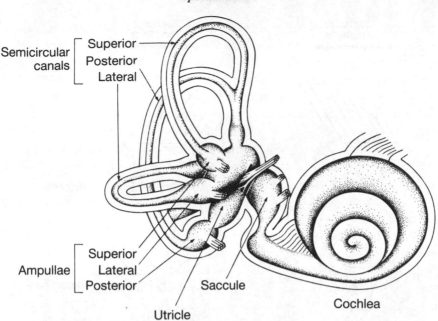

Fig. 5.25. Vestibular apparatus of the right side in three dimensions. (Adapted from Melloni, B.J. (1957) *The Internal Ear, An Atlas of Some Pathological Conditions of the Eye, Ear and Throat*, pp. 26–31. Abbott Laboratories, Chicago.)

each cell. These are packed hexagonally and increase progressively in length towards the side of the bundle from which the *kinocilium*, a specially long cilium with a different structure, projects (Fig. 5.26). Hair cells are polarized, so that movement of their cilia towards one side of the kinocilium causes depolarization and to the other side hyperpolarization. As they are depolarized, the hair cells release progressively more transmitter resulting in an increased rate of firing of the afferent axons with which they are in contact. Hair cells in the resting state are typically in the mid-range of polarization so that there is a steady release of transmitter and tonic firing of sensory nerve impulses. Any slight movement of their hairs results in an increase or a decrease in the rate of firing.

The semicircular canals

The three semicircular canals lie in nearly orthogonal planes, i.e. each is at right angles to the other two. With the head erect the lateral ('horizontal') canal is raised anteriorly at about 30° to the horizontal while the superior (anterior) and posterior canals are at 45–55° to both sagittal and frontal planes, respectively. The two lateral canals on the left and right side of the body are in a single plane, and the posterior canal of one side is in a plane nearly parallel

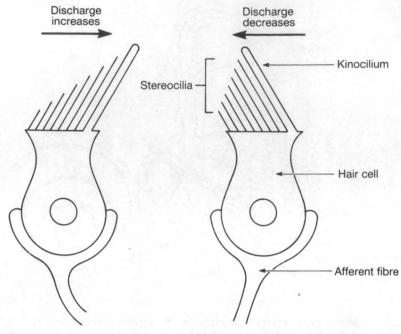

Fig. 5.26. Diagram of the vestibular hair cell. Bending the cilia towards the side of the kinocilium increases the discharge frequency; bending the cilia away from that side decreases it.

to the superior canal of the other side, so that these respective pairs are stimulated similarly and at least one pair is affected by any given angular acceleration. The canals of either side alone can generate afferent impulses signalling movement in any direction.

The labyrinths are filled with endolymph, a watery fluid with ion concentrations similar to that of intracellular fluid, with $K^+ \approx 150$ mM and $Na^+ \approx 20$ mM. The vestibular epithelium contains *dark cells*, responsible for the ion transport which maintains these concentrations. Each canal is connected at both ends to the utricle, so that the endolymph is free to circulate. At one end of each canal is an *ampulla* (Fig. 5.25) containing receptors which transduce circulation of endolymph into nervous impulses. The receptor apparatus, the *crista ampullaris* (Fig. 5.27) consists of the *cupula* (or cupola), a gelatinous wedge-shaped structure running fully across the cross-section of the ampulla to form a diaphragm which effectively blocks any bulk flow of endolymph. As it has nearly the same specific gravity as the endolymph it neither floats nor sinks and thus is almost unaffected by the direction of gravity. The cupula is mounted on a ridge carrying many hair cells which have their cilia embedded in the base of the cupula. The hair cells in each ampulla all have the same orientation; those in the ampullae of the horizontal canals

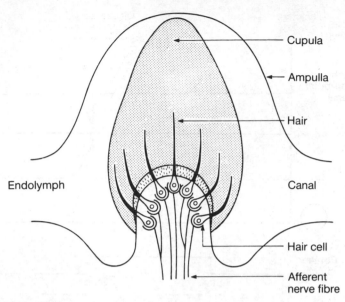

Fig. 5.27. Diagram of the crista ampullaris.

have their kinocilium facing towards the utricle, while in the vertical canals they face away from the utricle. Thus, movement of fluid towards the utricle depolarizes (stimulates) the hair cells of the lateral canals, but hyperpolarizes (inhibits) those of the vertical canals.

During any angular acceleration of the head, the endolymph tends to circulate through the semicircular canals as a consequence of its inertia, and the consequent fluid pressure causes deflection of the cupula and bends the cilia projecting from the hair cells. Movement of the fluid is soon damped and during steady motion the cupula rapidly returns to its resting position and produces no further sensation.

Figure 5.28 illustrates the changes in frequency of impulses recorded from axons in the vestibular nerve when the head is rotated to the right at a constant speed. The sensation of rotation persists beyond the period of firing in vestibular axons, demonstrating that this information is integrated and stored, probably in the brainstorm. There is a spontaneous resting discharge which changes with any movement of the head. When the head accelerates to reach constant velocity, there is a sudden increase in frequency; the discharge then returns to its frequency at rest before the sensation of movement ceases in 25–30 s. When the rotation is stopped quickly, the opposite sequence takes place. The discharge frequency of the receptors changes in the opposite sense to that at the beginning of rotation and there is a feeling of rotating in the opposite direction until the cupula again returns to its resting position.

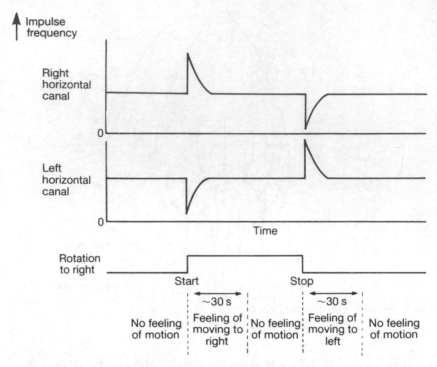

Fig. 5.28. Impulse frequency in vestibular afferent fibres when the head is rotated to the right at a constant angular velocity and then stopped.

The discharge patterns from the ampullae of corresponding semicircular canals on the two sides of the body are mirror images, so that the brain receives both positive and negative sense information on any movement.

The utricle and saccule

Both these organs have *maculae*, thickened regions containing hair cells, and vestibular nerve terminals. The hair-cell cilia are embedded in a gelatinous mass called an *otolith* which contains a mass of calcium carbonate crystals called *otoconia* (Fig. 5.29). The specific gravity of the otoconia is about 2·9, much higher than that of the endolymph. If the head is still, the otolith tends to fall to the lowest possible point, and gives rise to sensation of the direction of the force of gravity. During movement, its equilibrium position is determined by the direction of the vector sum of all accelerations of the head— linear, centrifugal and gravitational. It should be noted that conscious awareness of an accelerating force depends also on cutaneous receptors and muscle and joint proprioceptors.

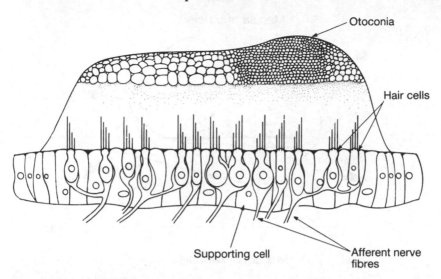

Fig. 5.29. Diagram of the otolith organ found in the utricle and saccule. (After Lindeman, H.H. (1969) *Ergeb. Anat. Entwicklungsgesch.* **42**, 1–113.)

Hair cells are arranged in each macula in a complex pattern of orientation (Fig. 5.30). There is a line of reversal in the orientation of the hair cells in the maculae of both utricle and saccule. In the macula of the saccule, the kinocilia are oriented away from each other on either side of the line of reversal rather than toward each other as in the utricle. As a consequence of these patterns of orientation of the hair cells, each different movement of the otolith causes a different spatially organized pattern of discharge in the afferent fibres. These highly organized afferent patterns generated in different parts of the vestibular system probably account for the direction of the compensatory eye movements which occur.

There is a small tract of efferent fibres to the vestibular apparatus whose function is probably inhibitory, as is the corresponding supply to the cochlea. These efferent fibres may play a role in *habituation* to repeated patterns of acceleration or during constant rotation.

Functions of the vestibular organs

In experimental animals, spontaneous rates of firing in different vestibular fibres vary from a few per second up to about 200 per second, averaging 90. The frequency of this discharge is altered by bending the hairs of the hair cells. To a first approximation, the ampullae of the semicircular canals are sensitive to angular acceleration and the maculae of the utricle and saccule to the

Macula of utricle

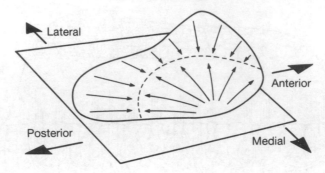

Macula of saccule

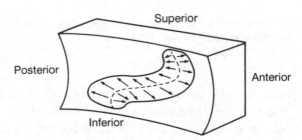

Fig. 5.30. Directions of orientation of hair cells in the maculae of the left utricle and saccule. The arrows indicate the direction of bending of the hairs for excitation. (From Spoendlin, H.H. (1966) Ultrastructure of the vestibular sense organ. In *The Vestibular System and Its Diseases*, ed. Wolfson, R.J., pp. 39–68. University of Pennsylvania Press, Philadelphia.)

direction of gravity and to linear acceleration. However, they can all respond to other modes of stimulation to some extent.

The vestibular apparatus evokes two interrelated categories of reflex response, *dynamic* and *static*. Oculomotor reflexes, important in stabilization of the visual image on the retina, depend primarily on the dynamic function, which is mediated principally by the semicircular canals. To maintain a stable retinal image during head movements, two quite different reflexes are involved in causing compensatory eye movements. If an object is being looked at, movements of the object or movements of the head are compensated by the *fixation reflex* which by pursuit movements of the eyes keeps the image of the object on the part of the retina giving the clearest vision, the fovea centralis. Experimentally, this reflex can be evoked by non-geniculate afferents to the superior colliculus, but it normally involves participation of the visual parts of the cerebral cortex to distinguish objects of interest. There is a limit to the amount by which an eye can be turned in the head to follow an object and to extend this it is necessary for the head to move, or for the eye to return quickly to roughly the straight-ahead position and to find another object of interest to follow as, for example, when looking out from a moving vehicle. This type of movement is called *optokinetic nystagmus*. The word 'nystagmus' refers to any oscillatory movements of the eyes, whether normal or pathological, and especially to this slow turn–quick return sequence just described. The direction of the nystagmus is conventionally named according to the quick phase.

The other mechanism by which the eyes are moved so as to keep an image as far as possible steady on the retina is driven by the vestibular apparatus and takes place to some extent whenever the head is moved, even with the eyes closed or in total darkness, and is called *vestibular nystagmus*. Here the slow drift is in the opposite direction to the rotation of the head and the quick phase is in the same direction. Even in the dark, this reflex may provide full compensation for head movements (between 60 and 100% compensation, depending on the methods of testing). In the light, this reflex collaborates with the fixation and optokinetic reflexes. The vestibular nuclei are connected with the ocularmotor neurones via the medial longitudinal bundle.

The conscious sense of movement depends also on the dynamic sensory components of the vestibular apparatus, and involves a part of the cerebral cortex close to the main auditory sensory area in the temporal lobe, tucked into the lateral or Sylvian fissure.

The static reflexes, mediated primarily by the utricle and saccule, are important for the maintenance of the upright position of the head and for body posture. The maintenance of balance depends on the distribution of tone in the body muscles, which is regulated via the vestibular nuclei in the medulla and their connections with the vestibulospinal tract and the cerebellum.

The central vestibular pathways

Vestibular axons on either side have their cell bodies in the vestibular ganglion (Scarpa's ganglion) which lies near the internal auditory meatus (Fig. 5.31). They project to the vestibular nuclei, which lie in the pons and the dorso-rostral part of the medulla. They are components of a number of reflex arcs responsive to movement of the head and to apparent changes in the direction of the force of gravity. Such reflexes are responsible for stabilization of the eyes, holding the head erect and maintenance of body stability. They are often not consciously perceived, and awareness of head movement or the direction of gravitational force is often secondary to the automatic reflex responses to acceleratory forces on the head.

Each vestibular nucleus has a particular pattern of projections to the

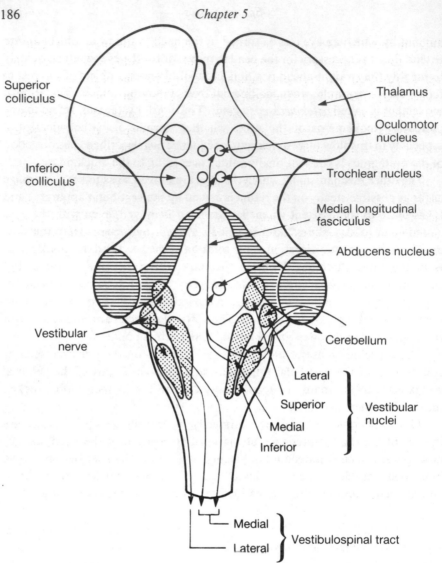

Fig. 5.31. A dorsal view of the brainstem (cerebral cortex and cerebellum removed) showing the central vestibular pathways.

oculomotor system, to the spinal cord and to the cerebellum. The *lateral vestibular nucleus* (Deiter's nucleus) receives axons from the macula of the utricle and from the cerebellum and spinal cord. Its neurones send their axons into the lateral vestibulospinal tract which terminates ipsilaterally in the ventral horn of the spinal cord, along its length. These axons powerfully facilitate α and γ motor neurones innervating antigravity muscles. Neurones in the lateral vestibular nucleus respond selectively to tilting of the head. Their resting discharge is increased by tilting in one direction and decreased by tilting in the other. The input from the cerebellum is inhibitory, and removal of this tonic inhibition gives rise to decerebrate rigidity. The *medial* and *superior vestibular nuclei* receive inputs from the ampullae of the semicircular canals. Outputs running in the medial vestibular tract terminate bilaterally in the ventral horn of the cervical region of the cord, making monosynaptic connections with neck-muscle motor neurones. These provide for the reflex control of neck movements to maintain the position of the head and provide a stable base for eye movements. Both nuclei also participate in vestibulo–oculomotor reflexes having outputs which run in the medial longitudinal fasciculus to the ocularmotor nuclei and give rise to rotatory nystagmus movements. The *inferior vestibular nucleus* receives excitatory inputs from the semicircular canals and from both the utricle and the saccule, and inhibitory inputs from the cerebellum. It has vestibulospinal and vestibuloreticular outputs and also has powerful effects on the cerebellum. Although its specific functions remains controversial, this nucleus integrates input from the whole vestibular apparatus and affects higher brain centres.

Disorders of vestibular function

Semicircular canal function can be tested by rotating the subject in a special chair whilst the head is held in such a posture as to stimulate one functionally associated pair of canals selectively. One can examine the consequences of stopping rotation on either eye movements or on body muscle tone. The latter is manifested by a tendency to deviate from a straight line when walking with the eyes closed. In these procedures, both right and left canals are inevitably stimulated together.

Unilateral stimulation can be produced by running water at above or below body temperature into the external meatus of the ear—the *caloric test*. The temperature difference gives rise to convection currents in the endolymph which excite the hair cells. This produces nystagmus which can be compared with a known normal response.

Diseases affecting the vestibule or its afferent fibres can cause abnormal discharges producing a sensation of *vertigo*, or giddiness, in which the external world may seem to move, the body may be felt to be moving, or the posture of the limbs, especially the legs, may be felt to be unsteady. Along with this there may be nystagmus, double vision, actual falling, and autonomic signs such as pallor, sweating, pulse rate and blood pressure changes, nausea and vomiting. *Menière's disease* is characterized by attacks of vertigo and progressive impairment of hearing. The cause is unknown but on post-mortem examination the endolymph-filled chambers of the inner ear are found to be enlarged.

When an often-repeated cyclical pattern of strong stimulation of the labyrinth occurs, as in sea travel or land travel in a car along a winding road, similar physiological consequences occur and are known generally as *motion sickness*. This is most marked when there is a lack of coordination between visual and vestibular information, and is often ameliorated or halted if it is possible to visualize the true horizon.

6 · Motor System

6.1 Spinal cord pathways
Motor neurones
Interneurones
Organization of function within the
   ventral horn
Organization of function within a motor
   nucleus
Organization between motor nuclei
Spinal reflexes
   Stretch (myotatic) reflexes
   Flexor (nociceptive) reflexes
   Reciprocal innervation
   Recurrent inhibition
   Regulation of muscle contraction
Autonomic reflexes

6.2 Supraspinal pathways to motor neurones
Corticospinal tract (pyramidal tract)
Rubrospinal tract
Lateral vestibulospinal tract
Reticulospinal tracts
Tectospinal tract
Medial vestibulospinal tract

6.3 Control of posture
Reflexes evoked by gravitational force
   (static reflexes)
   Local static reflexes
   Segmental static reflexes
   General static reflexes
   Righting reflexes
Reflexes evoked by linear acceleration
Reflexes evoked by angular acceleration
Effects of spinal and brainstem
   transection

6.4 Supraspinal organization of movement
Role of the cerebral cortex
Basal ganglia
   Functions of the basal ganglia
Role of the cerebellum
   Cerebellar cortex
   Deep intracerebellar nuclei
   Cerebellar afferents
      Functional zones of the cerebellum
   Functions of the cerebellum
Role of the ventrolateral thalamic nuclei
Motor cortex

6.5 Lesions of the motor pathway

A movement can occur as the result of stimulation of receptors (reflex) or it can be willed (voluntary). The chain of events involves the whole nervous system. It is convenient, however, to deal separately with sensory and motor systems and this chapter will deal with those parts of the nervous system directly involved in producing movement. It must be remembered though that this is an arbitrary division. For instance, many of the fibres of the corticospinal tract are concerned with regulating the traffic to and from sensory nuclei, rather than in directly exciting motor neurones.

In man, most overt activity is the result of contraction of skeletal muscle fibres. These fibres normally contract as the result of transmission of action potentials down the axons of motor neurones to motor endplates and the generation of endplate and action potentials in the muscle fibres they serve. The timing and force of contraction of a particular motor unit (the muscle fibres innervated by a particular motor neurone) depends on the pattern of action potentials discharged by that motor neurone. The problem of understanding movement thus reduces to the problem of the control of motor neurones.

6.1 Spinal cord pathways

Motor neurones

There are two classes of motor neurones—large (up to 70 μm) α cells with axons of 12–20 μm diameter that innervate skeletal extrafusal muscle fibres, and smaller diameter γ cells with axons of 1–8 μm diameter that innervate the intrafusal fibres of the muscle spindles in skeletal muscles. The cell bodies of both classes of neurones are located in the ventral horns of the spinal cord and in analogous positions in the brainstem.

Alpha motor neurones are multipolar cells (many dendrites, one axon), their dendrites spreading from the ventral horn to the dorsal horn and even across the segment and also up and down a particular cord segment, thus giving many opportunities for synaptic contact. The cell bodies of motor neurones of trunk and proximal limb muscles lie in groups (*motor nuclei*) in the medial part of the ventral horn (Fig. 6.1a), the ventromedial group forming a long column running almost the entire length of the spinal cord (C_1–L_4). In the brainstem, the column becomes discontinuous but the cell bodies of the motor neurones of the cranial nerves XII, VI, IV and III can be found in an analogous position. In the cervical and lumbar enlargements of the cord, the cell bodies of motor neurones of all distal limb muscles lie together in nuclei in the lateral part of the ventral horn (Fig. 6.1a). In general, the more lateral motor nuclei belong to more distal muscles. Electronmicroscopy has shown that 50% of the vast

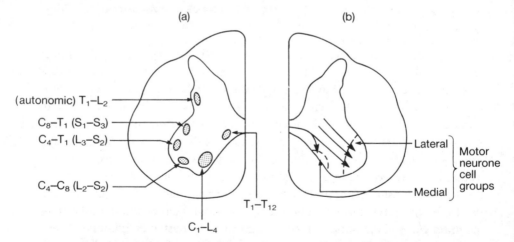

Fig. 6.1. Transverse section of the human spinal cord illustrating: (a) the cellular columns of the motor neurone groups. The longitudinal extent of each column of cells is indicated in figures and letters by the cord segments that it spans (adapted from Brodal, A. (1981) *Neurological Anatomy*, 3rd edn. Oxford University Press, Oxford); (b) the lateral and medial groups of interneurones (arrows) projecting mainly to the more lateral and more medial motor neurone cell groups (enclosed in broken lines) respectively (adapted from Sterling, P. & Kuypers, H.G.J.M. (1968) *Brain Res.* **7**, 419–443).

dendritic tree of any motor neurone is covered with synaptic terminals originating from sources within and above the spinal cord and brainstem. This extreme convergence occurs because motor neurones are the final common pathway from the central nervous system to muscle.

Gamma motor neurones are scattered amongst the α motor neurones of each motor nucleus in the medial and lateral columns of the ventral horn. Their axons pass out of the ventral roots with the larger diameter axons of the α motor neurones.

Interneurones

Lying dorsally at the base of the ventral horn and projecting to the motor nuclei are *excitatory* and *inhibitory* interneurones involved in the regulation of motor-neurone discharge. Their topography is similar to that of motor neurones, the more medial interneurones projecting mainly upon the medial motor nuclei and the more lateral interneurones upon the lateral motor nuclei (Fig. 6.1b).

Organization of function within the ventral horn

The trunk muscles, together with the proximal limb muscles, are involved in functions such as balance, posture and walking which require body or whole-limb movement. The muscles moving the fingers and wrists do not have postural functions but are involved in manipulatory movements. The motor neurones and interneurones in the ventral horn are organized to carry out these diverse functions, the medial group being involved in posture and the lateral group directing manipulatory activity. The descending inputs to the ventral horn similarly divide into those primarily influencing the medial neurones and hence posture, and those influencing the lateral neurones and hence manipulatory activity.

Organization of function within a motor nucleus

The motor neurones of a group innervating a given muscle share much the same afferent input and therefore their excitability tends to change in much the same way at the same time. In general the smallest diameter motor neurones have the lowest threshold and the smallest motor units. Their discharge therefore generates a small increment of tension. In graded contractions of a given muscle the recruitment order remains stable, each motor neurone firing in turn as its threshold is exceeded and increasing its firing rate as the excitation increases. Initially, as each small motor neurone discharges there is a small increment of tension, larger increments being added later in the contraction when larger diameter cells are excited. With this mechanism it is possible to vary tension over a wide range with a relatively small number of motor units.

Organization between motor nuclei

The information that reaches the spinal cord from the brain is coded in terms of movement not in terms of muscles. It is distributed to interneurones at the base of the ventral horn which in turn excite motor neurones in the motor nuclei required for the movement. Complex connections exist, walking, for instance, being generated following the delivery of a simple excitatory signal to the initiating interneurones. Swallowing is another example of a complex pattern of movements which may be begun with a simple afferent input provided by contact of food with the back of the pharynx. Normally, programmes are modified by incoming information from the periphery, superimposed on the basic pattern. For example, when walking the pattern is adapted to cope with irregularities on the surface.

Spinal reflexes

In muscle there are mechanoreceptors providing information about muscle length and tension, involved in the subconscious control of posture and movement and also contributing to sensation. These are associated with specialized structures in muscle known as *muscle spindles* and also with receptors on tendons known as *Golgi tendon organs*. In addition, there are, in muscle, receptors sensitive to severe pressure and stretch. These receptors, which give rise to cramping pain, are probably located in blood vessels and in connective tissue of the muscle and are sensitive to products released from muscle working without an adequate blood supply. As in the skin, the different types of endings are supplied by nerve fibres of different diameter but they are usually referred to by the I, II, III, IV classification (p. 128).

Muscle spindles. These structures (Fig. 6.2) are made up of 2–12 small-diameter striated muscle fibres termed *intrafusal* muscle fibres, the main muscle fibres being known as *extrafusal* fibres. The whole bundle of intrafusal fibres is enveloped in a capsule, lies in parallel with the main muscle fibres and is attached to them at its ends. The middle third of the intrafusal fibre, known as the equatorial region, is non-striated and largely non-contractile. The outer thirds, known as the polar regions, are striated and contractile. One muscle contains many muscle spindles.

There are two types of intrafusal muscle fibres:

1 *nuclear bag fibres*, commonly two per spindle, which are longer and thicker and have a group of nuclei clustered around the centre of the fibre;

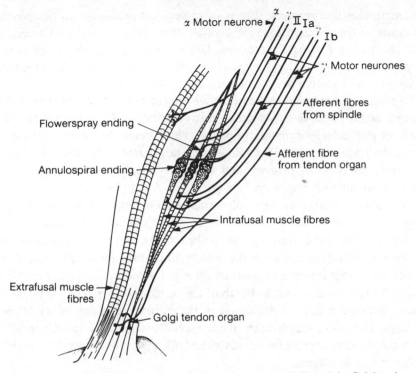

Fig. 6.2. The innervation of extrafusal muscle fibres, the muscle spindle and the Golgi tendon organ.

2 *nuclear chain fibres* which are shorter and thinner and have fewer nuclei arranged in a chain along the centre of the fibre; there are commonly four to five per spindle.

Group Ia fibres coil around the equatorial region of the nuclear bag and chain fibres to form *primary* receptor endings (*annulospiral endings*). Group II fibres (smaller and slower-conducting) end mainly on nuclear chain fibres to form *secondary* receptor endings (*flowerspray endings*).

The receptor endings of both group Ia and group II afferents are mechanoreceptors and they are stimulated when the equatorial region of the spindle is stretched, such as happens when the whole muscle is stretched, as this pulls on the ends of the spindle. This pattern of responses is known as '*in parallel*' behaviour as it is due to the fact that the spindles are in parallel with the extrafusal muscle fibres. The response from the spindle endings is reduced or silenced when stretch on the equatorial region of the spindle is removed, either by removing stretch from the whole muscle or by contracting the extrafusal muscle fibres, thereby slackening the intrafusal fibres.

Such a response to change in the length of the muscle is known as the *static* or length-sensitive response and is in some ways analogous to the position response of cutaneous mechanoreceptors. Both primary (Ia) and secondary (II) endings give a static response. During static firing, the frequency of impulses is proportional to the muscle length. The static response is probably associated with stretch of both bag and chain fibres.

Spindles also respond to the rate of change of length of the muscle. This is known as the *dynamic* or velocity-sensitive response. Only the primary (Ia) endings give a large dynamic response. The greater the velocity or rate of length increase, the higher the frequency. During release from muscle stretch group Ia fibres, both primary and secondary endings, commonly show a decrease in impulse frequency.

In addition to the two types of sensory ending, muscle spindles also have their own motor innervation. This innervation is supplied by *fusimotor* or γ motor neurones, the extrafusal muscle fibres being supplied by *extrafusal* or α motor neurones. Stimulation of the muscle fibres does not alter tension in the whole muscle but it produces contraction of the polar regions of the intrafusal fibres. This, in turn, causes a stretch of the equatorial region of the intrafusal fibres and thus results in excitation of the primary and secondary afferent endings. The motor supply to the muscle spindles allows the spindle length to be regulated thus altering its sensitivity and allowing it to function over a wide range of muscle lengths.

Golgi tendon organs. These receptors occur on tendons usually near the muscle–tendon junction (Fig. 6.2). Their afferents are Ib fibres which branch over several tendon fascicles, covered by a capsule which, at its ends, becomes continuous with the connective tissue of the muscle or tendon.

A Golgi tendon organ responds to the tension in its muscle fascicle (and hence in the tendon) with an increase in frequency of impulses both when the muscle is stretched passively and when it contracts actively. This is known as *'in series'* behaviour as the tendon organ is on the tendon, in series with the muscle. The increase in response during extrafusal muscle contraction distinguishes the Golgi tendon organ from the muscle spindle receptors whose response is reduced.

The Golgi tendon organ responds to both the velocity of tension development and to constant or maintained tension. It has thus both dynamic and static response characteristics. It is less sensitive to passive muscle stretch than is the muscle spindle as passive stretch lengthens the muscle rather than the tendon. However, during active muscle contraction the response of the tendon organ is greater than for passive stretch of the same degree because active contraction causes greater stretch of the tendon. Thus, Golgi tendon organs do not respond only at extreme tension as originally thought but respond also during normal tension development.

Stretch (myotatic) reflexes

Muscle spindles are the starting point for the simplest of all reflexes found in mammals—the *stretch reflex*. When a muscle is stretched the elongation leads to a discharge in the muscle spindle afferents that is conducted into the spinal cord via the dorsal roots. Within the cord the sensory fibres branch and some terminals make *monosynaptic* excitatory contacts with α motor neurones to the extrafusal muscle fibres of the *same* muscle (Fig. 6.3). Impulses generated in these motor neurones leave the cord in the ventral roots and conduct impulses back to the muscle causing it to contract in opposition to the change in length. These reflexes may be elicited by rapid or slow changes in muscle length and tend to maintain a constant muscle length with a change in load.

These reflexes persist after all cutaneous nerves have been cut and any tendon organs anaesthetized. They are specific to stretched muscle, hence the name *stretch* or *myotatic reflexes*. The degree of stretch need only be 1–2 mm and is normally the result of gravity. These reflexes have both phasic and tonic components, are of short latency, and, although the tonic component is well-nigh indefatigable, are very susceptible to inhibition (see below: reciprocal innervation). They are present to some extent in all muscles but are best seen in the antigravity muscles, which for most species means the extensors.

In recent years, analysis of stretch reflexes has shown an increasingly complicated picture. Briefly, it can be said that primary endings respond both to the velocity of stretch and to maintained length, secondary endings only to the latter. The length of the spindle can be modified independently of that of the extrafusal muscle fibres by impulses travelling along the γ efferent fibres. There are in fact two types of γ efferent fibres, static and dynamic. Stimulation

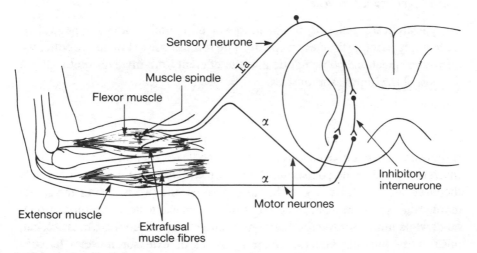

Fig. 6.3. Circuit diagram of a monosynaptic stretch reflex showing reciprocal inhibition.

of static fibres increases the static response of both groups Ia and group II afferents. Stimulation of the dynamic fibres increases both the dynamic and static responses of group Ia fibres. In addition, there are two types of ending on the intrafusal muscle fibres:

1 *plate endings*, on both nuclear bag and nuclear chain fibres at the ends or poles of the fibres, possibly associated primarily with dynamic γ efferent fibres;
2 *trail endings*, mainly on nuclear chain fibres just next to the equatorial region of the fibres and associated mainly with static γ efferent fibres.

Simple *phasic* stretch reflexes (tendon jerks) are evoked by a brief muscle stretch which excites the dynamically sensitive spindle afferents. When a neurologist flexes and extends, for example, the patient's elbow, that is, when he alternately stretches the biceps and triceps muscles, he normally encounters a certain resistance. This resistance or *'tone'* is in fact reflex and not due to any elasticity of the muscles. This reflex is more complicated than a tendon jerk. An electromyogram shows a series of waves. The first, of short latency, is generated by the phasic action already described. The origin of later waves which provide the tonic-maintained aspects of this effect is not well understood. On the one hand, it is claimed that there is a long-loop reflex with primary spindle activity routed to motor neurones by way of the motor cortex, with the additional 10–15 ms delay imposed by the cortical loop. On the other hand, it is claimed that the later components result from spinal reflexes originating perhaps from spindle secondary endings, the delay being imposed by their smaller diameter afferent fibres and largely polysynaptic connections.

Flexor (nociceptive) reflexes

A typical flexor reflex is the withdrawal of a limb following painful or potentially painful stimulation (p. 40). The receptors are skin nociceptors, the central connections polysynaptic and the efferent limb involves motor nuclei of most of the limb flexor muscles.

Reciprocal innervation

Built into the circuitry of stretch and flexor reflexes is an inhibition of motor nuclei of muscles antagonistic to the movement (Fig. 6.3). It is seen not only between the muscles at a joint but also between the two halves of the spinal cord. Thus an incoming nociceptive stimulus can excite flexor motor neurones and, while inhibiting the ipsilateral neurones of the antagonistic muscles, can also excite the contralateral motor neurones of extensor muscles (*crossed extensor reflex*).

Recurrent inhibition

In addition to the influence of the spinal reflex mechanisms, α motor neurones are influenced directly by the activity of interneurones which they themselves activate. In this feedback circuit, the interneurones, *Renshaw cells*, are inhibitory and are excited by transmitter released from collateral branches of the motor neurone (Fig. 6.4). *Renshaw inhibition* appears to play an important role in dampening the activity of motor neurones; in particular, it appears to limit the discharge frequency of tonically active α motor neurones.

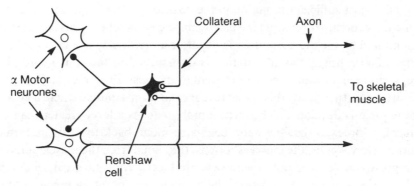

Fig. 6.4. Circuit diagram for recurrent inhibition of a motor neurone by a Renshaw cell.

Regulation of muscle contraction

The properties of muscles are such that a constant signal will not always produce the same increase in tension or change in length. This variability arises because the contractile responses of muscles depend on their initial length and because muscles fatigue. To compensate for this behaviour and the varying loads imposed on muscles, they and their nervous reflex loops are arranged as servo-mechanisms, that is, mechanisms that modify their output to adapt to varying loads. Such systems have distinct advantages as they can quickly and accurately follow command signals and automatically compensate for changes. Muscle spindles, Golgi tendon organs and the various neural pathways associated with these receptors are involved in such processes.

The myotatic (stretch) reflex pathway is involved in both the initiation of contractions and the control of muscle length. Contractions are initiated through this pathway when activity in γ motor neurones causes muscle spindles to increase their rate of discharge. Thus both α and γ motor neurones can be used to initiate muscle contraction. The pathway involving α fibres is faster, but the pathway involving γ fibres probably gives more sensitivity of control and is uniformly effective over the whole range of muscle lengths.

The myotatic reflex pathway is also a feedback system in the control of muscle length. Stretching a muscle beyond its desired length activates the muscle spindles so that they produce action potentials at an increased rate. Their signals activate α motor neurones supplying that muscle, so that the muscle contracts until the desired length is restored. If a muscle contracts so that it is shorter than the desired length then the spindle receptors are silenced, activation of motor neurones is reduced, and the muscle relaxes.

As an example, consider the effects on the biceps muscle of placing an extra load on the hand when the forearm is outstretched. Before the load is placed on the hand, the α and γ motor neurones to the biceps discharge at a rate that is just sufficient to maintain the position of the limb. As the load is applied, the forearm drops and the biceps muscle is stretched. This lengthening is monitored by the muscle spindles and, after a brief delay due to conduction in the afferent fibres, the information is transmitted to the spinal cord. The biceps motor neurones almost immediately increase their rate of discharge and, after a further delay, this signal reaches the biceps muscle where a greater muscle force is developed. The force rapidly reaches a level greater than the disturbing force, so that the hand begins to move back towards its former position. However, as the muscle shortens, the signal from the muscle spindles is progressively reduced until the muscle produces a force no greater than the loading force and movement stops. The muscle does not return precisely to its former length because a small steady-state error is required to produce enough output from the spindles to overcome the extra load.

In some circumstances, two opposing muscles can operate to fix the position of a limb. In this case, the pathways producing reciprocal inhibition become active. For example, stretching the biceps muscle spindles produces inhibition (via an interneurone) of triceps motor neurones and, similarly, stretching triceps causes inhibition of biceps. Thus, pairs of opposing muscles become organized into a single working unit. This unit may also be brought into play in normal limb movement when the contraction of antagonistic muscle will help to dampen movements and prevent overshoot.

In addition to the system regulating muscle length, there is a negative feedback loop regulating muscle tension. This system depends on signals from the Golgi tendon organs of a muscle and leads to inhibition of the motor neurones that control that muscle. This inhibitory influence is present even when the muscles are contracting normally.

The role of this loop can best be considered when it operates in isolation from the length system so that the force generated by a muscle is maintained, despite changes in length or the development of fatigue. For example, if the muscle force is altered by fatigue, then inhibition of the motor neurones is decreased and the muscle is excited more strongly. If the force becomes too great, for example as a result of a change in muscle length, the inhibitory effect will be increased and the force correspondingly reduced.

However, it should be noted that the above is an oversimplification. Both the muscle length and the muscle tension servo-mechanisms normally operate together and both employ the same α and γ motor fibres. They are both arranged so that they reciprocally innervate antagonist muscles. Thus, they allow muscles to exert a constant tension while changing length, and to adapt quickly to changing variations in load.

Autonomic reflexes

Activation of nociceptors can give rise to reflexes which have their connections with autonomic preganglionic neurones in the intermediolateral column of the spinal cord (Fig. 6.1a). These neurones have axons which excite the cells of the adrenal medulla and the postganglionic neurones innervating sweat glands and smooth muscle of blood vessels.

Other autonomic reflexes bring about the emptying of the bladder and rectum. These hollow organs have stretch receptors in their walls which signal the wall tension and thus the organ volume. The signals are appreciated consciously as a sensation of fullness and they also activate reflex connections in the spinal cord. *Defaecation* is brought about by distension of the rectum as faeces pass into it from the colon. This causes the *anal sphincter* to open reflexly, and the diaphragm and muscles of the abdominal wall to contract, with explusion of the faecal mass (p. 579).

The smooth muscle of the bladder wall (the *detrusor muscle*) and neck (the *internal sphincter*) is innervated by sympathetic fibres (inhibitory) from the lumbar segments of the spinal cord and by parasympathetic fibres (excitatory) from sacral segments 2 to 4. The *external sphincter* is a striated muscle with a somatic innervation. When bladder-wall tension reaches a certain level, the detrusor contracts reflexly and the internal sphincter relaxes. It should be noted, however, that the bladder wall can accommodate increasing volumes of urine with very little alteration in tension. Only when the volume reaches (usually) 300–400 ml is there an appreciable degree of discomfort associated with a steeper rise in tension and triggering of the *micturition reflex*. An indication that these empyting reflexes can be facilitated from the brain is that the bladder can be emptied at any volume. More often, however, these spinal reflexes are inhibited by cerebral activity until the sensations become insistent. The reflex activity is then augmented and the external sphincter relaxed. In the case of the bladder, the flow of urine begins and is facilitated by ancillary reflexes from the urethra. These reflexes are stimulated by flow of urine and the efferent activity reinforces bladder muscle contraction and sphincter relaxation. Contraction of the abdominal wall and the pelvic floor also aid the complete emptying of the bladder.

Following transection of the spinal cord and the onset of spinal shock the

bladder wall is inert and the sphincter closed, resulting in urinary retention
with overflow. As the shock wears off and detrusor muscle tone returns, reflex
emptying can be brought about by, for example, cutaneous stimulation, which
may at the same time cause defaecation and other evidence of widespread
autonomic activity (*'mass' reflex*).

6.2 Supraspinal pathways to motor neurones

Through these pathways the brain evokes movements and their postural
concomitants, by exciting directly or indirectly the motor neurones. We can
distinguish four descending pathways extending the length of the spinal cord
which have these functions. These are the cortico-, rubro-, lateral vestibulo-
and reticulospinal tracts (Fig. 6.5). A further two tracts extend only through
the cervical region and are primarily concerned with the motor nuclei of neck
muscles. These are the tecto- and medial vestibulospinal tracts (Fig. 6.5).

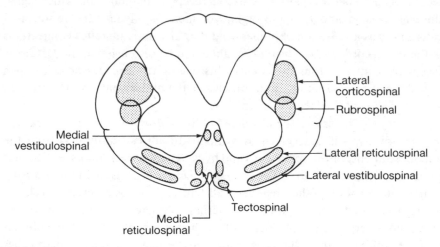

Fig. 6.5. Diagram of the descending pathways of the spinal cord in transverse section.

Corticospinal tract (pyramidal tract)

This tract is the most important pathway in motor control. It comprises axons
of pyramidal cells in layers III and V of the premotor area, precentral gyrus
and postcentral gyrus in almost equal proportions. The tract descends through
the posterior limb of the internal capsule and the cerebral peduncles to pass
through the pons and medulla. In the midbrain and medulla, axons (bulbospinal
fibres) leave the tract and pass to the cranial nerve motor nuclei. These are
crossed pathways but no definite tracts can be seen as the axons pass across
the brainstem in small separate bundles. Axons also pass to the red nuclei and

to the reticular nuclei, all of which are themselves the source of spinal pathways, sometimes spoken of as the *indirect* corticospinal pathway.

In the caudal medulla, the pyramids are formed mainly by the decussation of the *direct* corticospinal fibres. The crossed fibres thereupon enter the lateral columns of the spinal cord forming the lateral corticospinal tract (Fig. 6.5). A few fibres (15%) do not cross at this level and pass down ipsilaterally. They are thought to cross in the spinal cord.

The fibres of the lateral corticospinal tract terminate on the lateral group of interneurones in the dorsal part of the ventral horn. In the lower cervical cord, about 10% of the fibres also terminate directly on motor neurones, probably in motor nuclei which serve the finger and hand muscles.

Rubrospinal tract

This tract arises from cells of the posterior part of the *red nucleus* in the midbrain. The large diameter fibres cross immediately and run down the contralateral side of the brainstem and spinal cord (Fig. 6.5). Their terminals are distributed similarly to those of the corticospinal tract. The red nucleus receives input from the cerebral cortex and the dentate and intermediate nuclei of the cerebellum.

Lateral vestibulospinal tract

This tract (Fig. 6.5) arises from the lateral vestibular nucleus, is ipsilateral and terminates in the medial part of the base of the dorsal horn. A few fibres terminate on medial motor neurones. The lateral vestibular nucleus receives inputs from the vestibular nerve and the fastigial nucleus of the cerebellum.

Reticulospinal tracts

Two pathways exist (Fig. 6.5). One originates in the pontine reticular system and runs ipsilaterally down the spinal cord. The other originates in the medullary reticular system and bifurcates to run down both sides of the cord. Both systems terminate on the medial group of interneurones and to a small extent on medial motor neurones. The reticular nuclei receive input from the cerebral cortex and the fastigial and intermediate nuclei of the cerebellum.

Tectospinal tract

This tract originates in the midbrain in the region of the superior colliculus. It is a crossed pathway and the fibres terminate on interneurones in the cervical cord. At its origin it receives input from the cerebral cortex, especially the occipital lobe, and from the superior colliculus. The pathway seems to generate head movements to help direct one's gaze at a particular point.

Medial vestibulospinal tract

The medial vestibulospinal tract (Fig. 6.5) arises mainly from the medial vestibular nucleus and is the caudal equivalent of the medial longitudinal fasciculus. The tract is uncrossed and terminates in relation to the same cells as the tectospinal tract. The function of the tract is to operate neck muscles in response to vestibular stimulation, so keeping the head stable during body movements.

This list does not include all known pathways. Recently, it has become evident that several other pathways from the brainstem are of importance in movement. In particular, there are pathways running in the ventral quadrant which originate first in the *raphé nuclei*, a narrow continuous collection of cells in the midline of the brainstem which contain 5-hydroxytryptamine (5HT or serotonin), and second in the *locus coeruleus*, which is located near the floor of the rostral part of the fourth ventricle and the cells of which contain noradrenaline. These pathways appear to facilitate the discharge of motor neurones. Both serotonin and noradrenaline when injected intravenously increase the discharge of α and γ motor neurones in experimental animals. Moreover, section of the ventral quadrant of the cord above the level of the lumbar enlargement interrupts these pathways and abolishes the response of lumbosacral motor neurones to stimulation of the contralateral precentral motor cortex, despite the continued functioning of the lateral corticospinal tract which runs in the lateral quadrant. It seems likely then that these monoamine systems may set the background level of excitation of motor neurones in conformity with a person's emotional state.

6.3 Control of posture

By *posture*, we mean the maintenance for a period of time of position (of head, limbs, trunk) in space, as a prelude or background to movement. Fundamentally, posture implies a particular distribution of muscle tone and the most basic of postural mechanisms is thus the *stretch reflex*.

When standing comfortably, the body is in a position in which the weight is taken by bones and ligaments with little muscle activity. Attempts to displace the body from this position by gravity or acceleration are actively resisted. This resistance is the result of various postural reflexes acting via the eye, neck, trunk and proximal limb muscles, and which operate at an unconscious level. The brainstem and spinal cord pathways to the motor neurones of the postural (antigravity) muscles are the vestibulo- and reticulospinal tracts.

There are three basic classes of postural reflex:

1 reflexes responding to gravitational force;
2 reflexes responding to linear acceleration; and
3 reflexes responding to angular acceleration.

Reflexes evoked by gravitational force (static reflexes)

Stretch reflexes, requiring the nervous machinery of only one or two spinal cord segments, can be thought of as units which can be combined into more and more complex patterns (static reflexes) and modulated in more and more ways as the involvement of the CNS increases. Thus, a spinal animal cannot remain upright for any length of time although it will exhibit stepping motions; after upper medullary section, an animal can stand for long periods but cannot right itself if laid on its side; while a midbrain section in quadrupeds, although not in primates, will permit such righting responses.

Local static reflexes

These reflexes are confined to the stimulated limb and the most important is the *tonic stretch reflex*, the elements of which are described on p. 195 and which is the basis of the maintenance of an upright posture. As an example, the muscles which extend the knee during standing are continuously stretched and therefore are in a state of reflex contraction.

Segmental static reflexes

These reflexes are evoked by, for example, nociceptor stimulation of one limb. The reflex response is seen in both limbs. For instance, flexion of a leg in response to a noxious stimulus to the foot is accompanied by extension of the opposite leg (the *crossed extensor reflex*). This response reflects the inbuilt circuitry of stretch and flexor reflexes (see reciprocal innervation, p. 196).

General static reflexes

In these reflexes, the stimulus is local but the response includes many muscle groups. Many such reflexes involve stimulation, by changes in head position, of the abundant muscle spindles in the neck muscles. Such movements also produce stimulation of the labyrinths. These reflexes implicate both limb and eye muscles and are most easily illustrated in quadrupeds. After destruction of the labyrinths, *tonic neck reflexes* depending only on neck muscle proprioceptors can be elicited by, for example, turning the head to the right relative to the body. This will cause the right limb to extend and the left limb to flex. Turning to the left has the opposite effect. Dorsiflexion of the head will extend the forelimbs and flex the hindlimbs, while ventroflexion has the converse effect. If the neck proprioceptors are eliminated, dorsiflexion will extend all four limbs and ventroflexion will flex them. These later reflexes must therefore be labyrinthine in origin. Normally, the two sorts work in a complementary fashion.

An example of the influence of neck movement on the eyes is the *doll's head phenomenon* seen in comatose patients whose brainstem is still intact. In such patients, rotation of the head to the left yields an eye movement to the right and vice versa. Dorsiflexion of the head produces eye movement downward and ventroflexion an upward movement. Presumably these reflexes normally contribute to the maintenance of visual fixation.

Righting reflexes

When an animal laid on its side awakes from sleep, it stands up. It adopts this standing posture as a result of the operation of a series of righting reflexes which involve signals from skin receptors, the labyrinth, and muscle proprioceptors, and which also include visual reflexes dependent on the cerebral cortex.

Reflexes evoked by linear acceleration

Receptors in the inner ear have been described (Section 5.5). Fibres project from the utricle and saccule to the lateral vestibular nuclei which give rise to the vestibulospinal tracts. An example of this type of reflex is the *vestibular placing reaction*. This can be evoked by holding a blindfolded cat by its pelvis, head down, whereupon its forelegs extend. The response is lost if the utricle is destroyed.

If the blindfold is removed, a cat will still extend its legs even if its utricles are destroyed—a *visual placing reaction*. In general, visual stimuli can substitute for labyrinthine in postural reflexes. Thus, animals and men with bilateral labyrinthine damage have normal posture if they can use their eyes.

Reflexes evoked by angular acceleration

Rotatory stimuli are detected by the receptors in the semicircular canals (p. 178). The position of the cupulae, and thus stimulation of the receptors, depends upon angular acceleration with inertial lag of endolymph movement. The receptors project to the medial (and other) vestibular nuclei which give rise to the medial longitudinal fasciculi and the medial vestibulospinal tracts. The output side of the reflex response therefore involves movements of eye, neck and proximal arm muscles.

The situation is most easily analysed in relation to the horizontal canals. If a person is rotated to the *right*, movement of the endolymph and right cupula is to the left, i.e. *ampullopetal*, while that of the left cupula is *ampullofugal*. Appropriate signals are sent to the medial vestibular nuclei. The movement of the eyes is in the same direction as the endolymph flow, retaining the fixation point as the head moves. As the rotation continues, the fixation point cannot be maintained and there is a quick flick of the eyes to the right to a new fixation point. The slow drift and quick flick will be repeated until there is no longer movement of the endolymph in the canal. The eye movements are called *vestibular nystagmus*. A nystagmus is always named after the quick phase so, in the example, the nystagmus would be to the right (see also p. 185).

If the head is rotating at constant velocity, there is no relative endolymph movement, the cupulae gradually take up the resting position and vestibular nystagmus ceases although reflex movement of the eyes (*optokinetic nystagmus*) continues depending on the visual input. On ceasing rotation, cupular movement is now in the opposite direction, returning to rest in 25–30 s.

Ampullopetal movement, absolute or relative, of the endolymph in the right horizontal canal occurs:

1   on moving the head to the right;
2   on ceasing rapid rotation to the left; or
3   on syringing the right ear with warm water.

Obviously the same effect is obtained by syringing the left ear with cold water and after destruction of the left labyrinth. The sensations which arise are those of movement to the right or of the external world to the left; thus the slow drift of the nystagmus and forced movements are to the left. These are commonly accompanied by vertigo and general autonomic disturbance. Nystagmus in the absence of appropriate stimuli may be a symptom of brainstem, labyrinthine or cerebellar injury or disease.

Effects of spinal and brainstem transection

The preceding sections outline the reflex mechanisms by which the motor neurones (both α and γ) of antigravity muscles are excited, particularly via the vestibulospinal and reticulospinal paths. This excitation is an important determinant of the reflex resistance of a muscle to stretching (muscle tone). It is therefore understandable that a spinal transection which severs these connections causes a severe loss of tone and reflexes in segments below the section. Indeed, immediately following a spinal transection, for example in the cervical region below the level of the phrenic motor neurones, there ensues a state of profound torpor. The muscles lie inert, the strongest stimulation of a sensory nerve evokes no response, reflexes are unobtainable, the blood vessels dilate, the blood pressure falls, thermal sweating is absent, the bladder distends with urinary overflow and the viscera in general are quiescent. This condition is known as *spinal shock*. It is not due to the trauma nor to the low blood pressure but to interruption to the normal flow of impulses down the long descending tracts of the cord that impinge on interneurones and motor neurones so that their excitability is lowered. The lateral vestibular and pontine reticulospinal tracts are most important in this respect and spinal shock follows any section caudal to the lateral vestibular nucleus. The course of recovery varies with the species from a few minutes in frogs to some weeks in humans. There is a gradual increase in excitability; first flexor withdrawal reflexes return with extensor inhibition, then bladder and rectum empty reflexly, the blood pressure rises, and after some months (in man) extensor activity returns and tendon jerks can be elicited.

The recovery of excitability of these neurones which underlies the reflex recovery may be due to sprouting of the terminals of their remaining inputs. This may enable the input fibres to excite the motor neurones more readily. There may also be an increase in sensitivity of the motor neurones to the transmitter released by the remaining inputs, so that the same quantity of transmitter is now more effective. This would be akin to the postdenervation supersensitivity of skeletal muscle.

Brainstem section just rostral to the vestibular nuclei causes the tone to be exaggerated instead of lost and was first described by Sherrington (1857–1952), a pioneer of neurophysiology. The exaggeration of tone is so extreme in antigravity muscles that a particular posture is maintained—*decerebrate rigidity* (Fig. 6.6a). Humans in this condition have arms and legs extended, the back arched and the head dorsiflexed. The feet are ventroflexed and the arms pointed. The wrists are flexed and the fingers little affected. The rigidity is due, essentially, to the removal of inhibitory influences with the maintenance of facilitatory influences, and can be made more extreme by making a more rostral transection, in the midbrain, leaving the pontine reticular nuclei intact. Neurones in this nucleus, like the neurones of the lateral vestibular nucleus, excite the motor neurones of antigravity muscles.

Decorticate rigidity occurs when the postural mechanisms are damaged above the brainstem. It differs from decerebrate rigidity in that it can be modified by reflex means and shows great variation in severity in different species. Cats, for instance, can walk about in a decorticate condition, but a man in this condition (Fig. 6.6b) is unconscious, with legs extended and arms

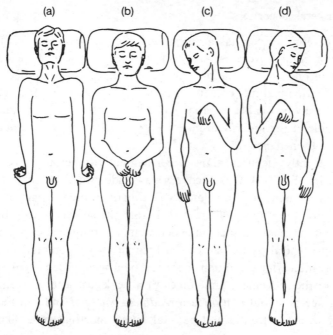

Fig. 6.6. The positions maintained in decerebrate rigidity (a) and decorticate rigidity (b to d). The tonic reflexes in the decerebrate condition are shown when the head is turned to the right (c) and to the left (d). (Redrawn from Fulton, J.F. (1955) *Textbook of Physiology,* 17th edn, p. 217. W.B. Saunders Co., Philadelphia.)

in a position determined by his head position (through neck reflexes). Thus, if the head is turned to the right, the right arm is extended and the left arm flexed (Fig. 6.6c) and if the rotation is to the left the opposite is seen (Fig. 6.6d).

6.4 Supraspinal organization of movement

The decision to move involves the cerebral cortex as a whole. The instruction is channelled through relatively direct connections to the basal ganglia and cerebellum. The instructions from these organs are then collected by the ventrolateral thalamic nuclei which relay to the motor cortex. The motor cortex is the source of those parts of the corticospinal and corticobulbar tracts which relay to the spinal cord and brainstem motor machinery and of the corticofugal pathways to the red and reticular nuclei which also relay to the spinal cord motor machinery.

At the spinal level there must be integration of these instructions with those that govern posture which are delivered via the medial pathway. The end result is alteration of firing rates of the appropriate neurones which leads to contraction or relaxation of muscles to produce the desired movement.

Role of the cerebral cortex

That the cerebral cortex has a prominent role in movement has long been suggested on clinical grounds. Actions in the usual sense of the word, like the taking off of one's hat or the turning on of a light, are composed of a large number of individual movements. A person who suffers a left-sided parietal lobe lesion may be unable to carry out such an action on command, though quite capable of the individual movements which compose the action. This condition is termed *apraxia*.

More recently, electrical signs of cortical activity before movement have been recorded (Fig. 6.7). This discovery was made by making subjects perform the same movement, i.e. the voluntary flexing of an index finger many times, and averaging the tiny signals recorded from the subject's scalp before and during the movement. In such an experiment when the electrodes are placed over the precentral gyrus, a slow surface negative potential is recorded, presumably indicating excitation of cortical neurones, which begins more than 0·5 s before the movement and comes to a peak just after movement. This initial surface potential is termed a *readiness potential* and can be recorded bilaterally over the frontal and parietal lobes. Starting some 80 ms before

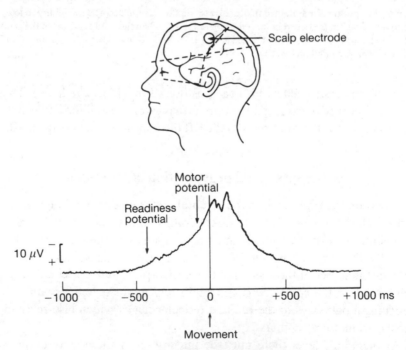

Fig. 6.7. Averaged cortical potentials recorded on the scalp over the precentral gyrus during voluntary flexion of an index finger. The premovement potential can be divided into an initial readiness potential and a later rapidly rising motor potential. (From Deeke, L., Sheid, P. & Kornhuber, H.H. (1969) *Exp. Brain Res.* 7, 158–168.)

movement, there is a rapid increase in potential called the *motor potential* which signals activity of motor cortical neurones. It is recorded only unilaterally over the appropriate part of the motor cortex.

Basal ganglia

The basal ganglia are found deep to the cerebral cortex in each hemisphere (Fig. 6.8). They comprise the *claustrum*, the *putamen* and the *caudate nucleus* and, separated only by white matter from the thalamus, the *globus pallidus*. Other smaller nuclei are for functional reasons usually considered along with the basal ganglia. The largest of these are the *subthalamic nucleus* and the *substantia nigra*, a darkly pigmented nucleus lying in the midbrain. The caudate nucleus and putamen are often referred to together as the *neostriatum*, while the putamen and globus pallidus are sometimes referred to together as the *lentiform nucleus* because of their combined shape.

All components of the basal ganglia are richly interconnected. Little is known of these connections but some at least of the interneurones in the neostriatum use acetylcholine as their transmitter and the *striatonigral pathway* from the neostriatum to the substantia nigra (Fig. 6.9a) is thought to utilize γ-amino butyric acid (GABA) as the transmitter. More importantly, the *nigrostriatal pathway* from the substantia nigra to the neostriatum (Fig. 6.9a) is *dopaminergic* and is disordered in Parkinson's disease.

Afferent pathways. The neostriatum receives important inputs from the cerebral cortex (frontal, parietal, occipital and temporal lobes), the intralaminar nuclei

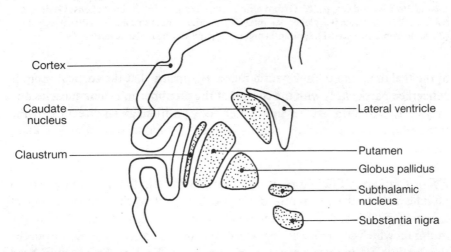

Fig. 6.8. Diagram of a coronal section of the right cerebral hemisphere showing the location of the basal ganglia (claustrum, caudate nucleus, putamen and globus pallidus) and associated structures (subthalamic nucleus and substantia nigra).

(a)   (b)

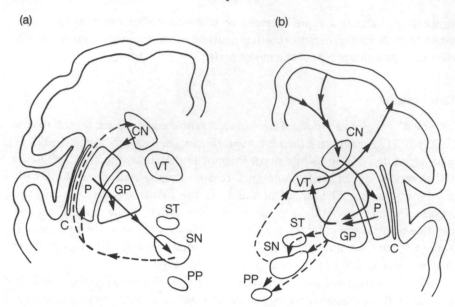

Fig. 6.9. Coronal sections of the cerebral hemispheres illustrating pathways thought to be important in the function of the basal ganglia. (a) Pathways thought to be disordered in Parkinson's disease. Note the striatonigral pathway (unbroken line) from the caudate nucleus and putamen through the globus pallidus to the substantia nigra; the nigrostriatal pathway (broken line) from the substantia nigra to the caudate nucleus and putamen. (b) Pathways by which the basal ganglia influence posture and movement. Note the circuit (unbroken lines) comprising the pathway, cortex–caudate nucleus and putamen–globus pallidus–thalamus–motor cortex, which is thought to play an important role in the supraspinal control of movement and posture and the pathway (broken lines) projecting to the pedunculopontine nucleus which influences locomotion. (Adapted from Brodal, A. (1981) *Neurological Anatomy in Relation to Clinical Medicine*, 3rd edn, p. 216. Oxford University Press, Oxford.) Abbreviations: caudate nucleus (CN), claustrum (C), globus pallidus (GP), pendunculopontine nucleus (PP), putamen (P), subthalamic nucleus (ST), substantia nigra (SN), and ventral thalamic nucleus (VT).

of the thalamus and the substantia nigra. It appears that the cortical input is concerned particularly with the control of the proximal and trunk muscles and may be used to form a representation of body position, given direction by the force of gravity and used in the early stages of the preparation of voluntary movement.

Efferent pathways. Efferent pathways from the neostriatum pass to the globus pallidus and to the substantia nigra. Both project to the ventral group of thalamic nuclei and to the pedunculopontine nucleus (Fig. 6.9b) in the midbrain which controls walking. Recently, an inhibitory projection from the substantia nigra to the neurones in the superior colliculus which control head turning has been described. Neostriatal stimulation, which excites globus pallidus neurones that inhibit the substantia nigra, induces contralateral head turning. The subthalamic nucleus appears to modulate these outputs for it has connections to the globus pallidus and to the substantia nigra and receives afferent inputs from the motor cortex and from the intralaminar thalamic nucleus.

Functions of the basal ganglia

Neurones in the globus pallidus of experimental animals discharge in relation to movement of the *contralateral* limb. The basal ganglia of one side then, like each motor cortex, are related to the opposite side of the body. Further, these neurones discharge *before* movement and presumably these signals pass to the ventral thalamic nuclei which in turn send signals to excite motor cortex cells. It is known that both the globus pallidus and the substantia nigra have an inhibitory effect on the ventral thalamic nuclei. The basal ganglia output then exerts its effect by selective inhibition of the ongoing activity of cells of the ventral thalamic nuclei. It is possible that the basal ganglia are concerned with the fixing of the body and proximal parts of limbs to provide a secure base for movements of the hands and feet. The basal ganglia also appear to play a role in the control of walking through their connection with the pedunculopontine nucleus (PP) (Fig. 6.9b). Stimulation of this nucleus in decerebrate animals will induce locomotion on a treadmill. The pattern of locomotion is not elaborated in the midbrain but in the spinal cord. The PP is connected to the cells of origin of the lateral reticulospinal tract (Fig. 6.5). This tract carries signals inducing locomotion to the interneurones and motor neurones in the spinal cord. The substantia nigra sends a tonically active inhibitory (GABA) input to the PP, while stimulation of the globus pallidus can excite locomotion by direct excitation of PP neurones. Stimulation of the subthalamic nucleus can also excite locomotion by inhibition of the substantia nigra.

The association of the basal ganglia with the control of movement was first proposed as the result of studies on patients with diseases of, or injuries to, this region. There are three main signs of such damage:

1 abnormal movements;
2 abnormal muscle tone (*rigidity*); and
3 *bradykinesia*, i.e. a slowness in initiating and changing movement.

An example of bradykinesia is where a patient with Parkinson's disease when asked to walk away from the examiner is slow to get going, and when asked to come back has great difficulty in turning around.

Abnormal movements are of four basic types:

1 *tremor*, which is characteristically involuntary, occurs at rest, and disappears during movement;

2 *athetosis* or slow writhing movements, which are particularly noted in patients with cerebral palsy whose basal ganglia were damaged at birth;

3 *chorea* or involuntary jerky movements of the extremities and facial muscles; and

4 *ballismus* or violent flailing movements usually of one limb, involving the proximal muscles.

Ballismus is a characteristic sign of damage to the subthalamic nuclei.

The most common disorder of the basal ganglia is Parkinson's disease, in which patients have a resting tremor, rigidity and bradykinesia. These signs are thought to be a consequence of the known loss of substantia nigral cells and thus of the dopaminergic pathway to the caudate–putamen complex (Fig. 6.9a). This view has led to treatment with a dopamine precursor, L-dopa, which crosses the blood–brain barrier and alleviates the symptoms. Surgery that interrupts the outflow of the basal ganglia (in globus pallidus or thalamus) can alleviate the tremor and rigidity, suggesting that these disorders arise from abnormal signals sent to the thalamus by the disordered basal ganglia. The bradykinesia is not attenuated, suggesting that it may be the primary deficit in the disease.

The rigidity in Parkinson's disease can be shown to arise from abnormal excitation of α and γ motor neurones of both flexor and extensor muscles around a joint. The abnormal degree of activity of both agonists gives rise to the rigidity. The excitation waxes and wanes with the tremor. Thus, the rigidity also momentarily increases and weakens giving rise to its name *cogwheel rigidity*.

Role of the cerebellum

The gross anatomy of the cerebellum has already been discussed in Chapter 2, (p. 81 & Fig. 2.30); its structure at the cellular level is fairly well understood, as are its neural circuits. It is thus particularly frustrating that an equivalent level of understanding of its function has so far not been achieved. All the lobes of the cerebellum have a three-layered cortex covering white matter in which are embedded the deep nuclei. Histologically, the cortex displays a remarkable degree of uniformity (Fig. 6.10).

Cerebellar cortex

The three layers comprise:

1 a *molecular layer* containing a large number of axons which run along each folium parallel to its long axis, and which intersect Purkinje dendritic trees at right angles;

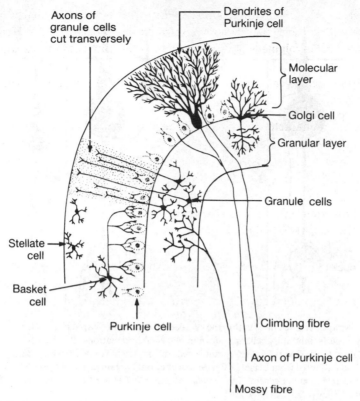

Fig. 6.10. Transverse section of a folium of the cerebellum. (Redrawn from *Gray's Anatomy* (1949), eds. Johnson, T.B. & Willis, J., 30th edn, p. 956. Longmans, Green & Co., London.)

2 a *thin Purkinje cell layer* containing the cell bodies of the large Purkinje cells; and

3 a *granular layer* containing 20 billion granule cells, afferent mossy and climbing fibres, and the elaborate synaptic complexes between incoming mossy fibre axons and granule cell dendrites.

There are five cell types—Purkinje cells, granule cells, Golgi cells, basket cells and stellate cells.

Purkinje cells are 15 million in number. Their dendritic trees pass up into the molecular layer to intersect with parallel fibres (Fig. 6.11a, b). Spines on their dendrites are the sites of excitatory synapses between granule cell axons and Purkinje cell dendrites. A second type of excitatory synapse is made by climbing fibres which wind round the Purkinje-cell dendrites. The axons of the Purkinje cells pass to other cell types and out of the cortex to the deep cerebellar nuclei. Everywhere the discharge of the Purkinje cells has an *inhibitory* effect.

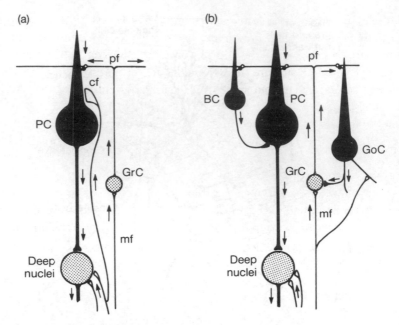

Fig. 6.11. Synaptic relationships within the cerebellum. (a) Input–output circuits (arrows). (b) Inhibitory circuits (inhibitory cells are shown in black). Abbreviations: BC, basket cell; cf, climbing fibre; GoC, Golgi cell; GrC, granule cell; mf, mossy fibre; PC, Purkinje cell; pf, parallel fibre. (Adapted from Eccles, J.C. (1966) Functional organization of the cerebellum in relation to its role in motor control. In *Muscular Afferents and Motor Control*, ed. Granit, R., pp. 19–36. John Wiley & Sons, New York.)

Granule cells are *excitatory* neurones which receive excitation from incoming mossy fibres and send their axons up to the molecular layer to form the parallel fibres (Fig. 6.11a, b).

Golgi, basket and stellate cells (Fig. 6.11b) are all *inhibitory* interneurones. *Golgi cells* have their cell bodies near the Purkinje cell layer. They inhibit granule cells and their dendrites receive excitation in both the molecular and granular layers. *Basket cells* have their cell bodies in the molecular layer close to the Purkinje cells. Their dendritic trees intersect the parallel fibres which excite them and their axons form a dense meshwork around the initial segment of each Purkinje cell. *Stellate cells* resemble basket cells but they lie more superficially in the molecular layer and their axons terminate on Purkinje-cell dendrites.

Deep intracerebellar nuclei

These are four masses of grey matter embedded in the white matter of each half of the cerebellum. They lie in line from lateral to medial. The largest, the *dentate nucleus*, which has a convoluted bag-like shape, lies most laterally and the *fastigial nucleus*, a solid grey mass, lies most medially under the vermis forming part of the roof of the fourth ventricle. Between the fastigial and dentate nuclei on each side are smaller masses of grey matter—the *globose* and *emboliform nuclei* in man, but which in experimental animals form the one *nucleus interpositus*.

The computations performed in the cerebellar cortex influence movement and posture via these nuclei. These neurones receive excitatory impulses continuously from branches of all types of incoming afferent fibres on their way to the cerebellar cortex (Fig. 6.11a). Superimposed on this activity are periods of inhibition of varying length produced by the inhibitory output of the Purkinje cells related to the particular nucleus. The functioning unit in the cerebellum comprises a group of neurones in a deep intracerebellar nucleus and the neurones in a related region of the cerebellar cortex.

Table 6.1 summarizes these relationships and indicates the final destinations of the nuclear axons. The vestibular nuclei are included because, although they are not intracerebellar, they do receive direct projections from Purkinje cells.

Table 6.1. Afferent and efferent connections of the cerebellar cortex.

Cerebellar zone	Related nuclei	Destination of nuclear axons
Posterior lobe (flocculus, nodule)	Vestibular	Ventral horn of spinal cord
Medial zone (vermis)	Vestibular, fastigial	Reticular nuclei, vestibular nuclei
Intermediate zone	Globose, emboliform	Ventrolateral thalamic nuclei (ipsi- and contralateral) and red nucleus (contralateral)
Lateral zone	Dentate	Same as intermediate zone

Cerebellar afferents

Mossy fibres send information from the *spinal cord* to the cerebellum by direct and indirect pathways. Of at least ten tracts four are sufficiently large and well known to be described here (Fig. 6.12). The *dorsal spinocerebellar tracts* in the dorsal quadrants of the spinal cord and the *ventral spinocerebellar tracts* in the lateral quadrants arise in the thoracolumbar segments of the cord and terminate in the cerebellar cortex as mossy fibres. The *cuneocerebellar* and *rostrospinocerebellar tracts* are the rostral equivalents of these tracts and also terminate in the cerebellar cortex as mossy fibres. The information carried by the dorsal spinocerebellar and cuneocerebellar tracts is about muscle length and tension, joint position and skin deformation (*proprioception*). Each half of the cerebellum

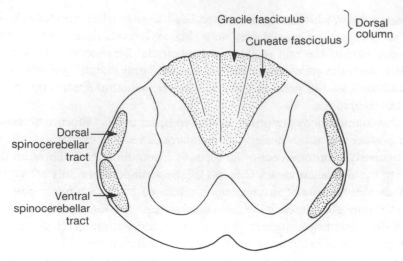

Gracile fasciculus
Cuneate fasciculus
} Dorsal column

Dorsal spinocerebellar tract

Ventral spinocerebellar tract

Fig. 6.12. Diagram of the ascending pathways of the spinal cord in transverse section.

thus receives information during limb movement about the phase and strength of contraction of individual muscles, the joint angles and the time at which the limb extremity touches the ground. In contrast, the ventral spinocerebellar and rostrospinocerebellar tracts are thought to convey information mainly about the activity of the anterior horn cells in the execution of motor commands, such as the rhythm generation which underlies walking. Other spinocerebellar tracts give rise to mossy fibres indirectly by terminating on the lateral cerebellar nucleus in the medulla. The axons of this nucleus form mossy fibres.

Information is sent from many areas of the *cerebral cortex*, including the visual and auditory areas and the motor cortex. The pathways run with the corticospinal and corticobulbar fibres and are indirect as they terminate on the ipsilateral pontine nuclei. The axons of the *pontine nuclei* form mossy fibres and are the largest source of such fibres. Collaterals from corticospinal and corticobulbar fibres also terminate here. These inputs are excitatory.

Vestibular input is both direct from the vestibular ganglion and indirect arising from cells in the vestibular nuclei which give rise to mossy fibres.

Climbing fibres are axons of neurones of the *inferior olivary nucleus*, a prominent nucleus in the medulla at the level of the pyramids. This nucleus has many subdivisions each with a particular pattern of input and each subdivision projects to Purkinje cells in a particular part of the cerebellar cortex. The largest inputs to the inferior olivary nucleus come from the premotor cortex, the vestibular nuclei and the spinal cord, but there are other inputs such as from the superior colliculus which conveys information about the direction of visual movement.

Functional zones of the cerebellum

Although the histology of the cerebellar cortex is uniform, it is divided functionally into longitudinal zones shown in Fig. 6.13. First, the division is made on the projection of the Purkinje cells to the cerebellar nuclei. In the most medial zone, the Purkinje cells all project to the fastigial nuclei; in the intermediate zone, they project to the globose and emboliform nuclei; and in the lateral zone they project to the dentate nucleus. Second, the longitudinally running climbing fibres and the Purkinje cell axons form zones which can be distinguished in the white matter by systematic differences in axon diameter with large-diameter axons toward the centre of each zone and smaller diameter axons at the boundaries where zones adjoin. Recently, it has been shown that in these longitudinal zones there are narrower longitudinal bands which are the ultimate functional microzones.

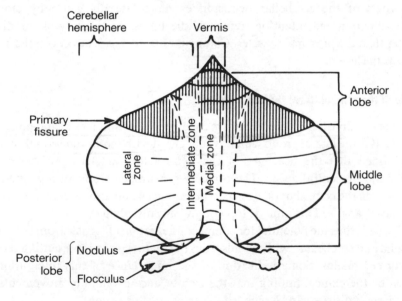

Fig. 6.13. Transverse and longitudinal subdivisions of the cerebellum. The surface has been unfolded and laid out flat. (From Noback, C.R. & Demarest, R.J. (1975) *The Human Nervous System*, 2nd edn, p. 290. McGraw-Hill Book Co., New York.)

Functions of the cerebellum

The cerebellum, like the basal ganglia, has long been thought to be concerned with movement because disorders of movement are associated with structural changes in the organ. There are three disturbances that follow damage to particular parts of the cerebellum:

1   *disorders of balance and gait*;
2   *hypotonia*; and
3   *incoordination* (ataxia).

With incoordination the rate, range and direction of movement are disturbed. An intention tremor, due to a timing disorder of the relations between agonist and antagonist muscles in a movement, is often present. Signs of incoordination are also seen in eye movements (nystagmus) and in speech muscles, producing peculiar speech patterns (staccato or scanning speech).

The region of the cerebellum affected determines the nature of the defect. Impaired balance without incoordination or hypotonia is a sign of flocculonodular lobe damage, e.g. in medulloblastoma, a tumour of children. Lesions of the vermis and intermediate area, particularly in the anterior lobe, are a feature of alcoholic cerebellar degeneration and are associated with incoordination of proximal muscles resulting in disorders of gait and of limb movement. Disorders of the cerebellar hemispheres more laterally cause hypotonia, incoordination and intention tremor of the limbs, particularly of the distal rather than the proximal muscles. Cerebellar disorders are always on the same side as the lesion.

Role of the ventrolateral thalamic nuclei

The ventral division of each thalamus consists of anterior, lateral and posterior nuclei. It is the lateral group which is concerned with motor function. It is well established that the neurones of the ventrolateral nucleus project to the ipsilateral precentral gyrus. The projecting neurones are in turn excited by axons from the contralateral deep cerebellar nuclei and also receive input from the basal ganglia. This input is topographically distributed.

There is thus the potential for complex integration of signals from the basal ganglia and cerebellum with signals from the motor cortex representing recent motor commands. Some indication of the importance of this integration is given by the clinical finding that the tremor and involuntary movements of Parkinson's disease can be relieved by lesions of this region.

Motor cortex

Stimulation of the cerebral hemispheres of man and experimental animals is known to evoke movements on the opposite side of the body. With reduced strength or stimulation or increased depth of anaesthesia, the regions from which movement can be evoked narrow down to three, all in the vicinity of the central sulcus.

Motor responses to stimulation are found on either bank of the central sulcus (Fig. 6.14) and this area is often referred to as the sensorimotor cortex.

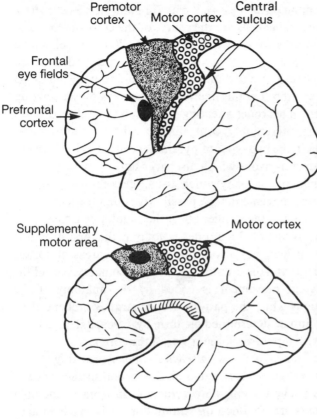

Premotor cortex Motor cortex Central sulcus

Frontal eye fields

Prefrontal cortex

Supplementary motor area Motor cortex

Fig. 6.14. The motor areas of the brain. *Top*: lateral surface of the left cerebral hemisphere showing the motor cortex (precentral gyrus) and the premotor cortex. *Bottom*: medial view of the surface of the right cerebral hemisphere (with the left hemisphere removed) showing the medial extension of the motor cortex and also the supplementary motor area in front of the motor cortex.

The *precentral gyrus* is called the *motor cortex* or *motor I (MI)* and is thought to be the region most immediately concerned with the generation of movements. However, the movements evoked by stimulation of MI are always fractional, that is, only part of a full movement—indeed it is impossible to elicit complete or complex movements.

Experiments in man and animals in which the precentral gyrus is systematically stimulated millimetre by millimetre from its medial to its lateral extremity (Fig. 6.14) indicate a representation of movements in MI such that leg movements are elicited from regions extending onto the medial surface of each hemisphere while movements of the face and scalp muscles are elicited most laterally. This somatotopic representation arises because each region is connected via the corticofugal tracts with the appropriate interneurones and motor neurones to produce muscle contraction of that part of the body. The somatotopic representation of the body is similar to that in the postcentral gyrus (p. 137). However, the size of the representation in the precentral gyrus is proportional not to the density of receptors in the body region but to the number of different movements which the muscles in that region can make.

Motor I and the adjoining *premotor cortex* contain giant nerve cells (Betz cells) which together with smaller cells are the source of the corticobulbar and corticospinal tracts. These cells are known as *pyramidal cells* from their shape and they are organized in columns which are oriented perpendicular to the surface. Each column is about 1 mm in diameter and contains many hundreds of pyramidal cells. Each column contains neurones influencing muscles acting on a particular joint. Cortical neurones affecting a particular muscle are found in several columns, each presumably controlling a different movement. Functionally, there appear to be two types of pyramidal cells in MI. There are larger cells with fast-conducting axons which discharge during movement and smaller cells with low conduction velocity which discharge continually though there is also a change during movement. One possibility is that the larger cells control the α motor neurones and the smaller affect the γ motor neurones.

The source of the excitation of MI neurones is not only the input from the ventral lateral thalamic nucleus, but also input from other areas of cortex, notably the postcentral gyrus and the premotor cortex. Presumably, part of the organization of muscles for movement is coded by the anatomy of the thalamocortical connections while the pattern of discharge is influenced by the pattern of thalamic input, determined in turn by cerebellar and basal ganglia input, together with input from other parts of the cortex.

Motor II (MII), also known as the *supplementary motor area* (Fig. 6.14), has only recently been investigated. It is found on the medial surface of each hemisphere in the gyrus above the cingulate gyrus and in front of the area from which leg movements are obtained on stimulation. It is much smaller than MI, extending for only about 2–3 cm. It was identified first in man and, on stimulation, complex synergistic movements of the contralateral limbs and body, often with vocalization, were produced. These responses were quite unlike the localized contralateral contractions evoked by stimulation of MI. Following lesions of MII, patients grasp reflexly at anything put into their hand. Ablation of MII in monkeys brings about a disturbance of bimanual coordination tasks and recording from cells in MII shows that they discharge with a particular movement whether it is carried out by the right or left hand. A role for MII in the planning and programming of movements appears possible, as recent studies of regional blood flow in man have shown that when thinking about a motor task there is an increased blood flow in both MII areas but no increase in the contralateral MI unless the movement takes place.

The third motor region, *motor III (MIII)*, which includes a region known as the frontal eye fields, is found in the prefrontal cortex in front of the premotor cortex (Fig. 6.14). Stimulation in the frontal eye field evokes movements of the eyes and sometimes turning of the head in the direction of movement. The commonest movement elicited is a conjugate deviation of the eyes to the contralateral side. Studies of neuronal activity in this region in unanaesthetized monkeys show that cells discharge during voluntary eye movements and destruction of the frontal eye fields prevents voluntary movements of the eyes though reflex movements are unaffected. Neurones of MIII contribute axons to the pyramidal tracts.

6.5 Lesions of the motor pathway

A complete transection of the spinal cord will cause permanent loss of all voluntary movement controlled by motor neurones caudal to the plane of section. People with a spinal-cord transection above the lumbar but below the cervical enlargement are paralysed in both legs and are said to be *paraplegic*. A section above or in the cervical enlargement below C_4 will give rise to paralysis of all four limbs. Such people are called *quadriplegics*. A transection above C_4 leads immediately to death as a consequence of respiratory paralysis.

Paralysis or weakness of one side of the body (*hemiplegia*) commonly results from a sudden interruption of the blood supply to an area of the nervous system, causing disruption of the cortical motor pathways to the spinal cord. This is commonly referred to as a *'stroke'*. Because of the *crossed* nature of this motor innervation, a hemiplegia signals a lesion of pathways on the opposite side of the brain.

After interruption at any level of the pathways from the motor cortex in man there is a period of shock, similar to spinal shock, in which there are no detectable reflexes. It takes 2–4 weeks to recover from the loss of reflexes. Recovery from the paralysis takes much longer and is rarely complete. In particular, finger and hand movements are seldom completely regained.

The signs commonly displayed by a patient with hemiplegia, on examination, include the following.

1 Muscle weakness without wasting, particularly of distal muscle groups, e.g. finger muscles are more affected than shoulder. Also the arm muscles are commonly more affected than the leg muscles. Because of their bilateral innervation, the muscles which move the eyes, the muscles of the upper third of the face and the trunk muscles are spared in unilateral lesions.

2 Changes in muscle tone. For weeks or months there is commonly reduced resistance to movement—*hypotonia*. In many patients, this is then replaced by *spastic tone* in which the resistance to movement typically increases as the movement continues.

3 Changes in reflexes. For weeks or months reflexes are lost or diminished in amplitude, but eventually in many patients the amplitude of reflexes such as the jaw, biceps, knee and ankle jerks is increased. *Clonus*, i.e. repeated responses to a single stimulus, may be present.

4   Absence of superficial reflexes. The abdominal reflexes—a brisk contraction of the abdominal muscles following stroking of the overlying skin—are
lost.

5   Extensor plantar response (*Babinski's sign*). Normally, the big toe and all
the other toes turn down when the lateral border of the foot is stroked. With
interruption of the corticospinal pathway the big toe dorsiflexes instead. This
is said to be the most important sign in clinical neurology.

This constellation of signs is known to neurologists as the *upper motor
neurone syndrome*. A fascinating and instructive account of this sequence of
events had been well described by a professor of anatomy who suffered from
thrombosis in his middle cerebral artery (Brodal, A. (1973) *Brain* **96**, 675–
694). In addition, the hemiplegic will display other signs which will indicate
the level of the lesion. For example, difficulties with speech with a right
hemiplegia suggest a left-sided cortical lesion, while the presence of field
defects and disorders of sensation may indicate a lesion in the internal capsule,
and the presence of a third-nerve palsy a midbrain lesion.

The neurological basis of the signs of hemiplegia is to some extent obvious.
For instance, the loss of movement and its distribution and the initial loss of
tone and reflexes reflect the interruption of the cortical output to motor
neurones. The spastic tone and brisk tendon jerks indicate an increase in
excitability of α and γ motor neurones. This may arise as the result of changes
in the spinal cord. These changes include a slowly increasing effectiveness in
the excitation of motor neurones by the primary endings of muscle spindles
and a slowly decreasing diminution of the effectiveness of the reciprocal
inhibition of antagonist motor neurones.

The abnormal reflexes can also be explained by the loss of cortical input.
In babies of a few months old, Babinski's sign is present, but as myelination
of nervous pathways proceeds and the corticospinal input takes control, the
reflex disappears. Its return in hemiplegia is thus no surprise.

7 · Higher Nervous Functions

7.1 **Consciousness**
The electroencephalogram
Variations in the level of consciousness

7.2 **Sleep**
REM sleep and dreaming
Neural basis of sleep
The need for sleep
Sleep disorders

7.3 **Disturbances of consciousness**
Epilepsy

7.4 **Learning and memory**
Characteristics of short-term (primary)
memory
Characteristics of long-term (secondary
and tertiary) memory
Neurological basis of memory

7.5 **Speech**

7.6 **Organization of behaviour**
Control of goal-directed behaviour
Emotion
Motivation
Subcortical control of behaviour
Role of the corpus callosum

7.1 Consciousness

Each morning we wake up—that is, we become conscious of the external world and of our own thoughts and emotions. This state of being conscious has so far defied psychological and physiological explanation. Indeed, all we can do is, first, define what is included in being conscious, realizing that each property is itself a topic for investigation, and then describe what is known of the relationship between brain states and consciousness.

Being conscious means being aware of our identity, of our past and present, and possessing the notion of the future. It evidently includes outward-looking aspects—awareness of others, aesthetic and ethical judgements, the ability to express one's thoughts and ideas—as well as inward aspects—our emotions and fantasies.

There is a variety of levels of consciousness. We can pay close attention (a heightening of consciousness), or daydream (a turning inward which to the observer is a diminution of consciousness). Finally, consciousness is regularly lost every night as we go to sleep and as miraculously regained as we wake up.

The electroencephalogram

It is possible, by recording the electrical activity of the brain, to define what conscious state a person is in. Such studies depend on the recording of the fluctuations in electrical potential, either from the scalp (*electroencephalogram*, EEG), or directly from the cortical surface (*electrocorticogram*). These electrical records are the algebraic summation, at the point on the surface at which the measurements are made, of the changes in potential resulting from current flow between dendrites and cell bodies in the underlying brain. This current is a consequence of the excitatory and inhibitory postsynaptic potentials. Note that the magnitude of the potentials recorded (in μV) is much attenuated, reflecting the fact that the electrical changes have been transmitted through a volume conductor. The EEG was first recorded by Berger (1873–1941) and the records are often called Berger waves after him. In conventional EEG recording, 8–16 leads are placed in a standard manner on the scalp over specific cortical areas.

The normal EEG in the awake and alert state shows irregular, low-voltage waves (30–80 μV) of high frequency. The trace, a typical example of which is shown in Fig. 7.1. (eyes open), is said to be *desynchronized*. If a subject is bored or closes his eyes, waves with a frequency of 8–10 Hz are recorded, particularly in *occipital* leads (Fig. 7.1, occipital trace), indicating *synchronized* activity of neurones. This is termed the *alpha rhythm*.

Alpha and other synchronized rhythms are thought to be induced by activity of the thalamic nuclei. Although decortication leaves thalamic activity unimpaired, lesions in the thalamus or deafferentation of a related cortical area abolishes the rhythm in that cortical area. Intracellular recording of

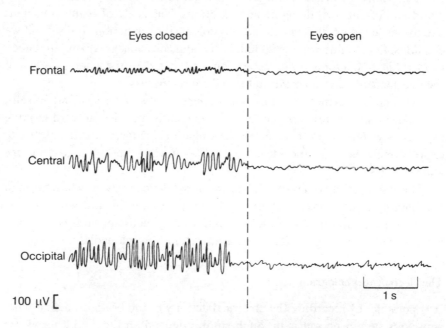

Fig. 7.1. Electroencephalogram (EEG) recorded from a human scalp. The traces show typical waveforms recorded over the frontal and occipital poles and in the central region between them. Note the appearance of the 8–10 Hz alpha rhythm in the central and occipital traces when the subject has his eyes shut and its disappearance (alpha blocking) when the eyes are opened.

electrical activity in the thalamus indicates the presence of multiple spontaneously active pacemaker neurones, which are influenced in their firing rates by brainstem signals (p. 226). The EEG finds clinical use mainly in the diagnosis of epilepsy (p. 232) but silent areas indicated by a flat trace in one or more leads may point to the presence of an infarction or tumour. A continuous flat trace from all leads indicates brain death. This occurs if the brain is deprived of blood for more than 3–8 min, although electrical activity in the brainstem may survive longer (7–10 min).

Variations in the level of consciousness

Consciousness is associated with brain activity. Anaesthetics, injury, metabolic poisons and disease regularly confirm that impairing brain activity impairs consciousness (Section 7.3). There is considerable evidence that brainstem activity helps to determine the level of consciousness and is essential for cortical function.

Modern research in this area started with the observation of patients who had a particular form of sleeping sickness in the aftermath of the influenza epidemic which followed the First World War. Victims of this disease, encephalitis lethargica, initially suffered from insomnia but as the disease progressed they fell into a deep sleep from which they could only briefly, and with difficulty, be aroused. Many died and at post-mortem there was a lesion in the midbrain in the grey matter adjacent to the nucleus of cranial nerve III and sometimes involving it; in those cases the patients had had a third-nerve palsy on the side of the lesion.

Repetition of this experiment of nature—by making lesions in the same region in animals—had the same effect as in man, the animals becoming somnolent and developing a synchronized EEG pattern. Furthermore, it was found that electrical stimulation in the midbrain grey matter would arouse normal sleeping animals, at the same time converting their EEG to the waking pattern.

Anatomical attention was then redirected to this region of the midbrain, which can be thought of as similar to the grey matter of the spinal cord. This, and similar regions in the pons and medulla, were collectively described by early brain anatomists as the *reticular formation*. Reticular means net-like, and refers to the finding of nets of cells (small brainstem nuclei) enmeshed in fibre tracts. The region has very many of these nuclei, only a few of which are large or distinctive enough to be named. One such is the *locus coeruleus* which is one of the noradrenergic-containing groups of cells found in the pontine reticular formation. Another recognized group comprises the *raphé nuclei* which lie in the midline in the pontine region.

Some reticular neurones lying largely in the medulla send descending axons to the spinal cord to excite or inhibit motor neurones. Others, largely in the pons and midbrain, send ascending axons to excite or inhibit thalamic cells. Some of the reticular neurones are very large, presumably because they have to support very long axons; some giant cells, for instance, have a bifurcating axon, one branch passing to the thalamus and the other to the spinal cord.

The afferent connections of the reticular formation are very widespread. It is possible to record from reticular nuclear cells and find that they are excited by skin, visual, auditory and olfactory inputs—a truly polysensory input. What can be the meaning of such a system? Clearly, it cannot convey precise information about a stimulus. It does however convey the message that a stimulus has occurred.

One modern interpretation of reticular function (the *reticular activation theory*) is that the signals from the reticular formation are used to vary the activity of the thalamic pacemakers, and thus cortical neurone excitability. The non-specific nuclei of the thalamus (including the intralaminar and anterior thalamic nuclei), some of which receive input from the reticular formation, project to wide areas of the cerebral cortex. Micro-electrode studies of the synaptic linkages show that the thalamic neurones of the non-specific system have two effects. First, through their axons, they excite cortical cells. Second, they have a complex interaction with the specific thalamic nuclei via axon collaterals to interneurones which in the drowsy state inhibit the neurones of the specific thalamic nuclei. Upon arousal or alerting, however, this inhibition is removed so that the specific thalamic nuclei can conduct their sensory information more easily to the cortex. One sign of this is that a bigger evoked potential is generated upon arousal. Incoming signals from arousing stimuli are also routed through the reticular formation to the hypothalamus.

7.2 Sleep

As one goes to sleep, the alpha rhythm, characteristic of the drowsy state, appears and is later replaced by low-amplitude theta waves of a slower frequency (4–6 Hz) which characterize stage 1 sleep (Fig. 7.2). As sleep becomes deeper (as judged by the difficulty of awakening the sleeper), occasional bursts of fast waves (12–15 Hz) called 'sleep spindles' are seen, signalling stage 2 sleep. Stage 3 is characterized by the presence of high-amplitude delta waves with a frequency of 1–2 Hz and the presence of K complexes (bursts of more rapid waves on top of delta waves). Stage 4 is characterized by a trace consisting almost entirely of large delta waves. Similar phenomena are seen in recordings from all mammals.

At times during sleep the EEG from the cortex shows a desynchronized pattern resembling that recorded in the waking state (Fig. 7.2). This shows

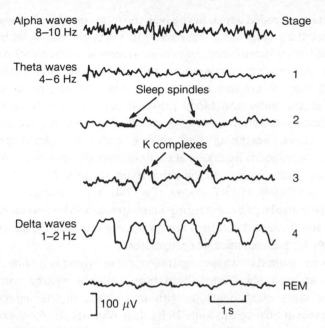

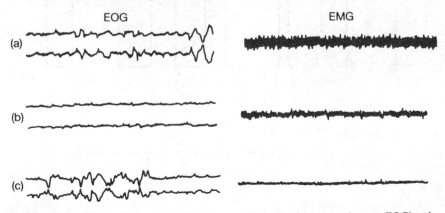

Fig. 7.2. Typical EEG traces recorded during the successive stages (1–4) of slow wave (SW) sleep, and during rapid eye movement (REM) sleep.

that the EEG alone is not a satisfactory indicator of the sleep state. It must be supplemented by recording muscle tone or eye movements. When people are awake their eyes are constantly moving (Fig. 7.3a) but during most of sleep eye movements are much reduced in amplitude and so too is muscle tone (Fig. 7.3b) though sudden jerks of limbs are common. In addition, during a particular phase of sleep, large-amplitude rapid (60–70 Hz) saccades in eye

Fig. 7.3. Records of electrical activity from extraocular muscles (electro-oculogram, EOG) and from limb muscles (electromyogram, EMG): (a) in the awake state, (b) during stage 2 of SW sleep and (c) during deep REM sleep. Note in REM sleep the characteristic activity in the EOG and the flat trace in the EMG.

movement are recorded which have given the name *rapid eye movement* (*REM*) *sleep* to this stage (Fig. 7.3c). Moreover, in sleep, the pupils of the eyes are constricted but sudden dilatation (*mydriasis*) may accompany REM.

Closer examination has shown that the REM state is accompanied by a number of signs of excitement and stress. For instance, there is a marked variation in the pulse and blood pressure and irregularity in respiration, accompanied by secretion of corticosteroids. Furthermore, a cycle of penile erections occurs, beginning and ending with each REM period, and testosterone secretion is increased at night in association with REM sleep.

It appears that we all spend some time every night in the two types of sleep, '*slow wave*' (SW) sleep and REM sleep (Fig. 7.4). Young adults normally spend some 16–18 h awake. The remaining hours are divided between the SW and the REM states, stage 4 coming early in the night and REM sleep becoming increasingly frequent as the night progresses.

Newborn animals, whose central nervous system is not completely developed at birth, show only alternations between waking and the REM state. Slow wave sleep develops with maturation of the nervous system. Human babies at birth spend some 18 h asleep of which 45–65% is in the REM state. A two-year-old is still spending 40% of his sleep-time in REM sleep, but a five-year-old approaches the young adult figure of 20%. For those over 50, the period spent in REM sleep forms about 15% of total sleep.

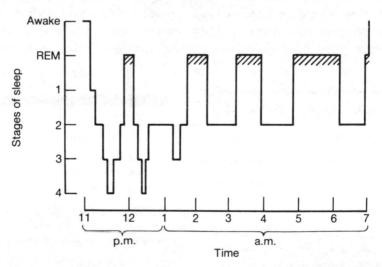

Fig. 7.4. The sleep cycle in a young man. The stages of sleep (ordinate) are judged from EEG criteria. Periods of REM sleep are indicated by cross-hatching. (From Hubbard, J.I. (1975) *The Biological Basis of Mental Activity*, p. 135. Addison–Wesley Publishing Co., Reading, Massachusetts.)

Sleep is a state of being, distinct from the waking state; it certainly does not resemble the unconsciousness resulting from brain damage.

REM sleep and dreaming

Subjects awakened from REM sleep almost invariably report that they have been dreaming. Apparently, we all dream although we do not always recall our dreams.

It is attractive to suppose that the eye movements, limb twitches and penile erections, together with the alternations of pulse, blood pressure and respiration, which occur during the REM state of sleep are associated with the acting-out of dreams. There is some evidence for this view largely drawn from correlation of the content of dreams and the observed signs. Dreams, of course, may be distressing and it is of interest that angina pectoris may also occur at this time while sufferers from duodenal ulcers and migraine, although not asthma, tend to have exacerbations of their symptoms during periods of REM sleep.

Mental activity also goes on during SW sleep. About 20% of subjects awakened from this state say they have been 'thinking' or even deny being asleep. Evidently, the mental activity is akin to everyday mental activity and lacks the hallucinatory aspect of dreaming. Though it has been suggested that learning is possible during sleep, research shows that this may occur only when the EEG shows alpha waves.

Neural basis of sleep

We might assume that if the reticular formation were inactive then an animal would go to sleep, and indeed radical deafferentation does make an animal somnolent with a synchronized EEG for a week or so following the operation. Thus, sleep has been attributed to the absence of arousing stimuli to the reticular formation and therefore to thalamus and cortex. However, sleep–waking rhythms reappear after a few weeks. Moreover, in man, removal of all external stimuli leads to hallucinations rather than sleep. More probably, sleep is produced by active inhibition of that part of the reticular formation responsible for arousal. The Swiss physiologist Hess (1881–1973) showed that there are a number of sites in the brain which, if stimulated through implanted electrodes in a conscious animal, will induce the animal to go to sleep in a natural manner, and will also evoke the appropriate synchronized EEG. The best characterized of these sites lies in the forebrain in the region rostral to the hypothalamus between the anterior commissure and the optic chiasma—the *preoptic region*. Lesions here make animals permanently sleepless and the region contains a population of neurones which discharge maximally in both SW and REM sleep.

The preoptic region has substantial efferent projections to the midbrain, particularly to its reticular formation. These are known to be predominantly inhibitory, suggesting that the preoptic region, as well as inducing the behavioural and EEG signs of sleep, also puts the machinery for wakefulness and attention out of action.

The need for sleep

The majority of people sleep 7–8 h a night but there are also short (5–6 h) and long (8–9 h) sleepers. When compared with long sleepers, short sleepers go to sleep more quickly, spend less time in stages 1 and 2 and in REM sleep but spend the same time in the deepest stages (3 and 4) of SW sleep. The working hypothesis of many investigators is that SW sleep is needed for bodily repair. There is evidence, for instance, that marathon runners have more SW sleep after their race than before. It is well established that growth hormone is only secreted during SW sleep. The idea has also been investigated that REM sleep, on the other hand, is concerned with repair of mental tiredness. There are interesting differences between the psychological profiles of long and short sleepers which support the hypothesis. For instance, psychological tests indicate that long sleepers tend to be significantly more introverted, anxious and depressed than short sleepers.

Sleep disorders

Sleep-walking (*somnambulism*) and bed-wetting (*nocturnal enuresis*) have been shown to occur during periods of arousal from SW sleep. They are not associated with REM sleep. Episodes of sleep-walking are more common in children than in adults and occur predominantly in males. They may last several minutes. Somnambulists walk with their eyes open and avoid obstacles, but when awakened they cannot recall the episode.

Narcolepsy is a not uncommon disease of unknown cause in which there is an eventually irresistible urge to sleep during daytime activities. In some cases, it has been shown to start with the sudden onset of REM sleep, whereas REM sleep almost never occurs without previous SW sleep in normal individuals.

Some adults have *sleep apnoea*—a condition in which breathing ceases completely a number of times during the night. When this happens, the subject wakes, takes a few breaths and falls asleep. The process may repeat itself, sometimes hundreds of times a night. Patients then complain of being very sleepy in the daytime. There are two sorts of patient: those in whom the lesion is in the central nervous system, and those in whom the lesion involves some obstruction in the airways.

Another interesting disorder is known as *nocturnal myoclonus*. Patients with this disorder show sudden repeated contractions of muscle groups, commonly of the legs, sometimes of the head, during sleep. The disease is thought to be akin to epilepsy.

7.3 Disturbances of consciousness

Damage to the midbrain, pons, medulla or cortex may result in unconsciousness due perhaps to disruption of the reticular activation system. Such unconsciousness may be transitory, in which case it may be a form of epilepsy or of concussion. *Concussion* implies a brief loss of consciousness due to head and brain injury. Often brain function is impaired for several hours after the patient apparently returns to consciousness so that he or she may later not remember anything that happened during that time (*post-traumatic amnesia*). Alternatively, unconsciousness may be prolonged, in which case it is known as *coma* and the patient is said to be *comatose*. The term 'coma' implies a loss of consciousness, which can be distinguished from sleep in that the subject cannot be totally aroused by strong external stimulation. Coma implies failure of either the association areas of both cerebral hemispheres, if the damage is very extensive, or of the ascending reticular formation of the brainstem and diencephalon, the structures which keep the cortex 'awake'. Therefore, damage to a single cerebral hemisphere does not produce coma unless it secondarily affects the brainstem reticular formation.

Disturbances of brain-cell metabolism, as in hyper- and hypoglycaemia, anoxia or following a drug overdose, may produce coma by affecting both the brainstem and the cerebral cortex. Most forms of coma are accompanied by a fall in the oxygen consumption of the brain. For instance, the blood supply may fall to 60% of normal during anaesthesia and to about 50% of normal in a diabetic coma.

Consciousness may be graded, as the recovery from concussion implies. If the patient can speak, it is usual to find out whether he knows who he is, where he is and what time it is. A patient who fails such tests is *disoriented*. Patients may be unconscious but still be aroused by a shout. The patient's eyes may then focus on the source of the stimulation. A very deeply unconscious patient may be unresponsive to anything save a painful stimulus to which he may react with grimaces, grunts, movements of arms and legs, and an increase in pulse rate. Such tests provide a rough measure of the level of failure of brain function.

Epilepsy

Epilepsy may be defined as a recurrent, paroxysmal, transitory disturbance of the central nervous system that is characterized by uncontrolled neural discharge. Epileptic seizures generally involve total or partial loss of consciousness and may be accompanied by uncontrolled motor reactions.

Epilepsy may be classified as generalized or focal in origin. *Generalized* or *centrencephalic* epilepsy is a term used to cover a very large group of patients in whom no cause for the disorder can be found. The term centrencephalic refers to the theory that generalized epilepsy has its origin in disorders of the activating system in the brainstem or thalamus. *Focal* epilepsy arises when the seizures begin in a localized area of cerebral cortex.

Where epileptic seizures start is of great interest in that a focal origin may be treatable. Both cortical and subcortical (limbic and reticular) foci have been found but in many patients there are no signs or symptoms indicative of a focus. However, the finding that in such patients the epilepsy may be successfully treated by cutting the corpus callosum suggests they may also have a focus in one hemisphere from which action potentials spread across the corpus callosum to the other hemisphere.

Two varieties of generalized epilepsy are relatively common—grand mal and petit mal. *Grand mal* seizures are characterized by an abrupt loss of consciousness and violent involuntary contractions of skeletal muscles. Together these phenomena are termed a *convulsion*. Many patients have mild symptoms which precede the attack, the *aura*. Many forms of aura have been described—tingling or numbness in the limbs, visual or auditory hallucinations or sudden emotional changes, such as fear. The aura is followed by the convulsion and is usually the last thing the patient remembers. The convulsion has several prominent phases. First, the patient stiffens and is apparently thrown to the floor, respiration stops and the pupils dilate. This is the *tonic phase* lasting 10–30 s in which the patient often lies with the legs fully extended and the arms flexed and abducted as if decorticate. It is followed by a *clonic phase* with severe jerking movements and, commonly, emptying of the bladder. After the convulsion, the patient appears relaxed, drowsy and may complain of headache. This stage lasts 30–60 min and is often followed by sleep. During a seizure, the EEG usually shows high-voltage spiking during the tonic phase which during the clonic phase becomes mixed with a high-voltage slow component (Fig. 7.5a). Electroconvulsive shock therapy (ECT) produces convulsions of a grand-mal type but the motor components are in practice prevented by anaesthetizing the patient and administering a muscle relaxant.

Petit mal usually affects children rather than adults; there is no warning, no aura, but brief periods of unconsciousness or altered consciousness during which the patient may stare blankly; the eyes may roll upwards until the pupils

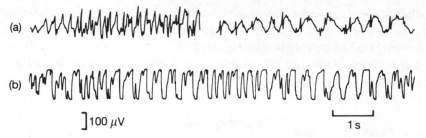

Fig. 7.5. EEG during epileptic seizures. (a) Tonic (left) and clonic (right) phases of grand-mal seizure; (b) petit-mal seizure.

are hidden under the lid. Episodes last 5–30 s and a patient may stop what he or she is doing and restart after the seizure without apparently being aware of what has happened. The EEG of petit mal shows a characteristic wave-form (Fig. 7.5b). There is an alternation between high-voltage waves of short and of long duration (spike and dome complex).

When seizures begin in a localized area of the cerebral cortex, *focal* epilepsy results. The commonest form in adults is temporal lobe or *psychomotor* epilepsy. Here the epileptic focus is located in the temporal lobe. The seizure is characterized by automatic purposeful reactions such as chewing, and smacking the lips. There are no convulsions and the patient may or may not lose consciousness. It is important to remember that focal seizures have recognizable underlying pathology whereas generalized seizures may not.

Jacksonian epilepsy, named after Hughlings Jackson (1835–1911) who first clearly described it, is a term used to describe focal motor seizures beginning in the motor areas of the cerebral cortex. Such seizures usually begin with a twiching, most often of the thumb or of a finger, a toe or the angle of the mouth. These regions are the most widely represented in the motor cortex. Thereafter, the 'march' of the seizure is that expected if the lesion had radiated out from these areas over the rest of the motor cortex. Focal sensory seizures are also known.

The classification of epilepsy has recently changed to take account of the finding that many patients with recurrent paroxysmal transitory disturbances of the CNS do not fall into the traditional categories. The international system now in use concentrates on the patient's signs and symptoms, grouping them as follows.

1 Generalized seizures. Patients in this class have bilateral signs and symptoms without local onset. Classical grand mal and petit mal both fall into this class.
2 (a) Partial seizures. Patients in this class have their signs and symptoms beginning locally. Jacksonian epilepsy falls into this class. (b) Partial complex seizures. Patients in this class have impaired consciousness, complex

hallucinations and may demonstrate automatism (a series of apparently purposive movements of which the patient is unaware) during the seizure. Temporal lobe epilepsy falls into this class.

3 Unclassified. Patients fall into this class if there are incomplete data for classification into classes 1 and 2.

7.4 Learning and memory

When we talk about memories or remembering we use the word in a variety of senses. We may say 'I remember how to ride a bicycle', 'I remember that picture', 'I recall that song'. Research has shown that each of these memories is of a different sort and relies on different brain mechanisms. 'Motor' memories, such as the ability to ride a bicycle, have not been much studied but it is clear that they may survive and be added to when other types of memory are badly impaired. The 'recognition' and 'recall' types of memory have been studied in human subjects mostly by means of verbal material and in animals by use of recognition tasks such as the learning of mazes. It has become clear that these types of memory are not akin to a photographic record but that what is remembered is a reconstruction based as much on what was expected as on what actually happened. Accordingly, witnesses are notoriously fallible.

People with so-called 'photographic memory' can recall at will an image of what they have seen, apparently localized in space in front of their eyes. The image is known as an *eidetic image*. Detailed investigation suggests that eidetic imagery is an alternative means of storing information, distinct from, but not more effective than, ordinary memory. Nothing is presently known of the physiological basis of this ability.

Studies of the normal type of memory for verbal material indicate that there are short-term 'primary' memories for snippets of information looked up immediately before use (e.g. telephone numbers), longer term ('secondary') memories of much-rehearsed material, and memories which last a lifetime ('tertiary' memory), such as one's name and important life-events. Recently, after-images have been brought into this scheme, it being realized that they are epiphenomena of the initial step in memory (sensory memory).

Sensory memory refers to that brief period when information has reached receptors and is about to be sent on to the central nervous system. Information presented to the eyes can be evaluated and gives rise to reflex responses before it is 'perceived'. The after-images seen when one looks at an object and then away from it may last about 250 ms, decaying over this period. Apparently, the memory trace is destroyed by fresh incoming information. The capacity of the visual sensory memory for after-images appears to extend to about sixteen or seventeen items. A sensory memory has also been demonstrated in the auditory system.

It seems obvious, although it is still a matter of much investigation and dispute, that for material to be remembered it must have been sequentially in sensory, primary, secondary and perhaps tertiary memory.

Characteristics of short-term (primary) memory

These are (1) small capacity, (2) short duration, and (3) storage as words.

1 Relatively few items can be recalled immediately after a short exposure (visual or aural). For instance, in a popular experimental method in which photographs, printed letters or drawings are flashed on a screen for very short, precisely controlled intervals of time (tachistoscopic experiments) subjects can recall only 5–9 items although the sample may be made up of material learnt for examination and remembered for a few days.

2 In this type of experiment, primary memory decays in a few seconds.

3 A study of errors in primary memory suggests that the material is coded in words. Volunteers in tachistoscopic experiments were found to be rehearsing the material they had learned, by mumbling either overtly or covertly. In remembering, mistakes were made between letters which sounded alike. The sounds were confused although the subjects were shown only the letters. The same mistakes were made, in fact, if the subjects were shown the material or if it was spoken. Thus, mistaking 'cat' for 'hat' was possible, but not 'cat' for 'kitten'.

Primary memory is thought to be represented in the nervous system by circulating nerve impulses, because an insult which disrupts the functioning of neurones prevents any memory of events which take place a short time before. Thus, after head injury causing loss of consciousness or after electroconvulsive thereapy, there is a short but persistent loss of memory (*amnesia*) for the period immediately preceding. A permanent reduction in the capacity of short-term memory may occur after lesions in the cerebral cortex. For instance a reduction in verbal and short-term memory has been described in a patient with a parietal lobe lesion.

Characteristics of long-term (secondary and tertiary) memory

These are (1) large capacity, (2) long duration, and (3) organization.

1 The storage capacity of secondary and tertiary memory is presumably very large though no quantitative estimates are available. There are accounts of people with phenomenal memories, perhaps the most sensational being that of a man observed by Luria (1902–1977), over a 30-year period. This man had no obvious limits to the capacity of his secondary memory for he could never be given a list of words so long that he could not recall it perfectly.

2 People vary greatly in their ability to retain secondary memories. For instance, Luria's subject could reproduce any lengthy series of words even if, as was several times the case, the test was held 15 or 16 years after the man originally learned the words. His memory was such, indeed, that he could recall the place where the test was first presented to him, how Luria was dressed, where he sat, and so on. It appears that tertiary memory may be permanent, in that some information is retained for the life of the subject. Often this information has a very strong emotional association.

3 When experiments of this type are done with secondary memory, confusion has a semantic rather than an acoustic basis; that is, words which have a similar meaning are confused rather than words with a similar sound, implying that the organization is semantic and relational. This was utilized by classical orators, training themselves in the art of memory. They were advised to imprint on the memory first a visual picture containing a series of identifiable places, commonly a building with its rooms and ornaments. In each place was then put, in imagination, a mark or pointer to a particular part of the speech. When it came to making the speech, the orator using his visual memory simply went through the building in an orderly fashion picking up each pointer as he went.

Secondary memory is notoriously difficult of access. There is much evidence that interference from material learnt before (*proactive inhibition*) or afterwards (*retroactive inhibition*) produces forgetting. Proactive inhibition appears to be the more troublesome. Tertiary memory appears to be easy of access.

Secondary and tertiary memories are thought to be represented in the actual structure or pattern of the CNS for they remain after neural insults which have destroyed primary memories. Nevertheless, damage to certain parts of the brain does affect long-term memory, perhaps by destroying the mechanisms by which material to be remembered is recovered or laid down.

Neurological basis of memory

Memory is disordered by injury to those structures seen as forming a fringe on the medial side of each hemisphere, around the brainstem, hence the name 'limbic' system (Fig. 7.6a, b). The structures concerned are many synapses away from the primary sensory or motor pathways and receive information from the overlying cerebral cortex which, after processing, is directed back to the cortex. The *limbic system* comprises parts of either frontal lobe, particularly the cingulate gyrus, which lies about the corpus callosum (Fig. 7.6a), the hippocampus and septum (Fig. 7.6a, b), the amygdaloid nuclei, the hypothalamus and anterior thalamic nucleus. In man, the hippocampus lies on the medial wall of the temporal lobe bordering the inferior horn of the lateral

(a)

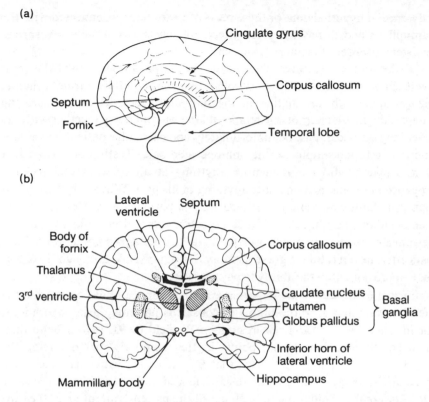

(b)

Fig. 7.6. The limbic system of the brain shown in (a), the medial surface of the right cerebral hemisphere, and (b) a frontal section through the junction of the 3rd and lateral ventricles. See text for further explanation.

ventricle. It is a layered structure like the cerebral cortex but of a simpler cellular pattern. The amygdaloid nucleus lies in the same region just in front of the hippocampus. The septal nuclei lie in the midline anterior to the hypothalamus between the lateral ventricles and under (ventral to) the corpus callosum, the genu of which forms their anterior boundary (Fig. 7.6a). A large fibre bundle known as the fornix takes origin from the hippocampus and overlying cortex on each side (Fig. 7.6a, b). This bundle arches forward underneath the corpus callosum accompanied by a similar but much smaller bundle of axons (stria terminalis) from the amygdaloid nucleus. The fornix and stria terminate in the septum and hypothalamus, and the fornix also terminates in the mammillary bodies of the hypothalamus which in turn project to the anterior thalamic nuclei. These in turn project back to the cerebral cortex.

The integrity of the limbic system is essential for long-term memory. Bilateral damage to certain parts of the hippocampus, fornix, mammillary body, medial hypothalamus or those parts of the thalamus connected with the mammillary bodies has the same effect, namely a remarkably severe and persistent disorder of memory (amnesia).

Of the limbic structures, damage to the hippocampus causes the most severe amnesia. There are reports of patients who suffered from epileptic seizures apparently originating in the temporal lobes. To try to cure the epilepsy, the medial part of one temporal lobe was cut away together with the underlying hippocampus, but unfortunately one particular patient had a non-functioning hippocampus on the non-operated side. Testing of this patient showed that he did not remember anything at all of what had recently happened nor could he remember anything of his past. When asked about his house, his family or the city he lived in, the patient did not even seem to understand these questions, as if their objective was entirely unknown to him. He appeared to feel completely isolated, with neither a past nor a future. Four years after the operation, a grave memory defect persisted although there was some improvement in the ability to pay attention.

Further evidence for the importance of the hippocampus comes from the effects of stimulation of the brains of conscious patients during operations. For instance, the Canadian neurosurgeon Penfield (1891–1976), found that, when he stimulated particular points in the temporal lobes overlying the hippocampus, his patients reported that fragments of past experiences seen, heard or felt long ago were being re-experienced and, moreover, at the same rate as originally. Things not in the focus of the patient's attention at the time of the original experience were not present in the hallucinations. Many of these patients suffered from epilepsy. Their attacks were preceded by 'flashbacks' consisting of the recall of experiences from their past similar to those evoked by brain stimulation. It is important to note that patients reported that the material either elicited by stimulation or coming into consciousness before an epileptic attack was much more distinct than anything they could normally recall. Penfield was convinced that his patients had records of their experiences in their brain which were normally not completely available to them.

Analysis of many case histories confirms that patients with damage to the hippocampal system show no loss of motor skills acquired preoperatively and that intelligence, as measured by formal tests, is unimpaired. But, with the possible exception of the acquiring of motor skills, they seem largely incapable of adding new information to their long-term store. This type of deficit is termed *anterograde amnesia*. The immediate registration of new input (short-term memory) appears to take place normally and material that can be rehearsed verbally is held for many minutes. Interruption of the rehearsal, however, produces immediate forgetting of what has gone before. Material already in long-term storage is unaffected by the lesion. A similar type of anterograde amnesia is shown by some chronic alcoholics (*Korsakoff's psychosis*).

These striking observations might suggest that each hippocampus is a site at which memories are stored. However, this cannot be completely true, because patients with bilateral hippocampal damage can acquire motor skills and also verbal long-term memories by special techniques in which they are given part of the information but have to supply the rest. In both types of learning, they are as effective as normal subjects. Memory formation is now thought to require concurrent activity in the hippocampus and other brain sites.

Hippocampal lesions may prevent that suppression of old information which is apparently needed for the storage of new information. After bilateral removal of the hippocampus, rats and monkeys react to new stimuli more sluggishly. Furthermore, they are very slow to reverse their responses in the type of study in which the subject has to respond to one of two situations, e.g. the right or the left. They also continue to respond for much longer in the absence of a reward than do normal animals.

Recently, a phenomenon known as *long-term potentiation* (*LTP*) has been described in the hippocampus, cerebellum and cerebral cortex which may be the cellular mechanism of memory. Long-term potentiation was first discovered following stimulation of one of the inputs to the granule cells of the hippocampus (*the perforant pathway*). This pathway normally excites a fraction of the granule cells. Following a number of trains of conditioning stimuli at frequencies of 20–100 Hz, the number of granule cells responding to single testing stimuli of this same pathway was increased from hours to days, depending on the number and duration of the conditioning trains.

Clinical experience also suggests that suppression of memory occurs, for example, following ECT. After treatment many patients forget their anxieties and are able to function normally again, although when interviewed later under the influence of a barbiturate they can recall their troubles, showing that these were not permanently forgotten but simply not recalled in the normal waking state.

Again, following head injuries, patients often lose their memory of events which occurred over a considerable period prior to the injury (*retrograde amnesia*), a much longer period than would normally be expected with failure of only short-term memory. The events before the accident are often very dramatic and must have been registered by the nervous system, yet quite often the subject has no memory of them for up to an hour before injury. During recovery, the length of the period of retrograde amnesia shrinks markedly, and, interestingly, association of ideas may aid this process as shown by the case of a soldier described by Russell in 1959, whose last memory was of setting out on a journey, driving a truck in the dark more than an hour before his accident. Some months later, in the cinema he saw a plane crash, with the usual noisy sound track. This noise suddenly brought back to him the noise his truck made as he crashed. We must assume then that, as well as the retrograde amnesia, due to failure of the nervous system to register as a result of injury, there was also amnesia due to suppression of presumably unpleasant information.

7.5 Speech

Speech, together with its associated activities (reading and writing), is a most complex phenomenon. Basically, it is the conveying of meaning by means of symbols, usually spoken, sometimes written. It is necessary to distinguish the production of voice (*phonation*), the shaping of voice into words (*articulation*), and speech proper, which implies meaning and understanding. In one sense, speech is a motor activity involving control of the labial, lingual, pharyngeal, palatal and respiratory muscles. Motor activities in general result from plans made in the frontal lobe, which are further elaborated by the basal ganglia and cerebellum and relayed to the ventrolateral thalamus. A final relay to the motor cortex brings appropriate motor neurones and thus muscles into action. The motor aspects of speech do not differ from other motor activities except in the planning stage, which is complex.

It has long been known that speech is unique amongst motor activities in that its higher control depends on the integrity of only one of the two cerebral hemispheres, the *dominant hemisphere*, which, in most people, is the left. This hemisphere also controls the right hand but cerebral dominance is not necessarily paralleled by appropriate handedness. Tests have shown that 95% of right-handed and 50% of left-handed people have their speech centre on the left. The production of speech thus involves subcortical and motor cortical components and also a component from association areas. Disturbance of any component will cause a difficulty in speaking. Such difficulties are termed *dysarthrias* if they involve the motor apparatus in the strict sense and *aphasias* if they involve the unique one-sided cerebral control.

A characteristic *dysarthria* occurs in many diseases, such as the scanning speech of cerebellar disease and the slow, slurred speech of Parkinson's disease. While there may be some temporary disturbance of articulation, a permanent dysarthria does not commonly occur with the hemiplegia following a cerebral vascular accident because the muscles of the palate, pharynx and vocal cords are bilaterally innervated through the vagus and glossopharyngeal nerves. Bilateral loss of cerebral control of the medullary cranial nerves, on the other hand, causes a severe dysarthria as well as numerous other signs and symptoms (*pseudobulbar palsy*).

Following brain lesions in the areas shown in Fig. 7.7, patients have a difficulty in speaking which differs from dysarthria. There is no impairment of the muscular apparatus or its control. The difficulty is in the higher processes connected with the selection and ordering of words. This type of disturbance is commonly called an aphasia, which literally means 'without speech', but, strickly speaking, such patients should be said to have a *dysphasia* since they can speak although with difficulty.

Two major and many minor forms of aphasia have been described. The commonest, Broca's or *motor* aphasia, is named after the French physician, Paul Broca (1824–1880), who noticed in 1861 that patients with a right-sided hemiplegia had a poorly articulated speech produced slowly and with great effort and abnormal in rhythm and intonation. For these reasons, it is sometimes termed a *non-fluent aphasia*. The speech content has been compared to a telegram in that patients can name objects and produce single words but have difficulty with sentences, coming to grief on the connecting words. These patients have trouble in reading and find writing difficult or impossible. Their comprehension appears unimpaired.

In 1874, a German neurologist, Wernicke (1848–1905), described patients with normal articulation, rhythm and expression, and correct grammar who had great difficulty in finding the correct words. The patients were unable to understand spoken or written language, that is, unlike the patients with motor aphasia, their comprehension was impaired. In severe cases, they might produce a grammatically correct but meaningless flow of language. This form of aphasia is term Wernicke's or *sensory aphasia*. At post-mortem, such patients are found to have lesions involving the posterior part of the temporal lobe (Fig. 7.7).

Wernicke's and Broca's areas are joined by a bundle of nerve fibres called the arcuate fasciculus. Patients with lesions in this structure have *conduction aphasia*. Such patients can comprehend the speech of others but cannot repeat it. Their own speech is fluent but, like Wernicke's aphasia, full of circumlocutions and of errors in word use.

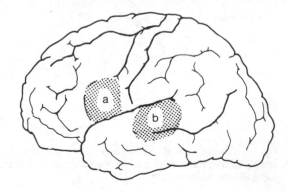

Fig. 7.7. A lateral view of a human brain showing areas important for the generation of speech. (a) Broca's area; damage here may cause motor aphasia. (b) Wernicke's area; damage here may cause sensory aphasia.

More restrictive language defects arise when the speech area is disconnected from the cortical representation of one particular modality. In these cases, a patient may be able to name an object presented in one sensory modality but not when it is presented in another (*anomic aphasia*). Speech itself is fluent and comprehension intact.

Emotional speech. The sounds made by animals as the result of strong emotion are akin to the expressions made by man in similar excited states. Such sounds are controlled by a different part of the brain from the part of the brain involved in true speech. A patient may lose true speech but retain emotional speech and vice versa. The production of such sounds is controlled from the limbic system in animals. Stimulation here in rhesus and squirrel monkeys can produce every sound these animals make. There is no sidedness to this phenomenon; one can evoke the same range of sounds from right or left limbic stimulation. In man, the control of emotional speech is not well understood. There is some evidence that lesions in the right (non-dominant) hemisphere may cause its loss.

7.6 Organization of behaviour

Control of goal-directed behaviour

That part of the frontal lobe anterior to the premotor area and extending to the orbital surface (prefrontal cortex, Fig. 7.8) is thought to be the site of integration of the information about the external world, sent to the cortex along the visual, auditory and somatosensory pathways, with the information about the internal world relayed from the hypothalamus and the limbic system. Behaviour in turn is thought to reflect the integration of this information within the prefrontal cortex.

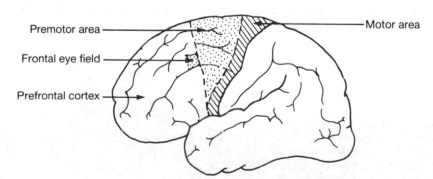

Fig. 7.8. A lateral view of a human brain showing the functional subdivisions of the frontal lobe, namely the motor area most caudally, the premotor area and frontal eye fields and the prefrontal cortex most rostrally.

This role of the prefrontal cortex was first suspected in the nineteenth century as a result of a study on Phineas Gage, a railroad foreman injured in an explosion when an iron bar penetrated his skull below the left orbit and emerged in the midline vertex. He survived despite (as a post-mortem later showed) the severing of all connections between the prefrontal cortex and the rest of the brain, but was described as being changed from being a careful, conscientious, mild-mannered man to a facile, careless and irritable drunkard. He was said to be impatient of restraint or advice when it conflicted with his desires. He was obstinate yet capricious, devising many plans for future activities which were no sooner arranged than they were abandoned for others that appeared more feasible.

In 1930, Fulton described the effects in the chimpanzee of cutting the connections of the prefrontal cortex with the rest of the brain. The animals showed a significant reduction in emotional responses to failure and stress, and, where previously aggressive, became tame. A similar operation (prefrontal leucotomy or lobotomy) thereupon became popular for treatment of severe depression and, initially, of schizophrenia. Following such lesions of the prefrontal cortex in man, speech, movement, vision, hearing and intellectual skills appear normal. Behaviour is, however, abnormal in subtle ways.

The more obvious changes are in mood. For instance, in one series of thirty-two patients, twenty became euphoric. Those patients in whom this trait was most prominent were spontaneously happy and contented no matter what their circumstances, laughing and joking upon the slightest provocation. In other cases, particularly when patients were very shy before operation, they became calmer, found it easier to meet other people and were less embarrassed. A few patients who were very depressed for years before operation did not become euphoric but became even more despondent.

Intellectual changes after prefrontal lobotomy are difficult to detect by ordinary psychological tests. Patients have difficulty in remembering anything for more than 2–3 h (Korsakoff's psychosis) and there are signs of increased proactive inhibition, that is, a memory trace is not replaced by the next trace at the usual speed. One such sign is inability to follow instructions involving constant changes. For instance, when told to draw a circle, the patient will do so, but when then asked to draw a cross he is likely to repeat the circle (*perseveration*). Another sign is the inability to carry out operations with both right and left hands, e.g. pushing a button with the left hand when a green light is seen or with the right hand if a red light is seen. Such tasks are done correctly for a few trials but the responses soon become random although the patient is found to understand the task. Reading and writing are not affected but there is a weakening of more complicated intellectual function as is demonstrated by the less satisfactory outcome of the operation in patients with more complex jobs.

In man, there is an inability to concentrate for long periods on a task, and animals become unable to perform delayed-response tests. This may be categorized as the loss of goal-directed behaviour. There is loss of social restraints, together with loss of the normal ability to evaluate one's own actions and to profit from mistakes. Superficially, the patients appear satisfactory after operation but, upon questioning in detail, reveal that they cannot feel deep happiness or deep sorrow, that they forget things, and that they have lost most of their interests because they cannot concentrate on any one topic. Moreover, close relatives are convinced that there has been a personality change. Interestingly, patients with seizures confined to the prefrontal cortex also show associated transient personality changes.

Emotion

Fundamentally, we do not know the mechanisms by which we become aware of emotional tone. It is a convenient working hypothesis to consider the *limbic system* as the site at which emotional tone and its expression are generated. In man, stimulation of limbic structures may enhance and lesions reduce emotional tone. For instance, agressive behaviour has been reported in a patient during stimulation of one of the amygdaloid nuclei through implanted electrodes, while bilateral amygdalectomy has resulted in an apathetic individual. Surgeons who attempt to treat psychiatric disorders nowadays use stereotactic techniques to make very localized brain lesions. Undercutting of the cingulate gyrus (Fig. 7.6a) (cingulotractotomy) has been used in severe chronic depression which does not respond to any other form of treatment. Enthusiasm for such operations should be tempered by the possibility that the improvement may not be permanent. In some studies, half of the patients initially improved by cingulotractotomy relapsed within 15–32 months of the operation and were left with permanent brain damage.

In contrast to the feeling of emotion, it is known in some detail how emotions are expressed through mechanisms involving parts of the frontal lobes of the cortex, the limbic system, the hypothalamus and the autonomic and somatic systems controlled by the hypothalamus. We can distinguish two basic states: excitement and alarm or worry. Superimposed on these basic states are additional groups of signs which mark off particular expressions of the emotions. Thus, in the first category, we can further distinguish anger, joy, lust, and in the second, worry, depression, anxiety or fear.

Darwin, in his classic book *The Expression of the Emotions in Animals and Man*, pointed out the special signs—the blushing of embarrassment, the pale face of anger and fear, the dilated pupils and the flared nostrils of excitement—by which we recognize excited or alarmed states not only in each other but in our closest relations in the evolutionary sense, the great apes.

Motivation

Animal studies have shown that electrical stimulation through implanted electrodes can be used as a reward for appropriate behaviour. Olds and Milner (1954) found that rats thus prepared could be trained to press a lever which turned on stimulating current. Regions of the brain could be classified according to the percentage of time the animals spent pressing the lever. Brain areas with high scores were found in the septal nuclei and to a smaller extent in the cingulate cortex and the hypothalamus. Self-stimulation of the hippocampus occurred less often but was still at rates above control levels. The most remarkable area was the *median forebrain bundle* in the lateral hypothalamic area which, among its other functions, connects the hypothalamic nuclei with the septal nuclei rostrally and with the midbrain caudally. If a rat had a stimulating electrode implanted there, it continually pressed the bar, suggesting that the ensuing brain stimulation was experienced as a reward. Indeed, given a choice between feeding and pressing, the rat would rather stimulate its median forebrain bundle than eat. Some brain areas were non-rewarding, in that animals avoided stimulating such areas.

A number of questions are raised by these studies, notably the meaning of 'reward' in this context. Similar results have been obtained in cats, dogs, monkeys and man. Studies on man are few but have particular value, for only in man can the meaning of 'reward' be investigated. The reports of Heath and his co-workers (1963) indicate that 'rewarding' brain stimulation is pleasurable. In a patient provided with a small stimulator connected to several electrodes implanted in his brain, the highest score was recorded from one particular point in the septal region, which the patient declared made him feel 'good'. He felt as if he was building up to a sexual orgasm although he was not able to reach the end-point, and often felt impatient and anxious.

Subcortical control of behaviour

Since 1897, when Goltz (1834–1902) demonstrated that a decorticate dog still showed its full range of emotional responses, it has been realized that emotional expression must, to some extent, be organized subcortically. The *hypothalamus* is the highest level of the nervous system needed for the display of emotional, sexual and gustatory behaviour, and hypothalamic stimulation can evoke this behaviour. Interestingly, acting-out of the behaviour outlasts the stimulation, which appears to be only a trigger. Furthermore, the hypothalamic neurones control not only autonomic neurones and the pituitary but must also have access to motor neurones, probably through reticulospinal pathways originating in the midbrain. Certainly, it is possible by midbrain stimulation to evoke behavioural acts but these displays differ from those evoked by hypothalamic stimulation in that they do not outlast the stimulation.

Limbic structures control and organize hypothalamic functions so that behaviours are expressed in appropriate situations and ways. For instance, the amygdaloid nuclei appear to have, as one function, control of what is eaten and when. Electrical stimulation of the medial amygdala in cats, dogs and monkeys facilitates eating, while stimulation of the lateral amygdala prevents eating even in starved animals. After lesions of the amygdala, animals and humans will put anything in their mouths. Rats, normally very conservative eaters, will for instance eat unfamiliar foods and even food which had made them ill in the past and which they had previously avoided. The septum, too, appears to have functions connected with coupling of food intake to metabolism and the regulation of taste preferences. Normally, when rats exercise, they eat appropriately to maintain body weight and drink enough to wash their food down. After septal lesions, increased activity leads to persistent weight loss and water intake becomes abnormally high, remaining so even if the animal is deprived of food. Further, the rats show an exaggeration of taste preferences as if some inhibitory control had been removed.

Role of the corpus callosum

The corpus callosum (Fig. 7.6a, b) is a large bundle of, for the most part, myelinated axons which have their origin in one cerebral hemisphere and pass to equivalent points as well as to other areas in the opposite hemisphere. All regions, except the primary visual, primary acoustic and foot and hand areas of the pre- and postcentral gyrus, are linked in this way.

The function of this largest of commissures was unsuspected even after its section became a treatment for otherwise intractable epilepsy. On clinical testing, the patients behaved normally. Sperry and his colleagues in the 1960s clarified the function of the corpus callosum when they devised techniques which tested each hemisphere separately. For instance, a cat, after section of the optic chiasma, had one eye blindfolded and was taught that, if it pushed a lever when it identified one pattern on a screen rather than another, it would be rewarded. If the blindfold was then put on the other eye, the cat performed as well as it did, after training, with the first eye. The information had clearly been 'transferred' between the hemispheres (Fig. 7.9). This conclusion was supported by repeating the experiment in another cat in which the corpus callosum had been sectioned together with the optic chiasma. Now when the blindfold was placed on the trained eye no transfer could be demonstrated.

The corpus callosum thus transfers information and this transfer can be of acoustic and somatosensory as well as of visual information. However, not all interhemispheric transfer is through the corpus callosum. Discrimination of brightness is transferred through midbrain commissures and the anterior commissure can also serve as a link for certain functions.

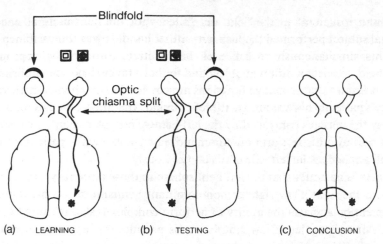

Fig. 7.9. Experiments demonstrating the role of the corpus callosum in transfer of visual information between the hemispheres. A visual discrimination task is learned through one eye and the animal retested through the naïve eye. (a) Animal with transection of the optic chiasm learns the task. (b) Testing with the naïve eye shows that the left hemisphere knows the task. (c) Conclusion is that information is transferred through the corpus callosum to the naïve eye. (From Cohen, D.H. & Sherman, S.M. (1983) The cerebral cortex. In *Physiology*, eds R.M. Berne & M.N. Levy, pp. 288–313. The C.V. Mosby Co., St. Louis, Mass.)

The question can be asked 'is the memory of a task laid down on the learning side only, so that the other side consults it through the corpus callosum as needed, or does the memory get laid down on both sides via the corpus callosum'? This can be investigated by teaching a cat a pattern discrimination, as before through one eye with the other blindfolded after section of the optic chiasma, but now only cutting the corpus callosum just before testing the other (naïve) eye. It is found that cats have laid down a memory on both sides so that the pattern is recognized through the naïve eye. However, when the identical experiment is carried out in a monkey, the hemisphere served by the naïve eye proves to be ignorant of the task. In this animal, memories are apparently laid down only on one side.

Sperry and his team next examined patients with a sectioned corpus callosum. These patients were examined by a procedure which enabled information to be confined to the visual field of one hemisphere. The sitting patients fixated on a marked point on a screen. In one type of experiment, a picture of an object was projected onto the screen to the left or right of the fixation point so that it fell within one hemifield and was thus represented only in the contralateral hemisphere. The patient's task was to reach under the screen and identify by touch, using one hand at a time, from a group of objects the object identical to that on the screen. Patients consistently succeeded with the hand ipsilateral to the field and failed with the contralateral hand. A normal subject performed the task with either hand. It was impossible in such patients simultaneously to ask each hemisphere, through the appropriate hemifield, to select a different object and for both tasks to be accomplished.

It is known that language functions are confined to one hemisphere and in Sperry's patients this was always the left. Verbal commands, then, were obeyed only by the muscles controlled by the left hemisphere. A patient, for instance, could not raise his left arm on command, nor could such a patient describe stimuli applied to the left side of his body.

There is no doubt that the left hemisphere in those patients was conscious. What of the right? The right or non-dominant hemisphere has memory, can pay attention and has the ability to perform complex tasks. It cannot control speech or writing but a few simple words or emotions can be expressed in

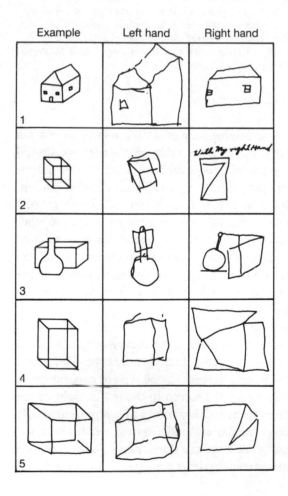

Fig. 7.10. Drawings with the right and left hand by a man with complete transection of the corpus callosum and anterior commissure. The patient had been right-handed but now his performance with his left hand appears to be better. (From Gazzaniga, M.S., Bogen, J.E. & Sperry, R.W. (1965) *Brain* **88**, 221–236.)

response to stimuli that are private to it. For example, if this hemisphere is shown pornographic pictures by means of the associated eye, the patient grins. Moreover, this hemisphere can be shown by non-verbal means to understand spoken or written words. Matching tests can be performed, for instance, in which a patient is shown a word through his right visual field and uses his left hand to pick out the matching object from a tray. It should be noted that in such tests the dominant hemisphere is ignorant of the whole transaction.

More recent investigations indicate that there is a fundamental difference in the way that the two hemispheres process information. The dominant hemisphere codes verbally and performs poorly if this coding is difficult. The non-dominant hemisphere, in contrast, appears to appreciate shapes as a whole. Figure 7.10 shows the drawings made with the right and with the left hand by a man with a complete transection of the corpus callosum and anterior commissure. The patient had always been right-handed and had never written or drawn with the left hand. As Fig. 7.10 shows, when he copied five sample figures his performance with the left hand was consistently better than with the right hand. Faces, which are difficult to describe in words, are recognized by the non-dominant hemisphere. Indeed, *prosopagnosia*, which is usually the result of a non-dominant hemisphere lesion, is a disorder in which familiar faces cannot be recognized. In man, the non-dominant hemisphere displays abilities which suggest a pictorial form of thought and a level of functioning which on its own terms is equivalent to the abilities of the hemispheres of other animals. It must be concluded that both hemispheres are capable of displaying attributes of consciousness and that consciousness is unified by the corpus callosum.

8 · Endocrine System

8.1 **Basic concepts**
Hormones
Hormone synthesis and secretion
Hormone transport and inactivation
Hormone actions
Control of hormone secretion

8.2 **The pituitary gland**
Anterior pituitary
  Hypothalamic neurohormones
  Control of anterior pituitary hormone
   secretion
  Growth hormone
   Growth and hormones
   Actions of growth hormone
   Control of growth hormone secretion
  Opioid peptides
  Melanocyte-stimulating hormone
Posterior pituitary
  Antidiuretic hormone
   Action of antidiuretic hormone
   Control of antidiuretic hormone
    secretion
  Oxytocin
   Actions of oxytocin
   Control of oxytocin secretion

8.3 **The thyroid gland**
Synthesis and secretion of thyroid
 hormones
Transport and inactivation of thyroid
 hormones
Actions of thyroid hormones
Control of thyroid hormone secretion
Disorders of thyroid gland function

8.4 **The parathyroid glands and calcium
metabolism**
Calcium metabolism
 Bone
Vitamin D
Parathyroid hormone
  Actions of parathyroid hormone
  Control of parathyroid hormone
   secretion
Calcitonin
Disorders of calcium metabolism

8.5 **The adrenal glands**
Adrenal medulla
  Actions of catecholamines
  Control of catecholamine secretion
Adrenal cortex
  Actions of glucocorticoids
  Control of glucocorticoid secretion
  Actions of mineralocorticoids
  Control of mineralocorticoid secretion
  Disorders of adrenocortical function

8.6 **The pancreatic islets**
Insulin
  Actions of insulin
  Control of insulin secretion
Glucagon
Control of energy utilization and storage
Effects of insulin deficiency

8.7 **Appendix**
Hormone assay
  Bioassay
  Radioimmunoassay
  Radioreceptor assay

8.1 Basic concepts

The endocrine system is one of the two coordinating and integrating systems of the body. It acts through chemical messengers or *hormones* carried in the circulation, and is involved in:

1 the maintenance of the internal environment (homeostasis);
2 the control of the storage and utilization of energy substrates;
3 the regulation of growth, development and reproduction;
4 the body's responses to environmental stimuli.

In contrast to the nervous system, the action of the endocrine system is slower in onset, more prolonged and generally more diffuse. The two systems are linked, however, through the *hypothalamus* which controls the secretion of many of the endocrine glands. Consequently, influences acting on or through higher centres of the brain, e.g. emotion, can affect endocrine secretion.

Evidence that an organ functions as an endocrine gland can be obtained by studying the specific effects of its removal from the body, by transplantation of the gland back into the body and by injections of extracts of the gland. The first experiment in endocrinology is attributed to Berthold, who in 1849 showed that transplantation of a testis into the abdomen of a castrated cockerel restored its secondary sex characteristics. The principal endocrine glands in mammals are the *hypothalamus, pituitary, thyroid, parathyroids, adrenals, pancreatic islets, gonads* and *placenta*. Hormones and hormone-like substances are also produced in the heart (p. 443), kidney (pp. 276 & 319), liver (p. 263), lungs (p. 540), pineal gland (p. 309), thymus (p. 342) and certain cells of the gastrointestinal tract (Section 14.4). Many substances which were first identified as hormones produced in endocrine glands are now also known to be synthesized in the nervous system, where they act as neurotransmitters or neuromodulators (see Table 2.2).

Hormones

A hormone can be defined as a chemical substance that is synthesized and secreted by a specific cell type, is transported by the circulation and at very low concentrations elicits a specific response in distant target tissues. Not all hormone-like substances meet these criteria. For example, some cells in the central nervous system and gastrointestinal system release substances which diffuse into surrounding regions and act locally. These are referred to as *paracrine* secretions. Sometimes cells secrete enzymes that act on plasma proteins to produce hormones, e.g. the renin–angiotensin system.

Hormones may be classified according to their chemical structure into:

1 *peptide hormones* (proteins, glycoproteins and polypeptides), e.g. growth hormone, insulin and antidiuretic hormone;
2 *steroid hormones*, e.g. aldosterone, oestrogen and testosterone; and
3 *tyrosine derivatives*, e.g. thyroxine and adrenaline.

The concentrations of hormones in the blood are extremely low, e.g. peptide hormones may range from 10^{-10} to 10^{-12} mol l^{-1} and steroid hormones from 10^{-6} to 10^{-9} mol l^{-1}. Because of the low concentration of hormones in the blood, their presence is difficult, if not impossible, to detect by chemical analytic techniques and so bioassays, radioimmunoassays and radioreceptor assays have been devised for measuring hormone levels (Appendix, p. 291). The development of the radioimmunoassay technique, which is highly sensitive and usually less cumbersome to perform than the bioassay, has led to rapid advances in endocrinology.

Hormone synthesis and secretion

Peptide hormones are synthesized on the ribosomes of the rough (granular) endoplasmic reticulum as part of larger precursor proteins called *pre-prohormones*, which are subsequently modified by deletion of peptide sequences. The 'pre' sequence appears to be a signal peptide involved in ribosomal attachment and transfer of the newly synthesized protein across the membrane of the endoplasmic reticulum (Fig. 8.1). From the rough endoplasmic reticulum the prohormone is transported to the Golgi apparatus where it is packaged into vesicles. Processing of the prohormone to the hormone occurs during these stages. With appropriate stimulation, Ca^{2+} enters the cell, causing the vesicles to fuse with the membrane and to extrude their contents into the extracellular fluid—a process known as *exocytosis*.

Steroid hormones are synthesized from cholesterol in steps which take place in the mitochondria, smooth endoplasmic reticulum and cytoplasm and require acetate, O_2, NADPH and other cofactors. Steroid hormones are not stored, to any extent, in vesicles and, being lipophilic, they readily cross the plasma membrane of the cell. Therefore, their rate of release is directly related to their rate of synthesis.

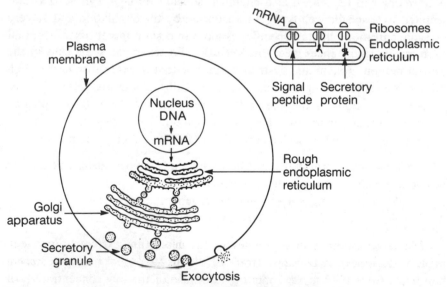

Fig. 8.1. Synthesis and secretion of peptide hormones. *Inset:* attachment of the single peptide and transfer of the nascent polypeptide across the membrane of the endoplasmic reticulum. During transfer the presequence is cleaved by a peptidase releasing the prohormone to the intracisternal space of the endoplasmic reticulum. (After Blobel, G. & Dobberstein, B. (1975) *J. Cell Biol.* **67**, 835–851.)

Synthesis of the hormones which are derivatives of tyrosine occurs by specific pathways in thyroid and adrenal medullary cells (see later).

Hormone transport and inactivation

Once released into the bloodstream, water-insoluble hormones, e.g. thyroid and steroid hormones, are bound to various plasma proteins. The free form is usually only a small fraction of the total hormone in the blood and exists in equilibrium with the bound form. Only the free hormone can affect its target cells. The concentration of the 'active' hormone in the blood is therefore determined by the dynamic relationship between its rate of secretion, its rate of inactivation and the degree to which it is bound to plasma proteins.

Hormones have a half-life in the body of minutes to days. Inactivation may occur in the blood, in the liver or kidney, or in some cases in the target tissues. Hormones may be inactivated by degradation, oxidation, reduction, methylation or conjugation to glucuronic acid, and excreted in the urine or bile. Peptide hormones generally have short half-lives (minutes) because they are rapidly degraded by peptide cleavage.

Hormone actions

Hormones affect the growth, development, metabolic activity and function of tissues. The responses are often the result of the actions of several hormones. Actions may be *stimulatory* or *inhibitory* and *additive* or *synergistic*. A hormone, which has no effect *per se* but is necessary for the full expression of the effects of other hormones, is said to have a '*permissive*' action. Hormones may alter:

1 membrane permeability;
2 activity of rate-limiting enzymes in reaction pathways;
3 protein synthesis (blocked by puromycin or cycloheximide); or
4 gene activation leading to the transcription of new messenger RNA species (blocked by actinomycin D).

These actions are not mutually exclusive and hormones may act in one or more of these ways.

Some hormones may act on several tissues; others only on a particular tissue or organ. In each case, the responsive cells have the appropriate *hormone receptors*. Several hormone receptors have been identified by binding studies with radioactively labelled hormones and shown to be discrete chemical entities, either protein or glycoprotein in nature. The first step in the action of a hormone is its binding to a specific cell receptor. Peptide hormones, which do not penetrate cells readily, act by binding to specific receptors in the plasma membrane; so too does adrenaline. Recently, it has been shown that some of the peptide hormone-receptor complexes may be internalized, i.e. taken up into the cell by endocytosis. The reason for this is not clear; it could be concerned with the regulation of receptor numbers or, in some cases, a necessary step in hormone action.

In contrast, radioactively labelled steroid hormones readily cross the plasma membrane and bind to specific *cytoplasmic* receptors in target tissues (Fig. 8.2). The receptor is a monomer or dimer that binds one or two molecules, respectively, of steroid hormone. The steroid hormone–receptor complex is then translocated to the nucleus, where it induces gene activation leading to the transcription of new messenger RNA species and, consequently, to specific protein synthesis. Thyroid hormones likewise cross the plasma membrane but bind directly to receptors in the nuclei of their target tissues.

Cyclic AMP. The effects of many hormones—adrenaline (acting on β-receptors), adrenocorticotrophic hormone, antidiuretic hormone (vasopressin) (acting on V_2-receptors), calcitonin, glucagon, follicle-stimulating hormone, human chorionic gonadotrophin, luteinizing hormone, melanocyte-stimulating hormone, parathyroid hormone and thyroid-stimulating hormone—are mediated by adenosine-3',5'-monophosphate (*cyclic AMP*). Hormone–receptor interaction at the membrane surface, via a *guanosine triphosphate (GTP)*-

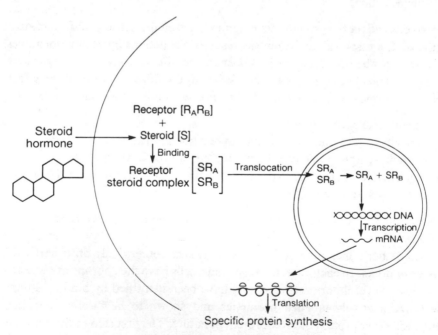

Fig. 8.2. Steroid hormone action at the cellular level. (After Buller, R. E., Schwartz, R. J., Schrader, W. T. & O'Malley, B. W. (1976) *J. Biol. Chem.* **251**, 5178–5186.)

regulatory protein, stimulates the enzyme *adenylate cyclase*, which catalyses the synthesis of cyclic AMP from adenosine triphosphate (Fig. 8.3). The increased level of cyclic AMP within the cell in turn stimulates *protein kinase A*. Protein kinase A is composed of two regulatory (R) units and two catalytic (C) units. Cyclic AMP binds to the R units causing them to dissociate and allows the free C units to phosphorylate cellular proteins. This may lead to specific changes in enzyme activity, protein synthesis or gene activation depending on the tissue involved. During this sequence of events, the initial hormonal signal is amplified many times. Eventually, the concentration of cyclic AMP is restored to its basal level by degradation to adenosine monophosphate, this being catalysed by the enzyme *cyclic nucleotide phosphodiesterase*. Sutherland (1915–1974), who discovered cyclic AMP, called it the '*second messenger*'.

In view of the specific nature of hormone action, it may appear odd that the action of so many hormones is mediated by cyclic AMP. However, only certain tissues contain receptors which will react with a particular hormone to produce an increase in intracellular cyclic AMP, and the responses of various tissues to raised levels of cyclic AMP are different.

Although the actions of many hormones are mediated by cyclic AMP, there are a number of hormones that do not increase cyclic AMP concentrations in their target glands. Some hormones, e.g. adrenaline (acting on α_2-receptors), may act by inhibiting adenylate cyclase and decreasing cyclic AMP levels, while others may act by increasing cytosolic Ca^{2+}. The mechanisms of action of some hormones, e.g. insulin and growth hormone, are not known. Other likely intracellular messengers are cyclic GMP (guanosine 3',5'-monophos-

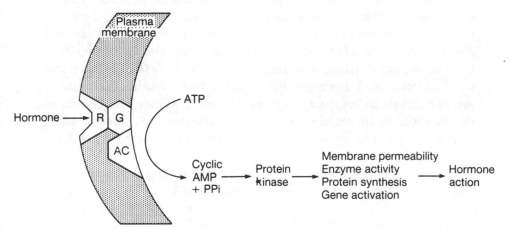

Fig. 8.3. Hormone action mediated by the 'second messenger', cyclic AMP. The hormone binds to its receptor (R) on the outer surface of the plasma membrane. At the inner surface are the GTP-regulatory unit (G) and the enzyme adenylate cyclase (AC) that catalyses the conversion of ATP to cyclic AMP. Binding of the hormone to the R–G complex allows the G unit to bind GTP and stimulate adenylate cyclase activity.

phate) and arachidonic acid derivatives (prostaglandins, thromboxanes and leukotrienes).

Calcium. Examples of hormones that increase intracellular Ca^{2+} are adrenaline (acting on α_1-receptors), antidiuretic hormone (acting on V_1-receptors) and angiotensin II. Some hormones, e.g. opioid peptides, may act by decreasing cytosolic Ca^{2+}. Hormones may increase cytosolic Ca^{2+} either by opening Ca^{2+} channels in the plasma membrane, which allows Ca^{2+} to flow into the cell down a steep electrochemical gradient, or by releasing Ca^{2+} stored in the endoplasmic reticulum. Increased cytosolic Ca^{2+} may in turn stimulate the activity of protein kinases or other regulatory enzymes in the cell either directly or through a Ca^{2+}-binding protein called *calmodulin*. Calmodulin, which binds up to four Ca^{2+} ions per molecule with high affinity, regulates the activity of many cellular enzymes, e.g. Ca^{2+}-calmodulin-dependent kinase, phosphorylase kinase, adenylate cyclase, cyclic nucleotide phosphodiesterase, Ca^{2+}-ATPase and myosin light-chain kinase. Not only does Ca^{2+} (via calmodulin) regulate enyzmes involved in the synthesis and degradation of cyclic AMP but the two intracellular messengers nearly always function together, although not necessarily in harmony.

Polyphosphoinositide system. It now appears that many of the hormones employing Ca^{2+} as an intracellular messenger operate as part· of the polyphosphoinositide system. Receptor activation leads to the stimulation of a phosphodiesterase (phospholipase C) in the plasma membrane (Fig. 8.4). This enzyme cleaves phosphatidylinositol 4, 5-bisphosphate on the inner face of the plasma membrane to form *inositol trisphosphate* ($InsP_3$) and *diacylglycerol* (DAG). $InsP_3$ migrates to the endoplasmic reticulum where it stimulates the release of Ca^{2+}, while DAG remains in the membrane and, in the presence of Ca^{2+} and phosphatidylserine, activates protein kinase C. Thus, in this system, Ca^{2+} acts as a third messenger and events in both branches determine the appropriate cellular response. Finally, $InsP_3$ and DAG are metabolized and the end-products are recycled to form phosphatidylinositol, which is used to resynthesize phosphatidylinositol 4, 5-bisphosphate as depicted in Fig. 8.4.

G proteins. A central feature in all these schemes is the transduction of the hormonal signal by GTP-regulatory proteins called G (or N) proteins. Such proteins are widely deployed in the transduction of extracellular signals (e.g. hormonal, neurotransmitter and sensory stimuli) into cellular responses. Receptor activation induces the G protein to exchange bound guanosine diphosphate (GDP) for GTP thus allowing it to interact with and control the activity of membrane-bound enzymes, such as adenylate cyclase (Fig. 8.3) and phospholipase C (Fig. 8.4). G proteins consist of α, β and γ subunits. Guanosine

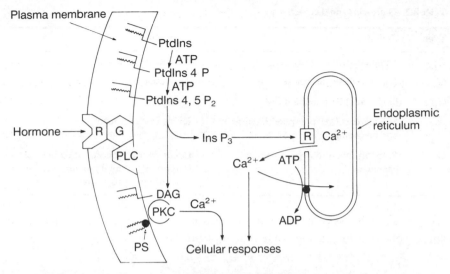

Fig. 8.4. Hormone action mediated by the polyphosphoinositide system. The binding of hormone to its receptor (R) in the plasma membrane activates the enzyme phospholipase C (PLC) via the GTP-regulatory protein (G). This enzyme cleaves phosphatidylinositol 4, 5-bisphosphate (PtdIns 4, 5P$_2$) to form the second messengers inositol trisphosphate (InsP$_3$) and diacylglycerol (DAG). InsP$_3$ mobilizes Ca^{2+} from the endoplasmic reticulum while DAG, in the presence of phosphatidylserine (PS) and Ca^{2+}, activates the enzyme protein kinase C (PKC), and these actions lead to specific cellular responses. InsP$_3$ and DAG are rapidly metabolized and their end-products recycled to form phosphatidylinositol (PtdIns) which is phosphorylated to form PtdIns 4 P and finally PtdIns 4, 5 P$_2$.

triphosphate binds to the α subunit causing the G protein to dissociate into α-GTP and $\beta\gamma$. The α-GTP subunit is the active component and its action is terminated by hydrolysis of GTP and re-association of the α with the $\beta\gamma$ subunits to form the inactive GDP-G protein.

A number of G proteins have been identified (Table 8.1). Bacterial exotoxins have been particularly valuable in identifying these G proteins: *cholera toxin* activates G$_s$ in the absence of hormones by ADP-ribosylation while *pertussis toxin* blocks the receptor-mediated activation of G$_i$ and G$_o$ similarly. The high concentration of G$_o$ in the brain (Table 8.1) no doubt reflects the important role of this protein in the actions of the various neurotransmitters and neuromodulators.

Control of hormone secretion

The immediate stimulus for the secretion of a hormone may be neural, hormonal, or the level of some metabolite or electrolyte in the blood. In the long term, secretion rates are usually maintained at fairly constant levels by *negative feedback mechanisms*, whereby increased levels of hormone in the blood lead to the inhibition of further hormone secretion. Positive feedback

Table 8.1. G proteins and their actions

Type	Action	Comments
G_s	Stimulates adenylate cyclase	Activated by cholera toxin
G_i	Inhibits adenylate cyclase	Activation blocked by pertussis toxin
G_p	Stimulates phospholipase C	—
G_t	Stimulates cGMP phosphodiesterase in photoreceptors	Known as transducin
G_o	Probably involved in gating of ion channels	Occurs in high concentration in brain (approx. 1% of total protein); activation blocked by pertussis toxin

control is much less common, but an example is to be found in the hormonal control of ovulation during the female reproductive cycle.

8.2 The pituitary gland

The *pituitary* or *hypophysis* is a small gland, weighing approximately 0·5 g in man, situated at the base of the skull and connected to the brain by the pituitary stalk. It consists of an *anterior lobe*, and a *posterior lobe* (Fig. 8.5). During embryonic development, the *neurohypophysis*, which includes the *pars posterior*, is derived from a downward evagination of the brain. The *adenohypophysis*, which includes the *pars anterior* and the *pars intermedia*, comes from an outgrowth of the roof of the mouth known as Rathke's pouch. In the adult human, the pars intermedia is only a remnant and the pars anterior and pars posterior may be equated with the terms anterior and posterior pituitary, respectively.

The pituitary secretes at least eight hormones. Four of these hormones regulate other endocrine glands and are referred to as 'trophic' (or 'tropic') hormones. The pituitary has been referred to as the master gland, but it is subject to control by the hypothalamus and by the secretions of its own target glands.

Anterior pituitary

The anterior pituitary synthesizes and secretes at least six hormones, namely, *growth hormone* (somatotrophin), *prolactin*, and the trophic hormones— *thyroid-stimulating hormone* (TSH, thyrotrophin), *adrenocorticotrophic hormone* (ACTH, corticotrophin), *follicle-stimulating hormone* (FSH) and *luteinizing hormone* (LH). FSH and LH are known collectively as the *gonadotrophins*. The hormones are produced by different cell types interspersed throughout the pars anterior and are secreted in response to stimulation by their corresponding

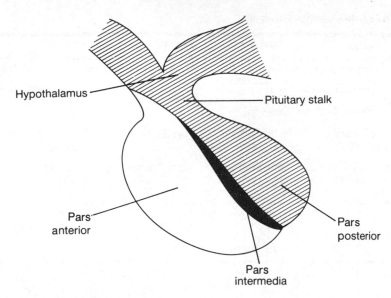

Hypothalamus

Pituitary stalk

Pars anterior

Pars posterior

Pars intermedia

Fig. 8.5. The pituitary gland.

hypothalamic neurohormones. Many of the hormones are synthesized as larger molecules or *prohormones*. The prohormone may contain the sequences of a number of hormones, e.g. the prohormone for ACTH also contains the sequences of melanocyte-stimulating hormones, of lipotrophins, and of endorphins (p. 264). The hormone secreted by a particular cell type will then depend on where the prohormone is enzymatically clipped.

The trophic hormones stimulate the growth and secretion of other endocrine glands in the body—TSH acts on the thyroid gland, ACTH acts on the adrenal cortex, and the gonadotrophic hormones, FSH and LH, act on the testes and ovaries (Table 8.2). Not only do the trophic hormones influence the secretion of their target glands, they also influence their size and development. Hypophysectomy therefore causes:

1 inability to grow due to lack of growth hormone;
2 atrophy of the thyroid gland and hypothyroidism due to lack of TSH;
3 atrophy of the adrenal cortex and corticosteroid deficiency due to lack of ACTH; and
4 infertility and deficiency of sex hormones due to lack of FSH and LH.

However, removal of the pituitary is not incompatible with life, although hypophysectomized animals have a low tolerance to cold and stress. (In hypophysectomized animals, the posterior pituitary hormones continue to be secreted because the neurones that synthesize these hormones have their cell bodies in the hypothalamus.)

Table 8.2. Anterior pituitary hormones

Hormone	Structure	Target tissue	Main action
Growth hormone	Protein (M_r 22 000)	Several tissues	Growth and metabolism
Prolactin	Protein (M_r 23 000)	Mammary gland	Milk production
Thyroid-stimulating hormone (TSH)	Glycoprotein (M_r 28 000)	Thyroid	Secretion of thyroid hormones
Adrenocorticotrophic hormone (ACTH)	Polypeptide (M_r 4500)	Adrenal cortex	Secretion of corticosteroids
Follicle-stimulating hormone (FSH)	Glycoprotein (M_r 32 000)	Testes and ovaries	Secretion of sex hormones and production of gametes
Luteinizing hormone (LH)	Glycoprotein (M_r 30 000)		

Hypothalamic neurohormones

Release of the anterior pituitary hormones is regulated by *neurohormones* which are elaborated in neurones in the hypothalamus. These neurohormones are released at the level of the median eminence. They diffuse into a primary plexus of capillaries and are transported down large portal vessels in the pituitary stalk to a secondary set of capillaries or sinusoids in the anterior pituitary—the so-called *hypophyseal portal system* (Fig. 8.6). They comprise both releasing and release-inhibiting hormones and some of the anterior pituitary hormones may be subject to dual control. It has been suggested that the name of the releasing hormone should end in '-liberin' and the inhibiting hormone in '-statin' but this has not yet been universally adopted and the most commonly used nomenclature is given below. The following releasing hormones (originally called 'releasing factors') have been identified—*growth hormone-releasing hormone* (GHRH), *thyrotrophin-releasing hormone* (TRH), *corticotrophin-releasing hormone* (CRH) and *gonadotrophin-releasing hormone* (GnRH). GnRH stimulates the release of both LH and FSH. The release-inhibiting hormones secreted by the hypothalamus are *growth hormone release-inhibiting hormone* (GHRIH, somatostatin) and *dopamine* (prolactin release-inhibiting hormone, PRIH), which inhibits the secretion of prolactin. Thus, secretion of growth hormone is subject to dual control by the hypothalamus. The structure of TRH was the first to be elucidated and it was found to be a tripeptide (pyroglutamyl–histidyl–proline amide). The other hypothalamic neurohormones which have been identified are also small peptides, except dopamine which is a tyrosine derivative related to noradrenaline.

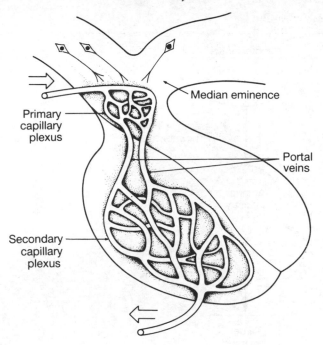

Fig. 8.6. Hypophyseal portal system.

Control of anterior pituitary hormone secretion

The secretion of releasing and inhibiting hormones by the hypothalamus may be influenced by emotional and environmental factors acting through the central nervous system. In the long term, however, regulation of the secretion of the hypothalamic neurohormones and anterior pituitary hormones is usually achieved through negative feedback mechanisms triggered by the blood level of the anterior pituitary hormones (short feedback loop) and the target gland hormones (long feedback loop). Figure 8.7 shows the possible negative-feedback pathways that could exist, and in many cases do exist, for the regulation of the anterior pituitary hormones.

Growth hormone

The conditions of *dwarfism* and *gigantism* are caused by disturbances in the function of the pituitary gland. Young animals cease to grow when their pituitary glands are removed but growth can be restored by injections of anterior pituitary extracts. Growth hormone was first isolated from bovine anterior pituitaries in the 1940s. It is a protein of M_r 22 000, there being some differences in the amino acid sequence between species. It has close structural

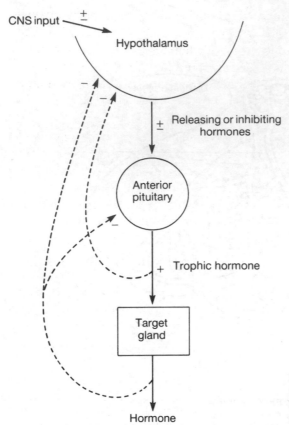

CNS input

Hypothalamus

Releasing or inhibiting hormones

Anterior pituitary

Trophic hormone

Target gland

Hormone

Fig. 8.7. Control of secretion of the anterior pituitary hormones. The broken lines indicate negative-feedback pathways.

similarities to prolactin and human chorionic somatomammotrophin, which suggests their evolution from a common progenitor molecule.

Growth and hormones. The major function of growth hormone is to stimulate the growth of bones and other tissues. The growth process is not simple and, in addition to an adequate food supply and to genetic endowment, a number of hormones are involved, namely, *growth hormone, sex hormones, thyroxine* and *insulin.* There are two periods of accelerated growth: the first occurs in the first two years of life and the second at the time of puberty. The period of accelerated growth at puberty is associated with increased levels of sex hormones, including androgens of adrenal origin. Paradoxically, the cessation of growth around 18–20 years of age is also due to the sex hormones, which cause fusion of the growing ends of bones (*epiphyseal closure*). Thus, the effect of excess growth hormone secretion depends on whether it occurs before or after closure of the epiphyses. Excess growth hormone secretion in young people causes gigantism, but in adults it leads to *acromegaly* in which there is a general coarsening of the features due to thickening of bone and soft tissue. On the other hand, a deficiency of growth hormone in children leads to dwarfism. Humans respond only to growth hormone of human (or other primate) origin. This is now synthesized using recombinant DNA technology. Synthetic human growth hormone is used to treat children with short stature due to growth hormone deficiency.

Actions of growth hormone. Growth hormone affects the metabolic activity of most of the tissues of the body. Its growth-promoting effect is due to its ability to stimulate the uptake of amino acids and their incorporation into proteins in muscle and bone. It is now known that the growth-promoting actions of growth hormone are mediated by one of the *insulin-like growth factors* (IGF-1) formerly known as *somatomedin*. IGF-1 is a polypeptide produced mainly in the liver. Growth hormone also has effects on carbohydrate and fat metabolism which in general are antagonistic to those of insulin, i.e. it is 'counter-regulatory' to insulin. Growth hormone increases blood glucose (after an initial decrease) by stimulating hepatic gluconeogenesis and by inhibiting glucose uptake by muscle. Large doses of growth hormone raise free fatty acid levels by increasing their mobilization from adipose tissue. The half-life of growth hormone is about 20 min but its actions continue for much longer.

Control of growth hormone secretion. The level of growth hormone in the blood as measured by radioimmunoassay is much the same in adults as in children and it fluctuates continuously. The physiological control mechanisms of growth hormone secretion are not well understood but secretion is increased during fasting and in stress. Its release from the anterior pituitary is under hypothalamic control by *growth hormone-releasing hormone* and *somatostatin* (GHRIH) which stimulate and inhibit secretion respectively. Factors that stimulate growth hormone secretion are low blood glucose and high blood amino acid concentrations. Bursts of hormone secretion also occur during stages 3 and 4 of sleep, and associated with the nocturnal rise in growth hormone is a characteristic early morning increase in blood glucose levels.

The other hormones secreted by the anterior pituitary, namely prolactin TSH, ACTH, FSH and LH, will be discussed later in the appropriate sections. Also found in the anterior pituitary are the opioid peptides. Their precise physiological role is not yet known but their potential importance is such that they warrant discussion at this point.

Opioid peptides

Endogenous peptides which interact with opiate receptors were first discovered in the brain and gut and later in the adrenal medulla. They were found to be pentapeptides and were named *met-enkephalin* (Tyr–Gly–Gly–Phe–Met) and *leu-enkephalin* (Tyr–Gly–Gly–Phe–Leu). Subsequently, larger opioid peptides were discovered in the pituitary and named *endorphins* and *dynorphins*. The former are found in the pars anterior and pars intermedia and the latter in the pars posterior. These opioids are also present in brain.

The family of opioid peptides has grown to at least nine members, all of which contain either the met-enkephalin or leu-enkephalin sequence at their N-terminals. Complementary DNA-gene cloning techniques have shown that the family has three branches which stem from (1) *proopiomelanocortin*, the precursor of β-endorphin and related peptides, (2) *proenkephalin*, the precursor for met- and leu-enkaphalin, and (3) *prodynorphin*, the precursor for dynorphins and neoendorphins. The proopiomelanocortin precursor contains the sequences of several active peptides including ACTH, β-lipotrophin, β-endorphin, and α- and β-melanocyte-stimulating hormones. This precursor molecule is processed differently depending on its location. In the pars anterior of the pituitary, ACTH and β-lipotrophin are produced (some of the β-lipotrophin is split to produce β-endorphin). In the pars intermedia, ACTH is further cleaved to α-MSH and β-lipotrophin is cleaved to produce β-endorphin.

Opioid peptides are powerful analgesics when injected into the ventricles of the brain (p. 144). There are at least three subclasses of opioid receptors:

1   the δ-receptor which binds enkephalins;
2   the κ-receptor which binds dynorphins; and
3   the μ-receptor which binds β-endorphin and the enkephalins, the activation of which correlate with analgesia.

The opioid peptides in brain appear to act as neurotransmitters or neuromodulators and play a role in the perception of pain and in behaviour. The role of the pituitary opioids is less clear but they appear to participate in the body's reaction to stress.

Melanocyte-stimulating hormone

The pars intermedia secretes *melanocyte-stimulating hormone* (MSH), which in amphibia and fish causes the skin to darken by dispersing the melanin granules within the melanophores and thus enables these animals to blend their skin colour with the environment. In the human adult, the pars intermedia occupies only 1% of the pituitary and its role is not known. There are two types of MSH in animals, α-MSH and β-MSH, both of which are polypeptides. The structural sequence of α-MSH corresponds to residues 1–13 of ACTH. Excess production of MSH (or ACTH) in humans can cause an increase in melanin synthesis and hyperpigmentation. MSH secretion is under dual control by both releasing and release-inhibiting hypothalamic neurohormones, the inhibiting hormone playing the dominant role.

Posterior pituitary

The posterior lobe of the pituitary secretes two peptide hormones, *antidiuretic hormone* (ADH) and *oxytocin*. These hormones are synthesized in discrete groups of neurones in the hypothalamus called the supraoptic and paraventricular nuclei (Fig. 8.8). Each hormone is synthesized as part of a prohormone, which also contains *neurophysin* (neurophysin I for oxytocin and neurophysin II for ADH). The prohormone is packaged into granules in the Golgi apparatus. Neurophysin is then cleaved from the hormone to which it binds non-covalently. This protects the hormone from degradation and also prevents it from leaking out of the storage granules. The granules are transported down the axons to their terminals in the posterior pituitary, where they are stored prior to their release into the circulation. Release of each hormone, together with its corresponding neurophysin, is triggered by nerve impulses originating in the hypothalamus. This process is known as *neurosecretion*.

Antidiuretic hormone

ADH is a nonapeptide (M_r 1102) in which two cysteine residues are joined by a disulphide bridge. It controls the reabsorption of water by the kidneys and also constricts arterioles, hence the alternative name *vasopressin*.

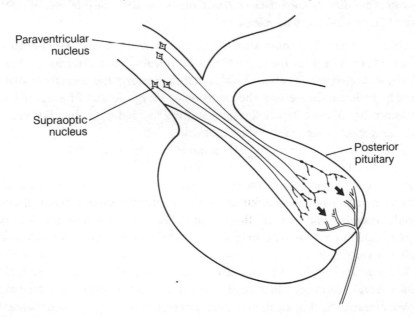

Fig. 8.8. Neurosecretion from the posterior pituitary.

In mammals, there are two types of ADH which differ by a single amino acid—arginine (man) or lysine. ADH has a half-life of about 5 min in the plasma and is metabolized by the liver and kidneys.

Actions of antidiuretic hormone. Two types of receptor for ADH have been identified: V_1-receptors found in vascular smooth muscle and V_2-receptors found in the distal tubules and collecting ducts of the kidney.

ADH increases the reabsorption of water by the kidneys and so reduces the excretion of water from the body. It acts on the distal portions of mammalian nephrons, increasing their permeability to water (p. 530). Water moves passively out of the nephrons along an osmotic gradient and so urine volume is decreased. This action of ADH (via V_2 receptors) appears to be mediated by cyclic AMP.

In addition to its role in renal function, ADH is a potent vasoconstrictor. It acts on vascular smooth muscle (via V_1-receptors) to increase the concentration of inositol trisphosphate (p. 256), which increases intracellular Ca^{2+} and triggers muscle contraction. Under normal conditions, this action of ADH does not significantly alter arterial pressure because it is counteracted by the baroreceptor reflexes. However, the vasoconstrictor effect of ADH is important in the maintenance of arterial pressure in potentially hypotensive conditions, such as haemorrhage, and may play a role in some types of hypertension.

Control of antidiuretic hormone secretion. Consistent with the dual role of ADH, several factors influence its release.

1 *Osmotic pressure.* Osmoreceptors in the hypothalamus respond to an increase in osmolality of the extracellular fluid leading to the release of ADH. Subsequent renal conservation of water thereby restores the normal osmolality of the body fluids. Conversely, the ingestion of a large volume of water reduces the osmolality of the extracellular fluid leading to a reduction in ADH release and an increase in the renal excretion of water.

2 *Blood volume and pressure.* Haemorrhage promotes ADH release in response to (i) decreased stimulation of atrial stretch receptors, (ii) decreased stimulation of arterial baroreceptors, (iii) activation of arterial chemoreceptors and (iv) increased release of angiotensin II. The decreased stimulation of atrial stretch receptors and arterial baroreceptors is also thought to underlie the transient effect on ADH release of changes in body position such as from a supine to a sitting position.

3 *Other factors.* Pain, exercise, stress, sleep and drugs such as morphine induce ADH secretion while alcohol strongly inhibits secretion. The well-known diuretic effect of alcoholic beverages results not simply from increased fluid intake but also from a direct suppression of ADH release.

Damage to the ADH-producing neurones in the hypothalamus may result in the condition known as *diabetes insipidus*, which is characterized by the voiding of large volumes of dilute urine (polyuria) and excessive thirst (polydipsia). This condition can be treated satisfactorily with synthetic ADH administered as a nasal spray.

Oxytocin

Oxytocin is a peptide (M_r 1025) that differs in structure from ADH in two amino acid residues. Its synthesis and metabolism are similar in many respects to those of ADH.

Actions of oxytocin. Oxytocin stimulates the myoepithelial cells of the mammary gland, causing milk let-down. It also causes uterine contraction in the oestrogen-stimulated uterus during parturition and is used clinically for the induction of labour. It has no known functions in males.

Control of oxytocin secretion. Milk let-down is a reflex action in response to suckling at the breast. A neuroendocrine reflex pathway is involved, in which impulses initiated by suckling are relayed to the hypothalamo–pituitary axis causing the release of oxytocin. This is carried by the circulation to the mammary glands. Pain, embarrassment and anxiety can cause inhibition of oxytocin release. There is a marked elevation of plasma oxytocin levels during parturition.

8.3 The thyroid gland

Hormones of the thyroid gland are essential for optimal metabolic activity in the body. They also have other important functions, particularly in the control of growth and the development of the nervous system. Thyroid hormone deficiency in the fetus produces the condition of *cretinism*, which is characterized by a failure to grow and severe mental retardation. This condition is caused by failure of the thyroid gland to develop properly. A closely related condition occurs in certain parts of the world, e.g. New Guinea, where the content of iodine in the soil and in the water is low, and in these areas it is associated with enlargement of the thyroid gland (goitre). This rarely occurs in coastal regions because of the high iodine content of fish and other sea foods. Endemic cretinism and endemic goitre have become less common since the introduction of dietary iodine supplementation, for example, in the form of iodized salt.

The thyroid gland is composed of two lobes joined by a narrow connection called the thyroid isthmus (Fig. 8.9). The gland is bright red because it has a very rich blood supply. It lies in front of the trachea just below the larynx.

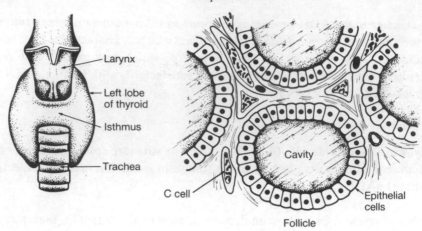

Fig. 8.9. The thyroid gland.

Each lobe is composed of many follicles which are spherical in shape and which are surrounded by a dense network of capillaries. Each follicle is composed of a single layer of epithelial cells surrounding a cavity in which the thyroid hormones are stored attached to a protein, *thyroglobulin*.

The main hormones secreted by the thyroid gland are thyroxine (T_4) *and triiodothyronine* (T_3), collectively known as the thyroid hormones (Fig. 8.10). The thyroid gland also secretes the hormone, *calcitonin*, which lowers plasma calcium concentrations and is discussed later (p. 278). Sufficient T_3 and T_4 are stored in the gland to last 2–3 months. The only structural difference between T_3 and T_4 is that T_4 contains four iodine atoms whereas T_3 has three. There is

Thyroxine (T_4)

3,5,3'-triiodothyronine (T_3)   **Fig. 8.10.** The thyroid hormones.

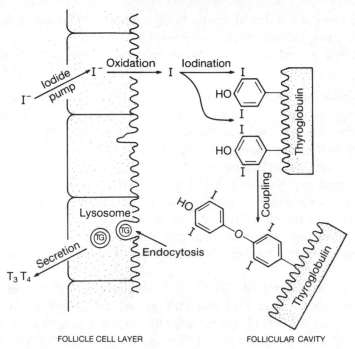

Fig. 8.11. Synthesis of thyroid hormones (T_3, T_4), their storage in association with thyroglobulin (TG) and their secretion.

approximately fifty times more T_4 than T_3 in the plasma. In the tissues, T_4 may be converted to T_3 or to reverse T_3 (3,3',5'-triiodothyronine). T_3 is the major hormone active in the cells; reverse T_3 is inactive.

Synthesis and secretion of thyroid hormones

The thyroid gland actively concentrates iodide to a level normally some twenty-five times that in the plasma. The iodide is oxidized by a peroxidase in the follicle cells to atomic iodine which immediately iodinates tyrosine residues contained in thyroglobulin (Fig. 8.11). Thyroglobulin is a large protein (M_r 670 000) which is synthesized in the follicle cells and secreted into the follicular cavity. The iodinated tyrosine residues in thyroglobulin undergo coupling to form T_4 and T_3. Iodination and coupling are thought to take place at the cell surface bordering the follicular cavity and the thyroid hormones are stored in the cavity conjugated to thyroglobulin.

The synthesis of the thyroid hormones can be summarized as follows.

1   Iodide uptake and concentration (blocked by thiocyanate and perchlorate).
2   Oxidation of $I^- \rightarrow I + e$ (blocked by propylthiouracil and carbimazole).

3 Iodination of tyrosine molecules in thyroglobulin by atomic iodine (also blocked by propylthiouracil and carbimazole).
4 Coupling of either two diiodotyrosine residues to form T_4 or of a diiodotyrosine (DIT) and a monoiodotyrosine (MIT) residue to form T_3.

As indicated above, the action of the antithyroid drugs—thiocyanate, perchlorate, propylthiouracil and carbimazole—is due to their interference with various steps in the biosynthetic pathway. Paradoxically, excess iodide also inhibits the biosynthesis of the thyroid hormones.

All the steps in the synthesis of the thyroid hormones are stimulated by *thyroid-stimulating hormone* (TSH) acting through cyclic AMP. TSH also stimulates secretion of the thyroid hormones. The first step in secretion is the uptake of small globules of colloid into the follicle cells by endocytosis. The globules then fuse with lysosomes and their contents are digested, thus liberating the thyroid hormones which diffuse out of the follicle cells into the blood.

The proteolytic degradation of thyroglobulin within the follicular cells causes the liberation of MIT and DIT as well as T_3 and T_4. A specific deiodinase enzyme in the follicular cells allows the iodine to be removed from MIT and DIT so that the iodine and the tyrosine can be recycled. This reaction is important for conserving supplies of iodine in the thyroid, and a genetic defect causing impaired deiodinating activity can cause iodine deficiency because of loss of iodotyrosine molecules in the urine.

Transport and inactivation of thyroid hormones

Most of the circulating T_4 is bound to plasma proteins, mainly to thyroxine-binding globulin and, to a lesser extent, to prealbumin and albumin. Total plasma T_4 and T_3 (protein-bound and free) can be determined by radio-immunoassay as measures of thyroid function. However, as the total concentration of hormone in the blood is influenced markedly by the concentration or binding capacity of thyroxine-binding globulin, an adjustment should be made for this, or specific measurement of free T_4 or free T_3 made. T_3 is less firmly bound to plasma proteins than is T_4 and this is reflected in the finding that, although the concentration of total T_4 in the plasma (about 100 nmol l^{-1}) is much higher than that of total T_3 (about 1·5 nmol l^{-1}), the concentrations of the free hormones are similar (about 0·04 nmol l^{-1} for T_4 and 0·02 nmol l^{-1} for T_3).

The thyroid hormones are broken down in several tissues, particularly the liver and skeletal muscle. T_4 has a half-life of 7 days, T_3 about 1 day. Much of the iodide that is released is reclaimed but about 150 μg of iodide is lost in the urine and faeces daily and must be replaced in the diet.

Actions of thyroid hormones

1 *Thermogenesis*. One of the principal effects of the thyroid hormones is to stimulate oxidative metabolism and thereby increase the production of heat in warm-blooded animals. They increase oxidative metabolism in all tissues of the body except the brain, lungs, spleen and sex organs. The increase in basal metabolic rate produced by a single injection of T_4 begins after a latency of several hours and lasts 9 days or more. The basal metabolic rate may increase by as much as 100% while after thyroidectomy it may fall by 50%. T_3 and T_4 may cause a slight increase in body temperature but increased secretion of hormone does not seem to occur in man as part of an acute response to cold. However, the hormones may act in a permissive manner to allow other mechanisms such as increased sympathetic activity to accelerate production of heat. It appears that the action of thyroid hormones on oxidative metabolism is due, at least in part, to an increase in the synthesis of Na^+–K^+-ATPase and hence in the activity of the Na^+–K^+ pump.

2 *Effects on growth and development*. As already mentioned, thyroid hormones are essential for normal growth in childhood. In amphibia, the thyroid hormones act as a trigger for metamorphosis and premature metamorphosis can be induced in tadpoles by feeding them pieces of thyroid gland. T_3 and T_4 appear to stimulate growth by a direct effect on tissues but they also have a permissive role on growth hormone action.

3 *Effects on the nervous system*. The thyroid hormones are essential for normal myelination and development of the nervous system in childhood. In the adult, a deficiency of thyroid hormones may lead to listlessness and blunting of intellect; an excess to restlessness and hyperexcitability.

4 *Other effects*. Often associated with excess production of thyroid hormones are an increased cardiac output and tachycardia. These symptoms are partly a direct response to the thyroid hormones, which increase sensitivity to catecholamines, and are partly secondary responses to increased demands for oxygen associated with their thermogenic action. The thyroid hormones have less well-defined effects on carbohydrate and lipid metabolism; they lower plasma cholesterol concentration. The metabolism of protein is also affected by thyroid hormones, and hyperthyroidism may lead to wasting of skeletal muscle and a negative nitrogen balance. The thyroid hormones influence calcium metabolism, and demineralization of the skeleton is common in severe hyperthyroidism. Adequate secretion of hormones is also necessary for gonadal function and lactation.

Thyroid hormones exert their varied effects upon their target tissues by binding of T_3 to the specific receptor for this hormone in the cell nucleus. Such binding modulates transcription of messenger RNA from a variety of genes coding for enzymes and structural proteins involved in the expression of thyroid hormone action but the details are not yet clear.

Control of thyroid hormone secretion

For normal thyroid hormone secretion, there must be adequate intake of iodine in the diet. The immediate stimulus for the release of thyroid hormones is TSH secreted by the anterior pituitary. TSH appears to stimulate every step in the production and secretion of T_3 and T_4. In addition, it controls the size and vascularity of the gland; if the pituitary is removed, the thyroid atrophies.

Secretion of TSH is stimulated by *thyrotrophin-releasing hormone* (TRH) which is secreted by neurones in the hypothalamus and transported to the anterior pituitary in the hypophyseal portal vessels. Its actions appear to be mediated by a rise in cytosolic Ca^{2+}, although some of them are mimicked by cyclic AMP. TSH secretion is also regulated by feedback inhibition by T_3 and T_4 (Fig. 8.12) which inhibit TSH secretion by a direct action on the anterior pituitary. The thyroid hormones may also act at the level of the hypothalamus to affect the output of TRH but the significance of this mechanism is unclear.

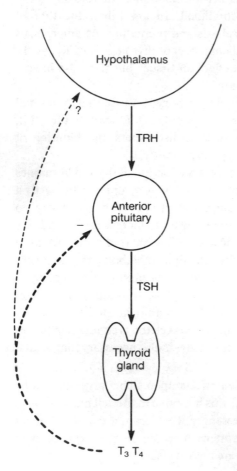

Fig. 8.12. Control of thyroid hormone secretion.

Disorders of thyroid gland function

Hypothyroidism may result from disease of the pituitary or thyroid gland (*autoimmune thyroiditis or Hashimoto's disease*), or from severe deficiency of iodine in the diet. Severe hypothyroidism in the adult is called *myxoedema* because of the puffiness of the hands and face due to an abnormal accumulation of mucoproteins in the subcutaneous layers. Other symptoms are low metabolic rate, bradycardia, cold intolerance, mental and physical lethargy and slow hoarse speech. Severe hypothyroidism in the fetus results in the condition of *cretinism*.

Hyperthyroidism or *thyrotoxicosis* results from the overproduction of thyroid hormones and is characterized by a high metabolic rate, tachycardia, heat intolerance, hyperexcitability, restlessness and weight loss. A common form of hyperthyroidism is *Graves' disease* which is also characterized by protruding eyeballs (exophthalmia) and goitre formation. Graves' disease is caused by abnormal thyroid stimulators in the blood. Long-acting thyroid stimulator (LATS) was the first of these to be discovered and it is now clear that this is one of a group of autoantibodies called *thyroid-stimulating antibodies* that are responsible for this condition.

Goitre formation is often associated with hyperthyroidism. However, it may also be a manifestation of hypothyroidism, in which there is a compensatory increase in TSH secretion, as occurs with iodine deficiency (endemic goitre) or in diseases of the thyroid gland which result in defective T_4 synthesis or secretion.

In the diagnosis of disorders of thyroid function, the physician is guided by estimates of plasma levels of free T_3 and T_4 and of TSH, by responses to test doses of TRH, or by the pattern of thyroid uptake of radioactive iodine measured externally. Radioactive technetium, given as the pertechnetate, is now commonly used instead of radioactive iodine for imaging of by the thyroid gland.

8.4 The parathyroid glands and calcium metabolism

In mammals, endocrine tissue important for controlling calcium balance is located in the region of the thyroid gland. Four parathyroid glands are usually embedded in the dorsal surface of the lobes of the thyroid gland (Fig. 8.13). They secrete *parathyroid hormone*, which raises plasma calcium and lowers plasma phosphate concentration. In addition, the parafollicular or 'C' cells (Fig. 8.9) of the thyroid gland secrete *calcitonin,* which lowers the plasma calcium concentration. Another important regulator of calcium metabolism is an active metabolite of *vitamin D* known as *calcitriol*.

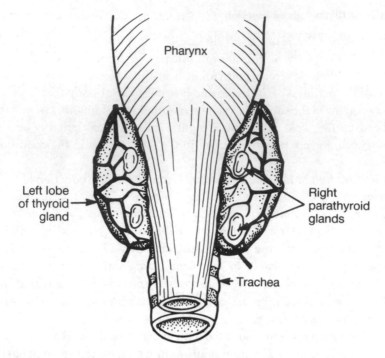

Fig. 8.13. The parathyroid glands viewed from behind.

Calcium metabolism

Calcium has a number of essential physiological functions in the body including:

1 maintenance of normal permeability of cellular membranes;
2 maintenance of normal excitability of nerve and muscle;
3 release of neurotransmitters, many hormones and exocrine secretions;
4 muscular contraction;
5 major constituent formation of bone and teeth;
6 coagulation of blood;
7 production of milk; and
8 activity of many enzymes.

About 99% of body calcium and 80% of body phosphorus is found in bone. The concentration of calcium in plasma is about 2·4 mmol l^{-1} (2·1–2·6 mmol l^{-1}), of which about 1·2 mmol l^{-1} is ionized, about 1·1 mmol l^{-1} is bound to protein and a small amount is present as complexes with anions, such as citrate, bicarbonate or phosphate. Phosphorus occurs in plasma in a number of organic constituents, including lipids and nucleotides, and is also present as inorganic phosphate at a concentration of about 1 mmol l^{-1}. The solubility product of calcium and phosphate is such that the product of the free ions remains constant.

On a typical diet, about 25 mmol (1 g) of calcium is ingested daily, but individuals vary greatly in their intake. Most of this calcium is in the form of calcium phosphate derived especially from milk and other dairy products. There is abundant phosphorus in the diet to meet daily needs. For calcium balance to be maintained, an intake of 12·5–20 mmol per day is recommended for infants and adults. A higher daily intake (25–37·5 mmol) is recommended for rapidly growing adolescents and for women during pregnancy and lactation, and after the menopause. Normally, about a quarter of the ingested calcium is absorbed in the gastrointestinal tract, but much of this is returned in the gut secretions while the rest is secreted in the urine (Fig. 8.14). Absorption occurs mainly in the duodenum by active transport (see p. 605). Both intestinal absorption and urinary excretion of calcium are under hormonal control, as is the exchange of calcium between the extracellular fluid and the skeleton.

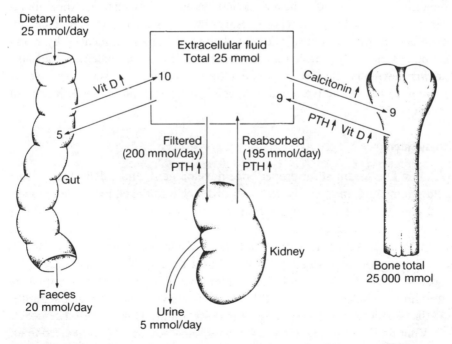

Fig. 8.14. Calcium balance in the body and its regulation by parathyroid hormone (PTH), calcitonin and vitamin D.

Bone

Bone is composed of an organic matrix of collagen in a 'ground substance' consisting largely of mucopolysaccharides and non-collagen proteins, on to which crystals of a complex salt of calcium and phosphate very similar to hydroxyapatite ($Ca_{10}(PO_4)_6(OH)_2$) are deposited. A more amorphous form of calcium phosphate is also present as well as small amounts of Na, Mg, Cl and F.

Three types of cells appear to function in the formation and resorption of bone, namely osteoblasts, osteoclasts, and osteocytes. The *osteoblasts* synthesize and secrete collagen fibres and promote the deposition of calcium phosphate crystals, while *osteoclasts* cause resorption of bone. Bone resorption depends on the destruction of collagen by lysosomal enzymes and phagocytosis, and on the dissolution of bone mineral by an increase in lactate and citrate production. *Osteocytes* are the most numerous cells in mature bone and are formed from osteoblasts. They appear to play an essential role in the exchange of calcium between extracellular fluid and bone. This role depends upon the activity of both parathyroid hormone and calcitriol. Only about 1% of the calcium and phosphate of bone is in equilibrium with the extracellular fluid through the activity of osteocytes, the so-called exchangeable pool of bone mineral which acts to buffer short-term changes in blood calcium ion concentration. The remaining 99% comprises the non-exchangeable pool, from which the mineral can be released only by osteoclastic resorption, as occurs in normal remodelling or in certain disease states. The remodelling of bone is influenced by many hormones, not only parathyroid hormone, calcitonin and calcitriol, but also growth hormone, thyroxine, gonadal hormones and cortisol. The role of vitamin D and its active metabolite calcitriol, has achieved prominence in recent years.

Vitamin D

Vitamin D is essential for proper bone development, and a deficiency of this vitamin in children causes *rickets*, a disorder characterized by stunted growth and bowing of the limbs due to failure of mineralization of bone. In adults, vitamin D deficiency can cause a failure of mineralization of bone (*osteomalacia*). Vitamin D is a steroid found in a limited number of foodstuffs, e.g. cod-liver oil, and is also synthesized in the skin by the action of ultraviolet light on 7-dehydrocholesterol. The vitamin occurring naturally in animals is vitamin D_3 (cholecalciferol) and that in plants is vitamin D_2 (ergocalciferol). As the two forms are equally active, the term vitamin D will be used hereon.

Vitamin D is now regarded as a prohormone because it is converted to an active metabolite *calcitriol* which controls calcium and phosphorus metabolism. It is first converted to a 25-hydroxy derivative in the liver and then, in the kidney, to *1,25-dihydroxycholecalciferol* (1,25-DHCC, calcitriol), if extracellular concentrations of calcium or phosphate are low. The renal enzyme 1 α-hydroxylase, which catalyses the final conversion, is stimulated by parathyroid hormone, low plasma phosphate, low plasma calcium (directly) and the gonadal steroid hormones. Synthesis of calcitriol is also controlled by negative feedback of calcitriol itself. If extracellular concentrations of calcium and phosphate are normal, most of the vitamin is transformed in the kidney to the 24,25-dihydroxy and 1,24,25-trihydroxy derivatives which are probably intermediates in degradative pathways.

Calcitriol acts on the small intestine to promote the absorption of calcium and phosphate which are necessary for bone formation and it facilitates bone mineralization directly. It has also been shown, together with parathyroid hormone, to have the reverse effect on bone mineralization *in vitro*, causing release of these ions from bone. This action may be important in bone remodelling. The precise mechanism of action of calcitriol is uncertain, although it appears to promote gene expression like other steroid hormones (p. 254), and in the intestine this leads to an increase in the synthesis of a calcium-binding protein that somehow promotes calcium absorption.

Parathyroid hormone

Parathyroid hormone is an 84 amino acid polypeptide (M_r 9500) secreted by the 'chief' cells of the parathyroid glands (Fig. 8.13). It may also be secreted by accessory parathyroid tissue which sometimes occurs in the neck region. Removal of the parathyroid glands (and accessory tissue) causes plasma calcium to fall by as much as 50% and leads to *hypocalcaemic tetany*. This is characterized by extensive spasms of skeletal muscle and can lead to asphyxiation due to laryngeal spasm.

The hormone is synthesized as the preprohormone containing additional amino acids which are removed within the gland before the hormone is secreted. Further cleavage of the hormone into two or more fragments occurs in the liver and kidney. This complicates the interpretation of radioimmunoassays used to measure the concentration of circulating hormone. As only the first 34 amino acids in the NH_2-terminal portion of the molecule are necessary for full biological activity, some of these fragments may retain biological activity.

Actions of parathyroid hormone

Parathyroid hormone *increases* ionized plasma calcium and *lowers* plasma phosphate concentration. It acts on the *bone*, the *kidney* and, indirectly, the *gastrointestinal tract*. Its actions on bone and kidney appear to be mediated by cyclic AMP. Parathyroid hormone increases the rate of bone resorption by stimulating the activity of osteocytes and osteoclasts. (The presence of vitamin D is needed for these actions.) This effect of parathyroid hormone is important in long-term regulation of plasma calcium but its actions on the kidney appear to be more important in compensating for short-term changes. In the kidney, parathyroid hormone increases the tubular reabsorption of calcium and decreases that of phosphate. The resulting fall in plasma phosphate stimulates calcium release from the bone and prevents the deposition of calcium phosphate. Parathyroid hormone also stimulates the formation of calcitriol in the kidney. Thus, the absorption of calcium in the gastrointestinal tract is increased as a consequence of an increase in calcitriol. In the long term, increased secretion of parathyroid hormone may result in a net loss of calcium from the body through the kidney. Under these conditions, the increased ionized plasma calcium increases the filtered load to an extent greater than that by which the reabsorption of calcium has been stimulated.

Control of parathyroid hormone secretion

Parathyroid hormone secretion is regulated solely by the level of *plasma calcium* acting on the parathyroid glands and varies inversely with plasma calcium levels (Fig. 8.15). A decrease in plasma calcium concentration causes an increase in secretion of parathyroid hormone and vice versa. Since parathyroid hormone increases extracellular calcium, further release is inhibited—a typical negative feedback mechanism.

Calcitonin

Calcitonin is a polypeptide (M_r 3500) secreted by the *C* cells of the thyroid gland. It *decreases* plasma calcium and phosphate levels by inhibiting bone resorption and possibly by decreasing the reabsorption of calcium and phosphate in the kidney. Its actions are mediated by cyclic AMP.

The release of calcitonin is stimulated by an *increase* in plasma calcium (Fig. 8.15) and it has been suggested that calcitonin protects animals against hypercalcaemia. Removal of the thyroid gland does not cause hypercalcaemia in man, whereas hypocalcaemia regularly follows removal of the parathyroid glands. Therefore, the physiological significance of the calcitonin is uncertain. It is used clinically in the treatment of Paget's disease to reduce the accelerated rate of bone turnover.

Disorders of calcium metabolism

Hypocalcaemia causes excessive neuromuscular irritability. A rapid decrease of ionized plasma calcium to below 1 mmol l^{-1} results in spontaneous firing of peripheral nerves. On the afferent side, this causes unusual sensations such as

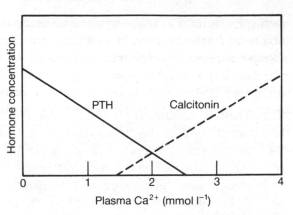

Fig. 8.15. Effect of plasma calcium on the concentrations of parathyroid hormone (PTH) and calcitonin in blood. (After Copp, D. H. (1968) in *Parathyroid Hormone and Thyrocalcitonin (Calcitonin)*, (eds R.V. Talmage & L.F. Belanger, p. 25, Excerpta Medica, Amsterdam.)

tingling (paraesthesia); on the efferent side, muscular twitching, spasm and cramps can develop, these motor manifestations being called *manifest tetany*. If ionized calcium decreases more slowly or to a lesser extent, stimuli such as localized ischaemia, hyperventilation or pressure over a nerve may be required to elicit the motor events. This is termed *latent tetany*. A low ionized plasma calcium may be consequent upon decreased levels of parathyroid hormone or calcitriol in the body. It may also result from increased plasma pH, e.g. in respiratory alkalosis, due to the release of hydrogen ions from plasma proteins making additional negatively charged binding sites available for calcium. Tetany may also occur when ionized magnesium concentration in plasma is reduced, and in potassium depletion associated with a metabolic alkalosis.

Osteomalacia, due to a deficiency of calcitriol, is characterized by a normal bone mass but the newly synthesized matrix fails to calcify. In *osteoporosis*, the bone is normally calcified but bone mass is decreased to a greater degree than in *osteopenia* which accompanies ageing. Oestrogen deficiency after the menopause is associated with osteoporosis which can be prevented by relatively small doses of oestrogen. Osteoporosis also occurs with prolonged immobilization and with excess glucocorticoid secretion or administration.

Hypercalcaemia is seen most frequently in hyperparathyroidism and in multiple myeloma (p. 348). Increased calcium mobilization from bone leads to painful softening and fractures of bones, and the increased calcium and phosphate excretion in urine may result in nephrocalcinosis and renal stones. Increased plasma calcium may also cause headaches and decreased tone (hypotonia) in skeletal and intestinal muscle.

8.5 The adrenal glands

There are two adrenal glands, situated one on top of each kidney. Each adrenal gland comprises two endocrine organs—the *adrenal medulla* and the *adrenal cortex* (Fig. 8.16). The two parts of the adrenal gland have different embryonic origins and are anatomically quite distinct. The adrenal medulla secretes *catecholamines* while the adrenal cortex secretes *corticosteroids*.

Adrenal medulla

The medulla is a modified nervous tissue derived from the neural crest and can be regarded as a collection of postganglionic sympathetic neurones in which the axons have not developed. The catecholamines are produced in chromaffin cells which are of two types, one secreting *adrenaline* (epinephrine) and the other *noradrenaline* (norepinephrine). In man, adrenaline constitutes about 80% of the catecholamines produced by the medulla. Small collections of chromaffin cells are also located outside the adrenal medulla, usually adjacent to the chain of sympathetic ganglia.

In addition to catecholamines, the adrenal medulla also contains enkephalins, dynorphin, neurotensin, somatostatin and substance P. The role of these adrenal peptides has not yet been elucidated, although it is known that met- and leu-enkephalins are located within the adrenaline-containing chromaffin cells and are co-secreted with catecholamines.

The synthesis of catecholamines from tyrosine has been dealt with in Chapter 2 (p. 69). The amines are stored in membrane-bound granules and their secretion is initiated by acetylcholine released from *preganglionic sympathetic fibres* in the splanchnic nerves. Acetylcholine depolarizes the chromaffin cells causing calcium to enter and trigger the release of the granular contents by exocytosis.

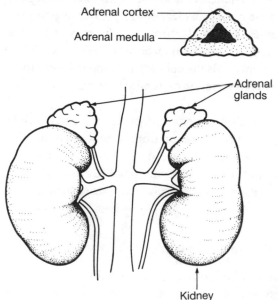

Fig. 8.16. The adrenal glands.

Once released into the bloodstream the catecholamines have only a short half-life (minutes). They are rapidly taken up into extraneuronal tissues and degraded by catechol-*O*-methyltransferase (COMT), or into nerve terminals and recycled or degraded by monoamine oxidase (MAO), the degradation products eventually appearing in the urine.

Actions of catecholamines

The actions of adrenaline and noradrenaline are complex and depend on their effects on the various subclasses of α- and β-receptors (p. 70) which differs in their distribution differ from tissue to tissue. Noradrenaline causes widespread vasoconstriction and a marked increase in peripheral resistance while adrenaline causes vasoconstriction in skin and viscera but vasodilatation in skeletal muscles. Both catecholamines increase heart rate and contractility directly, but in the intact animal the increase in peripheral resistance and mean arterial pressure caused by noradrenaline administration leads to reflex bradycardia. Adrenaline has a more pronounced effect on metabolic processes and increases the basal metabolic rate, stimulates glycogenolysis and mobilizes free fatty acids. Catecholamines also cause bronchodilatation and relaxation of the gastrointestinal tract.

Control of catecholamine secretion

The secretion of catecholamines is initiated by sympathetic activity controlled by the hypothalamus and occurs in response to such stimuli as pain, excitement, anxiety, hypoglycaemia, cold and haemorrhage. Increased secretion is part of the 'fight or flight' reaction described by Cannon (1871–1945). In an emergency, catecholamines released by the adrenal medulla are disseminated in the bloodstream while noradrenaline released from sympathetic nerve terminals acts at discrete points in the body. In frightened or stressed animals, there is a general increase in sympathetic activity in which the sympathetic nerves appear to play the dominant role because removal of the adrenal medulla does not seriously impair an animal's ability to cope with stress.

Catecholamine-secreting tumours of the adrenal medulla, one type of which is known as *phaeochromocytoma*, can result in severe hypertension and hyperglycaemia.

Adrenal cortex

The adrenal cortex secretes steroid hormones (corticosteroids) which can be classified as follows.

1 *Glucocorticoids*, i.e. *cortisol* and *corticosterone*, which affect the metabolism of carbohydrates, fats and proteins.

2 *Mineralocorticoids*, mainly *aldosterone*, which are essential for the maintenance of sodium and potassiom balance and extracellular fluid volume.

3 *Sex hormones*, mainly *androgens*, which may play a minor role in reproductive function, particularly as a source of androgens in the female, and are involved in growth at puberty.

The cortex is organized histologically into three zones—the outer *zona glomerulosa*, which secretes aldosterone, the middle *zona fasciculata*, which secretes mainly glucocorticoids, and the inner *zona reticularis*, which secretes mainly androgens. Removal of the pituitary causes the fasciculata and reticularis zones to atrophy but has little effect on the zona glomerulosa, which shows that the cells in the latter zone are regulated differently. During fetal life, the cortex has a large inner fetal zone which produces precursor steroids for the synthesis of oestrogens by the placenta (fetoplacental unit, see p. 312).

The adrenocortical steroids have the same basic structure as other steroids—the glucocorticoids and mineralocorticoids contain 21 carbon atoms and the androgens 19 (Fig. 8.17). They are derived from cholesterol which may be synthesized either directly by the adrenocortical cells or taken up from low-density lipoproteins in the circulation and stored as cholesterol esters in lipid droplets in the cytoplasm. Cholesterol is released from the lipid droplets by the action of cholesterol esterase which is stimulated by adrenocortico-trophic hormone (ACTH). It is then converted to pregnenolone in the mitochondria. This step, which is also regulated by ACTH, is the rate-limiting step in steroid biosynthesis. The actions of ACTH appear to be mediated by cyclic AMP. Pregnenolone is then transferred to the smooth endoplasmic reticulum. It undergoes further modifications here and in the mitochondria to form the three main classes of steroids (Fig. 8.17).

The steroid hormone produced depends on which hydroxylases and other enzymes are present in the tissue concerned. For example, formation of cortisol proceeds via hydroxylation of pregnenolone to 17-hydroxypregnenolone or of progesterone to 17-hydroxyprogesterone (Fig. 8.17). These steps are catalysed by the enzyme 17 α-hydroxylase and require NADPH and O_2. On the other hand, the synthesis of corticosterone (and aldosterone) proceeds via progesterone, and 17 α-hydroxylase is not required. Thus the activity of this enzyme determines the proportion of the two steroids secreted. There is no appreciable storage of these hormones in the adrenal cortex and the rate of release corresponds to the rate of synthesis. In man, approximately 20 mg of cortisol, 3 mg of corticosterone and 0·2 mg of aldosterone are secreted per day. The adrenal cortex also secretes significant amounts of androgens, particularly dehydroepiandrosterone and androstenedione. These are only weakly andro-

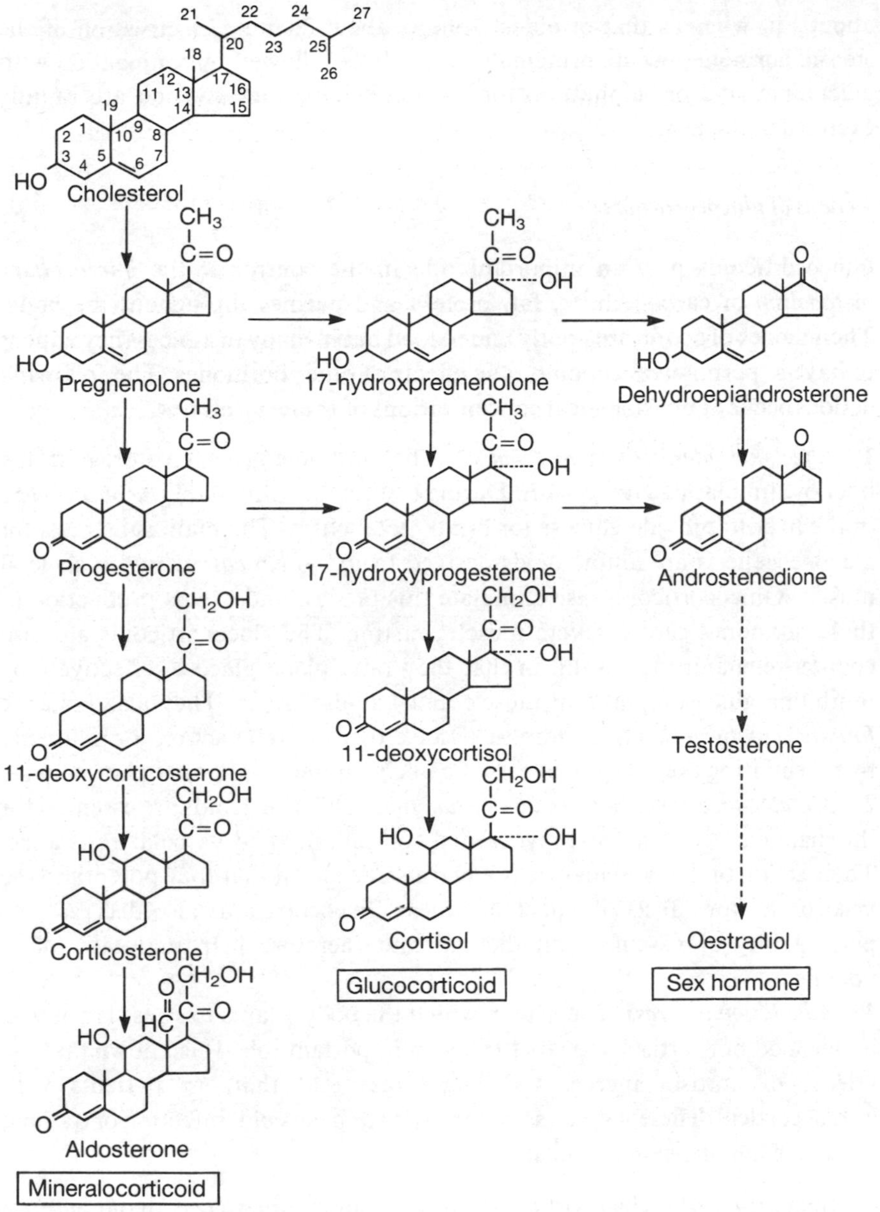

Fig. 8.17. Synthesis pathways for steroid hormones.

genic but can be converted in other tissues to the more potent androgens, testosterone and dihydrotestosterone (p. 298).

The corticosteroids are transported in the circulation mostly bound to plasma proteins such as *corticosteroid-binding globulin* (transcortin). About 90% of cortisol and 60% of aldosterone exists in the bound form, which acts as a reserve and protects the steroids from degradation. The half-life of cortisol is about 1 h, whereas that of aldosterone is about 20 min. Inactivation of the steroid hormones occurs principally in the liver followed by conjugation with glucuronic acid or sulphate to form water-soluble forms which are rapidly excreted in the urine.

Actions of glucocorticoids

Glucocorticoids play an important role in the control of the *intermediary metabolism* of carbohydrate, fat, protein and purines throughout the body. Their modes of action are poorly understood but in many instances they appear to have a 'permissive' action on the effects of other hormones. The following actions occur at physiological concentrations of these hormones.

1 *Effects on intermediary metabolism.* They promote glycogen storage in the liver by stimulating *glycogenesis*. During fasting they stimulate *gluconeogenesis* in the liver to provide glucose for brain metabolism. The main substrates for gluconeogenesis are amino acids derived from *protein catabolism* in skeletal muscle. Glucocorticoids also stimulate this process and excess production of these hormones causes severe muscle wasting. The glucocorticoids are also counter-regulatory to insulin in that they raise blood glucose, effectively by inhibiting glucose uptake in muscle and adipose tissue. They also enhance *fatty acid mobilization* from adipose tissue either by direct actions or indirectly by potentiating the lipolytic effects of catecholamines and growth hormone.

2 *Maintenance of normal circulatory function.* Glucocorticoids are essential for the maintenance of normal myocardial contractility and vascular resistance. Their action on the vasculature is a permissive one in that they potentiate the vasoconstrictor effects of catecholamines. Glucocorticoids also decrease the permeability of vascular endothelium and therefore help maintain blood volume.

3 *Adaptation to stress.* The way in which the body adapts to stress is not well understood but cortisol appears to play an important role. It is known that the release of cortisol increases during stress and that, in patients with adrenocortical deficiency, stress factors, such as heat, cold, infection or trauma, can cause hypotension and death.

In addition, the glucocorticoids do have some mineralocorticoid activity and this may be of significance because of their comparatively high secretion rate. In large doses, the glucocorticoids suppress the *immune response*. They decrease the number of circulating lymphocytes and eosinophils, cause involution of the thymus and lymph nodes, and may depress the antibody response. Therefore, synthetic corticosteroids are used therapeutically to suppress rejection of transplanted organs and to treat allergies. They also have *anti-inflammatory* properties and are used in the treatment of rheumatoid arthritis and related diseases. They may also be used to increase the synthesis of surfactant in fetal lung.

Control of glucocorticoid secretion

The secretion of cortisol and corticosterone (and androgens) is controlled by *adrenocorticotrophic hormone* (ACTH) produced by the anterior pituitary. The secretion of ACTH is regulated by hypothalamic secretion of *corticotrophin-releasing hormone* (CRH) and *antidiuretic hormone* (ADH) into the hypophyseal portal system. In this instance, the ADH is secreted by the same cells as those that release CRH and these are distinct from the ADH-secreting cells that project to the posterior pituitary. Cortisol exerts a negative-feedback effect on the hypothalamus and on the pituitary (Fig. 8.18). The level of plasma cortisol follows a diurnal pattern, peak levels occurring in the morning just before waking. This variation is related to the sleeping and waking pattern and interruption of this pattern may be responsible for the fatigue (jet-lag) experienced by air travellers. The effect of stress can increase cortisol secretion at any time.

Actions of mineralocorticoids

Aldosterone is the main mineralocorticoid produced by the adrenal cortex. It acts chiefly on the distal tubules of the *kidney* to promote the *reabsorption of Na^+* in exchange for K^+ and H^+ ions which are excreted (p. 541). Excess production of aldosterone with retention of Na^+ leads to expansion of ECF volume and hypertension. Adrenalectomy leads to a fall in extracellular Na^+, hypotension and eventually death.

Control of mineralocorticoid secretion

ACTH is not the major regulator of aldosterone secretion, in contrast to the other corticosteroids, but it does play a supportive role. The primary regulator of aldosterone secretion appears to be angiotensin II produced by the renin–angiotensin system (p. 540). An increase in plasma K^+ concentration also stimulates the release of aldosterone by the adrenal cortex.

Disorders of adrenocortical function

In *Addison's disease,* there is deficient secretion of all adrenocortical hormones usually due to autoimmune destruction of the gland. This disease usually develops slowly and is characterized by lethargy, weakness, weight loss and hypotension. A sudden stress can precipitate a crisis requiring emergency

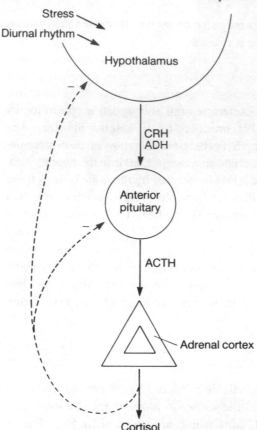

Stress

Diurnal rhythm

Hypothalamus

CRH
ADH

Anterior
pituitary

ACTH

Adrenal cortex

Cortisol

Fig. 8.18. Control of glucocorticoid secretion.

medical treatment. A common feature of this disease is hyperpigmentation of the skin due to excessive secretion of ACTH leading to stimulation of melanocyte activity in the skin (p. 264). The elevation in ACTH levels is brought about by removal of the negative feedback provided by cortisol. Patients with Addison's disease require replacement therapy with both a glucocorticoid and a mineralocorticoid.

Cushing's syndrome is a condition associated with excess plasma glucocorticoids resulting from excess ACTH secretion, tumours of the adrenal cortex or overadministration of glucocorticoids in the course of therapy. This disease is characterized by redistribution of body fat, a 'moon face', severe muscle wasting, osteoporosis, a predisposition to diabetes and hypertension. A diagnosis of Cushing's syndrome is indicated if plasma cortisol is elevated throughout the day and if ACTH secretion is not suppressed by low-dose administration of the potent glucocorticoid drug dexamethasone.

Conn's syndrome, or primary aldosteronism, is due to excess mineralocorticoid secretion caused by a tumour of the adrenal cortex. This leads to K^+

depletion and Na$^+$ and water retention, resulting in hypertension, muscle weakness, tetany and hypokalaemic alkalosis.

Adrenogenital syndrome is associated with excessive androgen secretion which may cause masculinization in the female and precocious puberty in the male. This may be due to an androgen-secreting tumour or it may be congenital. The latter is known as *congenital adrenal hyperplasia* in which one of the enzymes involved in cortisol synthesis is deficient. This leads to increased ACTH secretion by the pituitary and hence excess production of adrenal androgens. Treatment with glucocorticoids corrects the deficiency and also suppresses excess ACTH secretion.

8.6 The pancreatic islets

The pancreas is both an endocrine and an exocrine gland. The endocrine portion of the pancreas is localized in the *islets of Langerhans* which constitute only 2% of the mass of the pancreas. *Insulin* is produced in the B (β) cells and *glucagon* in the A (α) cells of the islets (Fig. 8.19). Recently, it has been discovered that a third hormone, *somatostatin* identical to growth hormone release-inhibiting hormone secreted by the hypothalamus, is also produced in the islets from D (δ) cells. Insulin was successfully extracted by Banting and Best in 1921 from the dog pancreas after they had first depleted it of proteolytic enzymes by ligating its exocrine ducts. The disease known as *diabetes mellitus* is due to a deficiency of insulin or to insulin resistance.

Insulin

Insulin is a small protein (M_r 6000) consisting of two peptide chains, called A and B, which are linked by two disulphide bonds. The A chain contains 21 amino acid residues and the B chain 30 residues. Beef and pig insulin differ from human insulin in only a few residues and both are used in the treatment of diabetes mellitus. Human insulin can now be made by chemical manipulation of porcine insulin and by DNA recombinant technology. The therapeutic potencies of these three types of insulin are remarkably similar. Insulin is synthesized as a larger single polypeptide pre-proinsulin, which is cleaved soon after synthesis to form *proinsulin* (Fig. 8.20). Proinsulin is packaged into vesicles and converted to insulin by cleavage of a connecting peptide to form two peptide chains. Insulin is released from the cell by exocytosis in response to an increase in blood glucose. Insulin is secreted into the portal vein and thus reaches the liver directly. It has a half-life in the blood of only a few minutes.

Actions of insulin

Insulin lowers blood glucose by facilitating glucose uptake in muscle and adipose tissue and by inhibiting hepatic glucose output. The most important

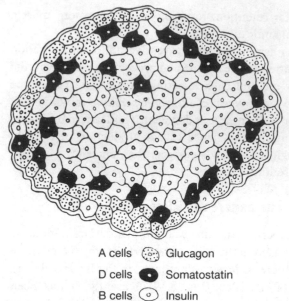

A cells ⊙ Glucagon

D cells ● Somatostatin

B cells ⊙ Insulin

Fig. 8.19. Islet of Langerhans. (From Orci, L. & Unger, R.H. (1975) *Lancet* **2**, 1243–1244.)

effects of insulin are in the liver, where it stimulates glycogen and fat synthesis and inhibits glycogen breakdown and ketone body formation. In muscle, it stimulates glucose and amino acid uptake, and glycogen and protein synthesis, and in adipose tissue it stimulates glucose uptake and triglyceride synthesis. Insulin also increases K^+ uptake into cells and consequently can lower plasma K^+.

The mechanism by which the binding of insulin to its specific cell membrane receptor is translated into these specific events is not known. The receptor is composed of two α subunits, which bind the hormone, and two β subunits, which are tyrosine-specific protein kinases. Binding of insulin results in autophosphorylation of the receptor, the role of which has yet to be elucidated. No specific intracellular second messenger has been identified, and it is probable that different mechanisms are involved for different kinds of cell response. Intracellular cyclic AMP levels decrease following insulin binding to its receptor, but the significance of this is not known.

Control of insulin secretion

An increase in plasma *glucose* concentration provides the major stimulus for insulin secretion. There is an initial rapid phase of secretion, due to release of preformed hormone, followed by a second slower phase of sustained secretion. Some *amino acids*, such as arginine, are also potent stimulators of insulin release. After feeding, the level of insulin may rise even before that of blood

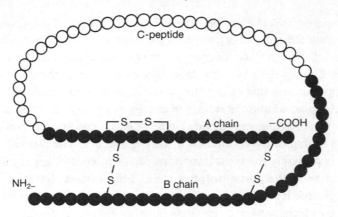

Fig. 8.20. Structure of proinsulin (insulin is shown in black).

glucose because *gastric inhibitory peptide* (GIP) can also stimulate insulin release. Other potent stimuli for insulin release are *glucagon* and the *sulphonylurea drugs*, such as tolbutamide, which are used in the treatment of mild cases of diabetes. *Adrenaline* inhibits insulin release, and so too does *somatostatin*. The latter which is produced in the islets of Langerhans probably plays an important role in the local control of insulin secretion. Neural control is also important; parasympathetic activity enhances insulin release while sympathetic activity inhibits it.

Glucagon

Glucagon is a polypeptide (M_r 3000) which is secreted by the pancreas in response to low blood glucose. In contrast to insulin, it *increases blood glucose* levels. Glucagon acts on the liver (via cyclic AMP) to stimulate *glycogenolysis* (breakdown of glycogen) and *gluconeogenesis* (synthesis of glucose from lactate, amino acids or glycerol). Glucagon secretion is also stimulated by amino acids, as is insulin release. This ensures that, when there is a high intake of amino acids, a precipitous fall in blood glucose due to elevated insulin is prevented by the action of glucagon. Glucagon is also a potent stimulator of fat mobilization but an infusion of glucagon does not normally increase free fatty acid levels, probably because its actions are offset by a concomitant rise in insulin.

Control of energy utilization and storage

The brain uses glucose almost exclusively as an energy source and it is necessary to maintain an adequate blood glucose concentration at all times or convulsions and coma will ensue. Blood glucose is maintained at fairly constant levels of 4–6 mmol 1^{-1} by the interactions of several hormones, namely *insulin, glucagon, growth hormone, adrenaline* and *cortisol*. Of these, insulin acts to lower blood glucose and is known as the 'regulatory' hormone, whereas the other four act to raise blood glucose and are known as the 'counter-regulatory' hormones. During the *absorptive state* following a meal, there are adequate glucose supplies and this causes the secretion of insulin which enhances the utilization of glucose and the storage of energy as glycogen and fat. During the *post-absorptive* or *fasting state* blood glucose falls and insulin secretion decreases in relation to that of the counter-regulatory hormones. The ratio of insulin to glucagon is probably the most important factor in controlling the shift from the absorptive to the post-absorptive state. This ensures that during fasting adequate glucose levels are maintained for the brain by gluconeogenesis in the liver and by glucose-sparing reactions in other tissues. A further account of energy storage and utilization is given in Chapter 15 (p. 611).

Effects of insulin deficiency

Diabetes mellitus is a major health problem which affects about 2% of the population. A predisposition to diabetes is inherited but the genetic factors are complex. Two types of diabetes are recognized clinically—juvenile-onset or insulin-dependent diabetes (*type I*) and maturity-onset or non-insulin-dependent diabetes (*type II*). Type I patients (10–12% of diabetics) have low plasma insulin and require injections of insulin. It is thought that type I diabetes is an autoimmune disease triggered in childhood by a viral infection. Type II patients may have normal or elevated levels of insulin but show decreased sensitivity to insulin, often correlating with a reduction in insulin receptor concentration. Type II patients are often obese and generally show improvement with weight reduction. Diabetes can be simulated in experimental animals by treatment with alloxan which destroys the B cells of the islets of Langerhans.

By 'diabetes mellitus' is meant the passing of sweet urine. The capacity of the renal tubules to reabsorb glucose actively is exceeded and glucose spills over into the urine. The loss of so much solute causes an *osmotic diuresis* resulting in polyuria and polydipsia. Large amounts of salts are consequently lost and this can lead to dehydration and hypotension. The blood concentration of free fatty acids is raised and their metabolism produces ketones which cause *metabolic acidosis*. If left untreated, the patient may become unconscious (*diabetic coma*). The administration of too much insulin can also lead to coma (*insulin coma*) because of the sudden lowering of blood glucose and the brain's dependence on glucose as an energy source.

There are also long-term effects of the disease: in particular, atherosclerosis in the coronary and leg arteries, and capillary abnormalities in the eyes, kidneys and nerves. Diabetics are prone to infection and injuries take longer to heal than normal.

In the diagnosis of diabetes, the physician is guided by the fasting blood glucose concentration and the presence of glucose and ketones in the urine. A *glucose tolerance test*, in which changes in the blood glucose are measured in response to oral administration of glucose, provides a more definite indication. In diabetics, the blood glucose rises higher in response to a glucose load and returns to baseline more slowly than in normal subjects.

8.7 Appendix

Hormone assay

There are three methods for assaying hormone concentrations in serum or urine—bioassay, radioimmunoassay and radioreceptor assay—although some hormones, e.g. catecholamines and steroid hormones, may be estimated by chemical analysis.

Bioassay

This involves quantifying the responses of tissues to various concentrations of a hormone (standard or unknown) either *in vitro* or *in vivo*. For example, human chorionic gonadotrophin was formerly assayed by its ability to induce ovulation in rabbits. A standard curve is plotted and the concentration of the unknown is extrapolated from it. Such assays are often cumbersome to perform and variable in their response. However, they do measure the biological activity of the hormone.

Radioimmunoassay

This assay, now widely used, depends on the availability of radioactively labelled hormone and on antibodies that react with it. Peptide, thyroid and steroid hormones can all be labelled with ^{125}I to a high degree of specific activity. Antibodies to the larger peptide hormones are prepared by immunizing experimental animals against these molecules. Antibodies to smaller polypeptides, thyroid and steroid hormones (and even to cyclic AMP) can also be produced if those molecules are first made antigenic by attaching them to proteins.

A competition assay is set up in which the standard or unknown hormone competes with the labelled hormone for a site on the antibody (Fig. 8.21); the more unlabelled hormone present the less labelled hormone is bound to the antibody. The hormone-bound complex is then separated from the free

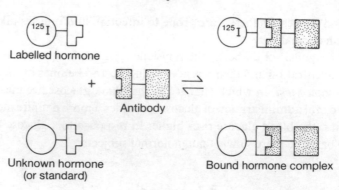

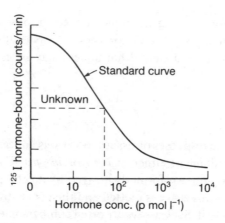

Fig. 8.21. Radioimmunoassay.

hormone by various physicochemical means (e.g. filtration, precipitation, charcoal exclusion) and the radioactivity of the hormone-bound complex determined. A standard curve is prepared and the concentration of the unknown is extrapolated from this.

One of the advantages of the radioimmunoassay is that it is extremely sensitive. However, it does not necessarily measure the biological activity of a hormone and it can detect precursors and degradation products of a hormone, thus leading to overestimation of hormone concentration.

Radioreceptor assay

This is similar to the radioimmunoassay but instead of an antibody for binding it uses a receptor of the hormone. Theoretically, such assays are ideal because they simulate the first step in the action of a hormone but in practice they are not as sensitive as radioimmunoassays.

9 · Reproduction

9.1 Sexual development

9.2 The male reproductive system
Spermatogenesis
Testosterone
Control of male reproductive activity
Semen
Erection
Ejaculation
Vasectomy

9.3 The female reproductive system
Oogenesis and follicular development
Gonadotrophic hormones
Menstrual cycle
Ovarian hormones: oestrogen and
    progesterone

Control of the female reproductive cycle
Hormonal contraception
Hyperprolactinaemia
Pineal gland

9.4 Pregnancy
Placental hormones
Parturition
Lactation

The sexual differences between male and female depend ultimately on differences in their chromosomes. In most mammalian species, the female has two X sex chromosomes while the male has one X and one Y sex chromosome, in addition to their autosomal chromosomes (22 pairs in humans). Both departures from the normal sex-chromosomal pattern and hormonal disturbances can result in defective development of the reproductive system. It is possible to determine the sex-chromosomal pattern by laboratory examination of cell smears, since the female pattern can be conveniently recognized by the presence of an extra piece of chromatin (an inactive X chromosome) called the 'sex chromatin' or 'Barr body'. Sex-chromosome abnormalities, such as XXY (Klinefelter's syndrome) and XO (Turner's syndrome), result in incomplete development of the gonads. The XXX ('superfemale') and XYY ('supermale') patterns, however, do not seem to be characterized by abnormal sexual development.

A fundamental difference between the female and the male is that the former undergoes obvious cyclic variations in reproductive activity. *Female* mammals have an *ovarian cycle* with a characteristic mean frequency for each species, e.g. 4 days in rats and 28 days in women. In women and in some other primates, the cycle is marked by a period of menstrual bleeding, which is due to the shedding of the endometrial lining of the uterus. Other female mammals do not menstruate although they show a phase called *oestrus*, or 'heat' as it is commonly known, when the female becomes receptive to the male just before ovulation. In some species, e.g. the cat and rabbit, ovulation is triggered by copulation. In most mammals, however, ovulation is regulated by an intrinsic rhythm controlled by interactions between the hypothalamus, pituitary and gonads.

9.1 Sexual development

In humans, the 'indifferent' gonads of both sexes are identical until differentiation begins at about the sixth week of fetal life. In the genetic male, the presence of the H–Y gene, a specific sex-determining gene located on the Y chromosome, causes differentiation of the gonads into the testes. Genetic females lack the H–Y gene and as a result the indifferent gonads develop into the ovaries. At this stage, the fetus has primordial genital ducts for both the male (*Wolffian ducts*) and the female (*Müllerian ducts*). In the male fetus, testosterone secreted by the developing testes causes the Wolffian ducts to develop into the internal genitalia (epididymis, vas deferens, seminal vesicles and prostate) and a peptide hormone called Müllerian-inhibiting factor (MIF), also secreted by the testes, inhibits the development of the Müllerian ducts. In the female fetus, the absence of testosterone and MIF allows the opposite to occur. The external genitalia are similarly bipotential at this stage. In males, androgenic hormones derived from the testes cause the external genitalia to develop into the male form, whereas external genitalia develop into the female form in the absence of these androgens.

The secretion of testosterone and MIF is not all that is required for the development of the male phenotype. The tissues must have the necessary androgen receptors to respond. Otherwise a defect known as *testicular feminization* (androgen-insensitivity syndrome) will result in which genetic males appear as phenotypic females with abdominal testes. Moreover, some fetal tissues require testosterone for differentiation and some, e.g. the prostate and the penis, require dihydrotestosterone. The conversion of testosterone to dihydrotestosterone requires the presence of the enzyme, 5 α-reductase. In *5 α-reductase deficiency*, the affected males at birth have testes, but lack a prostate gland, and their external genital organs resemble those of the female. Another defect caused by an inborn error of metabolism is *congenital adrenal hyperplasia* which is associated with masculinization of the female fetus (p. 287).

Animal studies indicate that the potential for developing cyclic reproductive activity is present in both sexes at the fetal stage. The secretion of testosterone in the male appears to abolish this intrinsic rhythm, because if testosterone is injected into the genetic female at a critical period the sexual rhythm is not generated. This critical period occurs within a few days after birth in the rat while in the monkey it occurs before birth.

After birth reproductive development is dormant until *puberty* when the reproductive organs in both sexes are reactivated. This coincides with accelerated growth of the body and the development of the *secondary sexual characteristics*. The onset of puberty usually occurs at about 11 in girls and 12 in boys, although this may vary considerably depending on genetic and environmental influences. Sexual maturity is signified by the *menarche* or first menstrual bleeding in the female and by the first ejaculation in the male. At about 50 years of age in women the *menopause* occurs: the ovary ceases to respond to gonadotrophins, the sexual cycles gradually disappear and menstruation eventually stops. In contrast, the production of sperm in males continues throughout life although it may gradually diminish.

9.2 The male reproductive system

The primary reproductive organs or gonads of the male are the *testes* which produce *spermatozoa* and also secrete the male sex hormone, *testosterone*. In addition, there are the *accessory* reproductive ducts (Fig. 9.1) and secretory glands (*seminal vesicles, prostate gland* and *bulbourethral glands*) which are

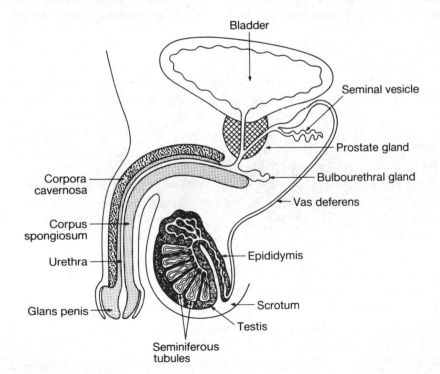

Fig. 9.1. The male reproductive system.

involved in the transport and delivery of spermatozoa to the female. A few weeks before birth the two developing testes pass out of the abdominal cavity into the scrotal sac. Failure of either one or both testes to descend into the scrotum (*cryptorchidism*) may result in infertility because production of spermatozoa (*spermatogenesis*) depends on a temperature about 4°C below body temperature. The scrotum can contract or relax to move the testes closer to or further away from the body so that this temperature is maintained.

Spermatogenesis

Spermatozoa are produced in the *seminiferous tubules* of the testes. These tubules are lined with maturing spermatozoa and also with *Sertoli cells* which appear to provide nutrients for the developing spermatozoa (Fig. 9.2). In response to gonadotrophic hormones released by the anterior pituitary, spermatogenesis is initiated at puberty and occurs continuously thereafter. Germinal cells called *spermatogonia* divide to form *primary spermatocytes* which undergo meiosis to form *secondary spermatocytes*. During meiosis, there is a reduction in the number of chromosomes from 46 to 23. The haploid secondary spermatocytes then divide to form *spermatids* which differentiate to give rise to the *spermatozoa*. The process of spermatogenesis from spermato-

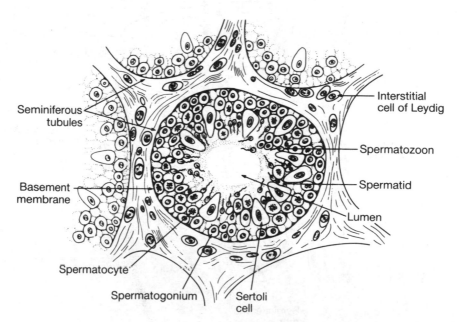

Fig. 9.2. Diagram of the testis in cross-section illustrating the different stages of spermatogenesis in the seminiferous tubules.

gonium to release of spermatozoa into the lumen of the seminiferous tubule takes about 75 days.

When spermatozoa are released into the seminiferous tubules, they are non-motile and incapable of fertilizing an ovum. From the seminiferous tubules, spermatozoa pass into the *epididymis* where they mature and are stored until ejaculation takes place. The production of spermatozoa is a continuous process and spermatozoa not ejaculated eventually deteriorate and are reabsorbed by phagocytosing epididymal cells.

The *mature sperm* consists of a head, middle piece and long tail (Fig. 9.3). The head is composed mainly of the nucleus and is covered by a cap known as the *acrosome*. The acrosome contains lytic enzymes which may enable the sperm to penetrate the ovum. The middle piece consists of a helical sheath of mitochondria surrounding a core of contractile filaments which extend into the tail. The mitochondria provide energy for motility of the spermatozoon which depends on a wave-like movement of the tail.

Testosterone

The principal androgenic hormone produced by the testis is *testosterone* (Fig. 9.4). This steroid hormone is secreted by the *interstitial cells of Leydig* which

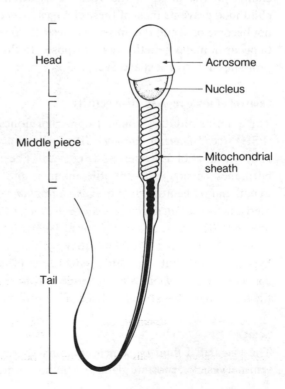

Fig. 9.3. The mature spermatozoon.

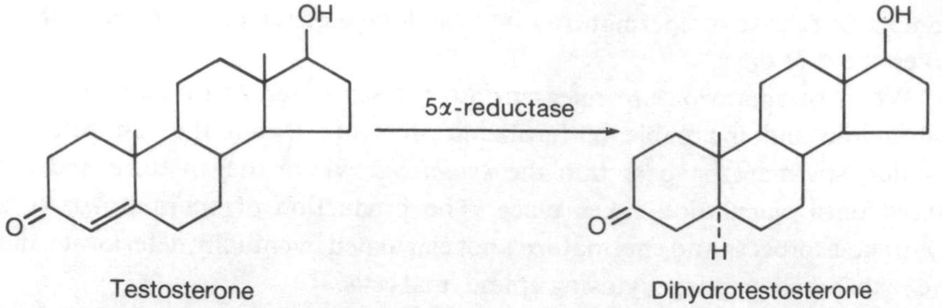

Fig. 9.4. Testosterone and its conversion to dihydrotestosterone.

lie scattered between the seminiferous tubules (Fig. 9.2). Its synthesis is similar to the formation of androgens in the adrenal cortex (see Fig. 8.17). Once released into the blood it is bound to a specific carrier globulin. In some of its target tissues, testosterone appears to be converted to a more potent androgen called dihydrotestosterone (Fig. 9.4).

Testosterone promotes the *development of the reproductive system* and the *secondary sexual characteristics* of the male and has important anabolic effects on skeletal muscle and bone. The most obvious effects of testosterone are seen at puberty, namely, enlargement of the penis and testes, increased rate of growth of muscle and bone, appearance of facial, axillary and pubic hair, and change in the pitch of the voice. Castration or removal of the testes in childhood prevents most of these changes from occurring. Castrated males do not become bald and the presence of testosterone seems necessary for baldness to occur in males genetically predisposed to this condition. Testosterone also promotes libido or sexual drive.

Control of male reproductive activity

The anterior pituitary gonadotrophic hormones—*follicle-stimulating hormone* (FSH) and *luteinizing hormone* (LH)—control respectively spermatogenesis and synthesis of testosterone (Fig. 9.5). The release of the gonadotrophic hormones occurs in a pulsatile manner and is controlled by a common hypothalamic neurohormone—*gonadotrophin-releasing hormone* (GnRH), formerly known as *luteinizing hormone-releasing hormone* (LHRH). Both testosterone and FSH are necessary for normal spermatogenesis. Testosterone inhibits secretion of LH in a typical negative-feedback manner, acting mainly on the hypothalamus, but has little effect on FSH secretion at physiological concentrations. *Inhibin*, a polypeptide hormone secreted by the Sertoli cells in the testes, has a negative-feedback effect on secretion of FSH.

Semen

The ejaculated fluid or semen contains spermatozoa and secretions of the seminal vesicles, prostate gland and bulbourethral glands. The average volume

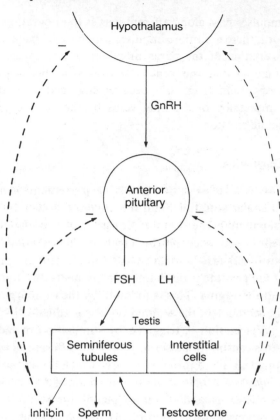

Fig. 9.5. Control of spermatogenesis and testosterone secretion.

of ejaculate in man is about 3 ml and this contains approximately 100 million spermatozoa per ml. The secretions of these accessory glands comprise the bulk of the semen and assist in the transport and nourishment of the sperm. The seminal fluid contains high concentrations of *fructose*, which serves as an energy substrate for the spermatozoa, and also high concentrations of *prostaglandins* which may increase motility of the uterus, thus promoting transport of spermatozoa in the female genital tract. Fertility depends on the quality of the semen, the two most important factors being the number and motility of the spermatozoa. A count of less than 20 million spermatozoa per ml is generally considered incompatible with fertility.

Erection

The erection of the penis, which is necessary for coitus and delivery of semen to the female, is due to the engorgement of the penis with blood. Erection may be initiated by psychic stimuli and by tactile stimulation of the glans penis. Failure of erection, i.e. *impotence*, may be due to psychological as well as the organic disturbances. In erection, a spinal reflex arc is involved in which impulses pass along afferent nerves to integrating centres in the *sacral* spinal cord, there to initiate impulses which travel back along *parasympathetic* fibres. Excitation of the parasympathetic fibres causes *arteriolar dilatation* in the penis so that the venous sinusoids of the corpora cavernosa and corpus spongiosum (Fig. 9.1) become engorged with blood. This, together with compression of draining veins by the ischiocavernosus muscle, producing penile erection.

Ejaculation

This is a reflex action involving movement (*emission*) of spermatozoa and glandular secretions into the *urethra* followed by the sudden ejection of the semen from the urethra. Emission of the glandular secretions occurs in a definite sequence. During erection, the secretion of the bulbourethral glands is discharged to lubricate the urethra. During ejaculation, the alkaline secretion of the prostate is discharged first to neutralize the acidity of the male urethra and the vagina. This is followed by the discharge of spermatozoa and finally the secretion of the seminal vesicles is added.

Ejaculation is triggered by stimulation of tactile receptors in the glans penis causing impulses to pass along afferent nerves to centres in the *lumbar* spinal cord and initiate impulses which return along sympathetic fibres. This *sympathetic activity* leads to contraction of the smooth muscle of the epididymis, vas deferens and secretory glands propelling spermatozoa and glandular secretions into the urethra. At the same time, the internal sphincter of the urethra constricts, preventing semen from entering the bladder. Contraction of the bulbospongiosus and ischiocavernosus muscles due to reflex activity in *somatic motor nerves* then leads to pulsatile emission of the seminal fluid from the urethra.

Vasectomy

This procedure, in which the vasa deferentia are cut and tied, is an effective means of contraception in the male. Spermatogenesis continues after vasectomy but, without an outlet, the spermatozoa degenerate and are reabsorbed in the epididymis. Since less than 10% of semen consists of spermatozoa, the volume of the ejaculate is little affected by vasectomy.

9.3 The female reproductive system

The primary reproductive organs of the female (Fig. 9.6) are the two *ovaries*, which produce ova and secrete the sex hormones, oestrogen and progesterone. The accessory reproductive structures comprise the two *oviducts* (Fallopian

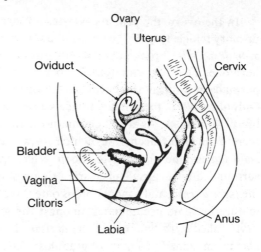

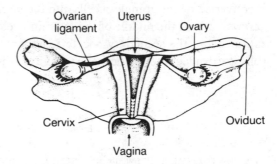

Fig. 9.6. The female reproductive system.

tubes), the *uterus*, the *cervix* and the *vagina*. At the opening of the vagina lie the external genitalia (the hymen, the labia and the clitoris).

Oogenesis and follicular development

The formation of germ cells or *oogonia* is completed during fetal development of the ovary. Many of the oogonia undergo the first stage of meiosis to form diploid *primary oocytes*; others degenerate. At birth, each human ovary contains approximately one million oocytes. The number surviving in both ovaries at puberty is less than about 400 000 and continues to decline so that at the time of menopause few viable cells remain. Although the first stage of meiosis occurs before birth, it is not until ovulation that meiosis is completed with the formation of the haploid *secondary oocyte*. In this process, most of the cytoplasm is retained by the oocyte and a smaller rudimentary cell called the first polar body is split off. After a second division during passage in the oviduct, a second polar body is eliminated and the mature *ovum* is formed.

In the ovary, the primary oocytes are arranged in *primary follicles*. Each primary follicle (Fig. 9.7) consists of an oocyte surrounded by a single layer of *granulosa cells*. From puberty onwards, in response to the gonadotrophic hormone, FSH, several follicles begin to develop during each cycle but normally only one of these reaches the stage of ovulation; the rest degenerate. Only about 400 of the primary follicles develop into ova during the reproductive life of a woman. As a follicle matures, the granulosa cells proliferate and secrete mucopolysaccharides which form a translucent halo called the *zona pellucida* around the oocyte. Soon after, the developing follicle becomes surrounded by a capsule of ovarian tissue made up of an inner cellular layer, the *theca interna*, and a more fibrous outer layer, the *theca externa*. With further maturation fluid accumulates amongst the granulosa cells to form a central cavity filled with fluid called the *antrum*. In the mature *Graafian follicle* the oocyte embedded in a mass of granulosa cells protrudes into the antrum (Fig. 9.7). The theca interna cells synthesize androstenedione (see Fig. 8.17), which is converted by the granulosa cells into the *oestrogen* hormones, *oestradiol* and *oestrone*. About the middle of the ovarian cycle, ovulation occurs: the follicle ruptures and the secondary oocyte together with its surrounding granulosa cells is extruded into the peritoneal cavity. It is then swept by the movement

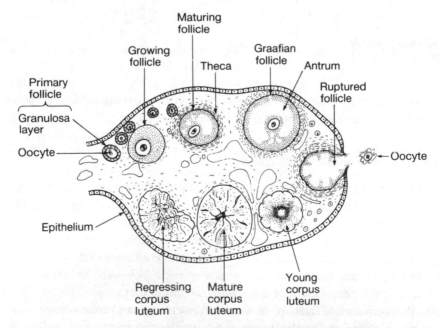

Fig. 9.7. Diagram of the ovary in cross-section illustrating the different stages in the development of a follicle and corpus luteum during one overian cycle. (The sequential arrangement is not an actual representation.)

of the cilia into the open end of the oviduct which is closely applied to the ovary.

After ovulation, the granulosa cells remaining in the ruptured follicle, together with cells of the theca interna, proliferate to form a new endocrine structure, the *corpus luteum*. This goes on secreting oestrogen but also produces the hormone *progesterone*. The corpus luteum is functional for about 12 days after ovulation. Thereafter, unless fertilization of the ovum and implantation have occurred, the corpus luteum regresses, causing a decline in the secretion of oestrogen and progesterone, the onset of menstruation and the initiation of a new cycle of ovarian activity.

Gonadotrophic hormones

Maturation of follicles in the ovary requires the presence of the *gonadotrophic hormones*, *FSH* and *LH*, which are secreted by the anterior pituitary. The secretion of both gonadotrophic hormones is stimulated by *gonadotrophin-releasing hormone* (GnRH) which is produced in the medial basal area of the hypothalamus and is released in a pulsatile manner with a periodicity of about 1 h. Secretion of the gonadotrophins is regulated by feedback actions of oestrogen and progesterone (see later) and of inhibin (Fig. 9.10). Inhibin, which is secreted by the granulosa cells of the follicles, acts on the pituitary to inhibit the release of FSH and on the hypothalamus to inhibit the release of GnRH.

FSH and LH are glycoproteins and each can be dissociated into two subunits called α and β. The α subunits of FSH and LH are identical in structure and specificity is determined by the β subunit. Their actions are mediated by cyclic AMP. FSH, together with LH, promotes *development of the follicle* and LH is needed for the *secretion of oestrogen* by the ovary. Oestrogen also promotes follicular development. The sudden peak in the LH secretion mid-cycle appears to *trigger ovulation*. Ovulation can be induced in infertile women by treatment with gonadotrophic extracts from the pituitary or urine. To be effective, such extracts must contain FSH for follicular development and LH for the production of oestrogen and ovulation. LH is also required for normal corpus luteum function.

Menstrual cycle

The average length of the menstrual cycle is 28 days but it may vary considerably. The first signs of bleeding signal the start of a new menstrual cycle and bleeding may continue for 3–5 days. During this period and until ovulation, the *ovarian follicles* begin to develop and secrete increasing quantities of *oestrogen*. Oestrogen acts on the uterus to stimulate regeneration

Chapter 9

and growth of the *endometrium* from the remnants left over from the previous menstrual cycle, causing a two- to threefold increase in the thickness of the endometrium. The first two weeks of the menstrual cycle are therefore referred to as the *follicular phase* with respect to the ovary and as the *proliferative phase* with respect to the uterus (Fig. 9.8). The variation that occurs in the duration of the menstrual cycle is usually due to variation in this first half of the reproductive cycle. Ovulation occurs at about the mid-point of the cycle, i.e. around day 14.

During the second half of the reproductive cycle, the *corpus luteum* develops and secretes both *oestrogen* and *progesterone*. Oestrogen continues to promote proliferative activity in the endometrium while, under the action of progesterone, the endometrial glands become distended with secretory products including *glycogen*, which is an important nutrient for the developing embryo should implantation take place. Endometrial blood flow increases and the spiral arteries become more tightly coiled and twisted. The second half of the cycle is therefore referred to as the *luteal phase* with respect to the ovaries and as the *secretory phase* with respect to the uterus. If implantation does not occur, the corpus luteum regresses, there is a rapid fall in secretion of oestrogen and progesterone and, for reasons which are unclear, the endometrium undergoes shrinkage due to the loss of extracellular water and constriction of the spiral arteries. This causes a reduction in blood flow to the endometrium with cell death and weakening of the walls of blood vessels. As the phase of vasoconstriction wears off, blood leaks from the damaged vessels to initiate

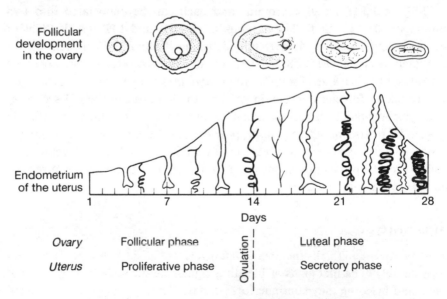

Fig. 9.8. Diagram showing the changes occurring in the endometrium and in an ovarian follicle during the menstrual cycle.

menstrual bleeding and eventually all but the basal layer of endometrium is detached from the uterus. A second phase of vasoconstriction of the spiral arteries minimizes loss of blood.

Ovarian hormones: oestrogen and progesterone

Oestradiol is the main oestrogen and progesterone the main progestin produced in the ovary (Fig. 9.9). Their synthetic pathways are outlined in Fig. 8.17. Oestrone is also secreted in significant amounts but its biological activity is less than that of oestradiol. These steroid hormones are transported in the blood bound to plasma proteins. They are degraded in the liver and their metabolites (particularly oestriol and pregnanediol) are excreted by the kidney. Like other steroid hormones, oestrogen and progesterone cause their effects in responsive tissues by combining with cytoplasmic receptors in cells to be translocated to the nucleus to induce increased production of messenger RNA with subsequent increase in synthesis of the effector proteins.

Oestrogen is secreted by the granulosa cells of the follicles and is also produced after ovulation by the corpus luteum. The peak of LH that occurs at ovulation is preceded by a rise in oestrogen secretion which stimulates LH secretion, i.e. a high level of oestrogen has a *positive-feedback* effect on LH (and FSH) secretion at this point in the cycle (Fig. 9.10). During the secretory phase of the menstrual cycle, oestrogen and progesterone exert a *negative-feedback* effect on the secretion of the gonadotrophic hormones. It is now believed that the primary site of the positive and negative feedback actions of oestrogen is the pituitary, although the hypothalamus also appears to be a site of action (Fig. 9.10). The negative-feedback effect diminishes as the concentrations of oestrogen and progesterone fall with regression of the corpus luteum.

In addition to its positive and negative effects on gonadotrophin secretion,

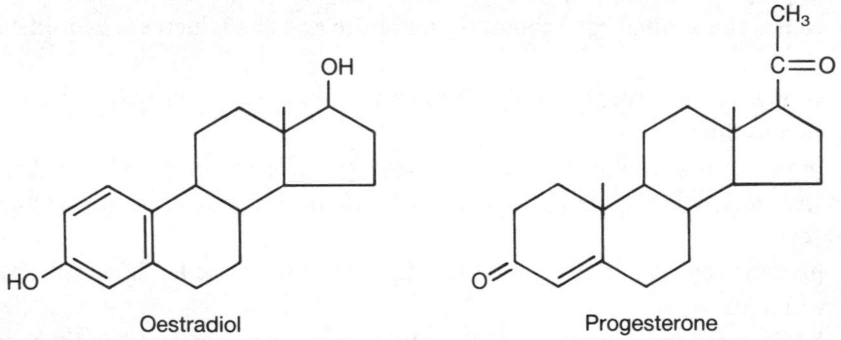

Fig. 9.9. Structure of oestradiol and progesterone.

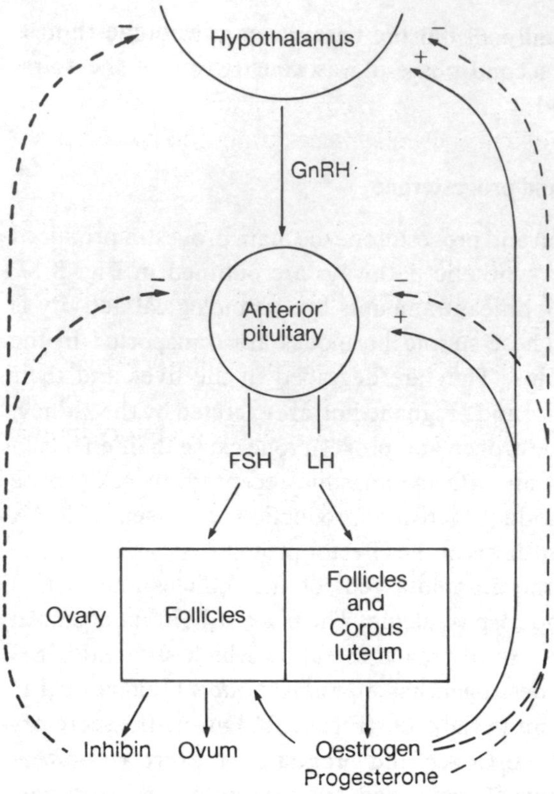

Fig. 9.10. Control of follicular development and oestrogen secretion.

oestrogen:

1 sensitizes the ovaries to the effects of gonadotrophins;

2 stimulates growth of the endometrium and contractility of the myometrium;

3 stimulates the output of mucus from the cervical glands and causes changes in the properties of the mucus which assist entry of spermatozoa;

4 causes the vaginal epithelium to proliferate and show increased cornification;

5 stimulates the growth and development of the breasts, particularly of the lactiferous ducts;

6 promotes the growth of bones and skeletal muscle and helps to bring about the characteristic female patterns of distribution of body hair and adipose tissue;

7 promotes closure of the epiphyses at the end of the period of linear skeletal growth; and

8 helps to conserve bone (p. 179); osteoporosis is a common condition in post-menopausal women.

Progesterone, which is present in significant amounts only during the luteal phase of each menstrual cycle, acts on tissues which have already been stimulated by oestrogen. In addition to having inhibitory effects on gonadotrophin secretion and follicular development during the luteal phase progesterone:

1 transforms the endometrium to its secretory phase and decreases the spontaneous electrical activity of the myometrium;
2 modifies the composition of cervical mucus, making it more viscous and resistant to penetration by spermatozoa;
3 causes further changes in the vaginal epithelium with regression of cornification;
4 promotes development of the breasts, particularly of the secretory units; and
5 causes an increase in basal body temperature after ovulation, which may be useful clinically to indicate that ovulation is occurring.

Control of the female reproductive cycle

The blood levels of the gonadotrophic and ovarian hormones throughout the ovarian cycle are shown in Fig. 9.11. A marked peak in the level of gonadotrophins occurs at the mid-point of the menstrual cycle and coincides with the time of ovulation. (Because it is difficult to pinpoint the time of ovulation, the peak in blood LH has been arbitrarily designated as day zero in Fig. 9.11, but in clinical practice the beginning of menstruation is taken as day 1 of the cycle.) Just before the beginning of each cycle there is a small rise in FSH. This rise in FSH stimulates follicular development and, together with LH, leads to an increase in oestrogen secretion. Oestrogen plays an important role in the maturation process because it sensitizes the granulosa cells to the effects of gonadotrophins and thus increases their capacity to produce oestrogen. The surge in oestrogen secretion that occurs just prior to ovulation has a positive-feedback effect on the anterior pituitary, and possibly the hypothalamus, causing a marked rise in LH and FSH secretion. The peak in LH secretion triggers ovulation. After ovulation, the increase in oestrogen and progesterone that parallels the development of the corpus luteum prepares the endometrial lining of the uterus for implantation. If implantation does not occur, the levels of oestrogen and progesterone fall in parallel with the demise of the corpus luteum. The fall in ovarian hormones causes menstruation to occur and also removes the negative-feedback influence on the secretion of the gonadotrophic hormones. The resultant rise in output of FSH triggers development of a new batch of follicles and the beginning of a new cycle of uterine and ovarian function.

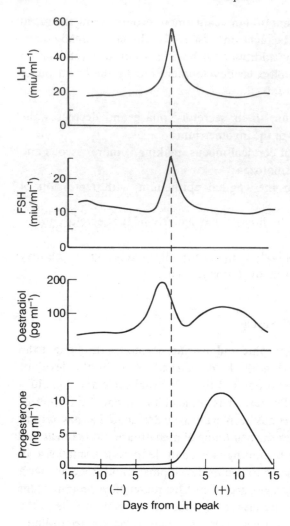

Fig. 9.11. Plasma levels of LH, FSH, oestradiol and progesterone during the ovulatory cycle centred on the mid-cycle LH peak (day 0). (After Thorneycroft, I. A. *et al.* (1971) *Am. J. Obstet. Gynecol.* **111**, 947–951.)

Hormonal contraception

The oral contraceptive pill is a very effective means of preventing pregnancy. Hormonal contraception is the outcome of research into the negative-feedback effects of the sex hormones. However, natural oestrogens and progestins are not effective when taken orally, because they are rapidly degraded in the gut and liver, and it was not until the development of synthetic oestrogens and progestins that hormonal contraception was widely adopted. The most commonly used contraceptive pill is a combination of synthetic oestrogen and progestin taken daily but withdrawn towards the end of the month. It acts primarily through feedback inhibition of FSH and LH secretion to suppress ovulation. Preparations containing progestin also produce changes in the cervical secretions making it more difficult for sperm to penetrate the uterus and they inhibit capacitation of sperm (p. 310). These appear to be the main actions of the progestin-only contraceptive which is not withdrawn and is administered continuously, either orally, or by the slow release of hormone from a vaginal suppository or intramuscular injection.

Hyperprolactinaemia

Prolactin is required in mammals for breast development and lactation (p. 313). In some rodents it prolongs the life of the corpus luteum and so has been called luteotrophic hormone. Although the production of prolactin is controlled principally by the inhibitory influence of dopamine, a stimulatory influence is exerted by thyrotrophin-releasing hormone and by oestrogen during pregnancy. Hyperprolactinaemia occurs when prolactin secretion is excessive, as for example in certain pituitary tumours. Elevated blood prolactin concentrations cause anovulation and amenorrhoea in women, and impotence and infertility in men. In these conditions, prolactin appears to inhibit the actions of LH and FSH on the ovary and testis, respectively.

Pineal gland

This gland, situated in the centre of the brain (see Fig. 2.26), secretes the hormone *melatonin*. In experimental animals, the administration of melatonin inhibits ovulation by preventing the release of GnRH and consequently LH. In some mammals, the secretion of melatonin fluctuates in relation to light (and hence day length) and so it is thought that the pineal gland plays a role in determining the seasonal breeding patterns of such animals.

Melatonin is synthesized from serotonin and its synthesis is controlled by the enzyme hydroxyindole-*O*-methyltransferase (HIOMT). Light suppresses the synthesis of this enzyme in those animals in which there is a 24 h (circadian) rhythm of melatonin release. Light energy is transduced to nerve impulses in the eye and relayed by a complex pathway to the spinal cord and thence to the pineal gland via sympathetic nerve fibres relaying in the superior cervical ganglion. The release of noradrenaline by these postganglionic sympathetic fibres during daylight suppresses the synthesis of HIOMT and hence of melatonin.

The pineal gland may also play a role in reproductive activity in man. Hypersecretion of melatonin due to a pinealoma can suppress gonadal activity and delay the onset of puberty. Conversely, hyposecretion of melatonin may lead to early sexual maturity.

9.4 Pregnancy

The duration of pregnancy (or gestation) in women is approximately 40 weeks from the last menstruation. Sperm can survive for a few days in the female reproductive tract; the ovum remains fertile for less than a day. Therefore, fertilization can occur if sperm are deposited in the female reproductive tract a few days before ovulation. Sperm need to spend some period of time in the female reproductive tract before they are capable of fertilizing the ovum. This process is known as *capacitation* of the sperm. Only one of the many hundreds of thousands of sperm deposited in the vagina can fertilize the ovum.

Fertilization occurs in the oviduct and the *zygote* begins to divide as it makes its way to the uterus. By the time (about 3 days) it reaches the uterus, it is a small mass of cells called a *morula*. Identical twins may arise if the morula separates into two parts during this stage. The morula develops into the *blastocyst* which is composed of an outer layer of trophoblastic cells separated from an inner mass of embryonic cells (Fig. 9.12). The *trophoblastic layer* forms part of the *chorion* which gives rise to the fetal part of the placenta. *Implantation* of the blastocyst in the uterine wall occurs about 7 days after fertilization.

At about 9 days after fertilization, the trophoblastic layer begins to secrete the hormone, *human chorionic gonadotrophin* (HCG). HCG prolongs the life of

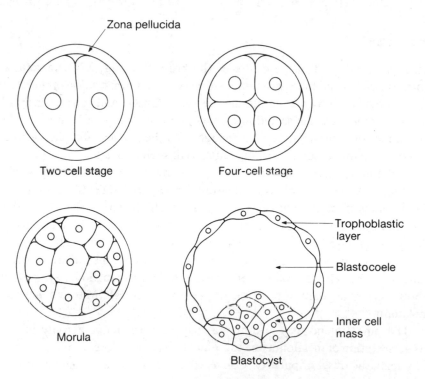

Fig. 9.12. Stages in the early development of the embryo.

the corpus luteum so that it continues to secrete oestrogen and progesterone which are necessary for the continuation of pregnancy during the first trimester. Thereafter, the role of the corpus luteum is supplanted by the placenta. The *placenta* is not only a means for exchanging respiratory gases, nutrients and waste products between fetal and maternal circulations, it also serves as an *endocrine gland* in its own right. The placenta secretes both protein and steroid hormones necessary for the continuation of pregnancy during the second and third trimesters.

Placental hormones

The placenta secretes at least four hormones. Two of these hormones are proteins, namely human chorionic gonadotrophin (HCG) and human chorionic somatomammotrophin (HCS), while two are steroids, namely oestrogen and progesterone. The maternal plasma levels of these hormones during pregnancy are shown in Fig. 9.13.

Human chorionic gonadotrophin. This is a glycoprotein (M_r 30 000) secreted by the trophoblastic cells of the placenta which is chemically and biologically similar to LH. It is secreted in large quantities during the first trimester

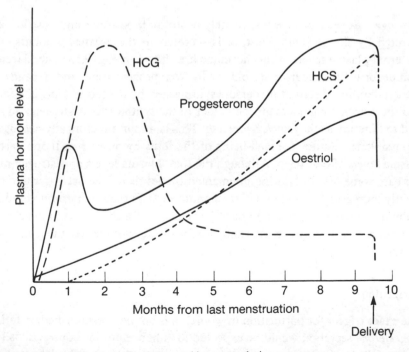

Fig. 9.13. Maternal plasma levels of placental hormones during pregnancy.

(Fig. 9.13) and its main role is to maintain the corpus luteum during the early part of pregnancy. Since HCG is secreted as early as 9 days after fertilization, its detection by radioimmunoassay provides a simple test for pregnancy.

Human chorionic somatomammotrophin. This was originally named human placental lactogen. It is a protein (M_r 18 500) secreted by the trophoblastic cells and is structurally similar to human growth hormone. The level of HCS in the maternal circulation increases steadily throughout pregnancy but its functions are not well defined. Like growth hormone it has a counter-regulatory action to insulin, promoting free fatty acid mobilization and inhibiting glucose uptake in the mother. As its name implies, it also promotes mammary development in preparation for lactation.

Progesterone. During pregnancy the maternal plasma levels of progesterone reach a peak 3–4 weeks after fertilization and then decline before increasing up until the time of parturition (Fig. 9.13). The initial peak reflects the development of the corpus luteum and the later rise is due to secretion by the placenta. Progesterone is required throughout pregnancy to maintain the endometrium and to suppress spontaneous contractions of the myometrium of the uterus. Progesterone also stimulates the development of the mammary glands.

Oestrogen. Several oestrogens, namely oestradiol, oestrone and oestriol, are secreted by the placenta. Oestriol is secreted in the greatest amounts; the maternal plasma levels of this hormone are shown in Fig. 9.13. The placenta cannot form the precursor steroid, 17-hydroxyprogesterone, and depends on the availability of steroid precursors synthesized by the fetus. Thus, the fetus and the placenta complement each other in the synthesis of oestrogen and are said to function as the *fetoplacental unit*. This is important clinically because it is possible to monitor the well-being of the fetus by measuring the maternal plasma levels of oestrogen. Although the total plasma levels of oestrogen may increase some fiftyfold during pregnancy, the levels of free oestrogen are not greatly increased because of a corresponding rise in the sex hormone-binding globulin. Oestrogen is required during pregnancy for the uterus to develop to accommodate the growing fetus and also for the development of the mammary glands.

Parturition

The exact trigger for parturition in women has not been established. A fall in *progesterone* secretion would be expected to enhance uterine contractions but such a fall does not appear to take place immediately prior to delivery. An increase in *oxytocin* release would likewise be expected to stimulate uterine contractions. Oxytocin is indeed released at parturition in response to stimulation of stretch receptors in the uterus and cervix and is used clinically to induce labour. Nevertheless, parturition can still be initiated in women with hypothalamic damage who lack oxytocin. *Prostaglandins* will also induce labour but again there is insufficient evidence that they provide the trigger for parturition in women. In sheep, however, it has been established that the fetus determines the time of delivery by increasing its release of ACTH and hence cortisol. The increase in fetal cortisol secretion causes a change in placental steroid synthesis that somehow initiates uterine contractions through the release of prostaglandins. The relevance of such a mechanism to parturition in women is not known.

Towards the end of pregnancy *relaxin*, a polypeptide hormone, can be extracted from the ovary, uterus and placenta. Relaxin and other similar polypeptides appear to play a role in parturition by promoting relaxation of the birth canal.

Lactation

Mammary glands provide both milk for nourishment and antibodies for protection of the young. Milk is produced by epithelial cells lining the alveoli which drain into slender ducts. Development of the alveoli and ducts require the presence of several hormones—oestrogen, progesterone, growth hormone, HCS, cortisol and insulin. During the latter part of pregnancy, prolactin is required for full maturation of the mammary glands and milk production. The initiation of copious lactation occurs shortly after parturition. Although the reason for this is not fully understood, it requires a fall in the concentrations of oestrogen and progesterone and a rise in prolactin. While prolactin stimulates milk production, milk let-down depends on the release of oxytocin which is induced by suckling (p. 267). Suckling is also necessary for the continued release of prolactin because it stimulates nerve endings in the nipples which in turn send impulses to the hypothalamus that inhibit the release of prolactin release-inhibiting hormone. In the absence of suckling, prolactin secretion is reduced and lactation ceases. The suckling stimulus also inhibits secretion of gonadotrophins and this presumably accounts for the suppression of ovulation that often occurs in nursing women.

10 · Blood

10.1 Blood cell production
Sites of production
Stem cells and haematopoiesis

10.2 The red blood cell
Erythropoiesis
Control of erythropoiesis
Synthesis of haemoglobin and
 erythropoietic factors
Iron
 Intake and absorption
 Iron exchanges and requirements for
 balance
 Deficiency of iron
Vitamin B_{12}
Folic acid
Megaloblastic erythropoiesis
Characteristics of red cells
 The red-cell membrane
 Haemoglobin
 Red-cell metabolism
Life-span and breakdown of red cells
Anaemia
 Causes of anaemia
 Red-cell indices and anaemias
 Compensatory adjustments in
 anaemia

10.3 The white blood cells
Granulocyte production and life-span
Abnormal white-cell counts
Acute inflammation

The immune system
 Antigens
 The effector cells of the immune
 system
 T-lymphocyte subsets
 B-lymphocytes and antibody
 production
 Functions of antibodies
 Active and passive immunity
 Tissue transplantation and the major
 histocompatibility complex
 Abnormalities of the immune
 response

10.4 Blood groups
Blood-group antigens and genes
Blood-group antibodies
ABO blood-group system
Rhesus blood-group system
Blood transfusion

10.5 Haemostasis
Reaction of blood vessels
Platelets
 Haemostatic reactions
 Prostaglandin synthesis and actions
 Other actions of platelets
Blood coagulation
 Intrinsic and extrinsic coagulation
 systems
 The coagulation factors
 Inhibitors of coagulation
 Anticoagulants
Fibrinolysis
Summary of haemostasis
Bleeding disorders
Tests of haemostatic function

Blood circulates through the body bringing O_2 and nutrients to the tissues and removing CO_2 and other waste products. As it moves around the body it aids interchange between the fluid compartments, dissipates heat and distributes hormones, thus helping to maintain homeostasis and to coordinate the activities of the various organs. By transporting various cells and antibodies (*immunoglobulins*) to the tissues, the blood promotes defence against foreign substances and pathogenic organisms, while it also has cellular and protein constituents which provide a means of self-defence to stop bleeding (*haemostasis*) after injury.

314

If a sample of blood is collected in a tube containing anticoagulant and the tube is centrifuged, the blood will separate into its main components (Fig. 10.1a). The supernatant yellowish fluid is *plasma*, in which are suspended the *cells* of the blood. Most of these are red cells (*erythrocytes*), with much smaller numbers of white cells (*leukocytes*) and *platelets* (occasionally called thrombocytes). If the blood is centrifuged in a tube of uniform bore, it is possible to estimate from the lengths of the columns of red cells and plasma, the fraction of a volume of blood occupied by the red cells (Fig. 10.1a). This is the *haematocrit* or *packed cell volume* (see Table 10.4).

The actual volume of circulating red cells (Table 10.1) can be estimated from the dilution of an intravenous injection of the subject's red cells labelled with a radioactive isotope (e.g. ^{51}Cr). This *dilution principle* (p. 535) can also be used to estimate plasma volume by measuring the extent of dilution in a subject's plasma of albumin labelled with ^{125}I. Blood volume can then be calculated by adding red-cell volume and plasma volume or it can be estimated from one of these two values by means of the haematocrit, e.g. blood volume = red-cell volume/haematocrit.

Some 90% by weight of plasma is water, about 8% is plasma proteins (Fig. 10.1b)—albumin, globulins (α, β, γ) and fibrinogen—while 2% consists of organic compounds and electrolytes (see Table 1.1, p. 3). When plasma coagulates, the soluble plasma protein, fibrinogen, is converted into insoluble fibrin (p. 361), with consumption of some other coagulation proteins. The liquid remaining is called *serum*.

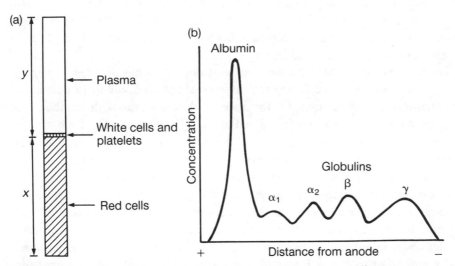

Fig. 10.1. (a) Fluid and cellular constituents of a sample of blood as seen after centrifugation. The fraction occupied by the red cells ($x/(x+y)$) is the haematocrit. (b) Separation of serum proteins by electrophoresis.

Table 10.1. Normal blood volumes (ml kg^{-1}) as the 95% range (mean $\pm$ 2 s.d.).

	Men	Women
Red-cell volume	25–35	20–30
Plasma volume	40–50	40–50
Total blood volume	60–80	60–80

Data from Dacie, J.V. & Lewis, S.M. (1984) *Practical Haematology*, 6th edn, p. 8. Churchill Livingstone, London.

The red cells of blood are central to the transport of O_2 and CO_2 between lungs and tissues (p. 490), while platelets are involved in haemostasis (p. 357). The white cells have important phagocytic and immunological functions (p. 338). In descending order of frequency in blood of adults (see Table 10.6), there are five types of white blood cell: neutrophil, lymphocyte, monocyte, eosinophil and basophil.

10.1 Blood cell production

Sites of production

The production of blood cells (*haematopoiesis*) begins in the first few weeks of gestation with the appearance of primitive cells in mesodermal tissues of the yolk sac. From about six weeks to seven months of fetal life, all types of blood cell are formed mainly in the liver but also in the spleen. During the fifth month, production of blood cells begins in the red marrow of the bones and at birth these sites take over completely from the liver and spleen. Lymphocytes are formed also in lymphoid tissue.

At first, red marrow occupies the cavities of all bones but from the age of 5–7 years, fat begins replacing the marrow of the limb bones. In adults, red marrow is normally confined to the bones of the trunk, the skull and the upper ends of the humerus and femur. Even at these sites about 50% of the marrow consists of fat but if there is increased demand for blood cells, the red marrow can expand into these fatty spaces and back into the limb bones.

In an adult, the red bone marrow consists of fat and developing blood cells in clusters held together by fine reticulin fibres and separated by a network of sinusoidal capillaries, which drain into a central venous sinus. New cells are released into the blood through gaps in the endothelium of the marrow sinusoids. Samples of red marrow can be obtained for cytological examination by aspiration with a syringe and marrow-puncture needle or by means of a *trephine*, with which a small core of marrow is cut from the bone for histological examination.

Stem cells and haematopoiesis

Blood cells are thought to develop in the bone marrow from precursors called *stem cells*. Stem cells can maintain their number by mitosis and can also develop along various lines leading to mature blood cells. The earliest stem cell in the marrow has the potential to develop into all the types of blood cell and is hence called *pluripotent*. Other later generations of stem cells have more limited proliferative capacity and also become *committed* to develop along some particular line(s). Thus the *myeloid stem cell* gives rise to other progenitor cells which eventually differentiate into red cells, granulocytes, monocytes or platelets. In films of bone marrow it is possible to identify and name fairly well-defined types of developing cell (Fig. 10.2). It is likely, although not certain, that pluripotent stem cells also give rise to *lymphoid stem cells*. These then develop into B-lymphocytes and also into other lymphoid cells which are processed by the thymus gland to become T-lymphocytes.

Although stem cells cannot be recognized in films of bone marrow by microscopy, they can be grown in culture where they give rise to colonies of descending cells. They are thus often referred to as *colony-forming units* (CFU). A stem cell appearing early in the development of erythrocytes is called a *burst-forming unit* (BFU-E) because it produces a large colony of several thousand cells. This sudden expansion requires a glycoprotein which can be formed by T-cells and monocytes. As the red-cell progenitors mature to the type called CFU-E, they become increasingly sensitive to a hormone, erythropoietin (p. 319), which induces their further progression along the red-cell pathway. Other specific glycoproteins appear to stimulate progenitors of granulocytes, monocytes and megakaryocytes to develop along the appropriate paths, while a hormone of unknown origin, thrombopoietin, stimulates production of platelets from megakaryocytes, probably in a manner analogous

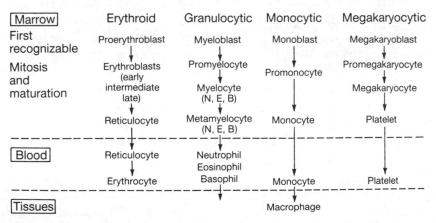

Fig. 10.2. Bone-marrow production lines. N, neutrophil; E, eosinophil; and B, basophil.

to the effect of erythropoietin on red-cell precursors. The study of haemato-
poietic stem cells is important because it is becoming clear that a number of
haematological disorders which involve more than one type of blood cell
probably arise because of defects in multipotent stem cells able to develop
along the affected pathways.

10.2 The red blood cell

Erythropoiesis

The process of blood cell production is well illustrated by the formation of red
cells, *erythropoiesis* (Fig. 10.3). In a Romanowsky-stained film of marrow, the
earliest recognizable red-cell precursor is a large nucleated cell, the *proerythro-*
blast (pronormoblast), with intensely basophilic cytoplasm reflecting its
content of RNA and ribosomal protein. This cell develops into an *early*
(basophilic) *erythroblast* (normoblast) and the cells subsequently become
smaller (*intermediate* and *late* erythroblasts), with increasingly acidophilic
cytoplasm as haemoglobin is synthesized, and the capacity for mitosis is lost
as nuclear pyknosis occurs. Extrusion of the nucleus then leaves cells,
reticulocytes, which contain remnants of RNA and ribosomes and continue

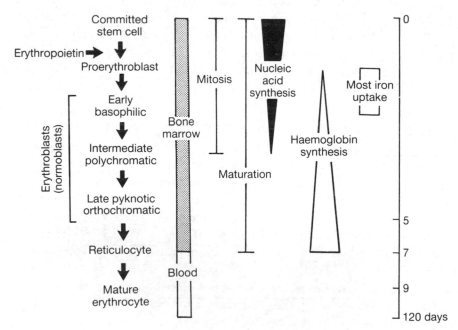

Fig. 10.3. Normal erythropoiesis.

making haemoglobin. Reticulocytes mature for 1–2 days in the marrow and are then released into the blood where, after a further 1–2 days, they lose their remaining ribosomes to become mature red blood cells.

The maturation of red-cell precursors in the marrow is accompanied by three or four mitotic divisions up to the stage of the intermediate erythroblast, so that one proerythroblast can give rise to about 16 mature red cells. Erythropoiesis is partly *ineffective*, however, because normally 5–10% of developing red cells do not survive to reach the circulation. This premature loss of cells is increased in certain disorders of red-cell production. Each mitotic division, and its effects, lasts about one day so that one erythropoietic cycle, from proerythroblast to release of reticulocytes, takes about six or seven days. This developmental cycle involves overlapping biochemical events, with early uptake of iron by erythroblasts, decreasing synthesis of nucleic acids and increasing synthesis of haemoglobin (Fig. 10.3). Up to 35% of the total content of haemoglobin in the mature red cell is made during the reticulocyte stage.

Reticulocytes can be identified in blood films stained with a dye such as new methylene blue which reacts with the ribosomes to form precipitates visible as dark blue granules or filaments. The *reticulocyte count* of blood (see Table 10.7) is a useful index of the erythropoietic activity of the bone marrow.

Control of erythropoiesis

In health, the rate of production of new red cells by the bone marrow balances the rate at which old red cells are destroyed. This steady state is controlled by a hormone, *erythropoietin*, which is secreted mainly by the kidneys in response to local hypoxia. Erythropoietin acts on the bone marrow to cause increased output of erythrocytes until the rise in haemoglobin concentration restores normal delivery of O_2 to the tissues.

Erythropoietin, a glycoprotein (M_r 33 000) which is normally present in plasma in low concentration (about 10 pmol l^{-1}), has a half-life of about 5 h. It is produced by the action of an enzyme, *renal erythropoietic factor* (erythrogenin), on a circulating plasma protein made by the liver but there is also evidence that preformed erythropoietin may be released directly by the kidneys. About 10–15% of the circulating erythropoietin appears to be made by the liver and this organ may be the source of the plasma erythropoietin detectable in anephric patients. The mode of action of the hormone is not known but it stimulates synthesis of RNA and of protein in red-cell precursors, increases their rate of mitosis and shortens their maturation time in the marrow.

Other hormones appear to stimulate erythropoiesis: corticosteroids, androgens, growth hormone and thyroxine. These hormones may act by increasing the sensitivity of red-cell precursors to erythropoietin.

Synthesis of haemoglobin and erythropoietic factors

Erythroblasts in the bone marrow undergo mitosis and begin making haemoglobin, first by synthesizing haem and the polypeptide chains of globin separately and then by their combination to form haemoglobin (Fig. 10.4). Certain dietary constituents are needed for normal erythropoiesis and deficiency of these may impair red-cell and haemoglobin production.

Synthesis of haem begins in mitochondria (Fig. 10.4) with the condensation of succinyl-CoA and of glycine to form δ-aminolaevulinic acid (ALA). This is the rate-limiting step in haem synthesis and is controlled by the enzyme ALA synthetase, which requires pyridoxine (vitamin B_6) as cofactor and is inhibited by haem. After several further reactions in the cytoplasm, the tetrapyrrole protoporphyrin IX is formed in the mitochondria and the enzyme ferro-chelatase catalyses the insertion of ferrous iron into the protoporphyrin ring to form haem. Iron is brought to the erythroblasts attached to a plasma protein, *transferrin*. It is probable that, after combination with a membrane receptor, the transferrin is taken up into the cytoplasm of the erythroblast where the iron is released and the transferrin is returned to the extracellular fluid.

While haem is being made in the mitochondria, *amino acids* are used for the synthesis on the ribosomes of matching quantities of the α and β polypeptide chains of globin. Synthesis of DNA by the dividing erythroblasts requires an

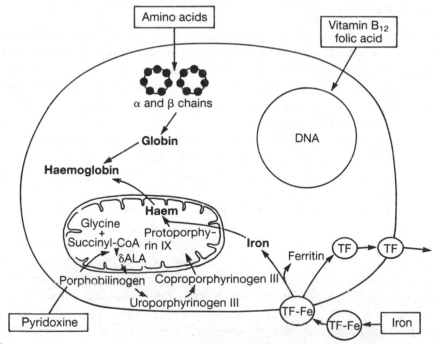

Fig. 10.4. Dietary requirements for synthesis of haemoglobin and DNA in erythroblasts. TF, transferrin.

adequate dietary supply of *vitamin B₁₂* and *folic acid*. Deficiencies of any of these dietary constituents can decrease erythropoiesis and haemoglobin production but of the various possibilities by far the most common is deficiency of iron. Synthesis of haemoglobin may also be *defective* because of a fault (often genetic) in the production of haem or of globin.

Iron

Depending on size, adults contain between 35 and 90 mmol (2–5 g) of iron, women having rather less than men (Fig. 10.5). Some 60–70% of this iron is in haemoglobin in the blood and 4–5% in myoglobin of muscle. Most of the rest is stored in hepatocytes and in macrophages of the liver, spleen and bone marrow, as *ferritin* or *haemosiderin*. Ferritin is a water-soluble compound consisting of a variable amount of ferric hydroxyphosphate enclosed by a protein shell of apoferritin. Synthesis of apoferritin is stimulated by the presence of iron in the cell. Haemosiderin, an insoluble complex of protein and iron, is probably formed by the partial lysosomal digestion of aggregates of ferritin. Ferritin is also present in the blood (Table 10.2) at concentrations which are usually higher in men than in women and which reflect the level of iron stores. Small amounts of iron are present in cellular enzymes (e.g. succinate dehydrogenase, monoamine oxidase) and in the cytochrome proteins of the electron-transport chain of mitochondria.

About 70 μmol of iron circulate in plasma bound to the β-globulin, transferrin. This protein is synthesized in the liver at a rate which varies directly with the level of stored iron and one molecule of transferrin can bind two atoms of iron in the ferric state. Transferrin is normally only one-third saturated with iron so that *serum iron* concentrations are lower than the total

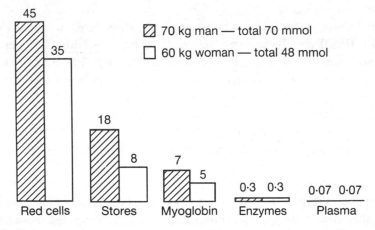

Fig. 10.5. Body iron in adults.

Table 10.2. Representative values of iron in adults.

	Men	Women
Haemoglobin iron (mmol l^{-1} blood)	8–11	7–10
Stored iron (mmol)	5–27	0–18
Serum iron (μmol l^{-1})	14–32	11–30
Serum iron-binding capacity (μmol l^{-1})	50–70	50–70
Serum ferritin (μg l^{-1})	40–340	15–140

iron-binding capacity of the serum (Table 10.2). Serum iron concentrations are lower in women than in men and decrease in pregnancy, iron deficiency and chronic infections. Serum iron-binding capacity increases in iron deficiency and in pregnancy but decreases with iron overload and chronic infections. The concentration of serum ferritin is usually a more accurate index of the state of the body iron stores than the serum iron and iron-binding capacity.

Intake and absorption

An average daily diet provides 180–270 μmol of iron. Red meats and sea foods are rich in iron, which is mainly in the ferric state and in organic form, e.g. as haem and as ferric–protein complexes. Iron is released from the food by acid and proteolytic enzymes in the stomach and small intestine and is reduced to the ferrous state in the presence of a low pH and reducing agents, such as ascorbic acid and sulphydryl groups of proteins. Ferrous iron is absorbed more readily than ferric iron because it is more soluble; haem is absorbed as such and the iron is released inside the mucosal cell. Absorption is limited by dietary substances which form insoluble complexes with iron, e.g. phosphate, phytate and the tannin of tea. Maximal absorption of iron occurs in the duodenum and upper jejunum and only very small amounts are taken up by the stomach and ileum.

Absorption rather than excretion of iron is regulated to maintain iron balance. Healthy adults absorb 5–10% of dietary iron although the proportion absorbed from different foods varies considerably, e.g. 20–30% of iron from meat, 5% of iron from eggs. In the case of inorganic iron (e.g. tablets of ferrous sulphate), the percentage absorbed decreases with increase in the total amount ingested. Absorption varies directly with the rate of erythropoiesis and inversely with the amount of iron in the body stores so that 20–30% of dietary iron may be absorbed in iron deficiency but less than 5% when the stores are full.

Iron not released to the blood after entry into the intestinal mucosal cell is sequestered as ferritin and is ultimately lost from the body when the mucosal cell is shed from the tip of the villus (see Fig. 14.2). This is the main route of iron excretion. How intestinal mucosal cells regulate the amount of iron absorbed is not understood. It appears that epithelial cells produced in the crypts of the villi when plasma levels of iron are low have a low iron content and a high absorptive capacity. Thus, the absorption of iron after a haemorrhage may be delayed 3–4 days until such cells are produced and migrate up the villi. After absorption, release to the extracellular fluid may depend upon how much 'exchangeable' iron is available to transferrin in other tissues, especially those storing iron.

Iron exchanges and requirements for balance

The iron in plasma, about 70 μmol in a 70 kg man, turns over between five and ten times a day. Of this turnover, some 80–85% arises from unidirectional movement of iron from extracellular fluid to bone marrow for the synthesis of haemoglobin and from phagocytic cells in the liver, spleen and marrow to the extracellular fluid, as iron is released from degraded haemoglobin (Fig. 10.6). The remaining movements of iron through the plasma are accounted for by exchanges of iron with non-erythroid tissues including absorption from the small intestine and loss in exfoliated intestinal cells. Small amounts are lost in sweat and urine. Iron from the breakdown of red cells is thus re-used for erythropoiesis, stored iron is available to meet extra demands and absorption normally balances loss from the body.

The inevitable daily loss of iron in adults, mainly from the gut, is between 9 and 18 μmol. Women lose, in effect, an extra 9–18 μmol a day due to menstruation and require an additional 18–36 μmol per day during pregnancy. Adolescents and infants require extra iron for growth. *Negative iron balance* occurs if these daily needs are not met by absorption of iron from the diet.

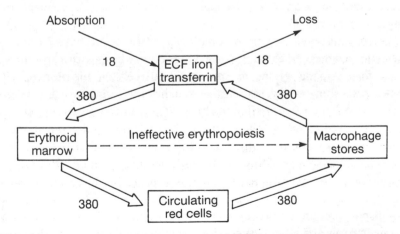

Fig. 10.6. Daily iron exchanges. Values (μmol) for a 70 kg man with a haemoglobin concentration of 150 g l^{-1} and red-cell life-span of 120 days. Iron exchanges from ineffective erythropoiesis are omitted. ECF, extracellular fluid.

Deficiency of iron

Continuing net loss of iron will lead to exhaustion of the stores and a fall in serum iron despite a compensatory rise in iron-binding capacity. When the saturation of transferrin is less than about 15%, delivery of iron to the marrow is impaired and an *iron-deficiency anaemia* develops.

There are three main causes of a negative iron balance and iron deficiency. *Increased physiological demands* for iron occur during periods of rapid growth especially in infants and young children, while the need for extra iron in pregnancy and for menstruation makes women liable to iron deficiency. *Inadequate absorption* of iron can arise after surgery (e.g. gastrectomy) and in diseases causing malabsorption. Low dietary intake in adults may contribute to iron deficiency due to other causes (e.g. pregnancy) and improper feeding may be important in infancy. The most common cause of iron deficiency in adults is pathological *loss of blood*, especially when the bleeding is chronic. Blood with a haemoglobin concentration of 150 g l^{-1} contains about 9 μmol of iron per ml. Therefore, an average daily loss of more than about 10 ml of blood becomes important since this volume of blood will contain more iron than can be maximally absorbed (about 90 μmol) from a normal diet supplying about 270 μmol of iron per day. Chronic loss of blood occurs most commonly from the uterus and the gastrointestinal tract.

Vitamin B$_{12}$

Vitamin B$_{12}$ (cobalamin), a cobalt-containing molecule, is synthesized by bacteria and the chief dietary sources are protein-rich foods of animal origin, especially liver and kidney, but also muscle meats, with smaller concentrations in eggs, cheese and milk.

The vitamin is released from complexes with food proteins by acid and proteolytic enzymes and combines on a mole-to-mole basis with a glycoprotein, *intrinsic factor*, made by the oxyntic (parietal) cells of the stomach. This complex resists enzymic digestion and reaches the terminal ileum where it attaches to specific receptors for intrinsic factor on the brush border of the mucosal cells. Mucosal uptake of the complex, possibly by pinocytosis, is followed by degradation of intrinsic factor and release of the vitamin to the portal blood. Absorption of vitamin B$_{12}$ is limited by the number of receptors available in the ileum and a normal diet supplies rather more of the vitamin than can be absorbed (Table 10.3). In the portal blood, vitamin B$_{12}$ is bound to a transport protein, *transcobalamin II*, which readily releases the vitamin to the bone marrow and other tissues. However, most vitamin B$_{12}$ in the blood is tightly bound to another protein, *transcobalamin I*, which gives up little of the vitamin to the tissues.

Table 10.3. Vitamin B_{12} and folic acid.

	Vitamin B_{12}	Folic acid
Daily diet supplies	4–22 nmol	900–2300 nmol
Absorption limit	2·5 nmol/day	50–100% of intake
Absorption site	ileum	duodenum, jejunum
Daily needs	1·5 nmol	220 nmol
Adult store	1500–2200 nmol	14–45 μmol
Time for depletion	3–4 years	2–7 months
Serum concentration	150–700 pmol/l*	4–20 nmol/l*
Red-cell concentration	—	300–1500 nmol/l*

* Radioimmunoassay; values for normal range depend on method.

An adult loses about 1·5 nmol of vitamin B_{12} each day, mainly in the urine and faeces, and needs to absorb a similar amount daily to stay in balance. Body stores, mainly in the liver, are enough to meet requirements for 3–4 years if dietary intake were to cease completely.

Deficiency of vitamin B_{12} is almost always the result of impaired absorption; only rarely does it arise from inadequate intake (e.g. in strict vegetarians). Impaired absorption may be due to lack of intrinsic factor (e.g. in pernicious anaemia or after gastrectomy), reduced absorptive capacity of the ileum (e.g. with ileal disease or after resection) and competition for the vitamin by intestinal parasites or bacteria.

Folic acid

Folic acid (pteroylglutamic acid) is a vitamin present in foods mainly as a polyglutamate. Folates occur in most foods, the highest concentrations being found in liver, yeast, green vegetables, nuts and fruit. The diet may provide as much as 2·3 μmol daily but the amount available varies widely depending on the type of food eaten and the method of preparation, since folic acid is easily destroyed by cooking. Folates are absorbed mainly from the duodenum and jejunum after reduction and methylation to methyl-tetrahydrofolate, which is the form in which the vitamin enters the portal blood. The small intestine has a large absorptive capacity for folate (Table 10.3).

An adult requires about 220 nmol of folic acid daily and this requirement increases to 660 nmol or more per day during pregnancy. Folate and its breakdown products are lost in the urine and sweat; faecal folate is probably derived largely from bacteria in the colon. Adults store between about 14 and 45 μmol of folate, mainly in the liver. In the absence of intake, body stores can become exhausted in several months (Table 10.3) so that severe deficiency of folate can develop rapidly.

Inadequate dietary intake of folate is probably not uncommon and will contribute to deficiency due to defective absorption (e.g. in mucosal disease), to drugs (e.g. anticonvulsants, alcohol) or to increased demand for the vitamin when cellular turnover is rapid, as in physiological states (e.g. pregnancy) or pathological conditions (e.g. malignant tumours).

Megaloblastic erythropoiesis

Deficiency of vitamin B_{12} or folic acid leads to decreased synthesis of DNA in proliferating tissues and produces characteristic morphological abnormalities in the bone marrow. The red-cell precursors are larger than normal (*megaloblasts*) and impaired synthesis of DNA leads to a delay in maturation of the nucleus relative to the cytoplasm. The nuclear chromatin has an open, stippled, immature appearance, at variance with the degree of haemoglobin formation occurring in the cytoplasm. Production of white cells and platelets is also affected and a macrocytic anaemia is often accompanied by low granulocyte and platelet concentrations in the blood.

It is thought that impaired synthesis of DNA due to deficiency of folate results from a slowing of a rate-limiting step, the methylation of deoxyuridylic acid to thymidylic acid. Tetrahydrofolate, as the 5,10-methylene derivative, is involved as coenzyme in this reaction so that deficiency of folate would decrease the rate of synthesis of thymidylate and hence of DNA.

It has been proposed that vitamin B_{12} is necessary for the synthesis of tetrahydrofolate from methyl-tetrahydrofolate which is the form in which it is released from the small intestine (see above). In this view, deficiency of vitamin B_{12} leads to reduced synthesis of DNA, and hence megaloblastic anaemia, because of inadequate supply of tetrahydrofolate.

Characteristics of red cells

The mature red cell is a biconcave disc, 7–8 μm in diameter, 2·5 μm thick near the rim and 1 μm thick at the centre. This curious shape enhances diffusion of the respiratory gases into and out of the cell since there is a greater surface area, and a shorter distance to central cellular regions, than would be present in a spherical cell of the same volume. The biconcave shape also increases the flexibility of red cells which allows them to bend easily as they squeeze through narrow capillaries and it keeps the tension on the membrane minimal when the red cells swell as they take up CO_2 from the tissues (p. 495).

Table 10.4 lists typical ranges of normal values of variables relating to the concentration of red cells in the blood: the *red-cell count, haemoglobin concentration* and *haematocrit*. These variables are maximal immediately after birth, fall to minimal values by about the age of six months, then rise slowly from the age of about one year to puberty, when the adult ranges are reached. In the elderly, there is a tendency for the values to fall. Changes in the

Table 10.4. Normal red-cell values as the 95% range (mean ± 2 s.d.).

	Men	Women	Children (1 year)
Red-cell count ($\times 10^{12}\, l^{-1}$ blood)	4·5–6·5	3·8–5·8	3·6–5·2
Haemoglobin (g l^{-1} blood)	130–180	115–165	105–135
Haematocrit (Packed cell volume, PCV)	0·40–0·54	0·37–0·47	—
Mean cell volume (MCV) (femtolitre, fl)	76–96	76–96	70–86
Mean cell haemoglobin (MCH) (picogram, pg)	27–32	27–32	23–31
Mean cell haemoglobin concentration (MCHC) (g l^{-1} cells)	300–350	300–350	300–350
Reticulocyte count ($\times 10^9\, l^{-1}$ blood)	25–85	25–85	25–85
Reticulocyte count (%)	0·2–2·0	0·2–2·0	0·2–2·0

Data from Dacie, J.V. & Lewis, S.M. (1984) *Practical Haematology*, 6th edn, p. 8. Churchill Livingstone, London.

haemoglobin concentration, red-cell count and haematocrit usually reflect changes in red-cell mass but this relationship can be distorted by variations in plasma volume, e.g. increase in the haematocrit with dehydration or decrease in the haematocrit as plasma volume rises in pregnancy.

The red-cell count, haemoglobin concentration and haematocrit can now be measured very accurately by electronic cell counters. These instruments also compute useful 'absolute values' or 'red-cell indices' which indicate the average size and haemoglobin content of the cells. The *mean cell volume* (calculated as haematocrit/red-cell count) reflects red-cell size, which may be normal (*normocytic*), increased (*macrocytic*) or decreased (*microcytic*). The *mean cell haemoglobin* (calculated as haemoglobin concentration/red-cell count) and the *mean cell haemoglobin concentration* (haemoglobin concentration/ haematocrit) indicate respectively the average amount and concentration of haemoglobin in the cells. Red cells are called *normochromic* when they have a normal concentration of haemoglobin; if it is reduced, they are *hypochromic*.

Variations in size and haemoglobin content of red cells are also detected by microscopic examination of a blood film stained with a Romanowsky stain. Normal erythrocytes appear as round cells, staining an even pink with sometimes an area of central pallor (corresponding to the biconcavity) which occupies about one-third of the area of the cell.

In diseases affecting red cells, morphological abnormalities are often apparent on the blood film and arise in general from three main causes: abnormal erythropoiesis, damage to circulating red cells and changes associated with compensatory increases in red-cell production. Thus, *anisocytosis* (variation in size) and *poikilocytosis* (variation in shape) are non-specific features of abnormal erythropoiesis, while *hypochromic* erythrocytes are seen when there is inadequate synthesis of haemoglobin because of impaired production of *haem* or of *globin*. The presence of an abnormal haemoglobin may cause a diagnostic abnormality, e.g. *sickle cells* in people with haemoglobin S. Fragmented red cells (*schistocytes*) are caused by damage in the circulation, e.g. by passage of blood through strands of fibrin. Increased release of reticulocytes from the marrow results in *polychromasia* due to greyish-blue staining of the reticulocytes.

The red-cell membrane

The membrane of the red cell is a lipid bilayer closely associated with proteins situated on one side of the membrane or passing right through it (p. 6). Carbohydrates occur mainly as glycolipids and glycoproteins on the external surface of the membrane and often have blood-group specificity (e.g. A, B, H antigens).

The membrane proteins form two main groups: *peripheral proteins* and *integral membrane proteins* (Fig. 10.7). Just inside the membrane, the peripheral proteins form an extensive filamentous network, called the *red-cell membrane skeleton* (cytoskeleton). This appears to be essential for maintaining the structural stability of the cell and probably also helps determine its shape and deformability. The major constituent of this cytoskeleton is *spectrin*, a dimer consisting of α and β subunits. The membrane skeleton also contains filaments of actin. The integral membrane proteins, e.g. *protein 3* and *glycophorin A*, are tightly bound to the membrane and extend through it, to act as anchors for the underlying membrane skeleton (Fig. 10.7). Defects of this protein cytoskeleton are thought to explain some inherited abnormalities of erythrocyte shape, e.g. hereditary spherocytosis and hereditary elliptocytosis.

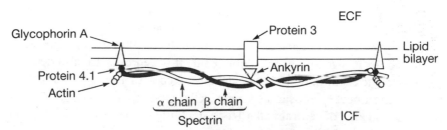

Fig. 10.7. Red-cell membrane skeleton. Peripheral spectrin dimers are connected to integral membrane protein 3 by ankyrin and to glycophorin A by protein 4.1.

The integral membrane proteins also subserve important cellular functions. Protein 3 has binding sites for several glycolytic enzymes and provides transport facilities for HCO_3^- and Cl^- ions entering and leaving the cell (p. 495). Other proteins include the Na^+–K^+-ATPase of the sodium pump and a Ca^{2+}–Mg^{2+}-ATPase, which mediates active efflux of calcium from the cell.

Haemoglobin

The red respiratory pigment of the erythrocyte, *haemoglobin*, is composed of four polypeptide chains of globin, each containing a molecule of haem. Haem consists of a protoporphyrin ring with a central atom of ferrous iron which can combine reversibly with a molecule of O_2. There are four types of globin chains, α, β, γ and δ, which differ in their constituent amino acids. In an adult, 96–98% of the circulating haemoglobin is *haemoglobin A*, the globin of which has two α chains and two β chains. Adults also have small amounts of haemoglobin A_2 ($\alpha_2\delta_2$) and of haemoglobin F ($\alpha_2\gamma_2$), which is the main haemoglobin of the fetus. Other globin chains (ε and ζ) and haemoglobins are synthesized in the large nucleated red cells which develop in the mesoderm of the yolk sac of the early embryo. The time-course of the change in synthesis from γ to β chains is shown in Fig. 10.8.

The tetrameric structure of haemoglobin allows conformational changes in the molecule which account for important characteristics of O_2 transport. One molecule of haemoglobin, with its four haem groups, can combine with up to four molecules of O_2 and uptake of one molecule of O_2 by a haem group facilitates uptake of the next. This 'cooperative' effect (the 'haem–haem interaction') gives rise to the sigmoid shape of the oxyhaemoglobin dissociation curve (see Fig. 12.13). Changes in the affinity of haemoglobin for O_2 are also brought about by variations in pH and P_{CO_2}, (the Bohr effect, p. 491) and by

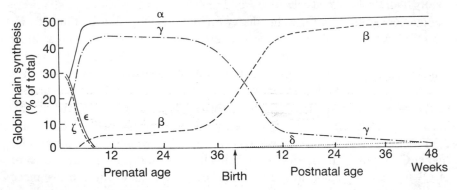

Fig. 10.8. Synthesis of globin chains.

the red-cell metabolite, 2,3-diphosphoglycerate (2,3-DPG). These various effects are mediated by interactions between spatially distinct sites in the haemoglobin molecule (*allosteric* interactions).

The molecule of haemoglobin A is almost spherical, four globin chains forming a tetrahedral array such that a pair of dissimilar chains (e.g. α_1 and β_1) make up a dimer. There is little mobility between the component chains of each dimer but there is freedom to rotate and slide along the contacts between the two dimers. These movements allow the molecule to exist in two different configurations, associated with deoxygenation and oxygenation. In the *deoxy* or *tense* (T) state, the globin chains are kept apart and immobile by ionic bonds ('salt bridges') between charged groups of amino acids and by 2,3-DPG wedged between the β chains (Fig. 10.9). In the *oxy* or *relaxed* (R) state, the salt bridges are broken, the 2,3-DPG is expelled and the β chains move closer together.

These changes in configuration are brought about by combination of O_2 with the iron atom of haem. Each haem group is contained within a waterproof crevice in a globin chain. In the deoxy state, the iron atom lies outside the plane of the protoporphyrin ring because of repulsion between the nitrogen atoms of the ring and a histidine residue of the adjacent globin chain (Fig. 10.9). The haem groups in the α chains are probably oxygenated first and the combination of a molecule of O_2 with iron enables the iron atom to move into the plane of the porphyrin ring, pulling the histidine residue with it. This causes movements of the α chains, the breaking of their linking salt bridges and the change in configuration of the molecule towards the oxy form. Movement of binding sites for 2,3-DPG on the β chains and narrowing of the gap between the β chains results in extrusion of the molecule of 2,3-DPG and eventual disruption of the remaining salt bridges between the α and β chains (Fig. 10.9). The haem crevices in the β chains widen allowing molecules of O_2 to enter so that oxygenation of haemoglobin is complete. Thus, the fourth molecule of O_2 taken up is able to attach to haemoglobin about 300 times more avidly than the first molecule.

The low affinity for O_2 of the deoxy form of haemoglobin is enhanced by 2,3-DPG which stabilizes the T structure by providing extra cross-links between the β chains (Fig. 10.9). Similar effects are provided by CO_2, which binds to terminal amino groups of all four globin chains to produce carbamates that form salt bridges. Hydrogen ions also bind to amino and carboxyl groups of the globin chains to stabilize the T form. These allosteric events are the basis of the Bohr effect (p. 491). Note, however, that the fetal γ chains bind 2,3-DPG weakly, producing the R state of the haemoglobin molecule with its higher O_2 affinity. This is vital to the fetus because of the low partial pressure of O_2 in maternal blood in the placental capillaries and diffusion distances that are greater than in the lung.

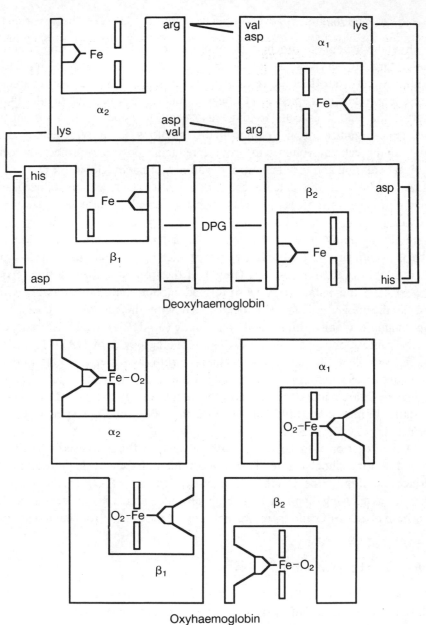

Fig. 10.9. Deoxy or tense (T) and oxy or relaxed (R) forms of haemoglobin. Lines connecting boxes indicate salt bridges. DPG, 2,3-diphosphoglycerate.

Red-cell metabolism

Without the benefit of mitochondria or of ribosomes for synthesizing protein, the erythrocyte survives for more than four months in the face of repeated oxidant stress from the high concentrations of O_2 and repeated mechanical stress from many daily passages through capillaries of diameter smaller than that of the cell. This cellular longevity seems to depend on a simplified metabolic organization with three main functions: to provide energy for maintaining cellular volume, to provide reducing power to protect the cell against oxidation and to help control the affinity of haemoglobin for O_2.

About 95% of the glucose consumed by red cells is metabolized by anaerobic glycolysis; 5% is used by the pentose phosphate pathway. As well as providing energy for the Na^+-K^+-ATPase, the glycolytic pathway also maintains a supply of reduced nicotinamide adenide dinucleotide (NADH). This is a coenzyme for a reductase enzyme which helps to maintain the iron of haemoglobin in the ferrous state (Fe^{2+}). If the iron is oxidized to the ferric (Fe^{3+}) state, the *methaemoglobin* so formed cannot combine reversibly with O_2. Methaemoglobin may also be reduced with the aid of the coenzyme nicotinamide adenine dinucleotide phosphate (NADPH), which is produced by the pentose phosphate pathway. The reducing power of NADPH is also made available to the cell through its linkage with the tripeptide, glutathione. This substance provides protection against the oxidation of sulphydryl groups of enzymes, globin and constituents of the membrane. It also counteracts autoxidation of membrane lipids and helps dispose of any hydrogen peroxide that forms.

A side-reaction of the glycolytic pathway leads to the synthesis of 2,3-DPG from 1,3-diphosphoglycerate. The 2,3-DPG combines with haemoglobin to reduce its affinity for O_2 thus promoting delivery of O_2 to the tissues. Conditions causing hypoxia (e.g. high altitude, anaemia) cause increased synthesis of 2,3-DPG and therefore increased release of O_2 from haemoglobin.

Life-span and breakdown of red cells

Mature red cells normally remain in the blood for between 100 and 130 days. Thus, a little under 1% of the circulating red cells are destroyed each day and must be replaced by reticulocytes released from the bone marrow. For a man of 70 kg, this turnover of cells amounts to more than 200×10^9 per day or, on average, 2·6 million every second.

As red cells age in the circulation, they become smaller and more dense and there is a decline in the activity of their glycolytic and other enzymes. At the end of their lives, probably in response to degenerative changes in the cell membrane, effete erythrocytes are removed from the circulation by macrophages in the bone marrow, spleen and liver. Within the macrophage, the red cell is broken down with release and degradation of haemoglobin (Fig. 10.10). The amino acids of globin are returned to the general amino acid pool of the body. The haem groups are degraded by microsomal enzymes, with release of iron to the extracellular protein transferrin and its transport to erythroblasts for insertion into new haem groups or to iron stores. The remainder of the haem group is converted to *biliverdin* and during this reaction a molecule of carbon monoxide is released. Biliverdin is reduced to *bilirubin* which enters the blood, becomes attached to albumin and is carried to the liver. Here microsomal enzymes in the hepatocyte make bilirubin more soluble in water by conjugating it with glucuronic acid and, mainly as the diglucuronide, bilirubin is secreted into the bile. In the intestine, reducing enzymes of bacteria convert the bilirubin into a number of products. One of these, *stercobilinogen*, is excreted in the faeces as *stercobilin*. Another, *urobilinogen*, is readily absorbed from the gut, passes back to the liver and is released into the bile (*enterohepatic*

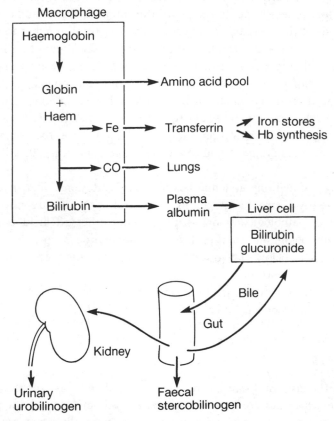

Fig. 10.10. Degradation of haemoglobin.

circulation). A small amount of the absorbed urobilinogen enters the systemic circulation and is excreted in the urine.

Jaundice, a yellow coloration of skin, sclera and mucous membranes, results when plasma levels of bilirubin are elevated above some 35 μmol l^{-1}. This can result from increased bilirubin formation (e.g. in haemolytic anaemia), decreased uptake by hepatocytes, disordered cellular conjugation, impaired secretion into the canaliculi or obstruction to biliary drainage. In duct obstruction, the level of conjugated bilirubin is increased and urinary bilirubin is increased. In contrast, increased production or failure of hepatocyte uptake results in increased unconjugated bilirubin, which, being bound to albumin, is not excreted in the urine.

Anaemia

In functional terms, anaemia is a disorder in which a decrease in the amount of circulating haemoglobin reduces the O_2-carrying capacity of the blood so that this is insufficient to meet the needs of the tissues for O_2. Various compensatory processes act to safeguard the supply of O_2 to vital organs and the demands of these organs for oxygen may alter from moment to moment. To encompass all these variables within a single useful definition is difficult and for clinical purposes it is usual to define anaemia on the arbitrary basis of the concentration of circulating haemoglobin. Thus anaemia is 'a reduction in the concentration of haemoglobin in the blood below the lower limit of the normal range for the age and sex of the patient'. A man is therefore anaemic if his haemoglobin concentration is less than 130 g l^{-1} and the corresponding value for women is 115 g l^{-1} and for children aged one year, 105 g l^{-1} (Table 10.4).

This way of defining anaemia has obvious limitations. Thus people may have apparently normal haemoglobin concentrations which are, nevertheless, suboptimal or pathologically low for them, while others, defined as anaemic, may in fact belong to that part of the population whose haemoglobin concentration is below the lower limit of a 95% range. Interpretation of haemoglobin concentration may thus require caution. However, as defined above, anaemias severe enough to cause symptoms are usually associated with unequivocally low concentrations of haemoglobin in the blood.

Causes of anaemia

In health, the erythroid marrow releases new red cells into the blood, while worn out erythrocytes are removed. Despite this continual cellular turnover, the total number of circulating erythrocytes normally stays remarkably constant. The rate of production of new cells, which is regulated by erythropoietin, precisely balances the rate of destruction of old cells. The result is that, provided the vascular system is intact, the mass of circulating red cells stays constant. Anaemia, and a decrease in red-cell mass, may arise from one (or more) of three fundamental disturbances: decreased production of red cells, increased destruction of red cells and bleeding.

Decreased production of erythrocytes (Table 10.5) may be due to a reduction in the rate of proliferation of precursors in the marrow, the *hypoproliferative anaemias*, or may arise from *defective* development of red-cell precursors, associated with an abnormal degree of ineffective erythropoiesis (p. 319).

Deficient iron supply for erythroblasts (p. 324) is a common cause of reduced erythroid proliferation and a similar defect occurs in many chronic inflammatory disorders, when release of iron from macrophages appears to be blocked. Diminished proliferation of red-cell precursors also occurs because of a reduced output of erythropoietin from diseased kidneys, or from normal kidneys in the presence of diminished requirements of tissues for O_2 (e.g. in hypothyroidism). The bone marrow may be infiltrated by abnormal cells to such an extent that erythropoiesis is reduced (e.g. in leukaemia) and this may also occur with bone-marrow failure (aplastic anaemia) due to drugs, chemicals, ionizing radiations (as occurred at Chernobyl) and for no discernible cause. The hypoproliferative anaemias are associated with a reticulocyte count disproportionately low for the degree of anaemia. The circulating red cells are

Table 10.5. Main causes of anaemia.

Decreased production
Reduced proliferation
   Iron deficiency, chronic disorders, low erythropoietin, marrow infiltration, marrow failure
Defective maturation
   Nucleus: vitamin B_{12}, folate deficiency
   Haemoglobin: defective synthesis of:
      globin: thalassaemias, haemoglobinopathies
      haem: sideroblastic anaemias

Increased destruction
Cellular defect
   Membrane, haemoglobin, enzyme
Extracellular defect
   Immune: autoimmune disease
   Non-immune: trauma, drugs, bacteria, etc.

Bleeding
Acute
Chronic

---

commonly normocytic and normochromic but microcytic and hypochromic cells are typical of iron deficiency. The bone marrow characteristically is hypocellular with evidence of depressed erythropoiesis.

Abnormalities of maturation affect primarily the nuclear function of erythroblasts or the synthesis of haemoglobin (Table 10.5). Deficiency of vitamin B_{12} or of folate, or both, interferes with synthesis of DNA (p. 336) with characteristic nuclear–cytoplasmic asynchrony seen in the abnormally large red-cell precursors (megaloblasts). Genetic defects can cause reduced rate of production of one or more chains of globin (the *thalassaemias*), or variations in structure of globin chains (*haemoglobinopathies*), with resultant instability of the haemoglobin molecule or abnormal functioning in the transport of O_2. Genetic and acquired anomalies may also lead to defective synthesis of haem with abnormal accumulation of iron in erythroblasts (the *sideroblastic* anaemias). These various disorders of maturation are commonly associated in the marrow with marked erythroid hyperplasia and premature destruction of many red-cell precursors. The reticulocyte count is low and the circulating red cells show evidence of the disturbance in production, such as macrocytic erythrocytes in megaloblastic erythropoiesis, and microcytic and hypochromic cells when synthesis of haem or of globin is defective.

Anaemia may also arise because of an abnormally high rate of destruction of erythrocytes (Table 10.5) either because of some defect of the cells (commonly genetically determined) or because the cells are damaged by something abnormal in the vascular environment (acquired disorders). Anaemias of this kind are called *haemolytic* and the characteristic feature of them all is that the life-span of the red cell is reduced. In very severe haemolytic anaemias, the red cells may survive for only a few days. The increased rate of destruction results in a raised output of bilirubin which may be visible in the tissues as *jaundice* (p. 334). The tissue hypoxia causes increased secretion of erythropoietin, which leads to a compensatory erythroid hyperplasia in the bone marrow and a raised reticulocyte count, often noticeable as a macrocytosis in the blood.

Acute loss of blood and oxygen-carrying capacity may result in the medical emergency of hypovolaemic shock (p. 461) but chronic bleeding, especially if it is intermittent, may cause an anaemia to develop very slowly. The anaemia may occur only after the iron stores in the body are exhausted (p. 324).

Red-cell indices and anaemias

Anaemias may also be classified in terms of the red-cell indices—MCV, MCH and MCHC (Table 10.4). This classification has the advantage that it immediately suggests a short list of likely disorders and hence any additional tests required. It may also allow identification of an important defect before overt anaemia has developed.

Macrocytic normochromic anaemias are characteristically those associated with megaloblastic erythropoiesis (B_{12} or folate deficiency) but also arise in alcoholism, liver disease and hypothyroidism. Increased numbers of circulating reticulocytes may also give a macrocytic blood picture.

Microcytic hypochromic anaemias are most often due to iron deficiency but also occur with thalassaemias and with sideroblastic defects. *Normocytic* and *normochromic* anaemias are commonly secondary to some chronic disorder (e.g. infection, renal failure, rheumatoid arthritis) but also occur in haemolysis, after acute bleeding, in bone marrow failure and as a result of infiltration of the marrow by abnormal cells (e.g. metastatic carcinoma).

Compensatory adjustments in anaemia

Delivery of O_2 from the lungs to the tissues involves a number of steps of which transport by haemoglobin is only one (Chapter 12). Ventilation, cardiac output, differential control of peripheral resistance to blood flow and diffusion from capillaries to cells are other links in the chain. In general, a decrease in effectiveness of any of these components of the O_2 transport system may be counteracted by increased activity of the others. Thus, in anaemia, various adjustments occur in order to compensate for the reduced oxygen-carrying capacity of the blood so as to maintain oxygenation of the tissues.

As the haemoglobin concentration falls, there is greater deoxygenation of haemoglobin in the capillaries. There is also increased synthesis of 2,3-DPG in the red cells which promotes deoxygenation of haemoglobin, shifting to the right the oxyhaemoglobin dissociation curve (see Fig. 12.13). While the increase in 2,3-DPG allows greater unloading of O_2 from arterial blood, this leaves a smaller venous O_2 content and thus a diminished reserve available for extra demands (e.g. in exercise). The total blood volume is maintained, despite the decreased red-cell mass, by an increase in plasma volume. There is also a redistribution of the flow of blood away from organs like the kidneys and the skin to more vital areas such as myocardium, brain and skeletal muscle.

When the blood haemoglobin concentration falls below about 70–80 g l^{-1}, cardiac output rises, at rest and during exercise. This is due to an increase of both stroke volume and heart rate. A net vasodilatation, with more rapid flow of blood of reduced viscosity (because of the decreased red-cell mass) gives rise to the characteristic *hyperkinetic* circulation of chronic anaemia.

The extent to which these adjustments allow adaptation to anaemia will depend upon a number of factors including the severity of the anaemia, its speed of onset and the adequacy of oxygenation and function of the myocardium.

10.3 The white blood cells

The white cells of the blood form two main groups both with defensive functions: *phagocytes*, which can engulf and destroy bacteria and other foreign matter, and *lymphocytes*, the effector cells of the immune system (p. 342). The phagocytic cells comprise *polymorphonuclear leukocytes*, in which the nucleus is often divided into several lobes, and *monocytes*, the precursors of *macrophages*, which are phagocytic cells, *fixed* or *free* in the tissues. All these phagocytic cells have cytoplasmic granules, which are often lysosomes. The granules of the polymorphonuclear leukocytes react with Romanowsky stains to give characteristic appearances in blood films and allow identification of three cell types: neutrophil, eosinophil and basophil, often known collectively as *granulocytes*. Normal ranges of total and differential white-cell counts in the blood are listed in Table 10.6.

The *neutrophil* is the most frequently occurring white blood cell in adults. It has a densely staining lobulated nucleus and fine dark-blue or purple (azurophilic) granules in the cytoplasm (Table 10.7). The cytoplasm often has a pinkish tinge because of other very small granules the size of which is at the limit of resolution of the optical microscope. Both types of granules are lysosomal in nature and liberate enzymes able to kill bacteria when the granules fuse with *phagosomes* (vacuoles containing bacteria or foreign material taken up by phagocytosis). Neutrophils act as phagocytes in acute inflammation (p. 341).

The *eosinophil* commonly has a nucleus with two lobes and the cytoplasm appears densely packed with large oval granules which stain a bright red or orange with Romanowsky stains. The granules contain lysosomal enzymes and a protein called *major basic protein* which can neutralize heparin. Eosinophils appear to have special roles in combating parasitic infestations, in phagocytosing antigen–antibody complexes and in modulating the effects of histamine and leukotrienes in allergic reactions.

The most infrequent white cell in the blood is the *basophil* (Table 10.6).

Table 10.6. Normal white cell counts ($\times 10^9 \, l^{-1}$ blood) as the 95% range (mean ± 2 s.d.).

Cell	Adults	Children (6 years)
Neutrophils	2·0–6·0	2·0–7·5
Lymphocytes	1·5–4·0	5·5–8·5
Monocytes	0·2–0·8	0·7–1·5
Eosinophils	0·04–0·4	0·3–0·8
Basophils	<0·01–0·1	<0·01–0·1
Total	4·0–11·0	5·0–15·0

Data from Dacie, J.V. & Lewis, S.M. (1984) *Practical Haematology*, 6th edn, p. 9. Churchill Livingstone, London.

Table 10.7. Types of white cell in Romanowsky-stained blood film.

Cell	Diameter (μm)	Nucleus	Cytoplasm	% of total (adults)
Neutrophil	12–15	2–5 lobes	Pink, granular; fine purple granules	40–75
Lymphocyte	6–8 (small) 12–16 (large)	Round, heavy chromatin	Thin rim, pale blue; occasional granule	20–45
Monocyte	12–20	Large, irregular; fine chromatin	Bulky, pale blue-grey	2–10
Eosinophil	12–15	Two lobes	Many large, oval, orange-red granules	1–6
Basophil	12–15	Large; irregular lobes	Few dark-blue granules; often overlie nucleus	< 1

This cell has an irregular nucleus, sometimes indistinctly divided into lobes and often obscured by the rather scanty dark-blue granules, which contain heparin and histamine. There are some similarities between basophils and the mast cells in tissue spaces but the precise relationship is not clear. Both types of cell have membrane receptors for IgE immunoglobulins and the combination of cell and antibody in immediate hypersensitivity reactions (p. 346) leads to release of the contents of the granules.

The *monocyte* is the largest white cell (Table 10.7), with a nucleus of irregular shape and fine chromatin. The bulky cytoplasm stains a blue-grey colour and has a characteristic 'ground-glass' appearance probably caused by the presence of many lysosomal granules beyond the resolution of most optical microscopes.

Most *lymphocytes* in the blood are small cells only a little bigger than red cells. They have a round nucleus with densely staining, condensed chromatin and a thin rim of featureless blue cytoplasm. About 10% of the circulating lymphocytes are larger (Table 10.7), with rather more cytoplasm which may contain a few azurophilic granules.

Granulocyte production and life-span

Granulocytes and monocytes develop in the red bone marrow from a common progenitor, the myeloid stem cell. The earliest recognizable granulocyte precursor is the *myeloblast*, which subsequently matures through the stages of *promyelocyte* and *myelocyte* (Fig. 10.2). The specific cytoplasmic granules first appear in the myelocyte and mitosis continues up to the end of this stage. Thereafter, with loss of mitotic capacity, the cells get smaller and there is progressive condensation of the nucleus in the *metamyelocyte* and *band* stages. Nuclear segmentation occurs in the mature granulocyte.

After mitosis, granulocytes spend up to a week maturing further and there are large numbers of metamyelocytes, band forms and mature cells kept in reserve in the marrow. This relative abundance of developing granulocytes is indicated by the ratio of white-cell to red-cell precursors, the myeloid to erythroid ratio. In normal marrow, this varies between 2·5:1 and 15:1, and a normal marrow stores between 15 and 20 times as many granulocytes as are present in the blood.

Control of granulocyte production is poorly understood but it seems likely that a feedback system operates between the marrow and mature cells in the blood and tissues, including macrophages and T-lymphocytes. Studies *in vitro* have identified various *colony-stimulating factors* which stimulate production and differentiation of granulocytes.

Following release into the blood, neutrophils are present in about equal numbers either freely circulating or rolling in a 'marginating' pool along the walls of capillaries and venules. These latter cells are not included in a white-cell count of blood but there is a rapid and free exchange of cells between the circulating and marginal pools. Neutrophils spend about 10 h in the blood before being lost at random into the tissues. Here they survive for probably 4–5 days before removal by macrophages after they have performed their phagocytic functions or have become senescent.

Rather less is known about transit through the blood of the other granulocytes and monocytes. Eosinophils remain in the circulation longer than neutrophils and show diurnal fluctuations in concentration which are related inversely to secretion of cortisol. Monocytes are stored to some extent in the marrow and spend 20–40 h in the blood before leaving to become macrophages in the tissues, where they may survive for months or years.

Abnormal white-cell counts

An increase in circulating neutrophils, a *neutrophil leukocytosis* (neutrophilia), is usually due to increased output from the marrow in response to bacterial infections, to inflammation and necrosis of tissues, or to acute haemorrhage. Very high counts, with precursor marrow cells in the blood, occur in *chronic granulocytic leukaemia*. A neutrophil count below the lower limit of normal, *neutropenia*, can occur due to the action of drugs on the marrow, in severe infections when output of neutrophils may fail to keep pace with demand, and as part of a *pancytopenia*, e.g. in bone-marrow failure.

An increased lymphocyte count, *lymphocytosis*, occurs in viral illness, in chronic bacterial infections (e.g. tuberculosis) and commonly in young children as a reaction to infections which produce a neutrophilia in adults. A high lymphocyte count in the middle-aged and elderly is commonly due to *chronic lymphocytic leukaemia*. A reduced lymphocyte count, *lymphopenia*, is uncommon but occurs in severe bone-marrow failure, in *acquired immune deficiency syndrome* (AIDS) and in patients taking immunosuppressive drugs. Raised concentrations of eosinophils, *eosinophilia*, are seen in allergic disorders (e.g. hay fever), parasitic infestations, reactions to drugs and in certain skin diseases (e.g. psoriasis). A *monocytosis* occurs in chronic bacterial infections while raised basophil counts are uncommon and are usually an accompaniment of a myeloproliferative disorder such as chronic granulocytic leukaemia.

Acute inflammation

Acute inflammation is the local response of living tissues to injury which may be due, for example, to infection, trauma, extremes of heat and cold, chemical agents, ultraviolet light, ionizing radiation and sometimes to antigen–antibody complexes. Tissue injury is also associated with other local reactions (chronic inflammation, repair and regeneration) and with general reactions (fever and leukocytosis) which are not discussed here. The acute inflammatory reaction is characterized by *local vasodilatation* causing redness and heat, and by *increased vascular permeability*, with the consequent increased accumulation of a protein-rich exudate and swelling. These reactions are mediated by a number of chemical mediators, such as plasma *kinins* activated in tissues and *histamine* released from mast cells and basophils. Kinins and histamine also stimulate nerve endings in the infected area producing the sensation of pain. Concomitantly with the exudation of fluid, leukocytes migrate from the circulation into the infected area in response to *chemotaxins* released by micro-organisms, by damaged tissues, or as products of complement activation (see below). The cells involved in acute inflammation are *neutrophils* ('polymorphs') and *monocytes*, the latter being transformed into *macrophages* on entering the tissues. These cells phagocytose micro-organisms and help to remove the tissue debris. The inflammatory exudate facilitates these processes and also serves to dilute noxious agents. The immune system (see below) also contributes to inflammation by virtue of:

1 antibacterial and antitoxic antibodies which may be present in the exudate;
2 derivatives of the complement cascade; and
3 the antiviral substance *interferon*.

The immune system

A further line of defence is provided by the immune system, the relative importance of which is well illustrated by the high mortality accompanying the major immunodeficiency diseases. Immune responses can be distinguished from the inflammatory reaction by virtue of the following characteristics.

1 *Specificity*—an immune system is tailored to deal with the initiating agent (*antigen*) through specific *antibody* and *cell-mediated* reactions. In contrast inflammation is non-specific.

2 *Memory*—a first encounter with an antigen is subsequently remembered, so that a second and subsequent encounters with the same antigen provoke a more effective response.

3 *Discrimination between 'self' and 'non-self'*—the host does not normally react with itself. Indeed specific immunological non-reactivity (*self-tolerance*) normally protects the host's own tissues against immunological attack. Thus immune responses are characteristically directed against chemical groupings recognized as foreign.

Antigens

Most antigens are proteins or polysaccharides of M_r above 5000–10000. Smaller molecules, called *haptens*, sometimes provoke an immune reaction if they become attached to body proteins or cells, and this accounts for such phenomena as nickel allergy. Only part of an antigenic molecule—the *antigenic determinant*—reacts specifically with the corresponding antibody. A determinant may be repeated several times on a single antigenic molecule or on a cell. Antigens important in clinical medicine include:

1 parts of the surface membrane of micro-organisms;

2 the toxic products of micro-organisms, e.g. tetanus toxin;

3 parts of the surface membrane of human cells (these antigens are important in blood transfusion and transplantation);

4 substances used therapeutically which sometimes act as haptens giving rise to drug allergies;

5 plant and animal antigens (e.g. pollen, animal dander, bee venom) which may provoke aberrant (allergic) immune responses.

Proteins from species other than man will in general be recognized as foreign. The ability to provoke an immune response is not an absolute attribute of a molecule; host factors ultimately determine the extent to which an antigen will be 'seen' as such.

The effector cells of the immune system

Lymphocytes are the main effector cells of the immune system and originate from bone-marrow precursor cells. They comprise two main populations: *T-lymphocytes* ($\sim 80\%$) and *B-lymphocytes* (15–20%). A small population of circulating cells apparently belongs to neither group (*null cells*).

T-lymphocytes depend on the thymus for their maturation (hence 'T-cells'), while B-lymphocytes probably mature in the bone marrow. B-lymphocytes were originally identified in birds, in which the Bursa of Fabricius (hence 'B-cells') is necessary for their maturation. This organ has no exact mammalian counterpart. While T- and B-cells are morphologically indistinguishable in routine histological preparations, the former differ from the latter with respect to:

1 several cell surface proteins which can be used as markers in clinical pathology;
2 the nature of their receptors for antigen;
3 their migration patterns and the sites which they occupy in the lymphoid organs; and
4 their functions.

All lymphocytes bear receptors for antigen. In B-cells, these are membrane-bound immunoglobulin (antibody) molecules. The T-cell receptor is made up of two polypeptide chains, and has structural features reminiscent of antibody molecules. Both the T- and B-cell receptors probably stem from a common ancestral molecule. One T- or B-lymphocyte and its mitotic progeny (i.e. one lymphocyte *clone*) can respond to only one antigen. The immune system as a whole, however, can cope with a vast array of different antigens. It will thus be apparent that the cells making up the immune system comprise an enormously varied population, which between them carry a huge repertoire of different receptors. An educated guess puts the number of different B-cell receptors at 5×10^8. In other words, there are at least 5×10^8 different clones of B-cells, each of which reacts with a different antigen. The number of T-cell clones is probably of comparable size. Thus the immune system contains lymphocytes bearing complementary (or near-complementary) receptors for virtually all of the foreign antigenic molecules likely to be encountered. 'Holes' in the repertoire are infrequent.

Lymphocyte circulation. Lymphocytes are found in large numbers in the lymph nodes and spleen, and in mucosal lymphoid aggregations in the gastrointestinal and respiratory tracts (e.g. tonsils, Peyer's patches). A substantial proportion of all lymphocytes constantly recirculates between the tissues, the lymph and the blood. If the immune system is to function efficiently, antigens entering the body must make contact reasonably quickly with the lymphocytes that express the complementary receptors. Before the immune system will recognize and respond to it, a new antigen must be displayed on either *macrophages* or *dendritic (interdigitating) cells*. Both cell types are nearly ubiquitous in their distribution. The antigen presented on the surface of these cells is, in effect, displayed before a stream of circulating lymphocytes. Those lymphocytes bearing receptors specific for the antigen will react with it and will proliferate; those bearing inappropriate receptors will not respond.

T-lymphocyte subsets

T-cells comprise several subsets, each with its own distinct function, that together result in *cell-mediated immunity*. About 80% of the circulating T-cells are regulatory, i.e. they facilitate or amplify other T-cell or B-cell responses (*T-helper cells*), or they dampen down or eliminate the same responses (*T-suppressor cells*). The regulatory mechanisms are still poorly understood. The ratio of helper to suppressor cells in the circulation is about 2:1 and is characteristically reduced in *acquired immune deficiency syndrome* (AIDS), in which the virus preferentially infects T-helper cells. Because T-helper cells are the first lymphocytes to interact with antigen displayed on macrophages and dendritic cells, they are sometimes referred to as T-helper/inducer cells. Another subset of T-cells (*T-effector cells*) subserves two main functions. One group (*cytotoxic T-cells*) is able to destroy cells infected with viruses. A second group mediates *delayed-type hypersensitivity* (DTH), an immune response which is important in the defence against those infections in which the causative organisms proliferate intracellularly (e.g. tuberculosis, leprosy), and also in the rejection of foreign tissue grafts. In DTH, the T-cells produce several different chemical substances, collectively known as *lymphokines*, which act non-specifically and whose main effect is to enhance the local inflammatory response. (One form of the antiviral substance interferon is a lymphokine produced by T-lymphocytes.) The term 'hypersensitivity' in DTH implies a pathological reaction. It is, however, with the exception of its extreme variants, a normal response.

B-lymphocytes and antibody production

B-cells proliferate on contact with antigen and ultimately develop into *plasma cells* which synthesize and secrete antibody. While B-lymphocytes make up 15–20% of blood lymphocytes, plasma cells are located in lymphoid and other tissues rather than in the circulation. The first encounter with an antigen (the *primary response*) is characteristically slow and relatively ineffectual. Antibody levels are low and the reaction takes 10–14 days to reach its peak (Fig. 10.11). The host's memory cells are however primed, with the result that a second and subsequent exposures to the same antigen will quickly (within 1–2 days) stimulate the production of high levels of antibody (the *secondary response*).

The antibodies secreted by plasma cells are found predominantly in the γ-globulin fraction of serum, with small quantities in the β-globulin fraction. Collectively they are referred to as *immunoglobulins*. The basic monomeric

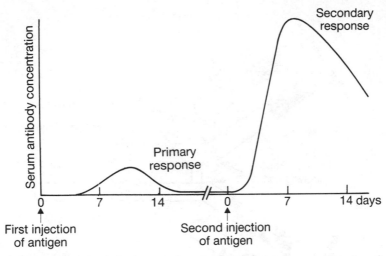

Fig. 10.11. Antibody concentration in serum in response to a first and a second injection of antigen.

form of all antibody classes consists of four polypeptide chains—two heavy chains and two light chains—held together by disulphide bonds (Fig. 10.12). The immunoglobulins comprise five structural classes of antibody which subserve different functions. These are named according to their heavy chains, both of which are identical in any one molecule. Thus the immunoglobulin (Ig) classes IgG, IgA, IgM, IgD and IgE (listed in descending order of their plasma concentrations) are characterized respectively by the heavy chains gamma, alpha, mu, delta and epsilon. There are two types of light chain, kappa and lambda, both of which are found in all five antibody classes. Within one antibody molecule, however, the two light chains are always the same.

The chemical structure of the antigen-binding site varies from one antibody (and one B-cell clone) to another, and forms part of the variable region of the molecule. The structure of the rest of the molecule is relatively constant. It is possible with papain to cleave antibody molecules into two antigen-binding fragments (*Fab fragments*) and a crystallizable fragment (*Fc fragment*) which contains part of the heavy chains. The Fc fragment is responsible for all the biological attributes of antibodies other than antigen binding (e.g. the ability to cross the placenta, to fix to mast cells and to activate the complement cascade). Because of differences in the structure of the variable region, relative molecular masses for the five different antibody classes given below are approximate.

IgG (M_r 150 000) makes up about 80% of the immunoglobulin in plasma. It is also found in extravascular tissues and can cross the placenta. It is the main antibody synthesized during the secondary response and is of major importance in the defence against micro-organisms and their toxins.

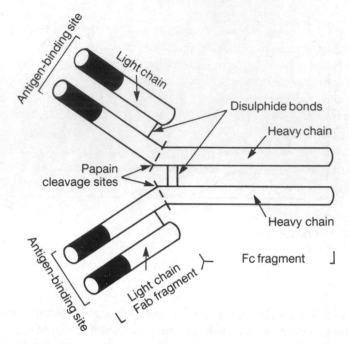

Fig. 10.12. Schematic representation of the IgG molecule showing the four polypeptide chains, the variable regions (shaded) of the chains and the fragments produced by enzymic digestion.

IgM (M_r 900 000) is a pentamer. It is the first immunoglobulin to appear in both phylogeny and ontogeny. It is also the first antibody to be produced in the primary response and the only antibody made by the fetus.

IgA (M_r 160 000) usually exists as a monomer in the plasma and as a dimer in surface secretions (tears, saliva, colostrum, gastrointestinal and bronchial secretions). It provides an important defence mechanism at these sites, particularly against viral infections.

IgD (M_r 185 000) is present mainly on the surface of B-lymphocytes where it acts as a receptor for antigen. (A given B-cell may have IgD surface receptors while secreting IgM and IgG. Even though the cell produces antibodies of more than one class, the antigen-binding site will be identical in all of them.)

IgE (M_r 200 000) is present at very low concentrations in plasma. It attaches to mast cells and basophils via receptors on the Fc portion of the molecule, i.e. it sensitizes these cells so that they release their granules if the attached antibody combines with antigen. This type of reaction exemplifies Type 1 hypersensitivity, a common example of which is hay fever. IgE levels increase significantly in response to infestation by certain parasites.

Functions of antibodies

The binding of antigen to antibody facilitates the removal of the antigen from the body. In the case of bacteria, antibodies either facilitate phagocytosis of whole organisms or neutralize their toxic products. Antibodies are capable of coating viruses, thereby reducing their pathogenicity. Certain antibody classes, such as IgM and some IgG subclasses, fix (or activate) *complement* on combining with antigen. Complement is a collective term encompassing a series of proteins which circulate in an inactive precursor form. They are activated sequentially giving rise to the production of several different factors, the functions of which include:

1 increasing vascular permeability;
2 enhancing phagocytosis; and
3 exerting a chemotactic effect towards polymorphs.

Complement is also capable of bringing about lysis of certain bacteria and cells. Complement is activated by IgM and IgG antibody (the *classical pathway*) and in a slightly different fashion by several other substances, e.g. bacterial endotoxin (the *alternative pathway*).

Active and passive immunity

Immunological memory is exploited in prophylactic immunization, which aims to provoke a primary response using vaccines (organisms or their toxic products treated to render them non-pathogenic). This is an instance of *active immunity* as the antibodies are made by the host's own cells. A subsequent encounter with the relevant live organisms or their toxic products evokes a secondary response, which will prevent or significantly attenuate the relevant disease. Memory is a function of modified T- and B-cells but the nature of the underlying changes is still uncertain. Its duration varies with the nature of the antigen and it can be lifelong for certain diseases. It is, for example, very uncommon to get measles twice.

It is possible to transfer antibodies from one person to another or even from animals to man. Immunity gained in this way is referred to as *passive immunity*. Antibodies transferred in the maternal milk provide significant protection to breast-fed infants (the antibodies are not extensively degraded in the gut at this age). Antiserum may also be useful as a stop-gap measure; for example, human serum containing anti-hepatitis A antibodies is sometimes given to short-term travellers to countries where the hygiene is poor. Passive immunity has a short duration of a few weeks at best. Animal serum is prone to cause hypersensitivity reactions and is now seldom used in clinical practice.

Tissue transplantation and the major histocompatibility complex

An individual will reject tissue or organ grafts from a genetically non-identical member of the same species. The cell-surface proteins that account for these individual differences are called *histocompatibility* (or *transplantation*) *antigens* and they are both numerous and complex. The most obtrusive of these antigens in transplant surgery belong to a genetic system known as the *major histocompatibility complex* (MHC), which in man is called the *human lymphocyte antigen* (HLA) *system*. The MHC, which is present in all higher species, is a recognition system which controls a variety of cell–cell reactions. It plays a crucial part in the interaction between T- and B-cells, and also between antigen-presenting cells and T-cells. Susceptibility to a variety of human diseases can be correlated with the HLA antigenic make-up.

Abnormalities of the immune response

Many disorders of the immune system have been described. Broadly these fall into three groups.

1 Immunodeficiencies affecting one or other component of the immune system. AIDS is the best publicized and now the commonest, but there are many others. The prevention and treatment of graft rejection involves the administration of immunosuppressive drugs (e.g. azathioprine and cyclosporin) and some measure of immunodeficiency is the price of a successful transplant. Immunodeficiency of varying severity also complicates other conditions, e.g. uraemia and severe burns.

2 Aberrant or excessive function of the immune system. In hypersensitivity or allergy (which occurs in several forms), the immune response to extraneous antigens is deleterious to the host. In autoimmune diseases the immune system reacts against the host's own tissues, e.g. in *Graves' disease* against the TSH receptor of the thyroid gland and in *myasthenia gravis* against acetylcholine receptors in skeletal muscle.

3 Neoplasms (cancers) arising in cells of the immune system. All lymphocytes can be triggered to proliferate in the uncontrolled fashion that characterizes cancer. The overproduction of one cell type at the expense of the others will sooner or later result in immunodeficiency. In *multiple myeloma* there is excessive synthesis of a single aberrant immunoglobulin arising from a malignant change in a single plasma cell and this *monoclonal antibody* appears in the serum and in some cases the urine. Characteristically, the levels of the normal immunoglobulins are markedly reduced in myeloma, resulting in increased susceptibility to infection.

10.4 Blood groups

Human red cells have on their membranes *antigenic* substances which permit the classification of blood into groups. If red cells with a particular antigen are mixed with the corresponding antibody, the red cells can clump together or *agglutinate*, a phenomenon which forms the basis of *blood-grouping* tests. If the combination of red cells and antibodies occurs in the body, the reaction can lead to dangerous breakdown of red cells within the circulation (*intravascular haemolysis*) or outside it (*extravascular haemolysis*). This is important in *blood transfusion* and in a disorder affecting the fetus and neonatal infant, *haemolytic disease of the newborn*.

If a patient receiving a blood transfusion has antibodies in the plasma which react with the transfused red cells, potentially lethal haemolysis may follow. The aim of the strict *cross-matching* procedures used in blood-banks is to ensure that such *incompatible* transfusions do not occur. During pregnancy, due to leakage of fetal red cells across the placenta into the maternal circulation, a woman may be stimulated to produce antibodies to antigens on the fetal red cells if her own red cells lack these antigens. This may occur if her lymphocytes have been sensitized to the antigen in a previous pregnancy. The antibodies (IgG) may cross the placenta into the fetal circulation and react with the fetal red cells causing them to be destroyed. This haemolytic anaemia, and its consequences, may be severe enough to cause death of the fetus *in utero* or to lead to severe complications after birth (haemolytic disease of the newborn). Many of these cases were caused by an antibody of the Rhesus blood-group system, anti-D (p. 352). This can now be largely avoided if, after transplacental haemorrhage of fetal red cells, anti-D immunoglobulin is injected into the woman to prevent immunization by the fetal cells.

Blood-group antigens and genes

The antigens on red cells which define blood groups are genetically determined. When it is shown that a set of red-cell antigens is inherited independently of others, these antigens are said to form a particular blood-group *system*. In people of European origin, there are, at present, fifteen well-defined, blood-group systems. Antigens in some of these systems are shown in Table 10.8.

The genes determining blood-group antigens are carried on pairs of autosomal chromosomes, except in the case of the Xg system, in which the gene concerned is on the X chromosome. The genes of blood-group systems generally behave as co-dominants, so that if a person has a particular gene, the corresponding antigen can be detected on the red cells. In such cases the genetic constitution of a person for the blood-group system concerned, the *genotype*, can be determined directly from the recognizable characteristics the

Table 10.8. Examples of antigens and antibodies in blood-group systems.

System	Antigens	Antibodies	Antibody type*
ABO	A, B, H	anti-A, anti-B	N
Rhesus	C, D, E, c, e	anti-C, -D, -E, -c, -e	I
MNS	M, N, S, s	anti-M, -N, -S	N
P	P, P$_1$	anti-P$_1$	N
Lutheran	Lu^a, Lu^b	anti-Lu^a	N & I
Lewis	Le^a, Le^b	anti-Le^a	N
Kell	K, k	anti-K	I
Duffy	Fy^a, Fy^b	anti-Fy^a	I
I	I, i	anti-I	N

*N = naturally occurring; I = immune

genes produce in the red cells, the *phenotype*. However, genes of some blood-group systems have no detectable effect on the red cells. These are *amorphic* genes (amorphs). When such genes are present, it may not be possible to determine a person's genotype directly from the phenotype but it may be possible to work it out by study of phenotypes in the family.

In those blood-group systems that have been investigated, the genes have been found to control synthesis of enzymes which modify the composition of glycoproteins or glycolipids on the red-cell membrane.

Blood-group antibodies

Blood-group antibodies are of two kinds, *naturally occurring* and *immune*. Naturally occurring antibodies are those found in the blood of people who have not been exposed to the red-cell antigens concerned (Table 10.8). The most common are those of the ABO system, which occur regularly in people whose red cells lack the corresponding antigens. Naturally occurring antibodies are usually IgM in type and, probably because of their large size, they cannot cross the placenta and thus cannot cause haemolytic disease of the newborn.

Immune blood-group antibodies arise as the result of an immune response to red-cell antigens not normally possessed by an individual but acquired by blood transfusion or by transplacental passage of fetal red cells during pregnancy. Thus most antibodies of the Rhesus, Kell and Duffy systems (Table 10.8) are formed in this way. Immune antibodies are most often IgG in type although IgM antibodies also occur, often early, in an immune response. The IgG antibodies can cross the placenta and can thus cause haemolytic disease of the newborn.

IgG antibodies are often *incomplete*, i.e. they fail to agglutinate red cells having the corresponding antigens when the cells are suspended in physiological saline. This is because the IgG molecule, in contrast to IgM (*complete* antibody), is too short to bridge the gap between adjacent red cells kept apart by a net negative surface charge. Agglutination will occur if these electrostatic repulsive forces between the red cells are reduced, e.g. by treating the cells with enzymes to remove charged groups from the membrane. Incomplete antibodies or complement attached to red cells can also be detected by means of the *antiglobulin test* (Coombs' test).

ABO blood-group system

The four main groups (phenotypes) of the ABO system, A, B, AB and O, are defined by the presence or absence of two red-cell antigens, A and B (Table 10.9). Either antigen may be present on the cells (group A or group B), both antigens may be present (group AB) or neither (group O). In addition, the naturally occurring antibodies, anti-A and anti-B, are present in the serum if the red cells lack the corresponding antigens (Table 10.9).

The A and B antigens are determined by the allelomorphic genes *A* and *B*. These genes, together with a third amorphic gene, *O*, are inherited as a pair, one from each parent. There are thus six genotypes but only four phenotypes (Table 10.9). The gene frequencies are not equal and, in people of European origin, groups O and A are much more common than B and AB (Table 10.9). Different phenotypic frequencies are found in other races. In some populations of Australian aborigines, for example, the groups B and AB do not occur.

The groups A and AB can each be divided into two subgroups because of two common variants of the A antigen, A_1 and A_2. About 80% of people of European origin with red cells of groups A and AB belong to the subgroups A_1 and A_1B, respectively. The remainder belong to the subgroups A_2 and A_2B. The practical importance of these subgroups is that A_2 and A_2B red cells react more weakly with anti-A antibody than A_1 and A_1B cells. A_2 and A_2B cells may react so weakly with some anti-A antibodies, that they can be wrongly grouped as being O and B, respectively.

The A and B antigens of red cells are formed from substance H, which reacts with the antibody, anti-H. The H antigen occurs on all red cells

Table 10.9. ABO blood-group system. Frequencies (UK) are approximate.

Phenotype	Genotype	Antigen on cells	Antibody in serum	Frequency (%)
A	*AA, AO*	A	anti-B	42
B	*BB, BO*	B	anti-A	8
AB	*AB*	A & B	none	3
O	*OO*	none*	anti-A & anti-B	47

*Almost always, red cells have H antigen, with maximal amount on group O cells.

irrespective of their ABO group but the amount present depends upon whether or not a person also has the *A* or *B* genes or both.

A, B and H substances are not confined to red cells: they also occur on white cells, platelets, epidermal cells and endothelium. In addition, in people called *secretors*, these substances are found in many body fluids and secretions, including saliva and sweat. About 80% of people of European origin are secretors. All have H-antigenic substance in their body fluids, while those of groups A, B or AB, also have the A, B or A and B antigens, as well as H. The ability to secrete A, B and H substances in body fluids is determined by a dominant gene, *Se*, inherited independently of the *ABO* and *H* genes, and active in people homozygous or heterozygous for *Se*. *Non-secretors* are homozygous for the allele, *se*.

Rhesus blood-group system

The five main antigens of the Rhesus system are known in Fisher's notation as C, c, D, E and e. According to Fisher's theory, these antigens are determined by three pairs of allelic genes: *C* and *c*, *D* and *d*, *E* and *e* (Table 10.10). These genes are inherited in sets of three, one set from each parent, and the *d* gene is amorphic.

The term 'Rhesus positive' is sometimes used for a person whose red cells have the D antigen and 'Rhesus negative' for a person whose red cells lack it. However, in transfusion practice, it is usual to determine the full Rhesus phenotype of all samples of blood and to reserve the term Rhesus negative for persons of genotype *cde/cde*, whose red cells lack the antigens C, D and E.

Naturally occurring Rhesus antibodies are very rare, e.g. some forms of anti-E. Immune antibodies are common following sensitization by transfusion or by pregnancy. Anti-D is responsible for many problems and causes haemolytic transfusion reactions and haemolytic disease of the newborn.

Table 10.10. Rhesus blood-group system. Frequencies (UK) are approximate.

Phenotype	Most common genotype		Frequency of genotype (%)
	Fisher notation	Short notation	
CcDee	*CDe/cde*	R_1r	32
CCDee	*CDe/CDe*	R_1R_1	17
ccee	*cde/cde*	*rr*	15
CcDEe	*CDe/cDE*	R_1R_2	14
ccDEe	*cDE/cde*	R_2r	13
ccDEE	*cDE/cDE*	R_2R_2	3
Others			6

Similar reactions are occasionally caused by antibodies with other Rhesus specificities, e.g. anti-c and anti-E.

Other blood-group systems are occasionally implicated in transfusion reactions and haemolytic disease of the newborn, e.g. the immune antibodies anti-Fy^a and anti-K (Table 10.8). These systems are, however, of much less clinical importance than the ABO and Rhesus systems.

Blood transfusion

Blood for transfusion can be kept for several weeks if it is collected from donors under aseptic conditions into sterile plastic packs containing a suitable preservative solution, and it is stored at 4°C. A commonly used preservative, citrate–phosphate–dextrose (CPD) solution, provides citrate as anticoagulant and glucose (dextrose) as metabolic substrate for the red cells. Adequate numbers (i.e. not less than 70%) of red cells remain viable after transfusion when previously stored in CPD solution for 3–4 weeks at 4°C. Adding adenine to the solution can increase this period to 5 weeks.

During storage at 4°C, the red cells show a progressive decrease in content of ATP and 2,3-DPG, while with decreased activity of the Na^+–K^+ pump, the cells gradually lose K^+ to the surrounding plasma and gain Na^+ from it. The concentration of K^+ in the plasma may reach values as high as 30 mmol l^{-1} after storage of blood for four weeks. The pH of the plasma also decreases with time of storage and its concentration of ammonia rises. These various changes can make stored blood dangerous for transfusion in certain patients, e.g. those with renal or hepatic failure. The decline in Na^+–K^+ pump activity makes some red cells spherocytic, with loss of deformability. These effects may be irreversible and after transfusion the abnormal red cells are destroyed very rapidly by macrophages in the spleen and elsewhere. Other constituents of blood do not withstand prolonged storage. Granulocytes begin to lose their phagocytic capacity within six hours of collection and they are functionally inert after 24 hours. Platelets lose their haemostatic effect within 48 hours at 4°C, while the labile coagulation factors, V and VIII, also rapidly deteriorate in chilled blood.

Before donated blood is made available for issue from a blood-bank, the ABO and Rhesus groups of the cells are determined and commonly the serum is screened for atypical antibodies. Serological tests are also done for syphilis, hepatitis B and AIDS. Before transfusion, the ABO and Rhesus groups of the patient's red cells are determined, the serum is checked for unexpected antibodies and red cells from the donor are tested against the patient's serum by *cross-matching* tests (compatibility tests). These cross-matching tests are essential for checking that there has been no error in ABO grouping of donor and recipient, and for ensuring that the recipient's serum does not contain naturally occurring or immune antibodies active against the donor's cells.

Transfusion of whole blood is sometimes necessary but over recent years the use of cell-separator machines and large-scale production of plasma constituents have made it increasingly possible to transfuse specific components of blood which the patient lacks. Thus *red-cell concentrates*, often resuspended in a small volume of electrolyte solution, are used to restore the haemoglobin concentration in an anaemic patient in whom the plasma volume may already be expanded. *Platelet concentrates* are of use in patients with severe thrombocytopenia. A variety of plasma fractions is also available to supply coagulation factors, e.g. *cryoprecipitate*, which is rich in factor VIII and fibrinogen, and for expanding plasma volume, e.g. *stable plasma protein solution*. A useful source of antibodies against common viruses is *pooled normal immunoglobulin* and various *specific* immunoglobulins are also available, e.g. anti-D and antibodies against tetanus, hepatitis B and diphtheria.

However much care is taken in cross-matching and administering blood; transfusion carries definite risks of unpleasant and even fatal complications. Major red-cell incompatibility can lead to lethal intravascular haemolysis or delayed extravascular breakdown of donor cells. Transfusion of blood contaminated with bacteria can cause profound shock with hyperpyrexia, while allergic reactions to transfused white cells, platelets and plasma proteins can also be severe. Circulatory overload, air embolism and changes in plasma electrolyte concentrations (e.g. hyperkalaemia) may occur and there may be direct transmission of disease, e.g. hepatitis, cytomegalovirus infection, malaria and AIDS.

10.5 Haemostasis

Haemostasis is the process by which bleeding from an injured blood vessel is arrested or reduced. Similar mechanisms appear to confine blood to the cardiovascular system by keeping vessels free of leaks, while other agents maintain the fluidity of the blood. A haemostatic defect, if it is severe enough, can cause a *bleeding disorder* and activation of haemostasis within the circulatory system can lead to the formation of a solid mass of blood constituents, a *thrombus*, within an intact blood vessel. In arteries, thrombosis can cause *ischaemia*—reduction or cessation of the supply of blood to a tissue—and this can lead to death of cells (*necrosis*), with areas of ischaemic necrosis (*infarcts*) often seriously, or fatally, impairing the function of an organ (e.g. myocardial infarction). In veins, such as the deep veins of the leg or the veins of the pelvis, fragmentation and onward movement of thrombi (*embolism*) can also have fatal consequences (e.g. pulmonary embolism).

Haemostasis involves interlocking reactions of blood vessels, platelets and the coagulation system, and the *haemostatic response to injury* comprises two main phases. In the *primary* phase, reactions of blood vessels and of platelets promote slowing of flow and formation of an aggregate of platelets at the site of injury. In the *secondary* phase, activation of coagulation by tissue factor released from damaged cells and by contact of plasma with disrupted vascular surfaces leads to the formation of *fibrin*, which stabilizes the platelet mass to yield a *haemostatic plug*.

Reactions of blood vessels

Many blood vessels respond to trauma by *vasoconstriction* and this reduces the rate of bleeding. After being punctured, capillaries appear to retract, with disappearance of the lumen. Endothelial cells seem able to contract in response to various stimuli. Contraction of smooth muscle in larger vessels, such as arterioles and venules, occurs promptly after injury as a result of smooth-muscle depolarization. It also appears to be mediated by local reflexes, possibly by vasoconstrictor substances of uncertain nature liberated from the vascular wall and particularly by *serotonin* and *thromboxane A_2* released by activated plaetelets. Intense vasoconstriction of large arteries in avulsion injuries can sometimes produce surprisingly effective, although temporary, haemostasis.

Vascular damage also promotes haemostasis as a result of the loss of endothelium and by the effect this has on endothelial functions which prevent or limit haemostatic activation in normal vessels. Thus the continuity of the basement membrane and of connective tissue in the walls of blood vessels, as well as the attachment of endothelial cells to these structures and to one another, is important for preventing leakage of blood. When injury removes endothelium, exposure of blood to subendothelial constituents, particularly collagen, allows adhesion of platelets to the damaged area and activation of the intrinsic pathway of coagulation. Damaged endothelial cells also release *tissue factor*, which triggers the extrinsic pathway of coagulation, and *von Willebrand factor* which is necessary for platelet adhesion.

The luminal surface of normal endothelium has a coating, about 50 nm thick, of *glycosaminoglycans*. These are polysaccharides made by endothelial cells and the main one, *heparan sulphate*, may act like the anticoagulant, heparin, in accelerating inactivation of coagulation factors by a plasma protein, *antithrombin III* (see later). *Protein C*, another inactivator of coagulation, is also activated by endothelial action. The endothelial glycosaminoglycans are strongly negatively charged and are thought to repel platelets, the surfaces of which are also negatively charged. Platelets may also be prevented from aggregating on normal vascular walls by *prostacyclin*, a prostaglandin synthesized by endothelial cells. When these various 'antihaemostatic' actions of endothelium are impaired by vascular injury, the balance will swing towards the haemostatic reactions generated by exposure of blood to subendothelial tissues.

Platelets

Platelets are cytoplasmic fragments, without a nucleus, which circulate as biconvex discs, 2–4 nm in diameter and with a volume of 6–8 fl. The normal platelet count is $150–400 \times 10^9$ per litre of blood. Platelets are derived from megakaryocytes in the bone marrow and survive in the circulation for 8–10 days. At any one time, up to a third of the platelets released from the marrow may be sequestered in the spleen.

Electron microscopy has revealed structural components of platelets (Fig. 10.13) which are essential for haemostatic function. An outer *glycocalyx*

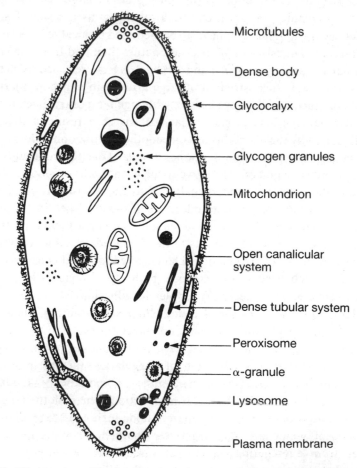

Fig. 10.13. Ultrastructure of the platelet.

consists of glycoproteins which are associated with the underlying plasma membrane and are receptors for agents which trigger platelet functions. The plasma membrane also contains *phospholipids* which provide a catalytic template for coagulation (*platelet factor 3*) and are the source of *arachidonic acid* for prostaglandin production. Extensive invaginations of the plasma membrane form a system of tubules, the *open canalicular system*, which greatly increases the reactive surface area of the platelet and provides conduits to the exterior for substances released from platelet granules during activation. A second group of intracellular tubules, the *dense tubular system*, serves as a store for Ca^{2+}, which moves into the cytoplasm to initiate contractile events. It also contains the enzymes for prostaglandin synthesis. The open canalicular and the dense tubular systems frequently come into close contact as interdigitating *membrane complexes*. These are reminiscent of, and may be equivalent in function to, the sarcoplasmic reticulum and T-tubules of skeletal muscle (p. 96). Just under the plasma membrane are *microtubules* which run as a circumferential band around the equator of the platelet and help to maintain the discoid shape of the unactivated platelet.

The cytoplasm of platelets contains contractile filaments of *actin* and *myosin* which enable activated platelets to change their shape and to release the contents of their granules. There are various organelles. The *electron-dense bodies* store adenosine diphosphate (ADP), 5-hydroxytryptamine (serotonin) and Ca^{2+}. The *α-granules* contain a number of proteins including *fibrinogen, heparin-neutralizing factor, β-thromboglobulin, thrombospondin* and *platelet-derived growth factor*, a mitogen for smooth muscle. *Lysosomes* (containing hydrolases), *peroxisomes* (with catalase), *mitochondria* and *glycogen granules* are also present. Energy for platelet function is obtained from oxidative phosphorylation in mitochondria and also from glycolytic degradation of glycogen.

Haemostatic reactions

Platelets take part in a sequence of actions in haemostasis (Fig. 10.14). After damage to the endothelium of a blood vessel, platelets adhere to subendothelial collagen fibres. Adhesion depends upon *von Willebrand factor*, part of the coagulation–factor VIII complex. This protein is thought to bind to collagen and to a glycoprotein (Ib) receptor in the platelet membrane. The adhering platelets then change shape from smooth discs to spiny spheres, with many projecting pseudopodia. The loss of discoid shape is thought to be due to depolymerization of the circumferential microtubules, while the formation of pseudopodia probably results from polymerization of actin. A gel–sol transformation of the cytoplasm makes the pseudopodia confluent so that the platelets spread out and flatten on the damaged area.

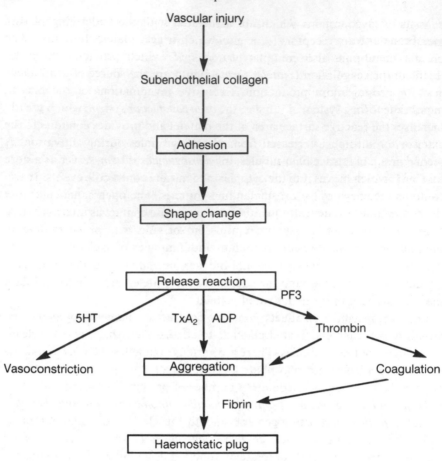

Fig. 10.14. Haemostatic reactions of platelets. 5HT, 5-hydroxytryptamine (serotonin); TxA$_2$, thromboxane A$_2$; ADP, adenosine diphosphate; PF3, platelet factor 3.

The sequence may stop at this stage but usually activation by collagen triggers the subsequent *release reaction*. In this Ca^{2+}-dependent step, the dense bodies fuse with the plasma membrane and the α-granules with the open canalicular system and their contents are discharged to the outside of the plasma membrane. Serotonin released in this way, reinforces local vasoconstriction, while ADP and also a prostaglandin derivative, *thromboxane A$_2$*, cause more platelets to stick to those already adhering, to form a clump. This process of *aggregation* depends upon Ca^{2+} and also upon fibrinogen, which is thought to link the platelets together by attaching to glycoprotein receptors IIb and IIIa. At this point, aggregation is reversible.

During these events, platelets also act as catalysts of coagulation, with local generation of thrombin and conversion of fibrinogen to fibrin. Negatively charged phospholipids, phosphatidylserine and phosphatidylinositol, exchange their places on the inside of the plasma membrane (during the inactive state) for sites on the outer surface, by a transmembrane, 'flip-flop' process. These phospholipids (platelet factor 3) act as catalytic surfaces to which complexes of the vitamin K-dependent coagulation factors are bound by Ca^{2+} bridges in optimal orientation for the generation of factor Xa and thrombin (see Fig. 10.16).

Thrombin causes further aggregation of platelets, which is irreversible (platelet 'fusion') and this effect appears to be mediated by the protein thrombospondin from the α-granules, which stabilizes the links of fibrinogen between platelets. Local production of fibrin reinforces the platelet aggregates. Eventually the mass of platelets and fibrin is compacted to form the definitive *haemostatic plug* by the action of actin and myosin filaments causing contraction of the platelets. A similar phenomenon can be seen as *clot retraction* in freshly collected blood allowed to coagulate in a test-tube.

Prostaglandin synthesis and actions

An important mediator of the sequence of haemostatic reactions of platelets is thromboxane A_2. This is formed in a series of steps (Fig. 10.15) from an

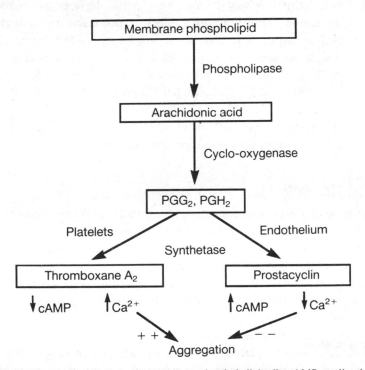

Fig. 10.15. Prostaglandin metabolism in platelets and endothelial cells. cAMP, cyclic adenosine monophosphate.

unsaturated fatty acid, *arachidonic acid*, which is a constituent of the phospholipids in the plasma membrane.

Stimulation of platelets by agents, such as collagen and thrombin, activates a Ca^{2+}-dependent phospholipase, which liberates arachidonic acid from phospholipid. In the dense tubular system, the membrane-bound enzyme *cyclo-oxygenase* (prostaglandin synthetase) converts arachidonic acid to the unstable cyclic endoperoxides, prostaglandin G_2 (PGG_2) and prostaglandin H_2 (PGH_2). PGH_2 is then transformed to thromboxane A_2 by the enzyme *thromboxane synthetase*. Thromboxane A_2 is very unstable, with a half-life of 30 s and it spontaneously breaks down to an inactive product, thromboxane B_2. Thromboxane A_2 is a powerful vasoconstrictor and one of the most potent natural platelet aggregators known. Its mode of action is uncertain but it causes release of Ca^{2+} to the cytoplasm from the dense tubular system possibly by acting as an ionophore. Increased concentrations of cytoplasmic Ca^{2+} activate actin–myosin complexes with contractile responses which are thought to drive the shape change, the release reaction and other events in platelet function.

A similar synthetic pathway produces prostaglandins in endothelial cells (Fig. 10.15) but PGH_2 is converted by a synthetase to *prostacyclin* (prostaglandin I_2). Prostacyclin is a vasodilator and a particularly potent *inhibitor* of platelet aggregation. This inhibitory effect is produced by activation of adenylate cyclase with rise in concentration of cyclic AMP in the platelet. This initiates a sequence of steps which reduces the concentration of cytoplasmic Ca^{2+} by stimulating active uptake of Ca^{2+} by the dense tubules. It has been proposed that prostacyclin acts to prevent deposition of platelets on normal endothelium, but the total amount produced per day is probably inadequate for this. However, release of prostacyclin by endothelium near an area of vascular damage may prevent unlimited growth of a platelet plug.

Other actions of platelets

Platelets take part in the process of *repair* after vascular injury. *Platelet-derived growth factor*, comprising a pair of polypeptide chains from the α-granules, promotes proliferation of smooth muscle. In ways not yet clearly defined, platelets may also be involved in inflammatory and allergic reactions. Neutrophils and macrophages release a powerful aggregating agent, *platelet-activating factor* but it is not known if this is of haemostatic significance.

Blood coagulation

The coagulation system consists of cofactors and a series of zymogens (proenzymes) which sequentially activate one another, leading to the formation of fibrin clot at a site of vascular injury. The coagulation factors (Table 10.11) are mostly globulins which are made by the liver and circulate in plasma at low concentrations. Most are known by capital Roman numerals, although the first four are usually referred to by name. The active forms of the zymogens are serine proteases, which activate the next factor in the sequence by splitting a limited number of specific peptide bonds to uncover the active enzymic site containing serine. The chain of proteolytic reactions in the coagulation system produces a *cascade* effect with amplification and acceleration at each step, so that even though the initiating stimulus may have been trivial, there is eventually an explosive production of large amounts of fibrin at the site of injury. The sequences of amino acids around the active sites and the mode of action of the coagulation enzymes are very similar to those of the pancreatic serine proteases—trypsin, chymotrypsin and elastase (p. 590). These enzymes probably all have a common evolutionary origin.

The end-stage of blood coagulation (Fig. 10.16) is conversion of the soluble plasma protein, *fibrinogen* (factor I) into insoluble *fibrin* by the protease, thrombin (IIa). Fibrinogen consists of three pairs of polypeptide chains, α, β and γ. Thrombin splits arginyl–glycine bonds near the N-terminus of each α and β chain to form *fibrin monomer* and two pairs of small peptides, fibrinopeptides A and B. Fibrin monomer molecules spontaneously polymerize to form a weak gel held together by electrostatic bonds. Thrombin activates factor XIII and, by transamidation in the presence of Ca^{2+}, this enzyme (XIIIa) causes strong peptide bonds to form between glutamine and lysine

Table 10.11. Coagulation factors.

Factor	Name	Approx. plasma concentration ($mg\,l^{-1}$)	Function
I	Fibrinogen	3000	Fibrin polymer unit
II*	Prothrombin	100	Protease
III	Tissue factor	—	Cofactor
IV	Calcium	100 (2·5 mM)	Cofactor
V	Proaccelerin	10	Cofactor
VII*	Proconvertin	0·5	Protease
VIII	Antihaemophilic factor, VIII:C	0·2	Cofactor
IX*	Christmas factor	5	Protease
X*	Stuart–Power factor	10	Protease
XI	Plasma thromboplastin antecedent	5	Protease
XII	Hageman factor	40	Protease
XIII	Fibrin stabilizing factor	10	Transamidase
—	High MW kininogen, Fitzgerald factor	80	Cofactor
—	Prekallikrein, Fletcher factor	35	Protease

*Hepatic synthesis requires vitamin K.

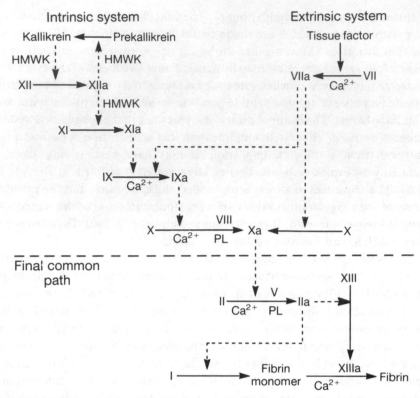

Fig. 10.16. The intrinsic, extrinsic and final common pathways of blood coagulation. HMWK, high molecular weight kininogen; PL, platelet phospholipid.

residues of the γ chains in the fibrin polymer, converting it to the insoluble, stable, fibrin clot.

Prothrombin is converted to thrombin by sequential cleavage of two peptide bonds in the prothrombin molecule by activated factor X, in the presence of Ca^{2+} and platelet phospholipid, and with factor V as cofactor (Fig. 10.16). By itself, factor Xa generates thrombin slowly but the reaction is about 100 000 times faster when factor Xa is bound by Ca^{2+} to the phospholipid surfaces of platelets, together with factor V and prothrombin.

The cofactor activity of factor V is initiated by splitting of peptide bonds in the molecule by thrombin, which thus has an autocatalytic action on its own production. This effect is limited, however, because thrombin also triggers inactivation of factor V by protein C (see later).

Intrinsic and extrinsic coagulation systems

Activation of factor X is the culmination of preceding enzyme reactions in two pathways, the *intrinsic* and *extrinsic* systems. The components of the intrinsic pathway are already present in circulating blood; the extrinsic pathway requires, in addition, *tissue factor* (thromboplastin), which is released by damaged cells.

The intrinsic system is activated when plasma comes into contact with constituents of subendothelial tissues, e.g. collagen fibrils or negatively charged, 'foreign' surfaces such as glass or particles of kaolin. Three zymogens, factor XII, factor XI and prekallikrein, and one cofactor, high molecular weight kininogen (HMWK), are involved in this phase of *contact activation*. The first stage (Fig. 10.16) is reciprocal proteolytic activation of factor XII and prekallikrein, in which factor XIIa, in the presence of HMWK, converts prekallikrein to kallikrein. Kallikrein, in the presence of HMWK, then activates factor XII. With HMWK as cofactor, factor XIIa then converts factor XI to its active form. It is uncertain what triggers this cyclic activation of factor XII and prekallikrein. Binding of factor XII to foreign surfaces may cause a conformational change in the molecule which could either expose the proteolytic site able to activate prekallikrein or make factor XII susceptible to a low level of protease activity in prekallikrein.

The next reaction in the intrinsic pathway is activation of factor IX in two proteolytic steps by factor XIa in the presence of Ca^{2+}. Factor IXa then activates factor X in a complex formed with factor VIII as cofactor, and with Ca^{2+} binding the reactants together on the phospholipid matrix of the platelet membrane. Thrombin acts on factor VIII, as it does on factor V, to increase the reactivity of the cofactor and eventually to cause its inactivation through the protein C system and by direct proteolysis.

Factor X may also be activated by the extrinsic pathway. Here, a complex formed of factor VII, tissue factor (III) and Ca^{2+} directly activates factor X (Fig. 10.16). Tissue factor is a lipoprotein found in microsomal preparations of various tissues and on the membrane of injured endothelial cells. The extrinsic pathway shows positive feedback with a reciprocal activating cycle between factor Xa and the activated form of factor VII.

The idea of separate intrinsic and extrinsic pathways leading to activation of factor X is useful as a basis for laboratory tests but from a functional point of view the distinction is probably artificial since there are cross-links between the pathways. Thus the factor VII–tissue factor complex directly activates factor IX (Fig. 10.16) and a product of the activation of factor XII can generate VIIa. Although the physiological importance of these reactions is uncertain, they may explain why both intrinsic and extrinsic systems are necessary for normal haemostasis. Moreover, it is also known that platelets can directly activate factor XI without the intervention of factor XII, HMWK or prekallikrein. This may explain why people with deficiency of these three coagulation factors do not have a bleeding tendency, whereas patients deficient in factor XI do.

The coagulation factors

The plasma protein coagulation factors fall into three main groups. The *contact factors* (XII, XI, prekallikrein and HMWK) are stable in blood or plasma stored at 4°C, do not need Ca^{2+} for activation to serine proteases and are also concerned in pathways other than coagulation. Thus factor XIIa, and kallikrein with HMWK, can activate the fibrinolytic system and HMWK is both a cofactor for the activation of prekallikrein and a substrate for kallikrein. Kallikrein splits off from HMWK the nonapeptide, *bradykinin*, one of a group of *kinins*, which increase vascular permeability, cause smooth muscle to contract and generate chemotactic activity in leukocytes.

The *vitamin K-dependent factors* (II, VII, IX, X) require Ca^{2+} for their activation, are stable in plasma kept at 4°C and, except for prothrombin, are not consumed in coagulation so that they are present in serum. Vitamin K is needed for a final post-ribosomal step in their synthesis in which an extra carboxyl group is added to glutamate residues to form γ-carboxyglutamic acid side-chains. These carboxyl groups allow the coagulation factors to bind to platelet phospholipid by calcium bridges.

The remaining protein coagulation factors, the *fibrinogen group*, are those which react with thrombin (fibrinogen, factors V, VIII and XIII) and they are all consumed or inactivated during coagulation. Factors V and VIII rapidly lose their activity in blood or plasma stored at 4°C and special blood products must be used for transfusing these factors. The concentrations of this group of plasma proteins increase in pregnancy, in women taking the contraceptive pill and in inflammatory states.

Factor VIII is a complex of two proteins with different actions. Factor VIII:C is the coagulation cofactor and its coagulant activity is deficient in patients with *haemophilia A*. It is probably a glycoprotein (M_r 300 000) made in the liver and its synthesis is controlled by a gene on the X chromosome. The other part of the factor VIII complex (VIII:WF) is von Willebrand factor which is required for the adhesion of platelets to collagen and also acts as a 'carrier' for VIII:C. It circulates as a population of multimers with M_r from 850 000 to more than 12×10^6. This protein is synthesized by endothelial cells under the control of a gene on an autosomal chromosome (number 12). This factor is deficient or defective in patients with von Willbrand's disease. It is not known how or where the two components of factor VIII become associated.

Inhibitors of coagulation

Thrombin is a very potent enzyme: it has been shown that there is potentially enough thrombin in 10 ml of blood to coagulate all the circulating plasma in less than 20 s at 37°C. Various inhibiting agents limit the action of thrombin. Phagocytic cells in the liver, and elsewhere, remove particulate thromboplastins from blood and hepatocytes can degrade activated coagulation factors (e.g. IXa, Xa, XIa). Flow of blood past an area of injury dilutes activated intermediates of coagulation and disperses loose aggregates of platelets. These actions limit the generation of thrombin. Thrombin is also removed from the blood during coagulation by adsorption to fibrin. This adsorption, which is called antithrombin I, inactivates only small amounts of thrombin. More important inhibitors are certain plasma proteins (Fig. 10.17). Antithrombin II was a term used for a plasma protein cofactor of the anticoagulant heparin. This is now known to be antithrombin III.

Antithrombin III is an α_2-globulin which reacts on a mole-to-mole basis with thrombin to form an irreversible complex in which both molecules are inactivated. Antithrombin III also inactivates factor Xa and the other serine proteases in the intrinsic pathway (IXa, XIa and XIIa). Two other proteins, *α_2-macroglobulin* and *α_1-antitrypsin*, also contribute to the antithrombin effect of plasma.

Thrombin is also bound by a specific receptor on endothelial cells, *thrombomodulin*. The result of this interaction is conversion of circulating *protein C* to its active form, Ca. Protein Ca is a serine protease which, in the presence of phospholipid, calcium ions and a cofactor, *protein S*, inactivates factors V and VIII and thus limits the generation of thrombin. Proteins C and S require vitamin K for their synthesis and protein Ca also enhances fibrinolysis. Protein Ca is itself inactivated by a specific inhibitor in plasma.

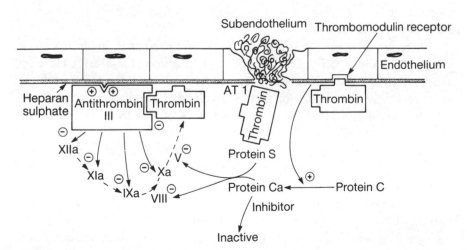

Fig. 10.17. Inactivation of coagulation by antithrombins and the protein C system. AT1, antithrombin I (the process of adsorption of thrombin by fibrin).

Anticoagulants

Ca^{2+} is needed at a number of steps in the coagulation pathways (Fig. 10.16) and agents which remove Ca^{2+} from solution keep blood fluid *in vitro*. Ethylenediamine tetra-acetic acid (EDTA) as the sodium or potassium salt and sodium oxalate are used for this purpose while trisodium citrate is the anticoagulant constituent of various solutions used for preserving blood for transfusion.

Heparin is a naturally occurring glycosaminoglycan which is used as an anticoagulant *in vivo* when rapid onset of action is desired. It acts by binding to antithrombin III, thereby inducing a conformational change in the molecule which greatly accelerates inactivation particularly of Xa but also of thrombin. Anticoagulant drugs of the coumarin-indanedione type, e.g. *warfarin*, produce an anticoagulant effect, which takes time to develop, by acting as inhibitors of vitamin K. They thus prevent the formation of the γ-carboxyglutamate residues in factors II, VII, IX and X. This effectively renders these proteins functionally inactive.

Fibrinolysis

During the repair of blood vessels and healing of wounds, fibrin deposited in haemostatic plugs and in extravascular sites is removed by the *fibrinolytic system*. Fibrin is broken down to soluble fragments (fibrin degradation products) by a serine protease, *plasmin*. Plasmin, a two-chain polypeptide, is derived by cleavage of a single peptide bond in *plasminogen*, a single-chain β-globulin in the plasma. This proteolytic reaction is brought about by *plasminogen activators*.

There are 'extrinsic' plasminogen activators present in many tissues, in endothelium and in various body fluids, including urine (*urokinase*), tears and saliva. 'Intrinsic' plasminogen activator activity arises in the blood itself and is generated by the coagulation factors involved in contact activation (factor XII, kallikrein and HMWK). The importance of this form of activator is probably slight since physiological fibrinolysis seems to be due largely to release of tissue plasminogen activator from endothelial cells. This activator has now been synthesized by recombinant DNA techniques and is being used to treat coronary thrombosis.

While fibrin is the physiological substrate for plasmin, this protease will also attack other proteins, especially fibrinogen, factor V and factor VIII. This non-specific proteolysis is normally prevented, since any free plasmin in the blood is rapidly inactivated by irreversible binding to α_2-*antiplasmin*, a circulating glycoprotein. The antithrombin, α_2-macroglobulin, neutralizes plasmin present in blood in excess of the inhibitory capacity of α_2-antiplasmin. Poorly defined *anti-activators* have also been reported.

Plasmin seems able to lyse fibrin selectively because plasminogen binds avidly to fibrin as the polymer forms. Tissue plasminogen activator, also incorporated into the fibrin as it forms, or diffusing in later from adjacent damaged tissue, lets plasmin be generated in close association with its natural substrate, fibrin. The plasmin bound to fibrin cannot be inactivated by antiplasmins in the blood but any plasmin leaking into the general circulation will be inactivated immediately. As the fibrin in a haemostatic plug is laid down, it is thus provided with an enzyme system for its subsequent dissolution.

Small amounts of plasminogen activator can be detected in venous blood and this seems to originate from endothelium of capillaries and venules in the microcirculation. Flow of blood through these vessels may at times be sluggish, and slowly flowing blood coagulates easily. A continuous slow release of plasminogen activator and the subsequent formation of plasmin may be important for maintaining the fluidity of blood in small vessels. Increase in the level of circulating plasminogen activator occurs in exercise, acute stress and in response to adrenaline. Fibrinolysis may also be important for maintaining the patency of other tubular systems in the body, e.g. the urinary tract.

Summary of haemostasis

For purposes of description, it is necessary to deal separately with each contributor to haemostasis—blood vessels, platelets and coagulation. In fact, all three work together in an interlocking manner (Fig. 10.18). Vascular injury promotes vasoconstriction which is reinforced by substances released by platelets. Platelets adhere to exposed subendothelial elements, then aggregate to form a platelet plug. Endothelial damage activates the intrinsic pathway of coagulation, which is catalysed by phospholipid in platelet membranes, while entry of tissue factor to the blood triggers the extrinsic pathway. Thrombin is generated causing formation of fibrin and the production of the haemostatic plug. Thrombin also reinforces aggregation of platelets, with free thrombin being rapidly inactivated by various antithrombins. Eventually, during repair, plasmin dissolves fibrin and antiplasmins inactivate any plasmin reaching the blood.

Bleeding disorders

People with haemostatic defects typically suffer from two kinds of bleeding: they bleed *for an abnormally long time* after injury and they bleed *spontaneously* without preceding trauma. In defects of platelets and small blood vessels,

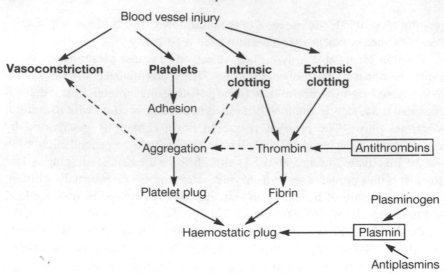

Fig. 10.18. Summary of normal haemostasis.

certain features of this bleeding tend to differ from those found in disorders of coagulaton (Table 10.12). Thus *purpura*, i.e. multiple small bruises and haemorrhagic spots (petechiae) in the skin, and bleeding from mucosal surfaces, are characteristic of the spontaneous bleeding seen with platelet and vascular defects. Severe coagulation defects are typically associated with single, spreading bleeds in deep tissues and joints. The responses to mild and moderate trauma also tend to differ in these two groups of haemostatic disorders (Table 10.12).

Purpuras due to acquired vascular defects may be benign local phenomena, e.g. simple easy bruising and senile purpura, or may be more generalized haemorrhagic lesions associated with, for example, severe infections, hypersensitivity reactions (e.g. Henoch–Schönlein syndrome) and drug allergies.

Table 10.12. Characteristics of bleeding in haemostatic defects. (Adapted from World Health Organisation (1972). *Inherited blood clotting disorders,* p. 17. World Health Organisation Technical Report Series No. 504).

Type of bleeding	Platelet/vascular defect	Coagulation defect
Spontaneous	Purpura Mucosal bleeding	Single deep bruises/bleeds Bleeds in joints
From superficial cuts	Profuse/prolonged	Not excessive
From deep injuries	Immediate onset	Delayed onset then prolonged/episodic
	Often responds to local treatment	Poor response to local treatment

Inherited malformations of small vessels (hereditary haemorrhagic telangiec-tasia) and defective perivascular connective tissue (e.g. scurvy) can also cause troublesome bleeding.

A low platelet count and *thrombocytopenic purpura* can arise from *decreased production* of platelets by the bone marrow, because of drugs, ionizing radiations, viral infections, failure of the marrow or infiltration of it by abnormal cells, e.g. in leukaemia. Thrombocytopenia may also be due to *increased destruction* of platelets by immune processes, as in immune thrombocytopenic purpura, after viral infections (e.g. measles) and as an allergic response to drugs. Purpura can also occur in people with normal numbers of circulating platelets but with functional defects, e.g. failure of aggregation in thrombasthenia.

Inherited deficiencies of coagulation factors are rare but spectacular disorders, when severe, and usually involve only one factor (e.g. deficiency of factor VIII in haemophilia A). In von Willebrand's disease, in addition to a deficiency of factor VIII, there is also an inherited lack of von Willebrand factor (VIII:WF), leading to defective adhesion of platelets and bleeding typical of that seen with platelet defects. Acquired disorders of coagulation factors are usually due to multiple deficiencies, commonly arising from deficient coagulant activity of the vitamin K-dependent factors because of lack of vitamin K. Since this vitamin is fat-soluble, deficiency, and defective vitamin K-dependent coagulation factors, can arise in disorders impairing absorption of fats. These deficiencies and bleeding also occur in liver disease, in premature babies and when induced intentionally by treatment with anticoagulants like warfarin (an overdose of which often causes bleeding). Disease of the liver can also be associated with other haemostatic abnormalities including thrombocytopenia, reduced synthesis of fibrinogen and increased fibrinolytic activity.

Diffuse intravascular thrombosis (disseminated intravascular coagulation) can arise because of release of procoagulant material (e.g. amniotic fluid) into the circulation or because of widespread endothelial damage (e.g. severe bacterial infections). There may be such gross depletion of coagulation factors and platelets that generalized bleeding occurs.

Tests of haemostatic function

Certain simple 'screening' tests are useful for assessing haemostatic function quickly. A *platelet count* will demonstrate a thrombocytopenia and examination of a *blood film* will often confirm a low platelet count and may reveal its cause, e.g. leukaemia. The *bleeding time*, the time taken for cessation of bleeding from small punctures in the skin (made by a standard technique), is an index of the integrity of platelets. In people with a normal platelet count, a prolonged bleeding time would suggest defective platelet function or von Willebrand's disease.

Two simple coagulation tests are used to monitor the extrinsic and intrinsic pathways (Fig. 10.19). The *prothrombin time* (PT) tests the extrinsic system and final common path by measuring the time taken for a sample of citrated plasma to coagulate when tissue factor and calcium ions are added. If a test plasma takes longer to coagulate than a known normal plasma, this suggests deficiency (or inhibition) of one or more of the factors VII, X, V, II and I.

In the *partial thromboplastin time with kaolin* (PTTK), the intrinsic system of a sample of citrated plasma is activated by incubating it with kaolin for several minutes, then a substitute for platelet phospholipid and calcium ions are added. If the test sample takes longer to coagulate than a control sample, this indicates deficiency (or inhibition) of one or more of the factors XII, XI, IX, VIII, X, V, II and I. The combined results of these two tests, PT and PTTK, will usually indicate if a coagulation defect is present and will suggest the likely possible defect(s), depending on whether either or neither test, or both, give abnormal results.

If thrombin and calcium ions are added to a sample of citrated plasma, the time taken for the plasma to coagulate, the *thrombin time*, is a useful test for rapidly assessing the concentration and reactivity of fibrinogen.

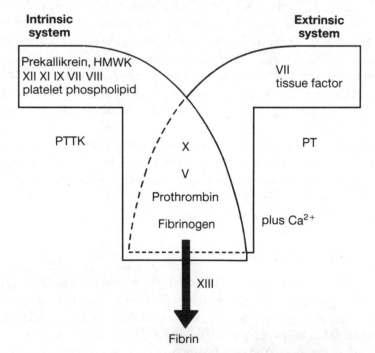

Fig. 10.19. Coagulation factors tested by the prothrombin time (PT) and by the partial thromboplastin time with kaolin (PTTK). HMWK, high molecular weight kininogen.

11 · Cardiovascular System

11.1 Introduction
Basic design and functions of the
    components of the cardiovascular
    system

11.2 The heart
Functional anatomy
Origin and conduction of electrical
    activity
Characteristics of cardiac potentials
Cardiac refractory period and the
    temporal relationship between
    electrical and mechanical events
Regulation of heart rate
Cardiac arrhythmias
Electrocardiogram (ECG)
Cardiac cycle and heart sounds
Abnormal heart sounds
Jugular venous pulse
Mechanical characteristics of cardiac
    muscle
Control of stroke volume
    Intrinsic control of stroke volume
    Factors influencing end-diastolic
        volume
    Extrinsic control of stroke volume
Law of Laplace and the heart
Cardiac output and its measurement

11.3 The vascular system
Physics of blood flow (pressure,
    resistance and velocity)
Deviations of blood flow from that
    predicted
Other properties of walls of blood
    vessels
Relationships between flow, pressure,
    resistance, cross-sectional area and
    volume throughout the vascular
    system
Characteristics of the systemic arterial
    circulation
    Measurement of arterial blood
        pressure
    Determinants of arterial blood
        pressure

Circulation through the arterioles and
    the regulation of blood flow in the
    systemic circuit
    Intrinsic regulation of blood flow
    Extrinsic regulation of blood flow
Circulation through the capillaries
    Ultrafiltration and reabsorption of
        fluid across the capillary wall
Characteristics of the systemic venous
    circulation
    Alterations in venous capacity
    Determinants of venous return
    Relationships between cardiac
        output, venous return and atrial
        pressure
Characteristics of the lymphatic system

**11.4 Integrated regulation of the
cardiovascular system**
Arterial baroreceptor reflexes
Reflexes initiated by cardiac stretch
    receptors
Reflexes initiated by other receptors
Cardiovascular centres

**11.5 Characteristics of the circulation through
different organs**
Pulmonary circulation
Coronary circulation
Cerebral circulation
Cutaneous circulation
Skeletal muscle circulation
Splanchnic circulation
Renal circulation

**11.6 Cardiovascular adjustments under
physiological and abnormal conditions**
Effects of posture
    Hydrostatic pressure
    Transmural pressure and venous
        capacity
    Reflex compensations
Hypertension
Hypotension, fainting and shock
Haemorrhage
Cardiac failure

11.1 Introduction

The cardiovascular system provides the mechanism by which the *blood* contained within it is *circulated* through the tissues of the body. It is composed of a *cardiac pump* and a series of distributing and collecting *vascular tubes* linked by very thin *capillaries*. The capillaries permit rapid *exchange* of substances over the short distance between the blood and the interstitial fluid bathing the cells of the tissues (Section 1.5). Thus the cardiovascular system is a *transport system* and, in the last analysis, it links the external environment to the tissues and distributes substances essential for metabolism. These are O_2 from the lungs and nutrients from the gastrointestinal tract. At the same time, the cardiovascular system removes from the tissues CO_2 and other byproducts of metabolism, carrying them to the lung, kidney and liver. Such actions of the cardiovascular system are essential for *homeostasis* of the plasma component of blood and the interstitial fluid, which both comprise the ECF or internal environment (Section 1.2), and hence ensure the even distribution of available water and electrolytes to all parts of the body.

There are additional homeostatic functions of the cardiovascular system. Not only does it circulate hormones but the heart itself produces a hormone called *atrial natriuretic factor* or *peptide* (p. 542). The cardiovascular system is also concerned with the distribution of heat and maintenance of body temperature which requires the control of blood flow to heat-exchanging areas. It also transports the agents involved in haemostasis and the cells and antibodies concerned with defence.

Basic design and functions of the components of the cardiovascular system

The heart consists of two pumps lying side by side but functioning in series: the right-hand one receives blood from the body and then propels it at low pressure through the lungs (the *pulmonary circulation*); the left-hand one receives blood from the lungs and then propels it at high pressure to all other tissues of the body (the *systemic circulation*) (Fig. 11.1). There is in adults normally no direct communication of blood between the two pumps. Thus, in one cycle, all the blood has to circulate through the lungs but only a small portion circulates through any one systemic tissue since these are arranged in parallel.

Approximate blood flows, at rest, to each of the systemic tissues are given in Table 11.1. This distribution of blood flow is related to the weight of the tissue, to its level of activity and, in the case of the kidney and skin, to the extra blood flow required for filtering excretory products and for temperature regulation, respectively. During moderate exercise, for example, the total blood flow increases and its distribution changes (Table 11.1). Heart muscle and skeletal muscle require increased blood flow to satisfy their increased metabolism and the skin has a higher blood flow to dissipate the extra heat production. Some of this blood is shunted away from the gut and kidney but it is imperative that a constant blood flow to the brain be maintained. These

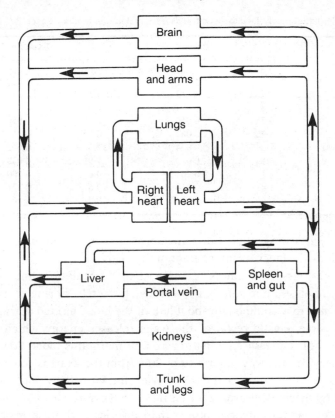

Fig. 11.1. Schematic representation of the cardiovascular system. For simplicity, bronchial vessels to the airways of the lung and coronary vessels to heart muscle have been omitted.

flow changes are controlled by local mechanisms acting on particular blood vessels (arterioles) within the tissue and by some hormones and the autonomic nervous system acting on both the heart and these blood vessels.

The distributing vessel leaving the right-hand side of the heart is the pulmonary artery and that leaving the left is the aorta. To supply a tissue or organ, these branch into smaller arteries and finally arterioles before entering the vast capillary network of fine tubes (Fig. 11.2). The capillary network is drained by venules which collect into veins and finally into the pulmonary veins or into the large superior and inferior venae cavae. The heart and each of these vessel types have special functions.

The *heart* fills with blood during its relaxation phase (*diastole*) and then generates pressure through muscular contraction to expel some of that blood (*systole*). Thus the pumps generate pulsatile pressure, 0 to ∼25 mmHg on the right and 0 to ∼120 mmHg on the left. The frequency of pumping (*heart rate*), the volume ejected at each stroke (*stroke volume*) and the product of these

Table 11.1. Distribution of blood to systemic tissues at rest and in moderate exercise.

	Rest		Exercise	
	$(l\,min^{-1})$	$(l\,min^{-1}\,100\,g^{-1}$ tissue$)$	$(l\,min^{-1})$	$(l\,min^{-1}\,100\,g^{-1}$ tissue$)$
Total flow	5·8	—	17·5	—
Brain	0·75	0·050	0·75	0·050
Cardiac muscle	0·25	0·083	0·75	0·250
Skeletal muscle	1·2	0·004	12·5	0·042
Skin	0·5	0·010	1·9	0·038
Kidney	1·1	0·367	0·6	0·200
Gut	1·4	0·050	0·6	0·021
Other	0·6	0·002	0·4	0·001

(cardiac output) can be altered by some hormones, by the autonomic nervous system and by the mechanism inherent in the heart.

The high-pressure conduits of the systemic circulation, the *aorta* and *arteries*, have elastic fibres in their walls which stretch, storing energy as the vessel distends to accommodate about half of the blood ejected during systole. During diastole, elastic recoil of the walls releases energy which helps to prevent the aortic pressure falling to zero and maintains blood flow towards the periphery. In this way, intermittent flow from the heart is converted into continuous pulsatile flow through the arteries. Backward flow into the heart is prevented by the valve guarding the entrance to the aorta.

Arterioles have a narrower lumen than arteries and are a major site of resistance to blood flow. The amount of blood flow is inversely proportional to the resistance which is determined mainly by the radius of the tube (Section 11.3). The high resistance of the arterioles is reflected in the considerable fall in blood pressure as blood flows through them. This is accompanied by a large

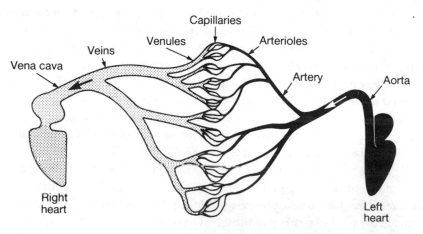

Fig. 11.2. Blood vessels of the systemic circulation.

damping of the pulsatile flow and its conversion to a steady continuous flow. Arteriolar resistance can be increased, and hence blood flow through to the capillary network decreased, by contraction of circular smooth muscle in the wall of the arterioles and vice versa. Contraction of this smooth muscle is controlled by local mechanisms, by some hormones and by the autonomic nervous system. Such arteriolar constriction will inevitably elevate the pressure in the arteries and decrease the pressure in the capillaries. The adjustment of arteriolar calibre regulates tissue blood flow, aids in the control of arterial blood pressure and alters capillary pressure and hence the net flow of water across the capillary wall.

Capillary networks provide a very large cross-sectional area through which blood flows slowly, giving ideal conditions for diffusional exchange between blood and interstitial fluid. Across the capillary wall, lymph plasma is formed from the imbalance between osmosis and ultrafiltration of water (Section 1.8). The lymph is returned slowly to the CVS by a secondary set of collecting tubes called the *lymphatic system*.

The primary set of collecting tubules, the *venules* and *veins*, are low pressure conduits returning blood from the capillaries to the heart. Systemic veins have a relatively large capacity. Indeed, in the resting supine position, veins hold four times as much blood as do the arteries. They are also very distensible and accommodate about 90% of any blood transfusion. Contraction of smooth muscle in the walls of veins, which is controlled by the autonomic nervous system and some hormones, causes a reduction in venous compliance (distensibility) and hence in venous volume (capacity). This shifts proportionately more blood into skeletal muscle capillaries during exercise or into the arteries during the reflex maintenance of arterial blood pressure after a haemorrhage.

At certain sites in the wall of the aorta and carotid arteries, there are nerve endings known as *baroreceptors* which respond to stretch and hence monitor the arterial blood pressure. This information, together with other sensory inputs, is relayed to the *cardiovascular centres* (the coordinating centres in the brain). The resulting motor output alters the *autonomic nervous activity* to the heart, arterioles and veins in order to maintain arterial blood pressure at a certain level whilst altering total blood flow and its distribution.

11.2 The heart

Functional anatomy

The heart consists of four chambers: the *right atrium* leading into the *right ventricle* and the *left atrium* leading into the *left ventricle* (Fig. 11.3). The aperture between each atrium and its respective ventricle is guarded by an

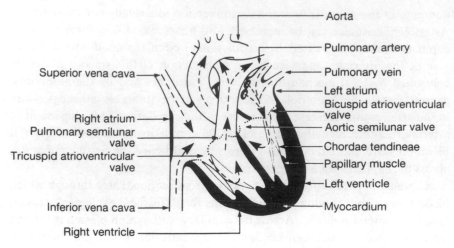

Fig. 11.3. Structure of the chambers and valves of the heart. The atrioventricular rings of connective tissue lie in the plane of the valves. Arrows indicate direction of blood flow.

atrioventricular (AV) *valve*. The right valve has three cusps (the tricuspid valve) and the left valve has two cusps (the bicuspid or mitral valve). Attached to the inner margins of these valves are tendinous cords—*the chordae tendineae*—which are attached to projections of ventricular muscle known as *papillary muscles*. When the pressure in the atrium is greater than in the ventricle, the AV valve opens and blood flows from atrium to ventricle. When ventricular pressure is greater than atrial pressure, the valve shuts while the chordae tendineae and papillary muscles prevent the valve from everting into the atrium. Blood therefore cannot flow from ventricle to atrium. The exits from the right ventricle into the pulmonary artery and from the left ventricle into the aorta are guarded by a pulmonary and aortic *semilunar valve*, respectively. The valve opens when the contraction of the ventricles generates a pressure higher than that in the pulmonary artery or aorta. The semilunar valves close when ventricular pressure falls during ventricular relaxation. This prevents reflux of blood back into the ventricles.

Venous blood continuously flows from the systemic circulation via the superior and inferior vena cava into the right atrium and from the pulmonary circulation via four pulmonary veins into the left atrium. Providing the AV valves are open, blood will flow onwards into the respective ventricles. When the ventricles are about 80% full of blood, the atria contract almost simultaneously to propel their enclosed blood into the ventricles to complete ventricular filling. After a very short pause, both ventricles contract almost simultaneously (systole) and the resulting increase in pressure immediately closes the AV valves. Continued contraction of the ventricles causes the ventricular pressure to become greater than the pressure in the pulmonary artery or aorta and thus the pulmonary and aortic valves open. Blood is then ejected from the right ventricle at low pressure into the pulmonary circuit and from the left ventricle at high pressure into the systemic circuit. When the ventricles stop contracting, the ventricular pressures drop below the pulmonary arterial and aortic pressures, causing the pulmonary and aortic valves to close, terminating blood ejection. As the ventricular pressures continue to fall and become less than atrial pressures, the AV valves open and the ventricles refill with blood. At rest, the period of ventricular filling (diastole) is about two-thirds of the total cycle.

The magnitude of pressure generated by each heart chamber during contraction is reflected in the thickness of its muscle wall—known as the *myocardium* (Fig. 11.3). The atrial myocardium is thin whilst ventricular myocardium is thick, especially that of the left ventricle which generates the highest pressure. The inner surface of the myocardium is lined with *endocardium* which is a continuation of the endothelium of blood vessels. The outer surface of the myocardium is covered with mesothelial tissue called *epicardium*. Enclosing the heart is a thin fibrous sac called the *pericardium*. This is relatively inelastic and limits excessive enlargement of the heart. The *pericardial space* between the epicardium and pericardium contains just enough interstitial fluid to act as a lubricant.

Right and left *coronary arteries* originating from the systemic aorta supply the arterioles and capillaries of the myocardium. Venous drainage is via the *coronary sinus* (75%) and *anterior cardiac vein* (20%) which both empty into the right atrium. Some venous blood drains directly via *Thebesian veins* and small venules into all heart chambers. Venous blood entering the left side of the heart in this manner will cause a small reduction in the oxygen concentration of systemic blood. Blood flow through ventricular myocardium is reduced (right ventricle), abolished or even reversed (left ventricle) during the contraction phase of the ventricles because the high pressures generated by the ventricles compress the coronary blood vessels (Section 11.5).

Origin and conduction of electrical activity

Although the heart is completely divided internally into right and left chambers, the cardiac muscle fibres of the myocardium of the two atria are continuous as are those of the two ventricles. However, atrial and ventricular muscle fibres are separated by rings of fibrous tissue called the *atrioventricular ring*. This acts not only as a fibrous skeleton for the origin and insertion of atrial and ventricular muscle and for the attachment of heart valves but also prevents electrical coupling of atrial and ventricular cells.

Cardiac muscle cells are shorter than skeletal muscle cells, are branched and abut end-to-end to form a network or syncytium and are rich in mitochondria; they have one nucleus situated in the centre of the cell (Fig. 11.4, cf. Fig. 3.2). Just as in skeletal muscle, actin and myosin filaments are found in each cardiac myofibril arranged into sarcomeres with their characteristic A, I and H bands and M and Z lines. The transverse tubular system is located, however, at the Z line in cardiac muscles not at the A–I junction. The apposition of one cardiac cell with another coincides with one of these Z lines and becomes specialized into a dense *intercalated disc*. Within this disc, adjacent areas of membrane are bridged by low resistance pathways called *gap junctions* which provide for the rapid conduction of electrical current between the two cells. This ensures that, when an action potential is generated in any part of the cardiac network, it is propagated rapidly so that the atria or ventricles contract as one. The discs also provide a site of adhesion between one cardiac cell and another and ensure that the tension developed by one cell is transmitted through to the next.

A denervated heart continues to beat in an orderly sequence of atrial then ventricular contraction followed by a passive filling phase. This ability to depolarize and contract rhythmically without innervation is called *myogenic rhythmicity*. All parts of the heart have this ability and the inherent rate in the atria and ventricles is about 70 and 40 beats per min, respectively. However, the initiation of each beat and its sequential coordination is brought about normally by a *specialized conduction system* of cardiac cells that have few myofibrils and can contract only weakly. Action potentials originate at a rate of about 100 per min in a small area called the *sinoatrial* (SA) *node* (the pacemaker) located in the right atrium near the entrance of the superior vena cava (Fig. 11.5). From there they are conducted along the plasma membrane

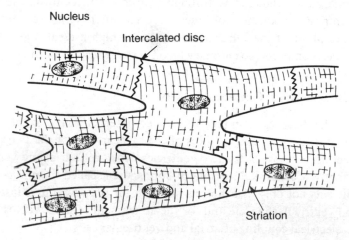

Fig. 11.4. Network arrangement of cardiac muscle cells.

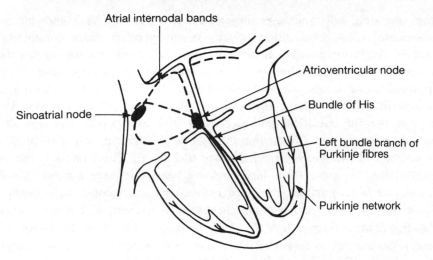

Atrial internodal bands

Atrioventricular node

Sinoatrial node

Bundle of His

Left bundle branch of
Purkinje fibres

Purkinje network

Fig. 11.5. The sinoatrial and atrioventricular nodes and the specialized conducting pathways of
the heart.

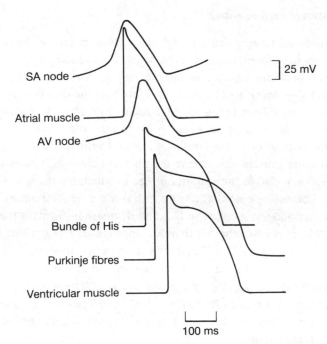

SA node

25 mV

Atrial muscle

AV node

Bundle of His

Purkinje fibres

Ventricular muscle

100 ms

Fig. 11.6. Shape, duration and sequence of cardiac action potentials. Note also the delays caused
by the anatomical sequence of depolarization and by the relative conduction velocities down the
conducting system. (After Hoffman, B.F. & Cranefield, P.F. (1960) *Electrophysiology of the Heart*.
McGraw Hill, New York.)

from one atrial cell to another through the low electrical resistance of the intercalated discs. Conduction velocity is enhanced by three conducting pathways (anterior, middle and posterior *internodal bands*) causing both atria to contract almost simultaneously. The internodal bands merge as they approach the next node, the *atrioventricular* (AV) *node*, which is located in the atrioventricular fibrous ring on the right side of the atrial septum. The AV node is the only electrical pathway from atria to ventricles through the insulating fibrous ring. Conduction through the AV node is slow (0.05 m s^{-1}) compared with atrial or ventricular muscle (0.5 m s^{-1}). This effectively delays transmission for about 0.1 s during resting heart rates (see Fig. 11.6) and ensures atrial contraction is finished before ventricular contraction begins. From the AV node, action potentials travel at a speed of 1 m s^{-1} in the Purkinje fibres of the *bundle of His* along the right and left branches down the ventricular septum to enter the *Purkinje network* which ramifies throughout the ventricles. The conduction velocity through the Purkinje network is very fast (5 m s^{-1}). Because of the extensive branching of the Purkinje network in the ventricles, excitation reaches all parts of both ventricles rapidly causing both to contract almost simultaneously.

Characteristics of cardiac potentials

The resting membrane potential and the action potential have different characteristics in different regions of the heart. Three patterns are found—one for Purkinje fibres and ventricular muscle, one for atrial muscle and one for the SA and AV nodes (Figs 11.6 & 11.7). Note that the duration of the action potential varies in different regions of the heart (Fig. 11.6). It is shortest in the SA and AV nodes and atrial muscle, longest in the Purkinje fibres and intermediate in duration in the bundle of His and ventricular muscle.

In ventricular muscle, the resting membrane potential (phase 4) is stable at about -90 mV due to the stability of the conductance to K$^+$ (G_K) and to Na$^+$ (G_{Na}). The action potential has an initial rapid depolarization (phase 0) resulting from a sudden increase in G_{Na} and decrease in G_K with the *net result* that more Na$^+$ ions enter the cell than K$^+$ ions leave. The peak of the action potential reaches about $+20$ mV and then has a rapid but short decline (phase 1) leading to a prolonged shoulder or plateau (phase 2). Phase 1 is mainly due to a rapid reduction in G_{Na}. Phase 2 is due to a slower reduction in G_{Na} and a delayed increase in conductance to Ca^{2+} (G_{Ca}) causing Ca^{2+} ions to enter the cell. The action potential then repolarizes relatively slowly (phase 3) as G_K, G_{Ca} and G_{Na} return to normal.

These events contrast with those in axons (Fig. 2.9) or skeletal muscle in which G_{Ca} is not involved in the action potential and G_K does not decrease during the initial depolarization. The involvement of G_{Ca} in cardiac muscle is

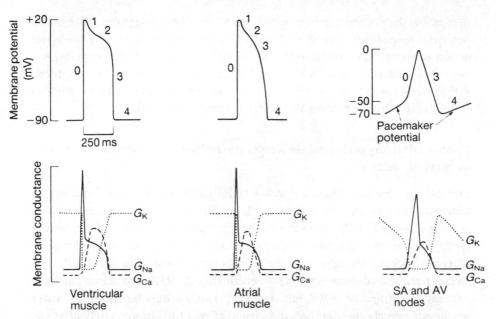

Fig. 11.7. The shapes and phase (0 to 3) of cardiac action potentials in different regions of the heart, the pacemaker potential in the SA and AV nodes and the underlying changes in membrane conductance to K^+, Na^+ and Ca^{2+}. Note membrane conductance is a logarithmic scale.

responsible for the long duration of the cardiac action potential (about 250 ms) compared with only a few milliseconds in skeletal muscles. The Ca^{2+} entering the cardiac muscle cells during each action potential contributes to the activation of the contractile mechanism and is also accumulated by the sarcoplasmic reticulum for release during subsequent contractions.

In atrial muscle, the action potential has a less obvious plateau (phase 2) and therefore a longer repolarization phase (phase 3) than ventricular muscle due to G_{Ca} decreasing more rapidly in these two phases.

In the SA and AV nodes, the cells have a high G_{Na} and hence a lower resting membrane potential than other heart cells. In addition, their membrane potential (phase 4) is unstable and slowly depolarizes from about $-70\,mV$ to about $-50\,mV$ in between action potentials. This slow depolarization is called the *prepotential* or *pacemaker potential*. The instability has been attributed to a slow decline in G_K, but an increase in G_{Ca} also contributes. When G_K is sufficiently low, the threshold potential of $-50\,mV$ is reached and this triggers the action potential. It has a relatively slow depolarization (phase 0) due to a slower increase in G_{Na} than in ventricular muscle. The peak of the action potential is about $0\,mV$, there is no plateau and hence the repolarization (phase 3) is long.

The rate of depolarization of the pacemaker potential in the SA node is faster than in the AV node. Therefore, the SA node triggers its action potential first and is the pacemaker from which each heart beat originates. The action potential propagates from the SA node through the atria, causing the atrial action potential, and arrives at the AV node before the pacemaker in the AV node has reached its threshold. Thus the action potential that occurs in the AV node is caused by the SA node. The action potential then propagates through the ventricle causing the ventricular action potential.

Cardiac refractory period and the temporal relationship between electrical and mechanical events

During the *absolute refractory period* (ARP) (about 200 ms) of the action potential the cardiac cell is inexcitable and during the subsequent *relative refractory period* (RRP) (about 50 ms) there is a gradual recovery of excitability (Fig. 11.8). A second action potential cannot be generated during the ARP whereas a strong stimulus can elicit an action potential in the RRP. The strength required decreases progressively during the RRP. An action potential generated during the RRP has a slower rate of depolarization, a lower amplitude and shorter duration than usual (Fig. 11.8). Immediately after the RRP, the next action potential is still of a shorter duration and results in less Ca^{2+} entry.

When measured from the beginning of the action potential there is a latency of about 10 ms (compared with about 2 ms in skeletal muscle) before the muscle starts to contract (Fig. 11.9). The peak of the developed tension occurs just before the end of the ARP and the muscle is half-way through its relaxation phase by the end of the RRP. The total duration of the mechanical event is about 300 ms. Thus the electrical and mechanical events in cardiac

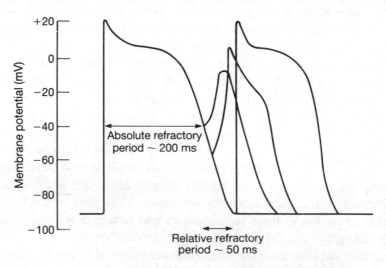

Fig. 11.8. The absolute and relative refractory periods of the ventricular action potential.

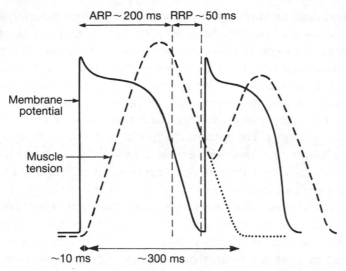

Fig. 11.9. The temporal relationship between the first action potential and the resulting muscle tension and between a second action potential (triggered immediately after the RRP) and its resulting muscle tension. Note summation of muscle contractions is impossible.

muscle overlap considerably in time in contrast with skeletal muscle where the short duration action potential is virtually over before the contraction begins (Fig. 3.7). This overlap means that when a second action potential is triggered at the very end of the ARP or during the RRP, the second contraction is superimposed on the semi-relaxed phase of the first contraction (Fig. 11.9). It is a relatively weak contraction because the shorter duration of the second action potential has resulted in the entry into the cell of less Ca^{2+} than usual. This means that the force developed by the cross-bridges between the actin and myosin filaments will be reduced.

Thus it is impossible to produce the summation and tetanus found in skeletal muscles during high-frequency stimulation (Fig. 3.7). The physiological importance of the prolonged cardiac refractory period is that it protects the ventricles from too rapid a re-excitation which would impair their ability to relax long enough to refill adequately with blood. Furthermore, since the total refractory period is longer than the time taken for conduction through the atria or ventricles, recycling of excitation in the muscular network is not seen in the normal heart.

Regulation of heart rate

The *cardiovascular centres* in the brain alter, via sympathetic and parasympathetic fibres and hormones, the intrinsic discharge rate of the SA node which is about 100 beats per min. The *sympathetic* nerves *increase* and the *parasympathetic* nerves *decrease* the heart rate. At rest, there is activity in both sympathetic and parasympathetic fibres but the latter are more active, reducing the heart rate to about 70 beats per min. Since the parasympathetic nerve fibres project to the heart in the vagus nerve (cranial nerve X), their normal activity is often referred to as *vagal tone*. Intense parasympathetic activity can actually stop the pacemaker of the heart for a brief period. Normally, a decreased heart rate (*bradycardia*), for example during sleep or in athletes at rest, is due to an increase in parasympathetic discharge and a decrease in, or even absence of, sympathetic discharge. The opposite changes in autonomic activity lead to an increase in heart rate (*tachycardia*) and in acute anxiety or severe exercise the heart rate can reach more than 200 beats per min. Alterations in heart rate are referred to as *chronotropy*.

The autonomic nervous system alters heart rate by its actions on the SA node. *Noradrenaline*, the transmitter released by the postganglionic sympathetic fibres, and *adrenaline*, the hormone released from the adrenal medulla in response to increased sympathatic activity, bind to β_1-*adrenoceptors* on plasma membranes of cardiac cells in the SA node. This increases the rate at which G_K declines (and G_{Ca} increases) in between action potentials, thereby increasing the rate of depolarization of the pacemaker potential (Fig. 11.10). The threshold potential is reached earlier and hence the action potential is triggered earlier allowing more action potentials to occur per unit time. *Acetylcholine*, the transmitter released by the parasympathetic fibres, acts on *muscarinic cholinoceptors* in the SA node causing a decrease in the rate at which G_K declines and hence prolongs the pacemaker potential (Fig. 11.10). The cardiac response to parasympathetic stimulation is quicker than it is to sympathetic and so vagal stimulation can alter the heart rate within the time required for a normal heart beat.

Other parts of the heart also receive sympathetic and parasympathetic

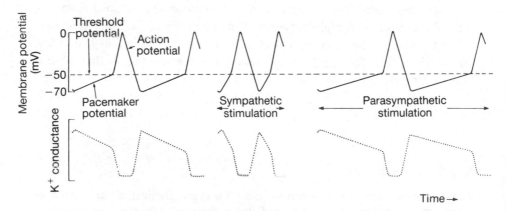

Fig. 11.10. Effects of sympathetic and parasympathetic stimulation on membrane conductance to K^+, and on the rate of depolarization of the pacemaker potential and the frequency of action potentials. For simplicity, the contribution of G_{Ca} to the pacemaker potential has been omitted.

fibres. The delay at the AV node is reduced by the sympathetic and prolonged by the parasympathetic since, via their effects on G_K, they increase and decrease, respectively, the slope of phase 0 of the action potential in the AV node. The duration of the atrial action potential is shortened by the sympathetic and lengthened by the parasympathetic. The Purkinje fibres and ventricular muscle receive mainly sympathetic innervation, the stimulation of which shortens the duration of their action potentials. Such alterations in the action potential duration are unique to cardiac muscle and shortening and lengthening are caused by a more rapid or a slower decrease, respectively, in G_{Ca} in phase 2 of the action potential.

The minimum duration of the cardiac action potential is about 120 ms in atrial muscle. Thus the theoretical maximum atrial rate of contraction is about 400 min^{-1}. However, the AV node cannot conduct more than about 220 action potentials per min, so that a coordinated heart cannot beat any faster than about 220 min^{-1}.

Cardiac arrhythmias

Arrhythmias are disturbances of the rhythm or sequence of depolarization due to disorders either at the site of action potential initiation or in the pathways for action potential conduction. Ischaemic heart disease, in which there is a reduced blood flow within the heart, is a common cause of arrhythmias.

If the discharge rhythm of the SA node is depressed the AV node, or more usually a *latent pacemaker*, assumes pacemaker function and drives the heart at a *lower frequency* than usual. These latent pacemakers are usually found in or close to the conduction system in both atria and ventricles. If this new pacemaker is in the AV node or in the ventricles, there could be occasions where both atria and ventricles are depolarized at the same time. The consequence of this will be that atrial contraction will occur during ventricular contraction when the AV valve would be closed and therefore cannot contribute to ventricular filling.

In the conducting system the AV node is the most susceptible to damage, developing a very slow initial depolarization and prolonged action potential. This decreases the speed of conduction through the AV node and, depending on the severity, causes a *longer interval* than usual between atrial and ventricular contractions (first degree AV block) or a partial or a complete heart block. In *partial heart block*, the atria beat normally but only some of the atrial action potentials reach the ventricle. The block may be regular, e.g. 2 to 1 or 3 to 1, in which case every second or third action potential, respectively, reaches the ventricle—or it may be irregular. The action potentials that did not reach the ventricles arrived at the AV node when it was refractory. A partial heart block may be evident in a normal heart when it is driven by sympathetic stimulation powerful enough to cause very high atrial rates. In *complete heart block*, the AV node fails to transmit impulses. The atria beat at the rhythm dictated by the SA node and, because of compensating reflexes, this may be faster than usual whilst latent pacemakers in the ventricle initiate their own rhythm somewhere between 20 and 40 beats per min. A transient complete heart block can occur during powerful parasympathetic stimulation which blocks the AV node.

Latent pacemakers may develop pacemaker potentials in the absence of damage to the SA or AV nodes and become sites known as *ectopic foci*. If these develop in the atria, their depolarization may occasionally reach threshold (when the atria are not refractory from a normal beat) resulting in an occasional extra atrial contraction and hence an *extra ventricular contraction (extrasystole)*. This excitation will rapidly depolarize the SA node which must repolarize and then depolarize before it can initiate the next normal beat. Consequently, the pause between the extra beat and the next normal beat is slightly longer than the usual beat interval. In *paroxysmal atrial tachycardia*, the atrial ectopic foci discharge at a higher frequency than the SA node (often only intermittently) and drive the heart at rates up to 220 min^{-1}. In *atrial flutter*, the ectopic foci discharge at rates of 200–350 min^{-1} causing rapid regular atrial contractions. However, the AV node cannot transmit every atrial action potential and thus some degree of partial heart block results. In *atrial fibrillation*, the multiple ectopic foci continuously discharge asynchronously at rates of 300–500 min^{-1} to give feeble atrial contractions. More importantly, the ventricle contracts at a totally irregular rate often before it is adequately filled with blood.

When the ectopic foci develop in the ventricles, the arrhythmias are more serious. If the ectopic focus reaches threshold at any time between the end of the refractory period of the usual action potential and the beginning of the next regular action potential, it will trigger an extra action potential causing an early *extra ventricular contraction*. This is a weaker contraction than usual because the action potential will be of shorter duration (Fig. 11.9) and because insufficient time will have been allowed for ventricular filling. The extra ventricular excitation will render the ventricular muscle refractory at the time the next SA discharge occurs. Hence, the next atrial contraction occurs but the next ventricular beat is missed. The *compensatory pause* resulting from the missed beat is nearly two normal beat intervals. The succeeding ventricular contraction will be large because of the extra time allowed for ventricular filling. The electrical event from the ventricles rarely propagates in a retrograde direction. The extra ventricular beats may be isolated and irregular or occur rather regularly. If the ventricular ectopic foci discharge at a high regular frequency, they cause *paroxysmal ventricular tachycardia*. If there are multiple ectopic foci in the ventricles discharging at a high rate, then rapid, uncoordinated and ineffective ventricular contractions (*ventricular fibrillation*) ensue causing heart failure. Sympathetic stimulation to the heart can increase the tendency of ventricular fibres to develop ectopic foci. Ventricular fibrillation can sometimes be treated successfully by electrical *defibrillation* in which the heart is exposed to a brief pulse of external electric current. This depolarizes instantly the entire myocardium and the myocardium may then repolarize as a coordinated unit without redeveloping fibrillation.

Arrhythmias can also occur because of unequal conduction velocities through branches of the conduction system (Fig. 11.11). One branch (B) conducts normally but the damaged one (C) has a long refractory period and a slow conduction velocity. When the action potential arrives at A, it can only propagate through B but because of the syncytial organization of the heart it can arrive again at C when C is no longer refractory and propagate in a retrograde direction through C to the original starting point, A. Because of this long pathway, the action potential arrives at A when it is no longer refractory and hence A is immediately re-excited. The action potential propagates back down B and again enters C setting up a circular pattern of depolarization and causes a frequency of contractions greater than normal. This *circus movement* can initiate atrial and ventricular fibrillation and possibly cause paroxysmal atrial and ventricular tachycardia and atrial flutter. Arrhythmias caused by circus movement are often referred to as *re-entry arrhythmias*.

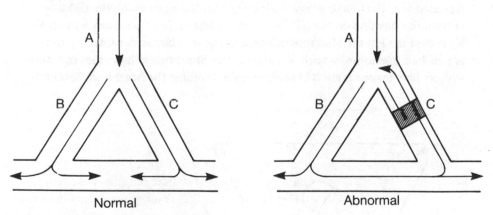

Fig. 11.11. Diagram showing 'circus movement' resulting from a long refractory period and slow conduction velocity in branch C.

Electrocardiogram (ECG)

The combined effect of the myocardial action potentials during each heart beat produces a voltage which is present throughout the whole body. By placing electrodes on the surface of the body, the differences in this voltage (about 1 mV) between electrodes can be measured and amplified. This record is the electrocardiogram (ECG). It is conventional to regard the heart, which is oriented to the left, as situated in the centre of an equilateral triangle, the *Einthoven triangle*, with the upper corners corresponding with the shoulders and the lower corner with the pubic symphysis. Recording electrodes, when attached to right and left arms (RA and LA) and to one or other foot, conventionally the left (LF), are effectively placed at the corners of this triangle (Fig. 11.12) and examine the heart vertically (coronal plane).

The recording electrodes (referred to as leads) fall into three categories: standard limb leads, augmented limb leads and chest leads. From *standard limb leads*, bipolar recordings are obtained between any two corners of the triangle (Fig. 11.12) either between LA and RA (lead I) or between LF and RA (lead II) or between LF and LA (lead III) with a fourth electrode on the right foot acting as an earth. From *augmented limb leads*, unipolar recordings are obtained, one of the three corners of the triangle acting as the active (positive) electrode, the remaining two leads connected together as the indifferent (negative) electrode and the right foot earthed (Fig. 11.12). When the active electrode is the RA, LA or LF, the lead is designated aVR, aVL or aVF, respectively. The term aV refers to the augmented voltage (about 50% more) obtained by these leads compared with standard limb leads.

Chest leads (Fig. 11.13) give unipolar recordings with the active electrode in one of six positions on the chest, the indifferent electrode consisting of the LA, RA and LF connected together and with the right foot earthed. Lead V_1 is placed over the fourth intercostal space near the sternum on the right, V_2 is in a similar position on the left, V_3 is on the left midway between V_2 and V_4, V_4 is over the left fifth intercostal space in the mid-clavicular line, V_5 and V_6 are in line horizontally with V_4 and in the anterior axillary line and mid-axillary line, respectively. These chest leads examine the heart horizontally.

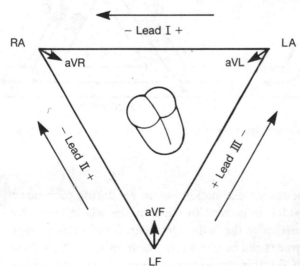

Fig. 11.12. Einthoven triangle for recording the electrocardiogram (ECG) from limb leads. RA, right arm; LA, left arm; LF, left foot. Arrows indicate the direction in which a particular lead 'looks' at the heart.

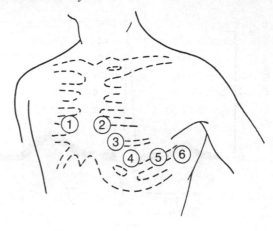

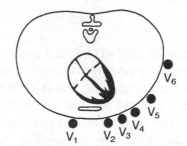

Fig. 11.13. Six positions for chest leads for recording the electrocardiogram (ECG).

The pattern of the ECG obtained varies with different leads but certain features are always present (Fig. 11.14). The *P wave* is produced by the spread of electrical activity during atrial depolarization. The *QRS complex* is produced by ventricular depolarization and the *T wave* by ventricular repolarization. When no depolarization or repolarization is occurring, there is zero voltage difference in the ECG (the *isoelectric line*). Atrial repolarization does not produce any detectable wave because it occurs during the QRS complex. Since ventricular repolarization is less well synchronized than ventricular depolarization the T wave is longer in duration but smaller in amplitude than the QRS complex. The *PQ* or *PR interval* (Fig. 11.14) is the time required for excitation to spread through the atria, AV node and bundle of His. The *QS interval* is the time required for excitation to spread through ventricles. The *QT interval* is a measure of the duration of the ventricular action potential (Fig. 11.15) and is closely related to the duration of ventricular contraction. Depending on the lead, the QRS complex may have one or two or sometimes three components. If the first deflection from the isoelectric line is negative (by convention

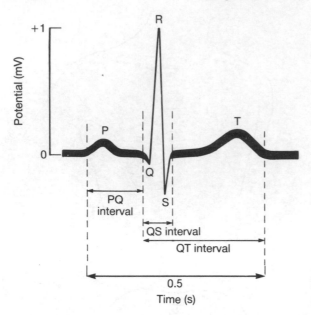

Fig. 11.14. Standard electrocardiogram (ECG) record for resting heart rate.

downwards), it is called a Q wave; if positive, it is called an R wave and if the next deflection falls below the isoelectric line it is called an S wave.

At a given electrode, the polarity of the potential change depends on the direction in which the wave of *de*polarization is moving in the heart. By convention, if the wave is travelling towards the electrode, a positive potential is recorded and if the wave is travelling away from the electrode a negative potential is recorded. A wave of *re*polarization travelling towards or away from the electrode causes a negative or positive potential, respectively. At any instant, the total electrical activity can be thought of as a vector having both direction and magnitude (Fig. 11.16).

Depolarization of the atria commences at the SA node and spreads through atrial muscle to the AV node. The net result is a vector directed downwards and to the left (P wave). Ventricular depolarization starts in the interventricular septum which depolarizes from left to right, resulting in a vector directed downwards and to the right (first wave of the QRS complex). Depolarization then spreads through the ventricular muscle from inner to outer surface. Since the left ventricle has more bulk than the right, the mean direction of the vector is downwards and to the left (second wave of the QRS complex). Activation of the last segments of the myocardium results in a vector directed upwards and to the right (third wave of the QRS complex). Ventricular repolarization spreads from the outer to the inner surface of the ventricular myocardium. Since it is opposite in both direction and sign from the second wave of the QRS complex, it produces a vector in the same direction, i.e. downwards and to the left (T wave).

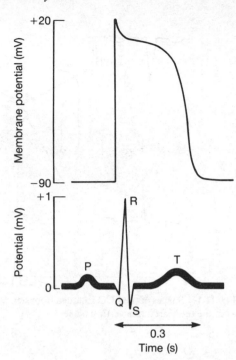

Fig. 11.15. Relationship between the ventricular action potential and the ECG.

The direction of these vectors and the position of the limb lead in relation to the Einthoven triangle determine the size and polarity of the ECG waves. Each lead can be regarded as 'looking' at the heart in a particular direction. Standard limb leads look from the positive to the negative electrode, augmented limb leads look at the heart from the limb by which they are named

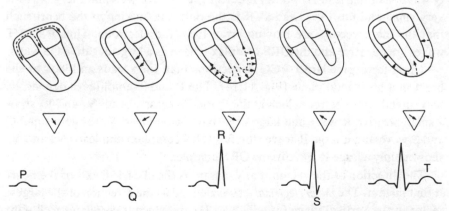

Fig. 11.16. Sequence of depolarization and repolarization of the heart, the resulting vector in the Einthoven triangle and the development of the ECG waves (as recorded by a limb lead 'looking' directly towards the vector arising from the middle stage of the sequence).

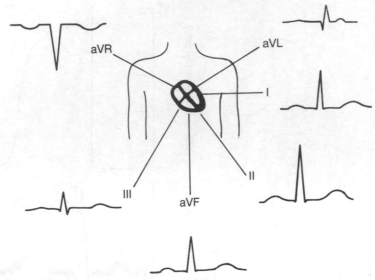

Fig. 11.17. Shapes of the ECG recorded from standard and augmented limb leads which examine the heart in the vertical plane.

and both types look in a vertical plane (Figs 11.12 & 11.17). Lead II normally looks most directly at the main vectors and therefore has the largest as well as positive P, R and T waves with small negative Q and S waves. In leads I and aVF, the deflections are all positive but of smaller amplitude than lead II because they do not look so directly at the main vectors. Lead aVL looks at the heart from the side and detects best the vector associated with interventricular septal depolarization. This vector moves away from the lead causing a negative Q wave. Neither lead III nor aVL detect the P wave well and the T wave is very small and positive. Lead aVR is the only one looking at the heart such that the main vectors are moving away from the electrode. Thus P and T waves are negative and the QRS complex is seen as a large negative QS wave.

   Chest leads give larger ECG deflections than limb leads and look at the heart in a horizontal plane (Fig. 11.18). The P and T amplitudes depend on how directly the electrodes look at the P and T vectors. Leads V_1 and V_2 show a small positive R wave and large negative S wave; lead V_4 shows a small Q and S wave and a large R wave (the full QRS complex) and leads V_5 and V_6 show mainly a large R wave in the QRS complex.

   The direction of the vector at any instant is the electrical axis of the heart at that instant. The *mean electrical axis* is defined as the direction of the largest vector in the vertical plane (usually lead II) and since it correlates well with the long axis of the heart (i.e. the interventricular septum) it can be used to infer the orientation of the heart. Tall, thin people tend to have a more vertical

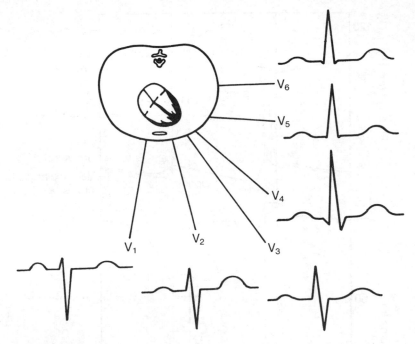

Fig. 11.18. Shapes of the ECG recorded from chest leads which examine the heart in the horizontal plane.

heart and the mean electrical axis lies between leads II and aVF. Hypertrophy of one ventricle with respect to the other will move the mean electrical axis to right or left of normal. The chest lead that gives equal amplitude R and S waves, usually V_3 (or V_4), overlies the anterior edge of the interventricular septum and is said to mark a *transition point*. Rotation of the heart in the horizontal plane also depends on body build and occurs with hypertrophy of one ventricle and can be detected as a right or left shift of the transition point.

Cardiac cycle and heart sounds

The interrelationships amongst electrical, mechanical (hydrostatic pressure and volume) and valvular events during one complete heart beat, referred to as the *cardiac cycle*, are illustrated in Fig. 11.19. Events on both sides of the heart are similar though slightly asynchronous. The organization of the electrical conduction system is such that the contraction of the right atrium precedes that of the left atrium and the left ventricular contraction precedes that of the right ventricle. However, right ventricular ejection begins before the left ventricular ejection because pulmonary arterial pressure is lower than the aortic pressure. Furthermore, since the pulmonary circuit is less resistant

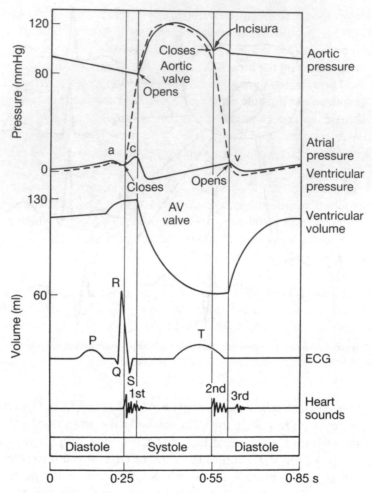

Fig. 11.19. The events occurring in the left side of the heart during one cardiac cycle at rest.

to blood flow than the systemic circuit, the right ventricular ejection goes on for longer and hence the pulmonary valve closes after the aortic valve.

At a resting heart rate of 70 min^{-1}, each cardiac cycle lasts 0·85 s and is composed of two periods: *systole*, when the ventricles contract and eject blood into the aorta or pulmonary artery, and *diastole*, when the ventricles are relaxed and fill with blood from the veins. The duration of ventricular muscle contraction and therefore by definition, systole, is 0·3 s (Fig. 11.9). Diastolic filling of the ventricles occupies almost two-thirds of the cycle (0·55 s) at rest. However, when the heart rate increases to its 200 min^{-1} maximum, the cycle is reduced to 0·3 s. A reduced duration of the action potential can shorten systole to about 0·15 s, leaving only 0·15 s for diastolic filling.

It is convenient to begin describing the cardiac cycle prior to atrial contraction, i.e. in *mid-diastole* (Fig. 11.19). At this point, both atrial and ventricular pressures are low but, since atrial pressure is slightly greater, blood from the veins entering the atria flows on into the ventricles through the open AV valves. The ventricles are already nearly 80% full and this venous inflow is small in volume. As diastole continues, both atrial and ventricular pressures slowly increase as the chambers are filled because, since the aortic and pulmonary valves are closed, blood enters but cannot leave the heart. At this stage, the aortic pressure remains high for reasons which will be explained later.

In *late diastole*, the P wave of the ECG occurs, reflecting atrial depolarization. Towards the end of the P wave, the atria begin to contract causing an increase in atrial pressure (the *a wave*, Fig. 11.19) which propels most of the blood within the atria into the ventricles. The addition of this final 20% to ventricular filling is accompanied by a small increase in ventricular pressure. As the atria begin to relax, atrial and ventricular pressures drop slightly. The volume contained in each ventricle at the end of diastole (about 130 ml when standing and about 160 ml when lying) is the *end-diastolic volume*.

Towards the end of diastole, the QRS complex of the ECG begins indicating ventricular depolarization and, by the end of the QRS complex, the ventricle starts to contract. In this *early systole*, the ventricular contraction generates an immediate increase in ventricular pressure which rapidly exceeds atrial pressure. This causes the AV valves to shut, setting up vibrations in these valves which are transmitted to the chest wall and can be heard with the aid of a stethoscope (*first heart sound*). With microphones and recording apparatus (phonocardiography), the waves composing this low-pitched sound can be displayed (Fig. 11.19).

As the ventricles continue to contract with their chambers closed, both the tension in the ventricular wall and the pressure in the ventricular lumen increase markedly. This is associated with vibrations in the ventricular muscle fibres which contribute to and prolong the first heart sound. Since blood can neither enter nor leave the ventricle, this period is referred to as the *isovolumetric* or *isometric phase of ventricular contraction*. Because of the high ventricular pressure, the AV valves bulge back into the atria causing a sharp rise in atrial pressure to about 10 mmHg (left atrium) or 5 mmHg (right atrium)—the *c wave*.

In *mid-systole* when the ventricular pressure first exceeds aortic and pulmonary arterial pressure, the aortic and pulmonary valves open and blood is accelerated and ejected very rapidly from the ventricles into the arteries. Their pressures follow very closely the ventricular pressures and in young adults rise from their diastolic minima of about 80 mmHg (systemic) or 8 mmHg (pulmonary) to their systolic peaks of about 120 mmHg (systemic) or

25 mmHg (pulmonary). During this rapid ejection phase, the shortening of the ventricles pulls the AV fibrous ring downward and, since the venous openings into the atria are in a fixed position, the effect is to lengthen the atria and increase their capacities. Thus there is a sudden fall in atrial pressure, often to negative values.

In mid-systole, the T wave of the ECG reflects repolarization. Towards the end of the T wave, in *late systole*, the ventricular muscle starts to relax and ventricular pressures drop (Fig. 11.19). Although ventricular pressure actually drops below aortic pressure, blood continues to be ejected slowly because of the momentum imparted to it during the initial acceleration earlier in systole. Note that the total energy causing blood to flow is the sum of the potential energy from hydrostatic pressure gradients and the kinetic energy from the mass of blood and its speed of movement (see p. 413). In the aorta, pressure does not drop as quickly as ventricular pressure because, owing to the resistance to flow through peripheral arterioles, only about half of the ejected volume has been propelled through the aorta. The remainder has been accommodated in the aorta by stretching its elastic elements. As the ejection into the aorta slows, the stretched aorta elastically recoils, propelling more blood onwards to the arteries. In addition, at the end of ejection, there is a small transient retrograde aortic flow towards the ventricles which causes the aortic valve to close. The volume contained in each ventricle at the end of systole is called the *end-systolic volume* and is about 60 ml when standing. Thus about 70 ml (the *stroke volume*) was ejected in systole. The proportion of the end-diastolic volume that is ejected (i.e. stroke volume/end-diastolic volume) is the *ejection fraction*.

Since all valves are now closed, no blood can enter or leave the ventricles and this period of rapid ventricular relaxation and rapidly falling ventricular pressure is called the *isovolumetric* or *isometric phase of ventricular relaxation*. During this period of *early systole*, the *second heart sound* (a quicker and higher pitched sound than the first) is heard. It is generated by the vibration from closure of aortic and pulmonary valves. This sound is often split, especially during inspiration, because the aortic valve closes slightly before the pulmonary valve. The pressure wave associated with the closure of the aortic valves is reflected along the aorta giving rise to the wave after the *incisura*. Throughout systole the atrial pressure has gradually increased from below zero to a peak of about 5 mmHg (left atrium) or 2 mmHG (right atrium) at the end of the period of isovolumetric ventricular relaxation—the *v wave*. This gradual increase in atrial pressure results from the continuous venous return accumulating in the atria gainst closed AV valves and from the return of the AV fibrous ring to its resting position in mid-systole.

When ventricular pressure has just dropped below that of the atria and the AV valves open. The ventricles fill rapidly with blood that has accumulated in the atria, setting up vibrations sometimes detectable as the *third heart sound*. During this rapid filling, both atrial and ventricular pressures decline. They then gradually increase in mid-diastole as ventricular filling continues slowly (*diastasis*) as a consequence of venous flow into chambers closed by aortic and pulmonary waves. Throughout diastole, atrial pressure is slightly greater than ventricular pressure. The rapid initial filling of the ventricles early in diastole is very important because it means that when heart rates are high and the diastolic period short, ventricular filling will lose only the relatively small contribution from the later diastasis. (Similarly the rapid large volume ejection early in systole prevents a short duration systole from markedly restricting stroke volume.) Throughout diastole, the aortic pressure remains high due to the continuing elastic recoil of aortic walls stretched during the previous systole and of course to resistance to flow downstream. However, there is a slow decline to a diastolic minimum as the blood flows from the aorta into the rest of the vascular system.

Abnormal heart sounds

A *fourth heart sound* can be heard just before the first heart sound if vibrations are set up during the rapid ventricular filling associated with atrial contraction. These vibrations may occur when the atrial pressure is high or the ventricle is stiff as in ventricular hypertrophy.

Abnormal sounds called *murmurs* can occur at any stage of the cardiac cycle. If blood passes through a narrowed orifice or regurgitates back into a chamber, its flow becomes turbulent and this may generate murmurs. Pathological narrowing of an orifice is referred to as *stenosis*. Stenosis of an atrioventricular orifice results in turbulent flow into the ventricles (*diastolic murmur*) and stenosis of the aortic or pulmonary orifice causes turbulent flow into the aorta or pulmonary artery (*systolic murmur*). If a valve is *incompetent*, this results in a backflow of blood. Incompetent AV valves allow systolic regurgitation into the atria (systolic murmur) and incompetent aortic or pulmonary valves cause diastolic regurgitation into the ventricles (diastolic murmur). Defects in the atrial or ventricular septum also cause systolic murmurs while a patent ductus arteriosus connecting the aorta with the pulmonary artery can be associated with a *continuous murmur* throughout the cardiac cycle. Frequency, character, duration, time of occurrence in the cardiac cycle and the site on the chest from which the murmur is heard most clearly frequently enable one to diagnose the underlying abnormality.

Jugular venous pulse

Since there are no valves at the entry of the venae cavae into the right atrium, the three positive a, c, and v waves in the right atrial pressure are transmitted into the large veins. They can be seen as pulsations in the jugular vein in the neck (p. 458) and are called the *jugular venous pulse* (JVP). Because they are transmitted backwards from the right atrium, each wave in the JVP occurs after the corresponding one in the atrium (Fig. 11.20). The *a wave* of the JVP occurs because of atrial contraction but it is not just a delayed reflection of the increase in atrial pressure. It is also contributed to by the damming of blood in the large veins resulting from the atrial contraction constricting the orifices of the venae cavae. The *c wave* is a delayed reflection of the increase in atrial pressure when the AV valve bulges into the atrium. The pressure pulse transmitted from the adjacent carotid artery during peak systole also makes a contribution. Hence the name *c wave*. The *v wave* of the JVP has the same origins as the v wave in the right atrial pressure. The relative amplitudes of the JVP waves are variable because the jugular vein is subjected to rhythmical pressure fluctuations induced by breathing (p. 436) which occur at a rate much slower than the heart beat. The shape and magnitude of the JVP can indicate the presence of cardiac arrhythmias, valvular incompetence and stenosis and particularly increases in atrial pressure.

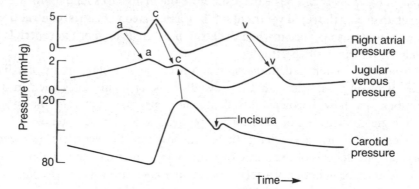

Fig. 11.20. Temporal relationships between right atrial, jugular venous and carotid arterial pressures.

Mechanical characteristics of cardiac muscle

Length–tension curves and *load–velocity curves* can be used to study the mechanical characteristics of strips of cardiac muscle and these curves are similar to those of skeletal muscle (p. 104). Muscles have both elastic and contractile elements which are stretched, generating a *passive tension* whenever the muscle is lengthened (Fig. 11.21a). When cardiac muscle contracts and shortens under a constant load, the contraction is referred to as *isotonic*. Here the contractile elements shorten without any change in tension (Fig. 11.21a).

If the muscle length is kept constant as it contracts (an *isometric* contraction,

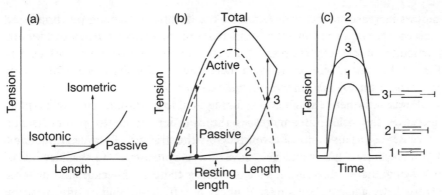

Fig. 11.21. Length–tension curves (a & b) for a strip of cardiac muscle. (a) Tension developed during passive stretch and during isotonic and isometric contractions. (b) Total and active tension developed during isometric contractions at different muscle lengths. (c) Tension developed during an isometric contraction at three different muscle lengths (identified also in b) and its dependency on the extent of overlap of actin and myosin filaments.

Fig. 11.21a), the elastic elements stretch as the contractile elements shorten and the muscle develops a *total tension* (Fig. 11.21b). This total tension is composed of the *active tension* generated during contraction plus the already present passive tension. The active tension developed depends on the initial length of the muscle. Unless the muscle becomes overstretched, differences in length which do not alter passive tension appreciably will result in large differences in active tension. While the increase in contractility has been attributed to the more effective formation of cross-bridges between actin and myosin filaments (Fig. 11.21c, cf. Fig. 3.9), it appears that the lengthening of cardiac muscle cells may also increase the sensitivity of troponin to Ca^{2+} and induce an increase in free cytosolic Ca^{2+}. In contrast to skeletal muscle, where the resting length is close to the optimal length, the resting length of cardiac muscle is below optimal and thus cardiac muscle is normally operating on the ascending part of the length–tension curve. This intrinsic ability of cardiac muscle to contract more powerfully whenever it is stretched over its physiological range is the basis of the *intrinsic control of stroke volume* considered later.

Since the myocardium is a synctium, the strength of contraction for a particular initial muscle length cannot be graded by recruitment of a variable number of motor units as occurs in skeletal muscle (p. 110). In cardiac muscle, the strength of contraction can be increased by the higher frequency of cardiac action potentials occurring when heart rate increases or, most importantly, by increasing the membrane conductance to Ca^{2+} (G_{Ca}). Both these increase the intracellular Ca^{2+} available for the contractile process. Changes in the strength of cardiac contraction are called positive or negative *inotropy*.

Sympathetic stimulation and circulating adrenaline acting via β_1 adrenoceptors increase the heart rate, reduce the duration of action potentials and increase the G_{Ca}. The considerable increase in total tension developed for any particular muscle length caused, for example, by adrenaline is illustrated in Fig. 11.22. This is referred to as an increase in *myocardial contractility* and is the basis of the *extrinsic control of stroke volume*.

Parasympathetic stimulation acting via muscarinic cholinoreceptors decreases the heart rate, increases the duration of action potentials and decreases G_{Ca}. This diminishes appreciably the strength of atrial contractions. Until recently, the ventricles have been considered to be devoid of parasympathetic innervation but, providing there is a background of tonic sympathetic activity, a depressant effect of parasympathetic stimulation on the strength of ventricular contractions can be demonstrated. Myocardial contractility is also depressed during cardiac failure (Section 11.6) and by the direct action of acidosis, hypercapnia and severe hypoxia.

The amount of passive tension in a muscle prior to its contraction is referred to as the *preload* and in the heart as a whole is related to the end-diastolic volume. If the muscle has to support an extra weight (called the *afterload*) after it has started contracting, it will first develop sufficient isometric tension to match the total load and then shorten, lifting the afterload in an isotonic contraction. As in skeletal muscle, if the afterload is increased, the velocity of this isotonic contraction for a particular afterload decreases in an exponential curve as illustrated by a *load–velocity curve* (Fig. 11.23b, cf. Fig. 3.10). In the heart as a whole the aortic pressure can be regarded for the

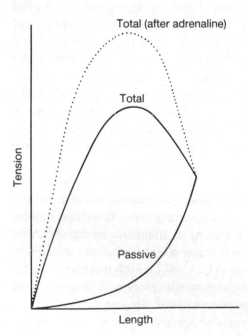

Fig. 11.22. The effect of adrenaline on the length–tension curve of a strip of cardiac muscle.

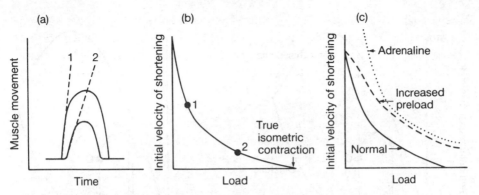

Fig. 11.23. (a) Measurement of velocity (distance/unit time, dashed line) from an isotonic contraction of a strip of cardiac muscle subjected to a small (1) and large (2) afterload. (b) A load–velocity curve. (c) Effect of increased preload or adrenaline on the load–velocity curve.

left ventricle as the equivalent of its afterload and it determines the total tension developed during systole.

When an isolated strip of cardiac muscle is stretched before its contractile responses to different afterloads are tested, this increased preload increases the velocity of shortening for any afterload, particularly with the larger afterloads (Fig. 11.23c) as well as causing a more powerful contraction (Fig. 11.21). Adrenaline increases the velocity of shortening for any afterload, particularly for the smaller afterloads (Fig. 11.23c), as well as causing a more powerful contraction (Fig. 11.22).

Control of stroke volume

By analogy with length–tension curves for strips of cardiac muscle, *volume–pressure curves* can be constructed for the heart. Length is now analogous to ventricular volume, passive tension to diastolic ventricular pressure and total tension to the maximum systolic ventricular pressure that could be developed at each volume (Fig. 11.24). During diastole, the heart fills with blood increasing in volume from an end-systolic volume (ESV) of 60 ml to an end-diastolic volume (EDV) of 130 ml with an increase in left ventricular pressure from about 5 mmHg at A to about 10 mmHg at B as shown in Fig. 11.24a. During the isovolumetric (i.e. isometric) phase of ventricular contraction, the pressure increases to C. If the aorta was clamped, since blood cannot escape, the pressure would continue to rise to C* on the curve for maximum systolic ventricular pressure. Note that for a normal EDV, C* is well to the left of the maximum peak.

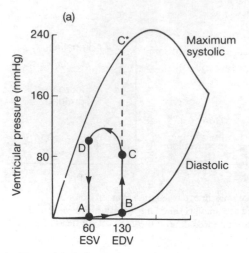 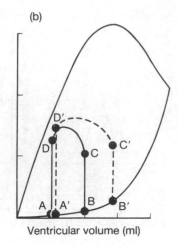

Fig. 11.24. Left ventricular volume–pressure curves. (a) The normal cardiac cycle of ABCDA (see text). C* is the ventricular systolic pressure achieved if the aortic valve remains closed. (b) The effect on the cardiac cycle of increasing the end-diastolic volume (EDV) from B to B'.

In a normal cardiac cycle, the isovolumetric phase of ventricular contraction terminates when the semilunar valve opens (at pressure C) and the volume begins to decrease. The pressure at C will depend on the diastolic pressure in the aorta or pulmonary artery. For the left ventricle, this is about 80 mmHg and during the ejection of blood the pressure increases to about 120 mmHg and then declines to about 100 mmHg at D. From C to D, the contraction is referred to as *auxotonic* occurring against the afterload of increasing aortic pressure. When the aortic valve closes, isovolumetric ventricular relaxation occurs and the pressure drops from D to A. The cardiac cycle is now completed. The *area* enclosed by this volume–pressure loop (ABCDA) is a measure of the *external work* done by the heart each beat.

Intrinsic control of stroke volume

The volume–pressure curves and the concept of the work loop were first described in 1895 by Frank using an isolated frog heart in which the aorta was clamped and the ventricles filled to different volumes between each beat. In 1914, Starling and his colleagues developed a mammalian heart–lung preparation in which the pulmonary circuit was intact and the lungs were mechanically ventilated (to keep the heart supplied with O_2) but the systemic circuit was replaced by a system of blood-filled tubes. The diameter of these tubes could be decreased to increase resistance and hence increase aortic pressure (afterload) and the height of the reservoir returning blood to the heart could be elevated to increase the filling pressure and hence increase end-diastolic volume (preload). Because the temperature of the blood was kept constant and the heart was denervated, heart rate remained constant.

In such a heart–lung preparation, an increase in EDV (preload) causes the heart to begin its isovolumetric contraction at a higher pressure and volume (position B′ in Fig. 11.24b). Thus the new volume–pressure loop is larger but with pressures at C′, D′ and A′ only slightly elevated. The ESV is slightly increased but the heart now operates at a larger EDV and ejects a larger stroke volume (Fig. 11.24b) at a higher ejection velocity (Fig. 11.23c). The relationship which is obtained between each EDV and the resulting stroke volume (Fig. 11.25) is often referred to as the *Starling curve* or ventricular function curve and it illustrates the so-called *Frank–Starling law of the heart*. According to Starling, 'the mechanical energy set free on passage from the resting to the contracted state depends . . . on the length of the muscle fibres'. The heart thus has an intrinsic ability to regulate its stroke volume, responding with a greater force of contraction to the stimulus of increased diastolic stretch. The upper limit to this intrinsic ability is reached when the enlarged EDV has resulted in the optimal myocardial length. (This intrinsic control of stroke volume is also referred to as *heterometric autoregulation*.)

An increase in aortic pressure (afterload) in the heart–lung preparation requires that an equal increase in ventricular pressure must occur before the aortic valve can open (position C′ in Fig. 11.26). Thus with the remaining energy of this contraction the stroke volume is smaller in the first ejection against the increased afterload (dotted line). Hence the ESV increases and with normal venous return the EDV subsequently increases, stretching the ventricle so that at the next contraction more tension is developed and so on

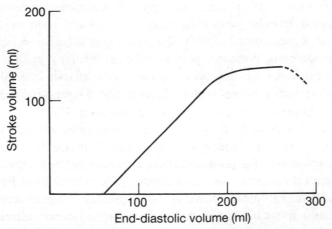

Fig. 11.25. The Starling curve—the relationship between end-diastolic volume and stroke volume (providing afterload is constant).

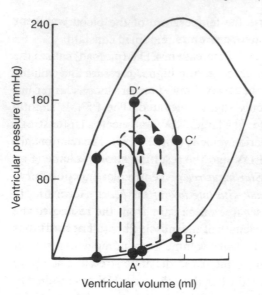

Fig. 11.26. Left ventricular volume–pressure curves showing final cardiac cycle A′B′C′D′A′ and intermediate steps (dotted lines) in response to an increased afterload.

until a new steady state is reached. The final volume–pressure loop is shifted to the right with A′, B′ and especially C′ and D′ pressures increased. The loop has a larger area, indicating more work, and the heart operates with a larger ESV and EDV but with an unchanged stroke volume. However, this intrinsic mechanism of compensation for an increased afterload is limited and with larger increases in afterload the eventual response is a decreased stroke volume, large EDV and ESV and a short ejection phase with decreased ejection velocity (Fig. 11.23b).

The Starling curve is often plotted with other variables on its axes. On the abscissa, diastolic muscle length and, providing ventricular compliance is unaltered, end-diastolic ventricular pressure or mean right or left atrial pressure are all used instead of EDV. On the ordinate, stroke work (i.e. stroke volume × mean aortic pressure), peak systolic ventricular pressure or cardiac output (if heart rate and afterload are constant) are all used instead of stroke volume. When cardiac output is plotted against mean right atrial pressure, the Starling graph is often called a *cardiac function curve* (p. 437).

The intrinsic control of stroke volume operates whenever there is a change in diastolic filling of the ventricles. Its primary function is to correct any *momentary imbalance of the cardiac outputs of the two ventricles*. Since the two ventricles beat at the same rate, their outputs can be matched only by adjusting their stroke volumes. Imbalance between the systemic and pulmonary circuits can occur as a result of change in the arteriolar resistance (afterload) or a change in the volume of blood returned to the heart (preload) by one circuit but not the other. Indeed, the latter normally happens throughout the respiratory cycle, with a larger right and a smaller left stroke volume during inspiration and the converse during expiration (p. 435).

Intrinsic control is also important when systemic *venous capacity* changes. For example, the volume contained in the systemic venous system is decreased when reclining compared with standing (Section 11.6). This results in a greater central blood volume in the heart and lungs and the increased EDV leads to an increased stroke volume (Fig. 11.27). Reflex contraction of smooth muscle in the veins also decreases venous capacity, for example in response to haemorrhage (Section 11.6) or in exercise where the effect is augmented by the skeletal muscle venous pump (p. 435). Decreased venous capacity contributes slightly to the increased EDV of exercise in the upright posture (Fig. 11.27). Acute increases or decreases in *total blood volume* (transfusion or haemorrhage) also increase or decrease the EDV and hence alter stroke volume accordingly.

Factors influencing end-diastolic volume

The EDV can be decreased by *very high heart rates* which reduce the time available for the slow diastolic filling phase (diastasis). However, since the increased sympathetic activity also causes a distinct increase in the rate of ventricular relaxation, much of the loss of diastasis is compensated for by the rapid *relaxation recoil of the ventricles*, increasing the volume achieved in the initial rapid diastolic filling phase. An increase in the power of *atrial contraction*, for example during exercise, increases EDV by increasing the wave of atrial pressure and hence the atrial contribution to ventricular filling (Fig. 11.27).

A large increase in *atrial pressure* at any other time in the cardiac cycle reduces the pressure gradient and hence blood flow from the capillaries to the

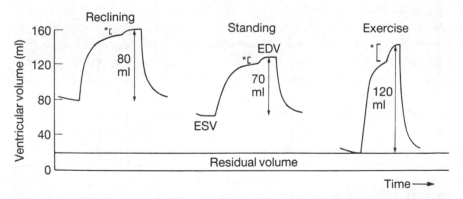

Fig. 11.27. End-systolic volume (ESV), end-diastolic volume (EDV) and the resultant stroke volume during reclining, standing and exercise in the upright posture. Note the residual volume that can never be ejected and the atrial contribution to ventricular filling (*).

atria and this will tend to limit increases in EDV. Similarly, an elevated
ventricular pressure during diastole will restrict further filling of the ventricles
(see diastolic pressure curve of Fig. 11.24). Reduced distensibility or *compliance*
of the ventricles caused by hypertrophy or fibrosis or by accumulation of fluid
in the pericardial space (cardiac tamponade) will result in a smaller EDV for
any particular diastolic ventricular pressure.

The surrounding *intrathoracic pressure* (or intrapleural pressure) influences
EDV. This pressure is subatmospheric (Section 12.2). At rest it is about −
3 mmHg in expiration and about −5 mmHg in inspiration but it becomes
positive in large expirations and more negative in large inspirations. The more
negative the intrathoracic pressure the greater is its dilating effect on the
ventricles and atria. Thus there is a greater diastolic filling during inspiration
than expiration. However, because there are valves in the veins preventing
retrograde flow (p. 435), large expirations do not reduce diastolic filling
appreciably.

Extrinsic control of stroke volume

In the 1950s, Sarnoff investigated the effect of *adrenaline* and *sympathetic*
stimulation on the heart–lung preparation and demonstrated that the maximum
systolic pressures which could be generated for any EDV were increased (Fig.
11.28a) and in proportion to the degree of stimulation. This is the positive
inotropic effect of the sympathetic nervous system, increasing *myocardial*

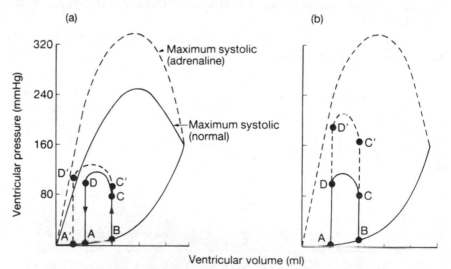

Fig. 11.28. Effect of adrenaline on the maximum systolic pressure of the ventricular volume–
pressure curve permitting in (a) an increased stroke volume with the cardiac cycle A′BC′D′A′ or
(b) a normal stroke volume maintained in the face of an increased afterload of the magnitude of
C′.

contractility (as shown earlier in the length–tension curve of Fig. 11.22) and increasing ejection velocity (Fig. 11.23). Parasympathetic stimulation has a negative inotropic effect but the extent of this and its importance is not clear. A reflex increase in sympathetic activity to the heart occurs during exercise, during circumstances when the mean arterial pressure has decreased, for example in haemorrhage, or when cardiac output has to be maintained against a high afterload.

The positive inotropic effect results in two types of volume–pressure loops (Fig. 11.28), one leading to an increase in the stroke volume and one maintaining a normal stroke volume despite an increase in afterload. In the first case (Fig. 11.28a), for the normal EDV and end-diastolic ventricular pressure (B) and only a slightly larger than normal aortic diastolic pressure (C′), the shift of the maximum systolic pressure curve to the left allows position D′ to move to a higher pressure and a very much lower volume. Thus the ESV at A′ is reduced and a large stroke volume (see also Fig. 11.27) has been ejected. It will occur at a higher ejection velocity (Fig. 11.23c). (This extrinsic control of stroke volume is also referred to as *homeometric regulation*.) If the effects of different degrees of sympathetic activity and hence of myocardial contractility on stroke volume are examined over a range of EDVs, a family of Starling curves can be plotted (Fig. 11.29).

In the second case (Fig. 11.28b), if sympathetic stimulation has also caused an increased afterload as a result of constriction of the arterioles, the increased myocardial contractility will permit the high pressure of C′ and D′ to be

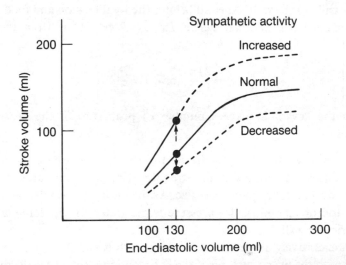

Fig. 11.29. Effect of alterations in sympathetic activity on the Starling curve and the resultant alteration in stroke volume for any particular end-diastolic volume (providing the afterload is constant).

reached without any change in EDV or ESV. Thus the stroke volume and its ejection velocity are normal. Usually, the compensation for an increased afterload is a mixture of intrinsic (Fig. 11.26) and extrinsic control of stroke volume.

Since the maximum systolic pressure curve would be difficult to measure clinically, an index of ventricular contractility is used. One such index is the measurement with a cardiac catheter of the maximal rate of pressure increase during the isovolumetric phase of ventricular contraction. A measurement of stroke volume itself would be an insufficient index as stroke volume depends on preload, afterload and the competence of valves, as well as on contractility.

Law of Laplace and the heart

Length–tension curves of strips of cardiac muscle and volume–pressure curves of the whole heart are not identical because volume is proportional to length³ (or radius³) and the tension within the muscle is not identical to the pressure in the chamber lumen. The relationship between wall tension and lumen pressure is described by the *law of Laplace*.

For hollow spherical organs, the total circumferential wall tension depends on the tension per unit length and thickness (T), the two equal radii of curvature ($2r$) and the thickness of the wall (u) and equals $T\pi 2ru$. The total luminal pressure depends on the transmural pressure per unit area (P_t) and the radius of the lumen (r) and equals $P_t\pi r^2$. *Transmural pressure* (P_t) is the pressure per unit area on the inside of the wall minus the pressure per unit area on the outside of the wall. At equilibrium, the total tension and total pressure counterbalance one another, hence $T\pi 2ru = P_t\pi r^2$. Thus from the law of Laplace:

$$P_t = \frac{2Tu}{r} \quad \text{or} \quad T = \frac{P_t r}{2u}.$$

With the law of Laplace a number of properties of the heart can be explained:

1 The rise in ventricular pressure during the ejection phase (C to D, Fig. 11.24) is due not to increasing the strength of muscle contraction but to the physical effect of a change in heart size. As the radius gets smaller and the wall thicker, additional ventricular pressure is generated for the same tension in the ventricular wall.

2 In an excessively dilated heart, the large radius and thin walls contribute, together with the overstretching of the actin and myosin, to the decline in the maximum systolic ventricular pressure (Figs 11.21 & 11.24).

3 An excessively dilated heart caused, for example, by congestive cardiac failure has to generate more total wall tension to develop the same systolic ventricular pressure than does a normal heart. This extra tension requires more O_2 consumption at a time when O_2 is being insufficiently transported by the whole cardiovascular system and when the high cardiac wall tensions are compressing the coronary arteries more and for a longer period within the cardiac cycle (Section 11.5).

4 In trained athletes or in the chronically volume- or pressure-loaded heart, hypertrophy of the heart develops. This increase in myocardial wall thickness can compensate for the increased chamber dilatation allowing a given pressure to be generated by the usual wall tension. However, since the cardiac hypertrophy in the loaded heart in contrast to that of the trained athlete is not accompanied by new cardiac capillary formation, this often leads to insufficient O_2 delivery to the cardiac muscle.

Cardiac output and its measurement

Cardiac output is the volume of blood pumped per minute by each ventricle and is thus the total blood flow through the pulmonary or systemic circuit. It is the product of *stroke volume* (volume per beat) and *heart rate* (beats per minute). Table 11.2 gives representative values for these variables at rest and near their maximum values. Note that increments in heart rate contribute more than stroke volume to the cardiac output of exercise, that the maximum heart rate and the minimum ESV are the same for non-athletes and trained athletes and that hypertrophy of the heart in the trained athletes allows their resting heart rate to be lower and their maximum cardiac output to be higher. Athletes can therefore supply more O_2 per minute to skeletal muscle and perform greater levels of exercise.

Cardiac output ($\dot{Q}$) can be measured by inserting flowmeters into the aorta or placing electromagnetic flowmeters or ultrasonic flow probes around the whole of the aorta. Clinically, two alternative methods are used. The *Stewart–*

Table 11.2. Representative cardiac variables at rest and at close to maximum exercise for non-athletes and trained athletes.

	Cardiac output ($l\,min^{-1}$)	Heart rate (beats min^{-1})	Stroke volume (ml)	EDV (ml)	ESV (ml)
Non-athlete					
Rest	5	70	70	130	60
Maximum exercise	22	180	120	140	20
Trained athlete					
Rest	5	40	120	200	80
Maximum exercise	36	180	200	220	20

Hamilton indicator dilution technique involves injecting rapidly a dye or radioactive isotope or cold saline into the venous circulation near the heart. The indicator quickly mixes with the cardiac contents and during the next few beats the entire blood–indicator mixture is pumped out of the heart into the circulation. The concentration of this mixture is measured by continual sampling from an artery in the arm for one complete circulation (about 30–40 s at rest). The mean concentration of the mixture for one complete circulation is determined and

$$\dot{Q} = \frac{\text{amount of indicator injected}}{\text{mean concentration of indicator} \times \text{duration of one circulation}}.$$

The second method uses the *Fick principle*. This states that the amount of a substance taken up by an organ (or the body) per unit time is equal to the blood flow multiplied by the arterial concentration minus the venous (mixed) concentration of that substance. In practice, the average steady-state oxygen consumption ($\dot{V}_{O_2}$) of the whole body is measured (Section 12.3) for about 15 min, during which time blood samples are taken from a systemic artery and the pulmonary artery (the latter contains mixed venous blood). The arterial (C_{a,O_2}) and mixed venous ($C_{\bar{v},O_2}$) O_2 contents are analysed (Section 12.4). Then:

$$\dot{Q} \text{ (litres min}^{-1}) = \frac{\dot{V}_{O_2} \text{ (litres min}^{-1})}{C_{a,O_2} - C_{\bar{v},O_2} \text{ (litres of } O_2/\text{litre of blood)}}.$$

Stroke volume can be determined from these cardiac-output measurements by dividing by the average heart rate (from an ECG record or arterial pulse). With new techniques, stroke volume itself can now be measured by non-invasive procedures of impedance cardiography, echocardiography and radionuclide imaging.

11.3 The vascular system

The design of the systemic circulation supplying tissues in parallel (Fig. 11.1) and the branching of the aorta into arteries, arterioles, capillaries, venules and veins (Fig. 11.2) have already been considered in Section 11.1. The functions of each type of blood vessel were also discussed briefly. Before examining these functions in greater detail, it is necessary to understand the fundamental physical concepts governing blood flow through a tube.

Physics of blood flow (pressure, resistance and velocity)

Liquid flows through a rigid tube from a higher to a lower pressure (but see pp. 396–413) and the *rate of flow* ($\dot{V}$, volume/unit time) is directly proportional to the *hydrostatic pressure gradient* (ΔP). In the vascular system, blood pressures are usually expressed in mmHg although other units are also used (1 mmHg $\approx$ 1·36 cmH$_2$O $\approx$ 133 Pa). The *resistance* (R) to blood flow depends upon the dimensions of the tube (length and radius) and the viscosity of the blood; the greater this resistance the less the flow. These determinants of flow can be summarized in an equation analogous to Ohm's law for electrical circuits:

$$\dot{V} = \frac{\Delta P}{R}.$$

If $\dot{V}$ and ΔP are in units of litres min^{-1} and mmHg, respectively, then the calculated values of R have the units mmHg l^{-1} min.

Resistance results from the friction (or viscous forces) between the molecules or particles of the liquid as they move and from the friction between this liquid and the walls of the tube. During steady streamline flow, an infinitely thin layer of fluid in contact with the wall does not move whilst the next layer moves slowly with the central axis layer moving at the fastest rate. Such layered or *laminar flow*, parallel to the axis of the vessel, has a parabolic velocity profile (see Fig. 11.32). The concept of viscosity (η) of a liquid expresses the fact that the adjacent layers interact rather than slip with infinite ease over one another. The greater the liquid's viscosity the greater the resistance to flow.

Since the friction is greatest between the liquid and the tube wall, the dimensions of length (l) and radius (r) of the tube affect the resistance to flow. The longer the tube or the smaller the radius of the tube the greater is the resistance offered. These determinants of resistance can be summarized in the following equation:

$$R = \frac{8\eta l}{\pi r^4}.$$

A twofold decrease in radius will cause a sixteenfold increase in resistance and resistance is affected more by changes in vessel radius than by changes in length or in viscosity.

In an unbranched tube, the relationships amongst flow, the pressure gradient and the determinants of resistance are described by the *Hagen–Poiseuille equation*:

$$\dot{V} = \frac{\Delta P \pi r^4}{8\eta l}.$$

In any blood vessel, the length is virtually constant, the viscosity is relatively constant and flow through it can be increased only by increasing its radius or the pressure gradient or both. Only arterioles show marked changes in radii which will affect resistance. An increase in flow through other types of blood vessels requires an increase in the pressure gradient. The Hagen–Poiseuille equation also indicates that at a constant flow an increase in radius of the arteriole will result in a decrease in the pressure gradient along the arteriole. This, as will be explained later, affects the pressure upstream (arteries) and downstream (capillaries).

In the systemic vascular system, it is important not only to consider an individual vessel but also an entire network. If the vessels are arranged *in series*, the total resistance to flow is the sum of all the individual resistances, whereas, if they are arranged *in parallel*, the reciprocal of the total resistance is the sum of all the individual reciprocal resistances (Fig. 11.30). Therefore, for vessels in parallel, the total resistance is considerably smaller than the resistance of each individual vessel. Arteries, arterioles, capillaries, venules and veins are in general arranged in series with respect to each other (Fig. 11.2). However the vascular supply to various organs (Fig. 11.1) and the vessels, e.g. arterioles, within any organ (Fig. 11.2) are arranged in parallel.

In a series arrangement (Fig. 11.31a), the same volume/unit time ($\dot{V}$) will flow sequentially through each of the vessels with a pressure gradient along each vessel that is inversely related to the fourth power of the radius. The liquid will also flow at a *mean linear velocity* (distance/unit time) which in each vessel will be proportional to flow ($\dot{V}$) and inversely proportional to its *cross-*

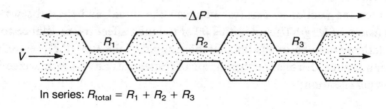

In series: $R_{total} = R_1 + R_2 + R_3$

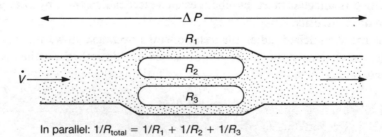

In parallel: $1/R_{total} = 1/R_1 + 1/R_2 + 1/R_3$

Fig. 11.30. *Upper*: resistance (R_1, R_2 & R_3) arranged in series. Addition of a fourth resistance would increase the total resistance. *Lower*: resistances arranged in parallel. Addition of a fourth parallel resistance would decrease the total resistance. (Assume for both diagrams that the resistance of the connecting tubes is negligible.)

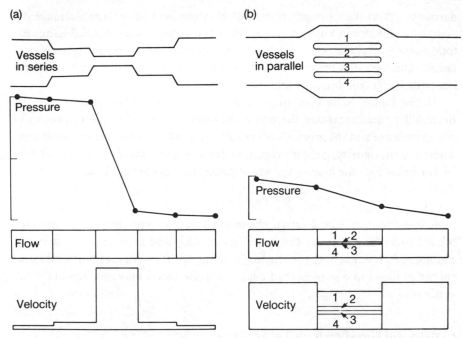

Fig. 11.31. Pressure, flow and velocity in vessels arranged (a) in series and (b) in parallel. Note that the scale of the velocity axis is expanded in (b).

sectional area (πr^2). Thus velocity $= \dot{V}/\pi r^2$ and the narrower the vessel, the smaller the cross-sectional area and the faster the velocity.

In a parallel arrangement (Fig. 11.31b), the pressure gradient across each of the parallel vessels will be the same and the flow through each vessel will be directly proportional to the fourth power of the radius. If the parallel vessels constitute an increase in total cross-sectional area, the total mean linear velocity through them will be slower. Since the flow through each parallel vessel is different, the velocity in each is derived by substituting from the Hagen–Poiseuille equation for $\dot{V}$ in the last equation. Thus velocity $= \Delta P r^2/8\eta l$ and in each individual parallel vessel velocity will be directly proportional to the square of its radius. In capillaries, extensive branching resulting in a large total cross-sectional area ensures that their combined resistance, the pressure gradient along them and the total and individual velocities within them are small.

A moving liquid has *kinetic energy*, which is proportional to the mass (m) of the liquid and the square of the mean linear velocity (v), i.e. kinetic energy $= \frac{1}{2}mv^2$. As described by *Bernoulli's equation*, the total fluid energy (E) in a moving liquid in a horizontal vessel is the sum of the kinetic energy and the *potential energy* (P_p), the latter being the hydrostatic pressure at any point along the length of the horizontal vessel. If the liquid flows suddenly from a narrower vessel into a much wider vessel, there will be a large decrease in velocity and hence in kinetic energy. However, as there will be little change in total energy, there is a relative increase in the hydrostatic pressure in the wider vessel. This becomes important in the cardiovascular system whenever there is a pathological dilatation in an artery (aneurysm).

If the liquid is moving in a vertical vessel, the total fluid energy in Bernouilli's equation is now the sum of the kinetic energy, the potential energy due to pressure and the *gravitational potential energy* (P_g). The latter equals $\rho g h$ where ρ is the density, g the acceleration due to gravity and h the height above $(+)$ or below $(-)$ the heart—the site of generation of energy. Thus:

$$E = P_p + P_g + \tfrac{1}{2}mv^2.$$

Note that the P_p of liquid raised above the heart becomes less by an amount related to its height above the heart (see p. 457) and vice versa. Hence the change in P_p is equal but opposite to the change in P_g so, in fact, the total fluid energy of blood in a vessel is not changed if the blood moves horizontally or vertically.

Deviations of blood flow from that predicted

The Hagen–Poiseuille equation was formulated for (1) laminar flow of (2) a homogeneous fluid with constant viscosity through (3) a rigid unbranched tube. By and large these characteristics do not apply to the vascular system.

Flow, especially in the arteries, is not always laminar. Flow is *pulsatile* in the arteries because of the rhythmic activity of the heart and this will result in a flat rather than a parabolic velocity profile and a lower mean velocity. Furthermore, under certain conditions, flow can become *turbulent*. This is characterized by eddies of liquid particles moving not only parallel to the vascular axis but also perpendicular to it, and this flattens the velocity profile (Fig. 11.32). The increased internal friction results in volume flow becoming proportional to the square root of the pressure gradient. In other words to double turbulent volume flow a quadrupling of the pressure gradient is required (Fig. 11.32). Turbulence always occurs transiently in the aorta or pulmonary artery during early systole and it can occur in large arteries if velocity is high and exceeds a critical value (e.g. in severe exercise). If viscosity is low (e.g. in severe anaemia) or there are pathological irregularities of the inner wall (sclerosis), turbulence occurs at a lower critical velocity. The noise of turbulent flow can be heard with a stethoscope.

Blood is not a homogeneous liquid with a constant viscosity inasmuch as it comprises cells and plasma; the former are responsible for most of the blood's viscosity. A greater proportion of the blood cells travel in the central or axial stream of flow resulting in a greater proportion of the less viscous plasma near

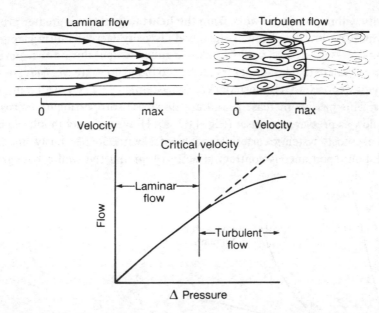

Fig. 11.32. *Upper*: velocity profiles of laminar and turbulent flow. *Lower*: relationship between pressure gradient and flow rate (or velocity) and the transition from laminar to turbulent flow once a critical velocity is reached.

the wall of the vessel. This flattens the parabolic velocity profile. Axial streaming is more evident as velocity increases. Thus the *effective viscosity* is less at high velocities and greater at low velocities. Under pathological conditions, low velocities can occur even in large vessels, either as a result of a failing heart or in a vessel distal to a pathological constriction. This results in aggregation of erythrocytes into rouleaux causing large increases in effective viscosity which will reduce the velocity still further. Since blood flows through capillaries at a low velocity, the effective viscosity would be expected to be high. However this does not occur because of an unexplained phenomenon in which the effective viscosity of any fluid suspension decreases considerably with decreasing tube radius as it flows through tubes smaller than 100 μm in radius. The effective viscosity of blood flowing through capillaries is also decreased because the erythrocytes travel through in single file, a phenomenon referred to as '*plug flow*'. Furthermore, in capillaries, the effective viscosity of blood depends on the *deformability* of erythrocytes since they are larger than the capillaries they traverse. In sickle-cell anaemia, the desaturated haemoglobin is crystalline and this reduces considerably erythrocyte deformability.

Total blood viscosity increases in hyperproteinaemia and when there is an increase in the concentration of red blood cells (increased haematocrit) as occurs, for example, in response to altitude (Section 12.6). Such an increased viscosity will require more work from the heart to generate a greater pressure to maintain normal blood flow. Total blood viscosity decreases during anaemia.

Blood vessels are not rigid but elastic tubes which, during the increased pressure gradient that accompanies increased blood flow, are *passively stretched* to varying degrees; this is most marked in systemic veins and in all pulmonary vessels. This passive increase in radius allows a disproportionate increase in blood flow as pressure increases (Fig. 11.33a). However, smooth muscle in the walls of most systemic arterioles exhibits, without nervously mediated stimulation, spontaneous contractile activity, giving the wall a background

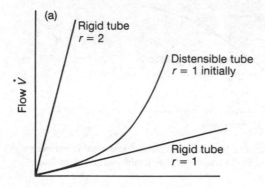

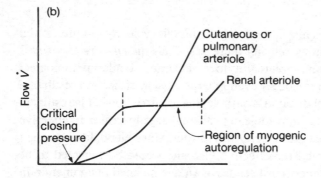

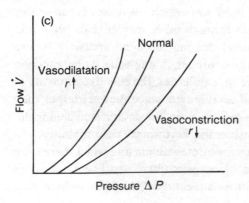

Fig. 11.33. Pressure–flow relationships (a) in rigid tubes of two different radii and in a distensible tube, (b) in arterioles which do and do not exhibit myogenic autoregulation, and (c) in cutaneous arterioles as a function of arteriolar radius. Note the critical closing pressure in (b) and (c).

tension called *myogenic tone*. When these arterioles are passively stretched slightly by an increased pressure gradient, their smooth muscle contracts causing constriction of the vessel. This reduction in radius in some arterioles can exactly match the increase in the pressure gradient such that there is no change in blood flow over a certain range of pressure (Fig. 11.33b). This is referred to as *myogenic autoregulation* of blood flow.

These passive and myogenic changes in radius that cause flow and pressure relationships to deviate from the predictions of the Hagen–Poiseuille equation must not be confused with the reflex autonomic nervous control of smooth muscle in systemic arterioles. This increases or decreases the radius altering flow accordingly. Therefore in most vascular beds, for a particular arteriolar radius, flow is relatively linearly related to the pressure gradient (Fig. 11.33c).

In small blood vessels, especially in systemic arterioles, the pressure–flow curves do not pass through the origin but intersect at a positive pressure (normally about 20 mmHg for an arteriole) called the *critical closing pressure* (Fig. 11.33b, c) below which flow ceases. The explanation for this phenomenon is unknown. The greater the smooth-muscle tone the higher is this critical closing pressure. Under pathological conditions, e.g. in shock, the pressure generated by the heart may not exceed the critical closing pressure of a vessel and, furthermore, concurrent constriction of the arterioles may actually increase the critical closing pressure, making the situation worse.

Other properties of walls of blood vessels

All blood vessels are lined by an endothelial layer and in addition, except in capillaries, there are varying amounts of elastin, collagen and smooth muscle (Fig. 11.34). The first component allows the vessel to stretch and the latter two tend to limit the stretch. This can be examined by taking an isolated segment of a blood vessel tied at both ends and determining the luminal pressure generated at various volumes (Fig. 11.35). As elastin is stretched it exerts *elastic tension* and the more elastin (for example, in an artery) the greater the luminal pressure generated for a given degree of stretch. Thus an artery has a high *elastic coefficient* ($\Delta P/\Delta V$) and a moderate *distensibility* or *compliance* ($\Delta V/\Delta P$). With aging, the major arteries become infiltrated with fibrous tissue and therefore less distensible.

Veins have only a small amount of elastin (Fig. 11.34) and hence in the volume range where the vein wall is actually being stretched (Fig. 11.35) it is less compliant than an artery. However, below this range, the vein is very distensible because the cross-sectional profile changes from the flattened ellipse of low volumes to the circular shape of higher volumes. This high distensibility explains why veins are often said to have a high *capacitance* for volume. The compliance of veins is decreased by sympathetic activity.

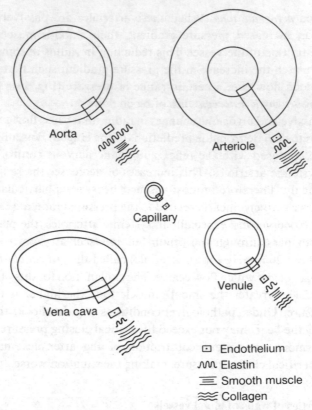

Fig. 11.34. Relative dimensions of internal lumen and wall thickness of blood vessels (arteriole, capillary and venule drawn to a scale 500 × that of aorta and vena cava). The relative proportions of components of the vessel wall are also shown.

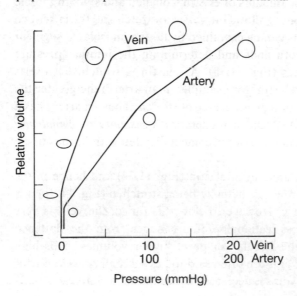

Fig. 11.35. Pressure–volume relationships in isolated segments of a vein and artery.

Systemic and pulmonary arteries are about ten- and twofold, respectively, less distensible than systemic and pulmonary veins. Arterioles and capillaries with little or no elastin have low distensibilities; they are therefore relatively rigid.

Strictly speaking, in Fig. 11.35, one should consider the *transmural* pressure, P_t, (i.e. the pressure on the inside minus the pressure on the outside of the wall) rather than the luminal pressure. Were the tube rigid the transmural pressure would be irrelevant but an elastic tube will distend if the inside pressure is higher or collapse if the outside pressure is higher. The outside pressure is the hydrostatic pressure of the interstitium and is usually close to zero (Section 1.8). Thus the transmural pressure alters the radius of the tube and this affects the *passive wall tension, T,* as described by the *law of Laplace.* For hollow cylinders, as opposed to spherical organs like the heart (p. 409), the other radius of curvature can be ignored since it is infinite for a cylinder and:

$$T=\frac{P_t r}{u} \quad \text{or} \quad P_t=\frac{Tu}{r}.$$

The tension of the wall will depend on the amounts of elastin, collagen and smooth muscle. These and the resulting tension are appropriate for the radius and wall thickness of the vessel and for the transmural pressure to which the vessel is usually subjected (Table 11.3). Note that a similar sized artery and vein have a fairly similar wall thickness but, since the vein is exposed to less transmural pressure, it has to develop less passive tension and appropriately has less elastin than an artery (Fig. 11.34). The thin endothelial cell wall of a capillary has to withstand a much larger transmural pressure than a vein but its very small radius ensures that only a tiny wall tension is required.

If a particular vessel is subjected to an increase in transmural pressure, its distensibility (Fig. 11.35) will result in an increase in the radius and a decrease in the thickness leading to large increases in wall tension. The stretched

Table 11.3. Passive wall tension and wall stress in blood vessels expressed in Nm^{-1} $(T=P_t r)$ and in kNm^{-2} $(T=P_t r/u)$ respectively.

Vessel	Internal radius	Wall thickness	Transmural pressure		Wall tension	Wall stress
			(mmHg)	(kPa)	$(N\ m^{-1})$	$(kN\ m^{-2})$
Aorta	13 mm	2·0 mm	100	13·3	173	86·7
Artery	2 mm	1·0 mm	85	11·3	23	22·7
Arteriole	15 μm	20 μm	65	8·7	0·13	6·5
Capillary	4 μm	1 μm	25	3·3	0·01	13·3
Venule	15 μm	2 μm	15	2·0	0·03	15·0
Vein	2 mm	0·5 mm	10	1·3	2·7	5·3
Vena cava	15 mm	1·5 mm	5	0·7	10	6·7

elastin, collagen and smooth muscle can generate these increased passive tensions within the physiological range of transmural pressures. However, an area of the aortic wall may weaken (for example, in severe arteriosclerotic disease) and become more distensible, resulting in a bulge or aneurysm. The increasing radius and progressive thinning of this bulge will require higher tensions (law of Laplace) from an already weakened wall. Furthermore, the slower velocity of flow at this point will result in a greater lateral pressure. Thus a point of rupture may be reached.

In arterioles and smaller veins, which possess sufficient smooth muscle for their contraction to decrease the radius of the vessel, the reduction in radius and increase in wall thickness will reduce the resulting passive wall stress (law of Laplace). Thus the active tension generated by contraction of the smooth muscle will not elevate the total tension as much as expected and the resultant effect of raising transmural pressure will be dampened.

Relationships between flow, pressure, resistance, cross-sectional area and volume throughout the vascular system

The relationships of these variables have been considered for series and parallel arrangements of vessels (Fig. 11.31) and will now be applied to the vascular system (Fig. 11.36). The *parallel branching* of the blood vessels is such that there is a rise in the *cross-sectional area*, greatest in the capillaries and of moderate proportions in the venules and small veins. The *percentage of total blood volume* that is accommodated in each set of vessels is determined by the cross-sectional area and the *length* of individual vessels. Total blood volume is about 5–6 and 4–5 litres, respectively, in average men and women (p. 316). In the supine position at rest, about 75% of the blood is in the systemic circuit, about 8% in the heart and about 16% in the pulmonary circuit. The volume in the heart and pulmonary circuit is referred to as the *central blood volume*. The aorta and systemic arteries contain about 12% of the total blood volume with only about 3% in the systemic arterioles and about 6% in the systemic capillaries for, despite their large cross-sectional area, these are nevertheless very short. Most of the blood (about 55%) is accommodated in the systemic venous system indicating its importance as a *blood reservoir*. In the pulmonary circuit, about 5% of the total blood volume is in the arterial system, 3% in the capillaries and 8% in the veins.

During standing at rest, the cross-sectional areas are slightly different (Section 11.6) and about 6% of the total blood volume is in the heart and about 9% in the pulmonary circuit, resulting in an increase to about 65% in the systemic venous system. During exercise (Section 16.3), there will be a greater proportion of blood in systemic capillaries, venules and, to a lesser extent, arterioles (the increase being in the blood supply of skeletal musculature), as

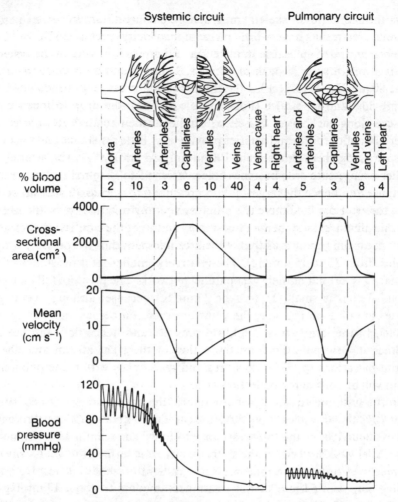

Fig. 11.36. Relationships amongst % blood volume, cross-sectional area, mean velocity and blood pressure in the systemic and pulmonary circuits at rest in the supine position.

well as in the heart and in the pulmonary circuit. The proportion in the systemic venous system will be reduced correspondingly.

The *velocity* in a particular vessel will be the volume flow rate divided by the cross-sectional area. Flow is *pulsatile* in the aorta and arteries and hence velocity changes but the mean velocity is about 20 cm s^{-1} at rest. Flow becomes *non-pulsatile* and very slow at a velocity of about 0·05 cm s^{-1} in the capillaries (Fig. 11.36). This low velocity is vital in that it allows time for adequate diffusion between blood and the cells across the capillary wall. Since large veins have a cross-sectional area about twice that of the aorta the mean velocity increases to only 10 cm s^{-1}. During exercise, increases in cardiac output can cause up to fivefold increases in velocity in the arteries and veins but in the capillaries of skeletal musculature a concomitant increase in cross-sectional area results in very little increase in velocity (Section 16.2).

Blood pressure is pulsatile in the aorta and arteries, having in the systemic circuit at rest a systolic peak of about 120 mmHg and a diastolic trough of about 80 mmHg (Figs 11.19 & 11.36). Since the pressure depends upon the volume flow and resistance (Poiseuille's equation), the drop in mean blood pressure along each set of vessels will indicate their relative *resistance*. The larger radii of the aorta and arteries provide very little resistance and the mean pressure along them drops from 100 to only 95 mmHg. As the arteries get smaller the pressure drop becomes bigger (from 95 to 75 mmHg). The greatest resistance and hence the largest fall in pressure (from 75 to 35 mmHg) occurs along the arterioles and there is a progressive transition from pulsatile to non-pulsatile pressure. Despite the even smaller radius of the capillaries, their total resistance is only about half that of the arterioles and thus the pressure drop is smaller (from 35 to 15 mmHg). The relatively smaller total resistance of the capillaries is a result of their vast parallel network. The pressure fall along the venous system is small (15 to 1 or 2 mmHg) because, although veins and arteries are of a similar size, the veins are more numerous and arranged in parallel. In the large veins, the blood pressure and hence flow and velocity becomes slightly pulsatile from the action of the right atrium and nearby arterial pulsations (p. 398). Pressures and resistances within the pulmonary circuit will be considered in Section 11.5.

In the systemic circuit at rest, therefore, the aorta and arteries constitute about 20%, the arterioles about 50%, the capillaries about 20% and the venous system about 10% of the total resistance to flow. The combined resistance of the parallel vascular beds of the systemic circuit is termed the *total peripheral resistance*. At rest, with a total pressure gradient of about 100 mmHg and a cardiac output of $6 \, l \, min^{-1}$, this resistance amounts to about $17 \, mmHg \, l^{-1}$ min. In the pulmonary circuit, by comparison, with a total pressure gradient of about 10 mmHg, there is a *pulmonary resistance* of about $1.7 \, mmHg \, l^{-1}$ min.

During exercise the profiles in Fig. 11.36 will change because the increased force of ventricular contraction and the consequent increase in cardiac output will elevate the mean arterial blood pressure (Section 16.3). There is also an increase in the radius of arterioles supplying skeletal musculature. This increase dominates the systemic response resulting in a decrease in total systemic arteriolar resistance and hence a smaller drop in pressure along the arterioles. This will elevate pressures in the capillaries and venous system.

Characteristics of the systemic arterial circulation

Since the blood enters the aorta from the heart only during systole, flow tends to be turbulent and the aortic and arterial blood pressures are pulsatile (Figs

11.19 & 11.36). Although two-thirds of the stroke volume is ejected during the first third of systole, which results in aortic flow and velocity reaching a peak early in systole, the peak of the blood pressure pulse occurs later (Fig. 11.37). This delayed peak in pressure is because the aorta is not a rigid tube and the pressure pulse is modified by the elastic stretching of the aorta. The aorta stretches in systole to accommodate about 50% of the stroke volume whilst the other 50% flows on into the peripheral blood vessels (Fig. 11.38). As the elastin of the aorta stretches, the kinetic energy of liquid motion is converted into potential energy. In diastole, as the elastic recoil converts potential energy back to kinetic energy, the aortic pressure does not drop to zero but falls gradually to a minimum of about 80 mmHg. This relatively high pressure during diastole ensures that the blood accommodated in systole is propelled out to the periphery during diastole. The pulsatile blood propulsion from aorta to the periphery results from the sequence of expansion followed by recoil of the next section of the aorta (Fig. 11.38) and later of the arteries. This elastic recoil converts the intermittent flow from the heart into a continuous, albeit pulsatile, flow through the arterial system.

The *pressure pulse* is transmitted through the arteries with a velocity considerably greater than the forward movement of blood itself (the flow pulse). The latter has a mean velocity of about 20 cm s^{-1} in the aorta and about 15 cm s^{-1} in a small artery (Fig. 11.36). In comparison, the pressure pulse has a very high velocity of about 4 m s^{-1} in the aorta reaching about

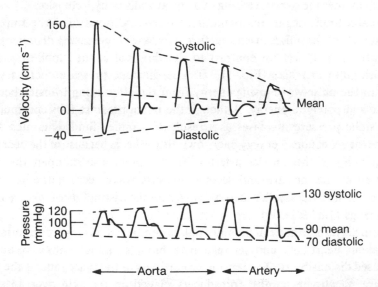

Fig. 11.37. Changes in velocity (flow pulse) and in pressure (pressure pulse) in the arterial system at increasing distances away from the heart. (After McDonald, D.A. (1974) *Blood Flow in Arteries*, 2nd edn. Edward Arnold, London.)

Aortic valve

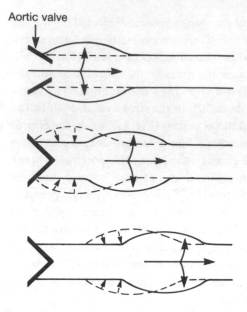

Fig. 11.38. Progression along aorta of the alternating sequence of elastic stretching, accommodating about 50% of the stroke volume, followed by elastic recoil propelling blood to the next segment. Solid arrows = movement of blood; dashed arrows = elastic recoil.

12 m s^{-1} in small arteries. This pressure pulse (pulse-wave) is transmitted through the column of blood and along the vessel walls and its velocity is higher the less viscous the blood, the greater the mean blood pressure, the more rigid or thicker the vessel wall and the smaller the lumen radius. With increasing age, the greater wall stiffness results in an increased pulse-wave velocity as does the overstretching of arterial walls in hypertension.

Because branching increases the arterial cross-sectional area, the amplitude and velocity of the pulses in arterial flow decrease at increasing distances from the heart (Fig. 11.37). In contrast to the flow pulse, the amplitude of the pressure pulse increases (Fig. 11.37). The diastolic pressure decreases as a result of the pulse-wave losing energy from the alternating transfer between kinetic and potential energy. Although the mean pressure falls continuously, the systolic pressure increases as a result of complex fluid dynamics which include the reflection of energy back towards the heart because of the decreased distensibility of the smaller arteries. Such factors also dampen the sharp vibration of the incisura and second pressure wave seen in the aorta and convert them to the smaller *dicrotic notch* and the distinct *dicrotic wave* of the artery (Figs 11.37 & 11.39).

Clinically, these arterial pressure pulses can be felt, giving information not only about heart rate and its regularity but also about stroke volume or decreased distensibility. The tension or hardness of the pulse reflects the pulse pressure. Electromechanical transducers placed on the skin over an artery yield details about the shape or contour of the pulse and provide more precise clinical information.

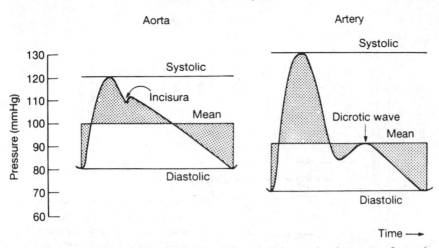

Fig. 11.39. Determination of mean arterial blood pressure by integrating the pressure fluctuation against time, i.e. the shaded areas above are equal to the shaded areas below the mean pressure.

Measurement of arterial blood pressure

The maximum to which the pressure rises is called the *systolic blood pressure* (at rest the range is 100–140 mmHg at 20 years of age) and the minimum is the *diastolic blood pressure* (range 50–90 mmHg). With increasing age, systolic, in particular, and diastolic pressures increase due to loss of arterial elasticity. The difference between systolic and diastolic pressure is the *pulse pressure*. The *mean arterial pressure*, which is the driving force for blood flow, is determined by integrating the change in pressure against time (Fig. 11.39). Different pressure contours in the arterial system mean that for the aorta the mean arterial pressure (about 100 mmHg) is approximately the arithmetic mean of the systolic and diastolic pressure or, expressed another way, the diastolic pressure plus half the pulse pressure, whereas in a peripheral artery it is approximately the diastolic pressure plus one-third of the pulse pressure (about 95 mmHg).

These pressures can be measured directly by inserting a fluid-filled cannula into the appropriate artery and recording with an electrical pressure transducer. Clinically, they are measured indirectly by a *sphygmomanometer* (Fig. 11.40). This comprises an inflatable rubber cuff covered by a layer of non-distensible fabric which is usually wrapped around the upper arm at the level of the heart and attached to a mercury manometer. The cuff pressure is altered by pumping air into the cuff or releasing it through a needle valve. A stethoscope is placed distal to the cuff over the brachial artery in the antecubital fossa. The cuff is then inflated to a pressure higher than the expected systolic pressure compressing the blood vessel so that no blood flows through the artery. When the cuff pressure is slowly released, varying sounds (*Korotkoff sounds*) audible

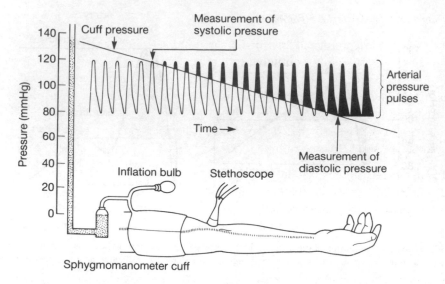

Fig. 11.40. Use of sphygmomanometer and stethoscope to measure indirectly systolic and diastolic pressure at the appearance and disappearance, respectively, of the Korotkoff sounds. (After Rushmer, R.F. (1970) *Cardiovascular Dynamics*, 3rd edn. W.B. Saunders Co., Philadelphia.)

with the stethoscope result from the intermittent and turbulent flow of blood through varying degrees of constriction of the brachial artery. Four phases of sound followed by a fifth phase of silence can be distinguished.

When the cuff pressure has fallen to just below the systolic pressure, a clear, but often faint, tapping sound suddenly appears in phase with each cardiac contraction. The tapping sound is produced by the transient and turbulent blood flow through the artery during the peak of each systole. The systolic pressure is defined as that cuff pressure at which the tapping sound is first heard. During the next ~15 mmHg fall in the cuff pressure, the tapping sound becomes louder (phase I). In the following ~20 mmHg fall, the sound becomes quieter with a murmuring quality (phase II) and may suddenly disappear in the latter part of this phase (the auscultatory gap). In the next ~5 mmHg fall in cuff pressure, the sound of the murmuring becomes very loud and thumping (phase III). In the following ~5 mmHg fall, the sound becomes muffled and rapidly grows fainter (phase IV) and finally the sound disappears (phase V). When blood flow velocity is high, for example in exercise, the beginning of phases IV and V may be separated by 40 mmHg or more. The beginning of phase IV (muffling) and of phase V (disappearance) are used to measure diastolic pressure. The diastolic pressure is usually defined as the cuff pressure at which muffling, not disappearance, occurs. However, if there is an obvious difference at rest between these, both values are reported.

Determinants of arterial blood pressure

The relative importance of any single factor in determining the arterial blood pressure is most simply examined if all other factors are held constant (Fig. 11.41). In general, systolic pressure is affected mainly by stroke volume and in particular by ejection velocity, and diastolic pressure by total peripheral resistance and the time allowed for blood to flow out of the arteries (i.e. the duration of diastole which is determined by the heart rate). In greater detail, systolic pressure is increased by:

1 an increase in the diastolic pressure of the previous pulse;
2 an increase in stroke volume;
3 an increase in ejection velocity (without a change in stroke volume); or
4 a decrease in aortic or arterial distensibility.

Diastolic pressure is increased by:

1 an increase in the systolic pressure of that particular pulse;

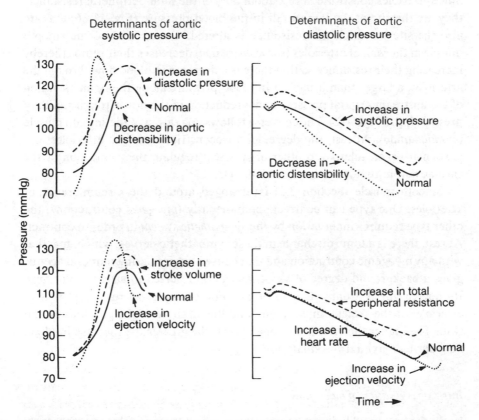

Fig. 11.41. Determinants of systolic and diastolic pressure in the aorta. Effect of each factor shown when all other factors held constant.

2 a decrease in ejection velocity;
3 an increase in aortic or arterial distensibility;
4 an increase in heart rate; or,
5 an increase in total peripheral resistance.

The resultant changes in pulse pressure and mean arterial pressure will depend on the direction and relative magnitude of the individual changes in systolic and diastolic pressures.

The numerical values given above for arterial blood pressure (and pressures elsewhere in the vascular system) are for the supine position when the vessels are at the level of the heart. In the vertical position the pressures are affected by gravity (Section 11.6).

Circulation through the arterioles and the regulation of blood flow in the systemic circuit

Since arterioles constitute at rest about 50% of the total peripheral resistance, they are the site of the largest fall in the blood pressure (Fig. 11.36) and are also the site at which the resistance is altered. Contraction of the smooth muscle in the wall of arterioles (*vasoconstriction*) decreases their radius thereby increasing their resistance so that there is a decrease in blood flow through the arterioles, a larger than usual fall in blood pressure along them, an elevation of the upstream arterial pressure and a reduction of the downstream capillary pressure (pp. 375 & 412). The converse follows relaxation of the smooth muscle (*vasodilatation*). At rest, the degree of vasoconstriction of the arterioles to a particular organ relative to other organs will regulate the proportion of the cardiac output that it receives (Table 11.1).

Smooth muscle (Section 3.2) is arranged around the circumference of arterioles. One type can contract spontaneously (*myogenic* contraction); the other type requires innervation by the *sympathetic system* in order to contract. At rest, there is a tonic discharge in these sympathetic nerves, which, together with any myogenic contraction and that caused by circulating humoral agents, gives a background degree of vasoconstriction referred to as *vasomotor tone*. Its degree varies from organ to organ. For example, at rest it is high in arterioles of the skeletal musculature and low in those of the gut, kidney and skin. The higher the vasomotor tone at rest, the greater the increase in blood flow during maximal vasodilatation.

Intrinsic regulation of blood flow

1 In the brain and kidney, the smooth muscle of the arterioles spontaneously contracts when the pressure increases and relaxes when the pressure decreases

(Fig. 11.33), which allows a blood flow that is independent of variations in arterial blood pressure. This *myogenic autoregulation* is less well developed, or overriden by other mechanisms, in the heart, gut and skeletal musculature. It does not occur in the skin.

2 All organs need to increase their blood flow in proportion to their metabolic requirements. During increased metabolism, there is a decrease in the partial pressure of O_2 (P_{O_2}), an increase in the partial pressure of CO_2 (P_{CO_2}) and an increase in H^+ concentration in the interstitial fluid. These changes cause the smooth muscle to relax to a degree that is appropriate to the increased metabolism. This is called *metabolic autoregulation*. Increases in local temperature and in the local concentrations of other metabolites, such as ATP, ADP, AMP, adenosine, lactate and pyruvate, as well as increases in the K^+ concentration and interstitial osmolarity of exercising skeletal musculature, have been said to cause vasodilatation. Metabolic autoregulation is well developed in the skeletal musculature, heart and brain, but in other organs it can be overridden by nervous control of the arterioles. It may contribute to the phenomenon of myogenic autoregulation inasmuch as an increase in pressure by increasing blood flow, in the face of constant metabolism, will lower tissue metabolites and result in vasoconstriction.

Small changes in the metabolism of an organ can be satisfied by metabolic autoregulation without affecting significantly the total peripheral resistance. However, the metabolism of an organ may increase so markedly that there will be a fall in total peripheral resistance. The resultant initial fall in mean arterial blood pressure will then elicit cardiovascular nervous reflexes (Section 11.4) to increase cardiac output and restrict the blood flow to other organs in order to maintain the mean arterial blood pressure.

When the blood supply to an organ is temporarily obstructed (for a period of seconds up to a few minutes), its restoration is accompanied by a large increase in blood flow (*reactive hyperaemia*) which depends on the duration of the obstruction and the metabolism of the organ over that time. The vasodilatating effects of the metabolites that have accumulated during the obstruction contribute to this reactive hyperaemia.

An increased P_{CO_2} (hypercapnia) or a decreased P_{O_2} (hypoxia) in arterial blood (Section 12.5) will cause vasodilatation of all systemic arterioles. However, hypercapnia and hypoxia, especially the former, will also elicit cardiovascular reflexes (Section 11.4) which oppose this vasodilatation.

3 In addition to metabolites, other chemicals released locally in an organ also affect its blood flow—*humoral control*. The salivary, gut and sweat glands when activated produce not only their exocrine secretions but also an enzyme kallikrein (p. 363). This converts plasma kininogens into active *kinins*, such as kallidin and *bradykinin*, which have marked vasodilating effects on the gland as well as locally within the skin or gut. These kinins are inactivated by other tissue enzymes. During inflammatory or allergic reactions, kinins are also liberated from tissues and the vasodilator *histamine* is released from basophils and mast cells (p. 341). When a blood vessel is cut, *serotonin* and the prostaglandin derivative *thromboxane A_2* are released from activated platelets and cause vasoconstriction (p. 355). On the other hand, some prostaglandins, e.g. prostacyclin released from endothelial cells, cause vasodilatation (p. 360).

Extrinsic regulation of blood flow

1 *Nervous control.* Since there is a resting vasomotor tone, an increase in *sympathetic* discharge to the arterioles causes further vasoconstriction whilst a decrease in discharge causes vasodilatation. The sympathetic neurotransmitter, *noradrenaline*, acts powerfully on α *adrenoceptors*, on the plasma membranes of vascular smooth muscles to cause contraction. (Noradrenaline also acts weakly on the β_2 adrenoceptors of vascular smooth muscle, the activation of which causes relaxation. However, the strong α response is dominant.) Arterioles of the skin, gut, skeletal musculature and kidneys receive a dense sympathetic innervation whereas those of the brain, and to some extent of the heart, are sparsely innervated. This indicates that there is little nervous control of the circulation to these latter organs (Section 11.5).

The vasoconstriction resulting from a certain level of sympathetic discharge is considerably greater in the skin, kidney and gut than in the skeletal musculature. Thus the blood flow to the skin, kidney and gut can be restricted in favour of other organs. This sympathetic pathway is controlled by the cardiovascular centres in the medulla oblongata and pons and operates, for instance, during the baroreceptor reflex to maintain mean arterial blood pressure (Section 11.4).

A special system of sympathetic nerves controlled from the motor cortex and hypothalamus innervates the arterioles of the skeletal musculature. It releases acetylcholine at the postganglionic nerve ending and causes vasodilatation. This *sympathetic cholinergic pathway* is usually silent but is activated during emotional reactions of alarm, rage or fear and during the initial phases of exercise.

The external genitalia receive a vasodilatating *parasympathetic* and a vasoconstricting sympathetic innervation and this dual innervation is integrated in the spinal cord. Parasympathetic nerves to the arterioles of the brain, heart and lung exist but their functional significance is not clear.

2 *Hormonal control.* The adrenal medulla continuously secretes *adrenaline* and a small amount of noradrenaline. The level of secretion by the adrenal medulla is proportional to sympathetic activity which is under the control of the hypothalamus (Section 8.5). Adrenaline activates to an equal extent the α-*adrenoceptors* resulting in vasoconstriction and the β_2-adrenoceptors resulting in vasodilatation. The net arteriolar response of a particular organ to adrenaline depends on the relative densities of α- and β_2-receptors. The density of the β_2-receptors is higher in skeletal musculature and in the heart. Thus during exercise the increased release of adrenaline contributes to the increased blood flow to the skeletal musculature and the heart whilst decreasing the blood flow to the skin, gut and kidney.

The hormones angiotensin II and antidiuretic hormone (vasopressin) also cause vasoconstriction, while atrial natriuretic peptide decreases the sensitivity of vascular smooth muscle to these and other vasoconstrictor substances. However, the cardiovascular function of these hormones is primarily concerned with the control of blood volume and the maintenance of a normal arterial blood pressure (p. 443).

Circulation through the capillaries

The entrances from the arterioles (or metarterioles) into the vast network of capillaries are guarded by a ring of smooth muscle, the *precapillary sphincter* (Fig. 11.42). These sphincters exhibit *myogenic rhythmicity* which results in intermittent and variable flow rates through any individual capillary. The

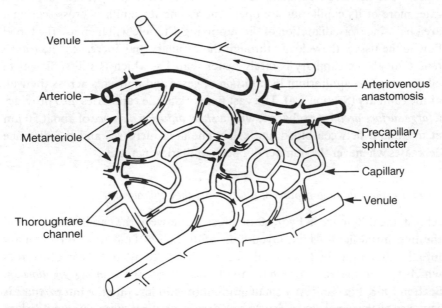

Fig. 11.42. Schematic diagram of a capillary network and its associated vessels. Thick walls denote considerable amounts of smooth muscle.

direction of flow may change in some capillaries depending on both their location within the capillary bed and the degree of constriction of nearby sphincters. This can be seen by examining, for instance, the capillary bed of the rabbit's ear. The net degree of constriction or dilatation of these sphincters is controlled by *metabolic autoregulation*. For example, in skeletal musculature at rest, at any one moment about 10% of the capillaries are 'open' and the rest contain blood that is stationary or they may even be completely empty. Blood bypasses the capillaries and flows through the *metarterioles* which contain little smooth muscle and through special capillaries called *thoroughfare channels* (Fig. 11.42). In the skin and gut, short, relatively large-diameter vessels called *arteriovenous anastomoses* act as shunts between arterioles and venules (Fig. 11.42). They have thick walls of smooth muscle controlled by sympathetic nerves acting on α adrenoceptors. Note that the capillaries themselves are not contractile nor are they distensible.

The parallel arrangement of the narrow (4 μm radius) capillaries, their contribution (20%) to the resting total peripheral resistance, the blood pressure drop along them, the absence of pulsatile flow, their large cross-sectional area, the low velocity of flow in the capillary bed and the low flow rate through an individual capillary have already been described (Figs 11.31b & 11.36). Under resting conditions, only about 25% of the capillaries are open giving a cross-sectional area of about 3000 cm^2. Since the average length of a capillary is about 0·1 cm and the velocity of blood flow is 0·05 cm s^{-1}, blood will take about 2 s to traverse a capillary at rest. When a tissue increases its metabolic rate, more of its capillaries are open, increasing its capillary cross-sectional area but, since vasodilatation of the supplying arteriole will increase the blood flow to the tissue, the velocity through the capillary may increase. The *transit time* through the capillary is rarely faster than 1 s, which is still sufficient to allow complete equilibrium by *diffusion* of gases and nutrients across the wall of the capillary (Section 1.4). The vast network of the capillaries also provides a *large surface area* for exchange, with a *short diffusion distance* of about 50 μm at maximum between blood and cells. The area increases and the distance decreases whenever the number of open capillaries increases.

Ultrafiltration and reabsorption of fluid across the capillary wall

At rest, the difference in hydrostatic pressure between the capillary blood and the interstitial fluid, which favours ultrafiltration of fluid from the capillary into the interstitial fluid, is virtually counterbalanced by colloid osmotic forces which favour the return of fluid to the capillary (the *Starling equilibrium*, Section 1.8 & Fig. 1.6). Any small amount of fluid lost into the interstitium is removed by the *lymphatic system*. In a variety of circumstances, this equilibrium may be disturbed, resulting in fluctuations in the circulating *blood volume*.

Excess ultrafiltration (Fig. 11.43) will occur with:

1 arteriolar vasodilatation;

2 an elevated arterial pressure (e.g. hypertension);

3 an elevated venous blood pressure (e.g. in the erect posture or in cardiac failure);

4 decreases in plasma osmotic pressure (plasma protein deficiency); and

5 increases in interstitial protein concentration (inadequate lymph drainage or the actions of histamine and kinins which increase the protein permeability of the capillary).

However, excessive increases in the volume of the interstitial fluid (*oedema*) are usually limited by factors that will increase lymphatic flow.

An excess of reabsorption (Fig. 11.43) will result mainly from:

1 arteriolar vasoconstriction;

2 a decreased venous blood pressure; and

3 dehydration which will increase plasma protein concentration.

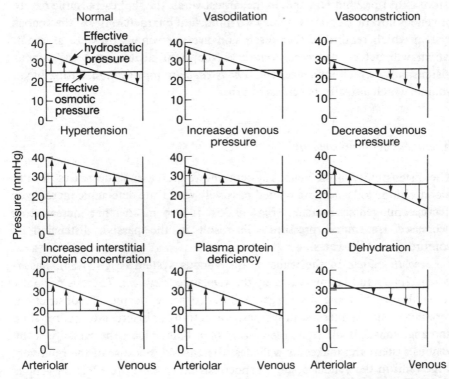

Fig. 11.43. The balance of ultrafiltration (↑) and reabsorption (↓) of fluid across the capillary walls under various conditions. Effective hydrostatic pressure is the difference in hydrostatic pressure between capillary blood and interstitial fluid and effective osmotic pressure is the difference in colloid osmotic pressure between capillary blood and interstitial fluid. See also Fig. 1.6.

In haemorrhage, increases in fluid reabsorption will occur indirectly as a response to the lowered capillary blood pressure and directly as a reflex response to haemorrhage via the vasoconstriction induced by increased sympathetic activity and increased circulating angiotensin II and antidiuretic hormone. Such responses will help to restore the blood volume by drawing on the reservoir (Section 1.2) of the extracellular fluid in the interstitium, principally from the skeletal musculature.

Characteristics of the systemic venous circulation

The following characteristics have already been described (Fig. 11.36) viz., the cross-sectional area of the systemic venous system in comparison with arteries and capillaries, the fact that the veins accommodate 55% of the blood volume when supine, the moderate venous velocity, the venous contribution of 10% to the total peripheral resistance, the resulting small pressure gradient (from 15 to 1 or 2 mmHg) and, in the largest veins, the slightly pulsatile nature of venous flow (a, c and v waves). Two important characteristics of the venous system which require further discussion are the venous pressure gradient, which will determine *venous return* to the right atrium, and the venous distensibility which influences the *venous capacity* for accommodating blood volume which also affects venous return.

Alterations in venous capacity

The distensibility of the venous system allows increases or decreases in total blood volume to be accommodated passively with little detectable increase or decrease in venous pressure (Fig. 11.35). It also means that increased or decreased transmural pressure will result in the passive distension or compression of the veins.

Smooth muscle in the walls of the venous system (Fig. 11.34), has α-*adrenoceptors* and is innervated by the *sympathetic system*. There is normally some degree of tonic discharge and hence some degree of constriction, *venomotor tone*. An increase in sympathetic activity or in circulating adrenaline and angiotensin II will cause *venoconstriction*. This reduces the capacity of the venous system and makes its walls less distensible. Because of the relatively large radii in the venous system, venoconstriction has little effect on venous resistance or total peripheral resistance. Conversely, a decrease in sympathetic activity to the veins results in *venodilatation* and an increase in venous capacity. Venoconstriction occurs as part of the reflexes coordinated by the cardio-vascular centres in response to haemorrhage or during exercise.

Determinants of venous return

The pressure gradient through the venous system is the primary determinant of venous return. This pressure gradient is the difference between *mean systemic filling pressure* (MSFP) and *mean right atrial pressure*.

The MSFP was defined by Guyton in the 1950s as the weighted average of the pressures in all portions of the systemic circulation; the weighting is in proportion to the volume capacity of the vessel. It is also the static blood pressure which prevails throughout the systemic circulation if the heart is suddenly arrested experimentally and rapid equilibrium of pressures in arterial and venous circuits has occurred. The MSFP is usually about 7 mmHg. In the dynamic situation, this pressure is reached in about the middle of the systemic venous circuit and thus the MSFP is approximately equal to the *mean venous pressure*. The MSFP is increased by an increase in total blood volume and by an increase in venomotor tone which reduces venous compliance and venous capacity. Although arteriolar vasoconstriction will decrease capillary and venous pressures (p. 375), reduce cardiac output and hence venous return, it will not alter MSFP. The maximum MSFP is about 20 mmHg and the minimum MSFP can be close to zero.

The mean pressure in the right atrium, often referred to as *central venous pressure*, is about 1 mmHg. It is elevated slightly in the supine position compared with the standing position and when total blood volume is increased. It can be elevated appreciably during right heart failure (Section 11.6).

Secondary factors also influence venous return. Because the right atrial pressure fluctuates in phase with the heart rate (Fig. 11.19), this results in fluctuations in venous return. The increase in atrial pressure during atrial contraction reduces venous return whilst both the downward movement of the atrioventricular ring during early systole and the rapid filling of the ventricle after the AV valve opens during early diastole decrease the atrial pressure, often to negative values, and hence increase venous return. This so-called '*suction effect of the heart*' therefore aids venous return and is greater during increased cardiac activity.

Since the atria are influenced by the subatmospheric pressure of the thorax (the intrathoracic or intrapleural pressure) which at rest is about -3 mmHg during expiration and about -5 mmHg during inspiration (Section 12.2), there is a further fluctuation in atrial pressure that is synchronous with respiration. Thus during inspiration atrial pressure decreases and causes a greater systemic venous return to the right atrium. Such an effect of increasing pulmonary venous return to the left atrium is opposed by the intrapleural pressure of inspiration causing a passive dilatation of pulmonary blood vessels. This increases the capacity of the pulmonary vessels and hence transiently during inspiration decreases venous return to the left atrium. The converse changes happen during expiration. This explains the respiratory-induced changes in preload to the right and left ventricles and hence in their stroke volumes (p. 405).

Venous return is also assisted by the effect of intrapleural and abdominal pressures on the intrathoracic and intra-abdominal veins, respectively, the so-called *respiratory venous pump*. Since veins have distensible walls, the fluctuations in transmural pressure in the thorax cause dilatation of intrathoracic veins during inspiration and their compression during expiration. The descent of the diaphragm during inspiration will, at the same time, raise intra-abdominal pressure and compress the abdominal veins; the converse will occur in expiration. Thus the respiratory pump, within limits, aids venous return since the relative dilatation of intrathoracic veins in inspiration will decrease their resistance to flow and cause some movement of blood towards the heart from the abdominal veins or upper extremities. The simultaneous compression of abdominal veins propels blood towards the heart since retrograde flow from the abdomen into the limbs is prevented by valves in the limb veins. *Venous valves* are thin cup-like structures whose cusps obstruct the lumen at the threat of any retrograde flow and thus ensure unidirectional flow. They occur only in veins of the limbs. In expiration, relative dilatation of the abdominal veins causes some movement of blood into the abdomen from the lower limbs. Whenever the depth of breathing increases, for instance during exercise, the effect of the respiratory pump on venous return is enhanced.

During exercise, a further mechanism, the *skeletal muscle venous pump*, aids venous return. The exercise must not be a sustained contraction but one of alternating contraction and relaxation. The contraction of the skeletal musculature alters the transmural pressure compressing the veins within it. This propels blood towards the heart since retrograde flow is prevented by the venous valves. When the skeletal musculature relaxes, the veins are dilated by inflow but only from below since the venous valves prevent retrograde flow. The further advantages of this pump during the gravity effects of standing will be considered later (Section 11.6).

Relationships between cardiac output, venous return and atrial pressure

The separate considerations of the control of cardiac output (in particular of stroke volume, p. 404) and of venous return can now be combined in a graphical approach designed by Guyton in the 1950s. It must be noted that these graphs represent only steady-state responses and not momentary changes and that the use of them requires some simplification of concepts and abstraction of ideas.

Earlier, in Fig. 11.25, the relationship between stroke volume and end-diastolic volume (the Starling curve) was described together with the concept that they could be considered as similar to cardiac output and mean atrial pressure, respectively. This relationship is referred to as the *cardiac or ventricular function curve* (Fig. 11.44a). The position of function (normally at a in Fig. 11.44a) moves up the curve when venous return increases since this results in a higher mean atrial pressure and hence a larger end-diastolic volume. Conversely, the position of function moves down the curve when venous return decreases. The curve becomes steeper during increased sympathetic stimulation of the ventricle (i.e. an increase in myocardial contractility, cf. Fig. 11.29) or during a decreased afterload in the arterial

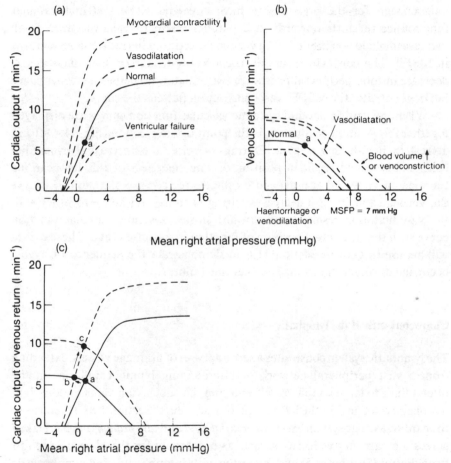

Fig. 11.44. (a) Ventricular function curves under different conditions; a = normal position of function with a cardiac output of 6 l min^{-1} and a mean right atrial pressure of 1 mmHg. This position moves with changes in venous return. (b) Systemic vascular function curves under different conditions; a = normal position of function, MSFP = mean systemic filling pressure. (c) Normal (resting) ventricular and systemic vascular function curves displayed together; a = normal position of function. Dashed lines and position of function (b & c) occur during moderate exercise (see text).

system (i.e. vasodilatation). A flatter curve occurs during the converse conditions or during ventricular failure or disease of the heart valves.

The equivalent relationship between venous return and mean atrial pressure is referred to as the *systemic vascular function curve* (Fig. 11.44b). As atrial pressure increases venous return is opposed whilst there is a limit to the possible increase in venous return caused by negative atrial pressures because they will also result in a tendency for collapse of the venae cavae in the chest. At a venous return of zero the atrial pressure in this graph is +7 mmHg, that is, the pressure (*mean systemic filling pressure*, MSFP) throughout the cardiovascular system when the circulation has been stopped experimentally long enough for the pressures to have equalized. The systemic vascular function curve shifts upwards in a parallel manner whenever there is an increase in blood volume or during venoconstriction. This indicates an increase in MSFP. The converse occurs during a haemorrhage or venodilatation. A decrease in total peripheral resistance (vasodilatation) causes a steeper curve but does not alter the MSFP; vasoconstriction flattens the curve.

When the ventricular and systemic vascular function curves are displayed together the position of function is the point of intersection (Fig. 11.44c) and at rest in health this is at a. During exercise, the increased myocardial contractility would result in position b. The steepness of this new cardiac function curve would also depend on the degree of vasodilatation. However the venous capacity also decreases during exercise and hence together with the vasodilatation results in an upward shifted systemic vascular function curve such that the actual position of function in exercise is at c. These curves will be used again in Section 11.6 to demonstrate the sequence of events occurring during haemorrhage and cardiac failure.

Characteristics of the lymphatic system

The lymphatic system constitutes a secondary set of drainage vessels, extending from a vast peripheral network of blind-ending lymph capillaries in the interstitium to large lymph vessels entering the subclavian veins via the right lymphatic duct and on the left via the thoracic duct (Section 1.8). It returns to the cardiovascular system *excess interstitial fluid* and any *protein* that has leaked across the capillary wall. It also acts as a pathway for the *absorption of fats* from the gut (Section 14.5) and, by virtue of its lymph nodes and lymphocytes, is involved in *immune responses* (Section 10.3).

Lymph flows sluggishly at a rate of about 2 litres/day. The pressure gradient driving this flow depends on the *interstitial hydrostatic pressure*, the magnitude of which is proportional to the amount of excess interstitial fluid. Lymph flow is aided by the *myogenic* rhythmic contractions of *smooth muscle* in the walls of lymph vessels, retrograde flow being prevented by *lymphatic valves* similar to those found in veins. During exercise, the rate of lymph flow can increase up to tenfold as a result of rhythmic contractions of skeletal musculature around the lymph vessels. Similarly, increased contractions of gut musculature will increase lymphatic return from the intestine. Furthermore, rhythmic alterations in the transmural pressure resulting from the cardiac and respiratory cycles will aid lymphatic return from the heart and lungs, respectively.

11.4 Integrated regulation of the cardiovascular system

The *short-term regulation* of the cardiovascular system is concerned with the adjustment of three variables—cardiac output ($\dot{Q}$), total peripheral resistance (*TPR*) and mean arterial blood pressure (*MAP*). Compensatory changes occur within seconds following perturbations such as a change in posture, haemorrhage or exercise. These compensations are dependent on a variety of sensory receptors, on the integration of the sensory information by the cardiovascular centres in the medulla oblongata and pons, which are also influenced by higher brain centres, and on the autonomic motor pathways to the heart and blood vessels. This autonomic (extrinsic) control interacts with the intrinsic control mechanisms operating within the heart and blood vessels.

The relationship $\dot{V} = \Delta P / R$, used previously in considering flow through a single vessel, must also apply to the systemic circuit as a whole. Here

$$\frac{\text{cardiac}}{\text{output}} = \frac{\text{mean arterial blood pressure} - \text{mean right atrial blood pressure}}{\text{total peripheral resistance}}.$$

Since mean right atrial pressure is close to zero, this equation can be condensed to

$$\dot{Q} = \frac{MAP}{TPR}.$$

Both $\dot{Q}$ and *TPR* are controlled by a combination of intrinsic and extrinsic mechanisms. In Section 11.2, the control of $\dot{Q}$ was considered in terms of (a) the control of stroke volume by the magnitude of the end-diastolic volume (Starling's law of the heart) and by sympathetic activity, and (b) the control of heart rate by parasympathetic and sympathetic activity.

In addition, extrinsic control of venous capacity by sympathetic nerves and circulating adrenaline and passive changes in venous capacity will alter end-diastolic volume and hence influence the intrinsic control of stroke volume. In Section 11.3, the control of *TPR* was considered, in particular, in terms of intrinsic control by metabolic autoregulation and extrinsic control by sympathetic nerve fibres and adrenaline. There is no direct control of *MAP*. It is a dependent variable, altered only indirectly by altering $\dot{Q}$ or *TPR* or both (Fig. 11.41). In contrast, there are no sensory receptors which directly monitor $\dot{Q}$ or *TPR* but there are sensory receptors, *arterial baroreceptors*, which directly monitor *MAP*.

Arterial baroreceptor reflexes

Baroreceptors are located in the walls of the aortic arch and in the enlarged part (the *carotid sinus*) of the internal carotid artery just after it arises from the common carotid artery (Fig. 11.45). The unmyelinated and encapsulated nerve endings of the baroreceptors are embedded in the elastic tissue of the arterial wall and are stimulated by stretching of the wall. The sensory myelinated axons from the *aortic baroreceptors* travel in the *vagus nerve* and those from the *carotid baroreceptors* travel in the *glossopharyngeal nerve*. Information from these receptors is relayed to the cardiovascular centres.

The aortic baroreceptors are less sensitive than the carotid baroreceptors. An increase in transmural pressure, usually brought about by an increase in

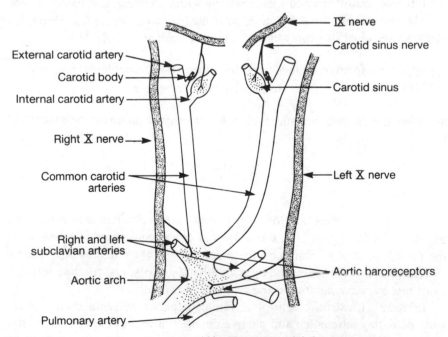

Fig. 11.45. Location of the aortic and carotid baroreceptors and their sensory nerves.

arterial blood pressure, will stretch the arterial wall and stimulate the baroreceptors. At normal *MAP*, carotid baroreceptors are tonically active as shown by streams of action potentials recorded from their afferent axons. The mean frequency of these action potentials changes in direct proportion to changes in *MAP* in the blood pressure range of 50–180 mmHg. Also the frequency of nervous discharge from the baroreceptors is in phase with the fluctuation from systolic to diastolic arterial pressure. Baroreceptors monitor the sudden increase in *MAP* that occurs after a change in posture from standing to supine and the sudden decrease in *MAP* that occurs after a haemorrhage or after a change in posture from supine to standing (Section 11.6).

The reflex changes initiated by stimulation of the arterial baroreceptors in response to a sudden increase in *MAP* are as follows:

1 an increase in parasympathetic and a decrease in sympathetic discharge to the SA and AV nodes causing a decrease in heart rate;
2 a decrease in sympathetic discharge to ventricular muscle causing a decrease in contractility and hence a reduction in stroke volume;
3 a decrease in sympathetic discharge to the veins causing an increased venous compliance and capacity and hence a reduced end-diastolic volume and subsequent stroke volume;
4 a decrease in sympathetic discharge to the arterioles causing a decrease in *TPR*. This vasodilatation results in a greater capillary pressure leading to an increased ultrafiltration across the capillary wall and hence, after about 5 min, to some reduction in blood volume.

All these changes except **4** will reduce $\dot{Q}$. Since $\dot{Q} = MAP/TPR$, the reflex reductions in $\dot{Q}$ and *TPR* will decrease *MAP* thus opposing the initial elevation and restoring the *MAP* to or towards normal. A decrease in *MAP* initiates the converse changes. In addition, a large decrease in baroreceptor stimulation, for example after a haemorrhage, will stimulate secretion of antidiuretic hormone (p. 266).

The baroreceptor reflexes adjust the balance amongst $\dot{Q}$, *TPR* and *MAP* whenever there is a small change in the metabolism of an organ. If one organ increases its metabolism, dilatation of its arterioles by local metabolic autoregulation rapidly increases its blood flow. This local vasodilatation will slightly decrease the *TPR* and hence the *MAP*. The smallest decrease in *MAP* will be detected by the baroreceptors and will result in a reflex increase in $\dot{Q}$ and, to some extent, an increased vasoconstriction in other organs, except the heart and brain. Thus, within a few seconds, the final decrease in *TPR* will be minimized and the extra blood flow required by that organ will be supplied by reducing blood flow to some other organs and by increasing the total blood flow. At the same time the decrease in *MAP* will be very small or even undetectable.

The precision of the baroreceptor reflex in regulating *MAP* at 100 mmHg on average depends on the magnitude of the initial deviation, on the influence of other receptor reflexes, on higher centre control of the cardiovascular centres and on blood volume. These other factors determine the final set-point for the *MAP* and the role of the baroreceptor reflex is to ensure that any particular set-point is maintained. If the baroreceptors are denervated the *MAP* is very variable around whatever the set-point, although initially the *MAP* will also be elevated.

Reflexes initiated by cardiac stretch receptors

Stretch receptors are also found in the walls of the venae cavae (as they enter the heart), the atria and the ventricles. Their afferent fibres are myelinated, except for those from the ventricles, and they travel in the *vagus nerve* to the cardiovascular centre. A physiological function is only well established for the reflexes resulting from stimulation of atrial and venae cavae receptors; both are referred to as *atrial 'B' receptors*.

Since increases in blood volume in the atria cause a lot of stretch for only a small increase in the low atrial pressure, these atrial 'B' receptors are low-pressure or *volume receptors*. Their peak stimulation occurs during the v wave of atrial pressure at the very end of systole. An increase in blood volume increases the total nervous discharge from the atrial 'B' receptors and reflexly initiates a decrease in sympathetic and an increase in parasympathetic activity. This complements the reflexes initiated by the baroreceptors which will also be stimulated by the increase in *MAP* that usually accompanies an increase in blood volume. These reflexes together result in a decrease in $\dot{Q}$ and *TPR*, an increase in venous capacity and an excess of ultrafiltration across the capillary wall. Thus *MAP* and blood volume decrease towards normal.

On the other hand, reduced stimulation of the atrial 'B' receptors increases sympathetic activity, in particular to the renal arterioles, and decreases parasympathetic activity. The former decreases renal blood flow, which reduces urinary *sodium excretion*, and increases the release of *renin* by the juxtaglomerular cells of the kidney (p. 540). The rise in plasma renin increases the circulating concentration of *angiotensin II* which (a) causes further vasoconstriction and hence increases *TPR*; (b) stimulates the release of *aldosterone* from the adrenal cortex, which increases reabsorption of sodium; (c) stimulates *thirst*; and (d) increases the secretion of *antidiuretic hormone* from the posterior pituitary, which not only reduces urinary loss of water but also contributes to systemic vasoconstriction.

In addition, there is a reduction in afferent impulses from the atrial 'B' receptors to the hypothalamus which further stimulates thirst and the release of antidiuretic hormone (pp. 266 & 536). Thus the urinary loss of water and salt is restricted and changes in blood volume are minimized.

These effects of renin, angiotensin II, aldosterone and antidiuretic hormone take from minutes to hours to develop. Such *long-term regulation* of blood volume and hence of *MAP* and mean systemic filling pressure involving the endocrine, renal and cardiovascular systems not only compensates for the decreased blood volume of a *haemorrhage* but also operates at all times in the background to maintain *total body water* and the *correct distribution of extracellular fluid* between the interstitium and the blood.

The hormones discussed above act to increase extracellular fluid volume but a recently discovered family of peptides, *atrial natriuretic peptides (ANP)*, can *decrease* extracellular fluid volume within a matter of minutes. ANP is released from secretory granules of the atria in response to stretch. It causes vasodilatation by decreasing the sensitivity of vascular smooth muscle to vasoconstrictor substances such as noradrenaline and angiotensin II. This decreases *MAP* and increases renal blood flow leading to increased urinary loss of water (diuresis) and sodium (natriuresis) and hence to restoration of normal extracellular fluid volume. ANP also directly decreases the release of antidiuretic hormone and aldosterone which may contribute to the diuresis and natriuresis, respectively.

Other cardiac stretch receptors are as follows. *Atrial 'A' receptors* are excited normally by atrial contraction (at the time of the a wave) and experimentally by a rapid infusion of a large blood volume. The reflex initiated is an increase in heart rate mediated by the sympathetic system (Bainbridge reflex). In the *left ventricle*, stretch receptors stimulated normally by the isovolumetric phase of ventricular contraction result in a decrease in heart rate and vasodilatation. These receptors appear to be stimulated also by alkaloids (e.g. veratrine), serotonin, nicotine and the accumulation of metabolites (Bezold–Jarisch reflex). Throughout the heart, afferent fibres which travel in sympathetic nerves are also stimulated by the above chemicals and by myocardial ischaemia, resulting in tachycardia, vasoconstriction and the pain of angina.

Reflexes initiated by other receptors

The receptors described below alter the set-point for *MAP*.

1 *Lung stretch receptors* (Section 12.5) are stimulated by the increase in tidal volume accompanying an increased ventilation, for example during exercise, and not only determine the pattern of breathing but also cause an increase in heart rate and a decrease in *TPR* and *MAP*.

2 *Nasal receptors* (Section 12.5) when stimulated by water in diving mammals cause a cessation of breathing, a decrease in heart rate and an increase in *TPR* and *MAP* (the diving reflex).

3 *Peripheral* or *arterial chemoreceptors* in the carotid and aortic bodies (Section 12.5) are stimulated by hypoxia, hypercapnia or acidosis and cause a decrease in heart rate and an increase in *TPR* and *MAP*. The increase in *TPR* and *MAP* will be counteracted to some extent by the local systemic vasodilatation (p. 429). This chemoreceptor effect contributes to the diving reflex and can be demonstrated only in a dive or breath-hold where there is no concurrent increase in ventilation. Stimulation of the peripheral chemoreceptors normally results in an increase in ventilation which will stimulate opposing reflexes from the lung stretch receptors. The final response to breathing an hypoxic gas mixture is an increase in heart rate and, in severe hypoxia, a decrease in myocardial contractility (p. 400), *TPR* and *MAP*.

4 *Central chemoreceptors* near the ventral surface of the medulla oblongata (Section 12.5) are stimulated by hypercapnia and reflexly increase heart rate, *TPR* and *MAP*. Most of the increase in *TPR* and *MAP* is opposed by local systemic vasodilatation and by the accompanying increase in ventilation stimulating the lung stretch receptors.

5 *Limb proprioceptors* believed to be stimulated during exercise send their afferent information to the hypothalamus and cause increases in $\dot{Q}$, *TPR*, *MAP* and ventilation.

6 *Peripheral and central temperature receptors* (Section 15.2) also have inputs to the hypothalamus and during an elevated body temperature will reflexly cause an increase in $\dot{Q}$, and possibly in *MAP*, and a decrease in *TPR* with preferential vasodilatation in the skin.

7 *Laryngeal and tracheal receptors* causing the cough reflex and *epipharyngeal receptors* causing the sniff reflex (Section 12.5) also cause an increase of short duration in *TPR* and *MAP*. In contrast, the *lung irritant receptors* of the lower airways have no known cardiovascular reflex.

8 *Juxtacapillary (J) receptors* in the lungs stimulated by local oedema (Section 12.5) not only cause rapid, shallow breathing but also cause a decrease in heart rate, *TPR* and *MAP*.

Cardiovascular centres

Cardiovascular centres are areas in the brain responsible for the integration of sensory information and the subsequent control of the autonomic nervous output to the cardiovascular system. They are found in the *pons* and *medulla oblongata* of the brainstem and in the *hypothalamus*. Modification of the autonomic nervous output also occurs as a result of activity in the *cerebral cortex* and *limbic system* descending to the hypothalamus, in *cerebellar pathways* descending to the pons and medulla and in the *spinal cord*.

Until recently, the centres in the pons and medulla were considered the principal integrating centres and the description below is a simple version of this classical model. However, it must be stressed that current research indicates that these centres are not clearly defined and are an important but by no means the only part of a very complicated integrating system extending from the cerebral cortex to the spinal cord.

In the classical model, the principal cardiovascular centres are located in the reticular formation of the pons and medulla rostrally and laterally as the *pressor area* and caudally and medially as the *depressor area*; there is a certain degree of overlap between them (Fig. 11.46). Electrical stimulation of the pressor area induces increases in heart rate, myocardial contractility, *TPR*, *MAP*, venomotor tone and adrenaline secretion. The converse effects result from stimulation of the depressor area.

Neurones in the pressor area send axons down the spinal cord to excite preganglionic sympathetic fibres. Those neurones in the pressor area controlling vasomotor tone have some tonic nervous activity and have been referred to collectively as the *vasoconstrictor area*. Those neurones controlling the heart have little tonic activity and have been referred to as the *cardioexcitatory area*. These areas are functional rather than anatomical since their neurones intermingle throughout the pressor area.

Neurones in the depressor area which have little tonic activity send axons to inhibit preganglionic sympathetic fibres and constitute the functional *vasodilator area*. Neurones in the depressor area which have considerable tonic

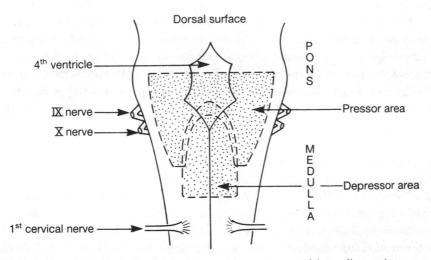

Fig. 11.46. Location in the brain of the pressor and depressor areas of the cardiovascular centres.

activity send axons to excite the dorsal motor nucleus of the vagus in the medulla and constitute the functional *cardioinhibitory area*.

The *sensory pathways* of the baroreceptors, atrial 'B' receptors and lung J receptors converge on the depressor area and those of the central chemoreceptors converge on the pressor area. The lung stretch receptors send impulses both to the cardioexcitatory neurones in the pressor area and to the vasodilator neurones in the depressor area, whilst the inputs from nasal, laryngeal and tracheal receptors and peripheral chemoreceptors converge on the cardioinhibitory neurones in the depressor area and on the vasoconstrictor neurones in the pressor area.

Local hypoxia, hypercapnia and acidity in the brain during *ischaemia* are believed to stimulate both the pressor area (particularly its vasoconstrictor area) and the preganglionic sympathetic fibres in the spinal cord.

The *respiratory centres* (Section 12.5) are also located in the medulla. When the dorsal respiratory group increases its nervous activity to cause inspiration, it also sends impulses to the pontomedullary pressor area (particularly to the cardioexcitatory area) resulting in increases in heart rate during inspiration (*sinus arrhythmia*).

The *cerebellum* not only coordinates movement and orientation (Section 6.4) but also assists in the postural control of *MAP*. Stimulation of the cerebellum excites the pontomedullary pressor area resulting in increases in heart rate, *TPR* and *MAP*.

Caudal *pressor* and rostral *depressor areas* can be identified in the *hypothalamus* and these project to the corresponding areas in the pons and medulla. Atrial 'B' receptors relay to the hypothalamic pressor area as well as to the supraoptic nuclei of the hypothalamus which control the release of antidiuretic hormone.

The *defence reaction centre* of the hypothalamus, stimulated electrically or via the motor cortex in alarm and defence responses, activates the hypothalamic pressor area. This results in increases in $\dot{Q}$ and *MAP* together with vasoconstriction of all organs except brain, heart and skeletal musculature. The defence reaction centre also activates, at synapses in the hypothalamus, axons which descend from the motor cortex to the preganglionic fibres of the cholinergic vasodilating sympathetic pathway in the spinal cord.

The *temperature regulating centre* of the hypothalamus in response to an increase in body temperature (Section 15.2) activates the hypothalamic pressor area causing an increase in $\dot{Q}$ and a sympathetically mediated vasoconstriction of all organs except the heart, brain and skin. At the same time the temperature-regulating centre decreases the activity in hypothalamic axons controlling the sympathetic adrenergic vasoconstrictor fibres to the arterioles, venules and arteriovenous anastomoses of the skin (Section 11.5). The resulting vasodilatation and consequent large blood flow through the skin aids in the restoration of normal body temperature. The balance between the degree of vasodilatation in the skin and vasoconstriction elsewhere coupled with the degree of increase in $\dot{Q}$ will determine the degree of decrease in *TPR* and *MAP*. The converse sequence of events occurs during a decrease in body temperature.

The *limbic system* integrates many emotional and behavioural reflexes (p. 244) and, via the hypothalamic pressor and depressor areas, will elicit the pressor response of excitement or mild pain and the depressor response of shock or severe pain. The *motor cortex* relays to the hypothalamic pressor area during the anticipatory and steady-state phases of exercise as well as activating the cholinergic sympathetic pathway.

11.5 Characteristics of the circulation through different organs

Pulmonary circulation

The pulmonary vessels have thinner, more compliant walls and are shorter in length and larger in diameter than the vessels in the systemic circuit. The entire cardiac output flows through the pulmonary circuit whose resistance is about a tenth that of the systemic. In the pulmonary artery, the systolic, diastolic and mean pressures are about 25, 8 and 15 mmHg, respectively, and the pressure declines gradually to about 10 mmHg in the capillaries and to a mean of about 5 mmHg in the left atrium (Fig. 11.36). Thus, in contrast to the systemic circuit, the arterial and venous parts of the pulmonary circuit contribute equally to the total pulmonary resistance. Since the pulmonary resistance is so low, flow remains pulsatile although the pulses decrease in amplitude as blood flows through the pulmonary vessels (Fig. 11.36).

The vast network of pulmonary capillaries in the walls of the alveoli is utilized for gas exchange between the alveolar air and pulmonary blood (Sections 12.1 & 12.3). The total cross-sectional area of the pulmonary capillaries and hence the velocity of capillary flow are similar to those of the systemic circuit (Fig. 11.36). However, since the pulmonary capillaries are shorter, the transit time through each capillary at rest is about 1 rather than 2 s. During severe exercise, the fivefold increase in pulmonary blood flow is associated with an increase in mean pulmonary arterial pressure which opens up non-perfused lung capillaries and thus increases their cross-sectional area but by only 1·5-fold. Hence velocity through the pulmonary capillary increases considerably and the transit time is as little as 0·3 s. Although this will impede equilibration of gases by diffusion it is not the limiting factor in the supply of O_2 during exercise, at least at sea level (Section 16.3).

In the erect posture, about 9% of the blood volume is in the pulmonary circuit. Since both pulmonary arterioles and veins are very compliant, the pulmonary circuit accommodates about 16% of total blood volume in the volume in the supine position (Fig. 11.36). Pulmonary blood volume also increases passively in exercise, generalized systemic vasoconstriction, left heart failure or mitral stenosis and decreases passively in generalized systemic vasodilatation and haemorrhage. Thus the pulmonary circuit functions as a blood volume reservoir (the *pulmonary reservoir*).

The compliant pulmonary vessels are also subjected to a varying transmural pressure induced by the fluctuating intrapleural pressure of respiration. Thus their volume increases during inspiration and decreases during expiration. Powerful expiration or positive pressure breathing can markedly reduce pulmonary (and cardiac) blood volume and pulmonary flow, as well as impeding systemic venous return.

The effect of gravity in the erect posture (Section 11.6) and the low pressure and high distensibility of the pulmonary circuit results in less, or even intermittent, flow to the apex relative to the base of the lung. The consequence of this for gas exchange and the total ventilation/perfusion ratio of the lung will be considered in Section 12.3 (p. 478).

Although pulmonary arterioles and veins are well supplied with sympathetic vasoconstrictor and parasympathetic vasodilator fibres, their roles are not clear. However, stimulation of the sympathetic system or administration of adrenaline can decrease the pulmonary blood volume by as much as 30%.

Neither myogenic nor metabolic autoregulation is demonstrable in the pulmonary circuit. In fact, the response of the pulmonary arterioles to local hypoxia, hypercapnia and acidity is the opposite of that of systemic arterioles. *Hypoxia* is the more potent of the three stimuli. In poorly ventilated regions of the lung, where alveolar O_2 is low and CO_2 is high, local pulmonary *vasoconstriction* diverts blood flow to better ventilated areas. This improves gas exchange by helping to equalize the regional ventilation/perfusion ratios in the lung. However, this hypoxic vasoconstriction has the disadvantage that during the hypoxia of altitude or of chronic obstructive lung disease all pulmonary arterioles are affected. The resultant increase in pulmonary resistance increases right ventricular and pulmonary arterial pressure (pulmonary hyperplasia).

Since it is essential for gas exchange that the alveoli do not accumulate liquid, the pulmonary circuit is accompanied by a lymphatic system more extensive than in any other organ. Furthermore, the colloid osmotic pressure of the plasma of 25 mmHg is greater than the difference in hydrostatic pressure between capillary (10 mmHg) and interstitium (−3 mmHg) and favours reabsorption of fluid across the entire length of the capillary. However, accumulation of fluid in the alveoli (pulmonary oedema) may occur and lead to dyspnoea (Section 12.6) when the left atrial pressure is increased during left ventricular failure or mitral stenosis or when the pulmonary arterial pressure increases during severe exercise.

The airways of the lung from trachea to bronchioles are supplied by the aorta via the bronchial arteries. This *bronchial circulation* constitutes only about 1% of the total cardiac output. Some venous drainage (about 25%) is via bronchial veins into the superior vena cava but the remaining 75% enters the pulmonary veins. There are also direct connections between the bronchial and pulmonary capillaries. This intermingling of the two circulations results in a very small degree of O_2 desaturation of the oxygenated pulmonary venous blood (p. 484). Pathologically, if the pulmonary arterial supply is inadequate, the bronchial arteries can provide some collateral circulation but with consequently more obvious O_2 desaturation.

Coronary circulation

The anatomical features of the coronary circulation have been described already (p. 377). Since the heart at rest receives about 5% of the cardiac output and has a very high O_2 consumption relative to its mass (Table 11.4), its venous O_2 content is low and there is thus a very large arteriovenous O_2 difference. In contrast to other active organs, during increased cardiac activity there is little further decrease in its venous O_2 content and the increased demand for O_2 is satisfied mainly by a large increase (up to fourfold) in coronary blood flow, a prime example of *metabolic autoregulation*. Hypoxia is a more potent coronary vasodilator than an increase in P_{CO_2} or blood acidity.

Although the smooth muscle of the coronary arterioles receives both sympathetic and parasympathetic innervation and contains α adrenoceptors, mediating vasoconstriction, and muscarinic cholinoceptors, mediating vasodilatation, the importance of these pathways is not clear. Any neurally induced vasoconstriction occurring during increased sympathetic stimulation is overridden by the metabolic vasodilatation resulting from the simultaneous

Table 11.4. Blood flow, oxygen consumption, arteriovenous O_2 difference and vascular resistance of tissues at rest.

Organ	Weight (kg)	Blood flow ($l\,min^{-1}$)	Oxygen consumption ($ml\,min^{-1}$)	a–v O_2 difference ($ml\,l^{-1}$)	Absolute resistance ($mmHg\,l^{-1}min$)	Resistance ($mmHg\,l^{-1}min$) per kg body wt)
Heart	0·3	0·25	30	120	400	120
Brain	1·5	0·75	45	60	133	200
Skin	5·0	0·50	10	25	200	1000
Skeletal muscle	30·0	1·20	55	60	83	2490
Gut	2·8	1·40	55	40	71	200
Kidney	0·3	1·10	20	15	91	27
Other	30·0	0·60	35	50	167	5010
Total	70·0	5·80	250	45	17	1190

increase in cardiac activity. Similarly, during increased parasympathetic stimulation, any neurally induced vasodilatation will tend to be overridden by the metabolic vasoconstriction. Note that adrenaline acting on β_2 adrenoceptors causes coronary vasodilatation.

The vascular bed of the heart has a very high resistance (Table 11.4) which is partly due to the *compression* of the blood vessels during cardiac contraction. In systole, the peak pressure is about 120 mmHg in the aorta, coronary arteries and left ventricle and about 25 mmHg in the right ventricle (Fig. 11.47). Within the myocardial wall, these pressures will be slightly greater. Thus, in systole, the coronary vessels of the left ventricle are compressed and there is little flow in the left coronary artery, whereas in the right ventricle the transmural pressure of the coronary arteries is reduced from 120 to only 95 mmHg and the right coronary flow is hardly affected. In diastole, the pressure in the coronary arteries is about 80 mmHg whilst that in both ventricles is about 0 mmHg. Hence there is no compression of blood vessels in diastole. Left coronary arterial flow is therefore intermittent, ceasing in early systole, and right coronary arterial flow is pulsatile, being slightly greater in systole than in diastole. Venous flow through the coronary sinus into the right atrium is greatest during the compression of systole and subsides during diastole.

Taking into account that diastole occupies about two-thirds of the cardiac cycle at rest, about 80% of the total, about 85% of the left and about 70% of the right coronary arterial flow occurs during diastole. Furthermore, left coronary arterial flow is about 60% of the total flow. With tachycardia, which reduces the period of diastole considerably (p. 394), and with an increased myocardial

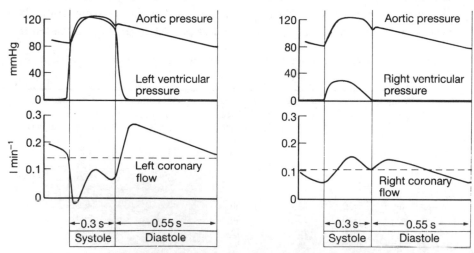

Fig. 11.47. Left and right coronary arterial flow in relation to aortic and ventricular pressures during a cardiac cycle of 70 beats/min.

contractility, a greater proportion of the coronary resistance arises from compression and coronary arterial flow is impeded for shorter but more frequent intervals.

Since about 80% of the total coronary blood flow occurs during diastole, the aortic diastolic rather than the mean arterial pressure becomes the primary determinant of the pressure gradient for coronary flow. Furthermore, coronary arterioles exhibit *myogenic autoregulation* such that coronary flow at rest is constant in the aortic diastolic pressure range of 60–180 mmHg. However, if the aortic diastolic pressure increases as a result of an increase in cardiac activity (Fig. 11.41), coronary blood flow will increase through metabolic autoregulation.

Reductions in coronary blood flow can be caused by atherosclerosis of the coronary vessels, by a low aortic diastolic pressure as in cardiac failure or haemorrhage, by a rise in atrial and ventricular diastolic pressures as in cardiac failure, or by a high left ventricular systolic pressure as in aortic valve stenosis. The resultant *myocardial ischaemia* affects mainly the left ventricle.

Cerebral circulation

The soft tissue of the brain, covered by a highly vascular arachnoid membrane and a tough dura mater, is encased in the rigid skull (Section 2.5).

Cerebrospinal fluid (CSF) fills the four ventricles and the subarachnoid space between the arachnoid and the pial surface of the brain (Fig. 2.32). The CSF is in continuity with the interstitial fluid of the brain and is formed from the plasma by the capillaries of the brain, particularly those in the choroid plexuses (Section 2.5). CSF flows in a specific direction around the brain and is reabsorbed from the subarachnoid space through the arachnoid villi into the venous sinuses of the dura. There is no lymphatic system in the brain; its drainage function is taken over by the CSF.

From the vertebral and internal carotid arteries are derived the arteries of the brain which, until they branch and penetrate the brain, lie close to the pia amongst the trabeculae in the subarachnoid space (Fig. 2.31). Venules drain into the superficial veins of the pia which in turn drain either into the venous sinuses in the dura before entering the internal jugular veins or into the vertebral veins of the spinal cord and thence to the external jugular veins. Unlike other capillaries, those of the brain (except in the circumventricular organs) are relatively impermeable to most substances except fat-soluble drugs, glucose, O_2 and CO_2. This impermeability is referred to as the *blood–brain barrier* and its significance is considered in Sections 2.5 & 12.5.

The brain receives about 13% of the cardiac output, has a moderately resistant vascular bed, a fairly high O_2 consumption and a large arteriovenous O_2 difference (Table 11.4). *Total blood flow* through the brain remains remarkably *constant* under physiological conditions. However, the normal variations in nervous activity within regions of the brain are associated with alterations in regional blood flow brought about by local metabolic autoregulation.

The control of cerebral blood flow (CBF) is highly developed. Although the cerebral arterioles are innervated by the sympathetic and parasympathetic nervous system, there appears to be no significant role for them in vascular control. In contrast, *myogenic autoregulation* maintains CBF remarkably constant in the *MAP* range of about 50–150 mmHg. Furthermore, *metabolic autoregulation* maintains a constant O_2 supply. Between the normal arterial P_{O_2} of 100 mmHg and an hypoxia of 50 mmHg P_{O_2} there is little increase in CBF. Below 50 mmHg P_{O_2}, the cerebral arterioles are sensitive to the vasodilating effects of hypoxia and CBF increases about twofold as P_{O_2} decreases from 50 to 25 mmHg. However, the cerebral arterioles are much more sensitive to the slightest increase or decrease in arterial P_{CO_2} and acidity. Considerable vasodilatation occurs when P_{CO_2} is elevated (40–100 mmHg) and considerable vasoconstriction occurs when P_{CO_2} is reduced (40–20 mmHg). Within such limits, a doubling or halving of P_{CO_2} results in about a twofold change in CBF. The dizziness felt after excessive hyperventilation is the result of the reduced CBF which accompanies the resultant hypocapnia. Since a reduced O_2 supply is usually accompanied by an increase in arterial P_{CO_2} and acidity (except at altitude), the CBF is regulated by hypercapnia rather than hypoxia to maintain a constant O_2 supply.

If the O_2 content of arterial blood declines sufficiently or if the *MAP* drops below the range of myogenic autoregulation such that CBF decreases, the O_2 delivery to the brain is impaired. This precipitates a sudden state of *unconsciousness* which if it is of short duration is a *faint* (syncope) but if prolonged is a *coma* (p. 231) and is often accompanied by brain damage. An obstruction or a rupture in a major cerebral artery leads to local areas of brain *ischaemia* (reduced blood supply) and often to local areas of brain damage (a *stroke*).

Intracranial pressure also affects CBF. Since fluid is essentially incompressible, the volume of blood, brain tissue and CSF within the rigid cranium must be essentially constant (the *Monro–Kellie doctrine*). An increase in intracranial pressure (i.e. the pressure of the CSF) caused by insufficient reabsorption of CSF or a blockage in its normal route of flow will thus compress cerebral vessels and reduce CBF. The consequent hypoxia and hypercapnia will stimulate directly the pressor area of the cardiovascular centre (Section 11.4) and the resultant rise in *MAP*, coupled with local cerebral vasodilatation, will tend to restore CBF. At the same time, the rise in *MAP* will reflexly decrease heart rate via the baroreceptors. These reflex responses to an increased intracranial pressure are called the *Cushing reflex* after Harvey Cushing (1869–1939) the pioneer of modern neurosurgery. There is obviously a limit to the increase possible in *MAP* and if intracranial pressure exceeds this in the clinical state of severe *hydrocephalus* the cerebral circulation ceases.

A further consequence of the Monro–Kellie doctrine is that increases in cerebral blood pressure caused by *gravity* during a head-stand, together with any tendency to an increase in cerebral venous volume, are transmitted to the CSF such that intracranial pressure also increases. Thus little change occurs in the transmural pressure and cerebral venous return decreases only slightly. Conversely, in the erect posture, the decreases in cerebral blood pressure are compensated for by a corresponding fall in intracranial pressure. Furthermore, the major cerebral veins do not collapse because they are held open by their association with the dura mater.

Cutaneous circulation

At rest the skin receives about 8% of the cardiac output through its relatively highly resistant vascular bed. Since it has a very low O_2 consumption, the arteriovenous O_2 difference is also small (Table 11.4). The skin has an extensive superficial network of arterioles, capillaries and venules and an extensive *deep plexus of veins*. The latter may contain up to 1·5 litres of blood. The arterioles and veins are richly innervated with sympathetic fibres acting via α *adrenoceptors* and hence the vascular resistance and capacitance of the skin can be altered by reflexes integrated by the cardiovascular centres in the pons and medulla. For instance during hypotension, haemorrhage, or exercise that does not generate a heat load, the blood flow to the skin can be restricted, making it feel cold, and the accompanying venoconstriction mobilizes a sizeable volume of blood from the deep venous plexus of the skin.

In the skin of the hands, feet and face (particularly the ears, nose and lips), there are also large numbers of *arteriovenous anastomoses* (Fig. 11.42). These are vessels about 50 μm in diameter with thick walls of smooth muscle which contain α *adrenoceptors* activated by sympathetic fibres. When *body temperature* is increased, sympathetic activity to the skin decreases under the control of the hypothalamus (Section 11.4), resulting in vasodilatation particularly of these arteriovenous anastomoses. This vasodilatation is enhanced by the *bradykinin* released by sweat glands when these are activated by cholinergic sympathetic fibres also originating from the hypothalamus. The large fall in total peripheral resistance, reflexly via the baroreceptors, triggers a corresponding increase in cardiac output. The massive increase in skin blood flow is accommodated mainly within the arteriovenous anastomoses and the venous plexuses which together provide a large area for heat exchange between the blood and skin, and hence the external environment. Thus the temperature of the body is regulated (Section 15.2) to a large extent by the amount of blood flowing through the skin which, from the thermoneutral point, can increase about thirtyfold in heat stress and decrease about tenfold in cold stress.

Skin tissue, to a much greater extent than other tissues, can tolerate reduced blood flow by *reducing its O_2 consumption* accordingly. However, too great a reduction during prolonged exposure to cold leads to skin necrosis (*frost-bite*). During less severe but prolonged exposure to the cold, the skin vasoconstriction changes to vasodilatation which explains the ruddy complexion on a cold day. This vasodilatation is not sympathetically mediated but is believed to be caused by *metabolic autoregulation*, which overrides sympathetic control, and by the damaged skin cells elicit an axon–axonal reflex (see below). Note that skin arterioles are incapable of myogenic autoregulation.

The α adrenergic sympathetic pathway to skin vessels is also activated via the hypothalamus by various *emotions* made manifest in blushing or in the pallor accompanying fear.

In the skin, the reactions of blood vessels can easily be observed. A pointed object drawn lightly over the skin causes the area of contact rapidly to become pale (the white reaction). The mechanical stimulation is thought to initiate local contraction of venules or precapillary sphincters. Greater pressure will cause the skin rapidly to become red (the red reaction), followed by a diffuse mottled reddening around the area of injury (the flare) and then a local swelling (the weal). The latter three reactions comprise the so-called *triple response*. The red reaction probably results from the damaged cells of the skin releasing substances like histamine which cause a local vasodilatation or venodilatation. The flare results from the mechanical stimulation of sensory endings causing antidromic conduction in axon branches (axon–axonal reflex) that release substance P (a vasodilator) near arterioles. The weal results from a histamine-induced increase in permeability of the capillaries and an increase in interstitial fluid.

Skeletal muscle circulation

Skeletal musculature receives at rest about 20% of the cardiac output through a highly resistant vascular bed, the tone of which is controlled by *myogenic autoregulation* and by sympathetic activity predominantly acting on α *adrenoceptors*. Because of its large mass (about 50% of body weight but about 90% of the total cellular mass), it consumes even at rest about 20% of the total O_2 consumption (Table 11.4).

During maximal exercise, skeletal musculature receives about 90% of the cardiac output and also consumes about 90% of the O_2 consumption (p. 629). The concomitant vasodilatation in the skeletal musculature results from *metabolic autoregulation*. The vasodilatation from adrenaline acting on β_2 *adrenoceptors* and from the *cholinergic sympathetic pathway*, which originates in the hypothalamus, plays little role once exercise is established. In rhythmic exercise, blood flow in the muscles concerned can cease during the contraction phase but because of the accumulation of metabolites it is then augmented during the relaxation phase. In non-rhythmic, especially isometric exercise, blood flow can cease entirely thus limiting the duration of such exercise.

Non-exercising skeletal musculature, because of its large mass, is an important site for sympathetic vasoconstriction mediated via α adrenoceptors during reflexes initiated, for example, by hypotension or haemorrhage. In contrast to splanchnic and cutaneous veins, the veins in skeletal musculature are practically devoid of sympathetic nerve endings and cannot alter their capacity neurogenically so as to contribute to the blood reservoir.

Splanchnic circulation

The splanchnic area receives through a low resistance vascular bed about 25% of the cardiac output at rest (Table 11.4). A number of arterial branches from the abdominal aorta supply the stomach, intestine, pancreas and spleen which are drained by the portal vein into the liver. The liver receives about 75% of its blood from this source, the remainder coming from the hepatic artery. Blood from the liver drains via hepatic veins into the inferior vena cava.

In the splanchnic and hepatic arteries, the mean blood pressure is about 90 mmHg whilst in the portal and hepatic veins it is 10 and 5 mmHg, respectively. The hepatic arterioles are highly constricted and consequently blood enters the relatively large liver capillaries or sinusoids at a pressure less than 10 mmHg. This vasoconstriction and hence the proportion of *blood flow to the liver* from the hepatic artery is controlled by myogenic and metabolic autoregulation. There is also a vasomotor tone from sympathetic fibres acting via α adrenoceptors. Increased metabolic activity in the liver or a decrease in portal flow will, via *metabolic autoregulation*, increase flow in the hepatic artery such that it can supply up to 50% of the blood flow to the liver. Conversely, an increase in portal flow will decrease flow in the hepatic artery via *myogenic autoregulation*. Note that the low hydrostatic pressure in the capillaries of the liver is matched by the high interstitial colloid osmotic pressure generated by the proteins which are being manufactured by the liver for the plasma. Hence the Starling forces for ultrafiltration and reabsorption across the capillary wall remain in balance (Section 1.8).

Blood flow to the gut is adjusted to the degree of activity in the smooth muscle layers by *metabolic autoregulation* and in the mucosal and submucosal layers by increased glandular activity releasing *bradykinin* which causes local vasodilatation.

The α *adrenoceptors* in the splanchnic arterioles and veins mediate vasoconstriction and venoconstriction in response to the generalized increased sympathetic activity and adrenaline secretion that occurs, for instance, during exercise, hypotension or haemorrhage. This serves to counteract a fall in mean arterial blood pressure and to redistribute blood to other vasodilatated regions. Furthermore, the reduction in *splanchnic venous capacity* can make about 300 ml of blood available to the rest of the body. Venous reservoirs of blood are also found in the spleen of some animals, for example the dog, but not in man. However, since prolonged splanchnic vasoconstriction would lead to cell damage in the gut, accumulation of metabolites will eventually override some of the vasoconstriction.

Renal circulation

The kidney receives through a very low resistance vascular bed about 20% of the cardiac output (Table 11.4) and its blood supply and blood pressures are described on p. 515. The arterioles of the isolated kidney exhibit well-developed *myogenic autoregulation* such that kidney perfusion is maintained constant at mean arterial blood pressures between 80 and 180 mmHg. Thus the renal functions of filtering waste products and regulating electrolyte balance are independent of fluctuations in blood pressure. However, *in situ*, this myogenic autoregulation can be overridden by sympathetic activity and adrenaline acting on α *adrenoceptors* of the kidney arterioles. Normally there is little sympathetic tone but, during exercise, hypotension, haemorrhage or systemic hypoxia and hypercapnia, the renal arterioles participate in the vasoconstriction that ensures adequate cerebral and coronary circulation. Furthermore, during exercise or heat stress, renal vasoconstriction helps to compensate for the vasodilatation in skeletal musculature or in the skin. Metabolic autoregulation in the renal circulation is poor.

11.6 Cardiovascular adjustments under physiological and abnormal conditions

The cardiovascular system has to adjust to the strains imposed by alterations in posture, by thermal stress and by exercise. The latter two are discussed in Sections 15.2 & 16.2. The cardiovascular system also has to compensate for the abnormalities induced by hypertension, hypotension, haemorrhage and cardiac failure.

Effects of posture

The pressures and volumes in the heart of Fig. 11.19 apply to the standing posture whilst the pressures and volumes in the vascular system of Fig. 11.36 apply to the supine posture. A change in posture alters the hydrostatic pressures in the blood vessels with resultant effects on transmural pressures and, in particular, on venous capacity and consequently on the distribution of blood around the cardiovascular system. This in turn affects end-diastolic volume.

Hydrostatic pressure

In the supine posture, all blood vessels are approximately at the level of the heart and the mean blood pressures in the aorta, the arteries and veins (of both the head and feet) and in the right atrium are about 100, 95, 5 and 1 mmHg, respectively. In the standing posture, the weight of the column of blood in the blood vessels results in an increase in arterial and venous pressure below the level of the heart and a decrease in these pressures above the level of the heart (Fig. 11.48). For every cm of blood above or below the level of the heart, pressure changes by 0·77 mmHg. Thus, in a man 1·8 m tall in whom the head is about 50 cm above the heart in the upright posture, there is a fall in blood pressure in the head of about 40 mmHg compared with the pressure in the supine position. Hence the arterial pressure at the entrance to the cranial cavity becomes 55 mmHg and, were it not for the fact that the veins in the

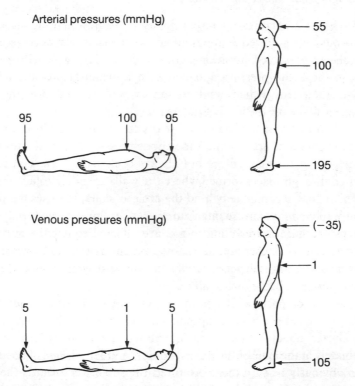

Fig. 11.48. Effect of gravity on arterial and venous pressures.

neck collapse, the venous pressure at the cranial exit would be -35 mmHg. In the feet, which are about 130 cm below the heart in the standing posture, there is an increase in pressure of about 100 mmHg and hence the arterial pressure becomes 195 mmHg and the venous pressure becomes 105 mmHg, provided that in the case of the veins the standing is motionless (see below).

It must be noted that blood flows from a point of high to low total fluid energy. The total fluid energy (p. 414) is the energy generated by the heart, i.e. the sum of the kinetic energy and the potential energy due to pressure plus that due to gravity. As blood flows towards the feet, it loses gravitational potential energy and gains potential energy due to pressure. Thus the total fluid energy in the arteries of the feet is the equivalent of 95 mmHg irrespective of a supine or standing posture. Therefore, although the changes in blood pressure on adoption of the standing posture mean that blood flows against a hydrostatic pressure gradient, for instance from the heart (100 mmHg) to the arteries in the feet (195 mmHg), the flow is not against a fluid energy gradient. Similarly, when total fluid energy is considered, it can be seen that it is not difficult for blood to return from the feet to the heart.

Transmural pressure and venous capacity

If blood vessels were rigid tubes, no cardiovascular adjustments to the standing posture would be required other than to compensate for decreased blood volume due to enhanced filtration across the capillary wall. However, the changes in the blood pressure and hence in transmural pressure affect, in particular, the veins which tend to empty and so collapse above and progressively fill below the level of the heart.

The collapse of veins in the neck breaks the column of blood and hence removes the effects of gravity and the venous pressure approximates to zero. The veins although collapsed are not always occluded as flow can continue unimpaired through spaces formed where the walls are not in total apposition. Blood which does accumulate behind the areas of collapse raises the pressure sufficiently above zero to cause intermittent venous flow. Intracranial veins do not collapse because the CSF and blood are affected equally by gravity and because the venous sinuses are maintained patent by the dura (p. 453). Thus the cerebral blood flow is dependent only on carotid artery pressure, the venous pressures at the exit being zero at all times.

Irrespective of posture, if the mean right atrial pressure equals about 1 mmHg, the height of the uncollapsed column of blood in the internal jugular vein is always about 1·5 cm *vertically* above the right atrium (which is approximately at the level of the sternal angle). A greater *length* of uncollapsed vein can obviously be seen the more tilted the subject is towards the supine position. The vertical height of this blood column is used clinically as a measure of central venous pressure (p. 435) and the pulsations of the a, c and v waves (p. 398) can sometimes be distinguished at the top of the column.

Initially, on assuming the standing posture, the passive distension of veins below the level of the heart results in an increase of about 500 ml in the volume of blood that they contain, most of this being displaced from the thorax. Consequently, in the absence of reflex compensations, there is a reduction in venous return and end-diastolic volume and hence in stroke volume and systolic blood pressure. Furthermore, the increased hydrostatic pressure in the capillaries below the level of the heart will cause considerably more ultrafiltration than usual and over a period of hours this will lead to loss of blood volume and, most noticeably, to local oedema in the ankles.

Reflex compensations

The decrease in end-diastolic volume is detected by the cardiac stretch receptors (p. 442) and the decrease in *MAP* (from both the direct gravitational effect at the level of the carotids and the decrease in cardiac output) is detected by the carotid baroreceptors (p. 441). In the space of about 5 s, there are thus reflex increases in sympathetic activity and in adrenaline release causing:

1 an increase in heart rate and in myocardial contractility tending to restore cardiac output towards the supine value;
2 vasoconstriction in the skeletal musculature, skin, kidneys and gut reducing blood flow to these organs, increasing *TPR* and decreasing capillary pressure; and
3 venoconstriction in all veins but in particular in the skin and gut which displaces some blood back to the thorax.

All of these contribute to the restoration of *MAP* at the carotid level.

Over a longer period of time, the cardiac stretch receptors will trigger an increase in antidiuretic hormone secretion. The reduced blood flow to the kidneys enhances the release of renin which increases circulating concentrations of angiotensin II and aldosterone. These responses, which are similar to those occurring after a mild haemorrhage, augment vasoconstriction and decrease urinary Na^+ excretion which helps restore blood volume.

One of the most important mechanisms counteracting the long-term effects of the standing posture is the use of the skeletal muscle venous pump. This, aided by the venous valves, reduces venous capacity and assists venous and lymphatic returns thus lowering the mean venous pressure in the feet from about 105 mmHg to as little as 30 mmHg. The normal muscular contractions associated with moving around in the standing posture are sufficient to achieve this and hence there is less of an increase in venous capacity and little local oedema. Prolonged standing, venous obstruction or pregnancy can overstretch

the veins and lead to incompetent venous valves. The consequent ineffective-
ness of the skeletal muscle venous pump results in the excessive distension of,
in particular, superficial veins which fill with static blood—the condition
known as *varicose veins*.

During motionless standing or in sudden standing of hypotensive patients
or of patients with an impaired sympathetic system or of subjects in a hot
environment where venodilatation in the skin predominates, restoration of
the *MAP* may be inadequate. If the *MAP* declines to a level (about 50 mmHg)
below which myogenic autoregulation of cerebral blood flow cannot occur, the
person becomes dizzy, has impaired vision and may even faint. *Fainting* has
the advantage of returning the person to the supine position!

Hypertension

Repeated measurements of arterial diastolic pressure above 90 mmHg in the
resting supine subject are taken clinically to indicate the condition of
hypertension. It is associated with increases in $\dot{Q}$, *TPR* or blood volume.
Clinically, hypertension is classified as primary or essential (95% of cases) and
secondary or symptomatic. The causes of *primary hypertension* are not clear
but may include:

1 hereditary factors;
2 a large Na^+ intake in the diet; and
3 psychological factors.

Secondary hypertension can be the result of:

1 a renal disease causing renal vasoconstriction, reduced renal blood flow,
the release of renin and a consequent increase in blood volume;
2 endocrine disorders leading to increased adrenaline secretion or to increased
aldosterone or glucocorticoid secretion resulting in Na^+ retention; or
3 toxaemia in pregnancy.

In hypertensives, the baroreceptor reflex modulates *MAP* around an
elevated set-point. Hypertension can lead to further renal damage, more severe
hypertension, ischaemic heart disease, congestive heart failure and strokes.

Hypotension, fainting and shock

Arterial systolic pressures below 100 mmHg can be taken to indicate
hypotension. Such pressures are associated usually with decreases in $\dot{Q}$ but
sometimes with decreases in *TPR* and occasionally in both. Hypotension may
be the result of:

1 endocrine disorders, for example Addison's disease (Section 8.5);
2 cardiovascular disorders such as aortic or mitral stenosis or cardiac failure;
3 loss of tone in resistance and capacitance vessels;
4 allergic or toxic reactions causing vasodilatation;
5 haemorrhage;
6 extreme pain;
7 adoption of the standing posture;
8 coughing or straining (e.g. in defaecation), if the intrathoracic pressure is raised sufficiently to impair venous return; and
9 insufficient $\dot{Q}$ during exercise to compensate for the vasodilatation in the skeletal musculature.

The commonest manifestation of acute hypotension is *fainting (syncope)*, which is the sudden loss of consciousness due to cerebral ischaemia. It is usually of short duration since the resulting supine position helps elevate the *MAP*. Fainting that results from sudden bradycardia and diffuse vasodilatation is referred to as *vasovagal syncope*. If this is precipitated by strong emotion the term *psychogenic syncope* is used.

Hypotension becomes critical only when inadequate blood flow disrupts organ function. This is referred to as *shock*. The signs of shock are pallor, coldness of the skin and sweating, collapse of superficial veins, hyperventilation, tachycardia, reduced urine formation, thirst, decreased body temperature, decreased metabolism and metabolic acidosis. If shock is untreated or sufficiently severe, it is likely to become irreversible with severe hypotension, unconsciousness, CNS damage, hypoventilation and further reductions in $\dot{Q}$ which will lead eventually to respiratory and cardiac failure and to death. Inadequate cardiac output can arise either from sudden failure of the heart to pump as in coronary occlusion or from a severe reduction in blood volume due to dehydration, haemorrhage or plasma loss resulting from burns.

Haemorrhage

Loss of 10% of the blood volume (about that given by a blood donor) is easily compensated for by a passive decrease in venous capacity and thus results in little change in *MAP*. Loss of up to 30% over a relatively brief period (e.g. 30 min) will cause a fall in *MAP* to about 70 mmHg and signs of mild shock. The cardiovascular reflexes initiated by such a fall eventually will compensate completely for the loss of blood volume. However, the rapid loss of a greater volume of blood will result in severe shock which may become irreversible if not treated by blood transfusion or by raising the blood volume by administration of a fluid with similar colloid osmotic pressure to plasma.

Following a *haemorrhage* the decrease in blood volume reduces end-diastolic volume and hence stroke volume, $\dot{Q}$ and *MAP*. Detection of these reductions by the cardiac stretch receptors and arterial baroreceptors will result in an immediate increase in sympathetic activity and adrenaline secretion and a delayed increase in the secretion of angiotensin II, antidiuretic hormone and aldosterone. The consequent responses (p. 442) are the same as those which follow a change from the supine to the erect posture (p. 459) but are of greater magnitude. These responses may serve to restore $\dot{Q}$ and *MAP* towards normal. Further restoration can occur only by restoring the blood volume. After about 15 min, at least 500 ml of plasma volume will have been provided by the transfer of interstitial fluid into the capillaries as a consequence of the reduced capillary pressure that has resulted from the hypotension itself and the vasoconstriction. In the next few hours, further restoration of circulating volume occurs because of the increased Na^+ retention by the kidneys and the resultant quenching of thirst. In contrast, replacement of the plasma proteins by synthesis requires 3–6 days and the formation of new erythrocytes over a period of 6 weeks or more.

The progress of these adjustments to haemorrhage can be examined using the graph of ventricular and systemic vascular function curves (Fig. 11.49), the basis of which was described on p. 437. In Fig. 11.49, position A is the normal pre-haemorrhage value, $\dot{Q}$ being about 6 l min^{-1} and mean right atrial pressure about 1 mmHg. Immediately after the loss of about 25% of the blood volume, the new position will be indicated by point B which lies on the normal

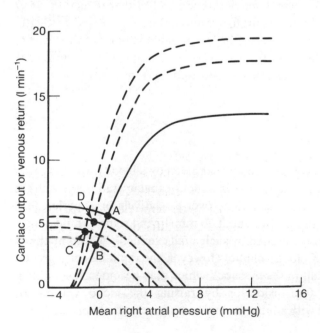

Fig. 11.49. Ventricular and systemic vascular function curves in response to haemorrhage. Position of function: A = pre-haemorrhage; B = immediately after haemorrhage; C = after reflex increase in myocardial contractility and venoconstriction; D = after restoration of some of blood loss.

ventricular function curve but on a systemic vascular function curve which is shifted substantially downwards. The reflex increase in myocardial contractility moves the ventricular function curve upwards and steepens it and at the same time the reflex venoconstriction moves the systemic vascular function curve upwards. Thus position C is reached. After about 15 min, restoration of some of the lost volume will move the systemic vascular function curve upwards still further and the improved *MAP* will result in less sympathetic drive to myocardial contractility and hence a flatter ventricular function curve. Thus position D will be reached. After a few more hours, the cardiovascular system will have returned to the original position of A.

Cardiac failure

Substantial *cardiac failure* (ventricular failure) most commonly results from impaired coronary perfusion (ischaemic heart disease) usually caused by coronary atherosclerosis (thickening and hardening of the arterial wall), from disease of the heart valves and from hypertension. However, a mild reduction in coronary perfusion may cause only *angina pectoris* which is the term applied to the pain arising from myocardial hypoxia and which may be triggered by exercise.

In heart failure myocardial contractility is insufficient to eject an adequate stroke volume. Thus $\dot{Q}$ and *MAP* are decreased, blood accumulates in the venous side of the circulation with consequent elevation of venous pressures. The heart is engorged with unejected blood and end-diastolic volume and right and left atrial pressures increase. Furthermore, inadequate organ perfusion and even shock may result. The decreased $\dot{Q}$ reduces renal blood flow so that Na^+ and water are retained. When fluid retention reaches 3–4 litres, oedema results as a consequence of the disturbances in the Starling equilibrium. Pulmonary congestion and oedema are exaggerated by left ventricular failure in which the large increase in left atrial pressure causes considerable passive expansion of the pulmonary vascular bed and a rise in pulmonary capillary pressure. Other respiratory symptoms that may accompany cardiac failure are cyanosis and hyperventilation often with dyspnoea.

The sequence of events in cardiac failure can be examined using the ventricular and systemic vascular function curves (Fig. 11.50). Acute myocardial failure results in position B on a flatter ventricular function curve. The accumulation of blood on the venous side shifts the systemic vascular function curve upwards and so position C is achieved. The next stages involve reflex adjustments. Arterial baroreceptors detect the hypotension and their reflexes dominate the opposing reflexes triggered by the increased end-diastolic volume detected by the cardiac stretch receptors. The increases in sympathetic activity and adrenaline release result in tachycardia, increased myocardial

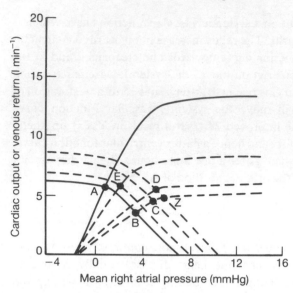

Fig. 11.50. Ventricular and systemic vascular function curves in response to cardiac failure. Position of function: A = pre-failure; B = onset of cardiac failure; C = as a consequence of blood accumulation in the venous system; D = after reflex adjustment providing heart can respond otherwise position Z achieved. E = after clinical treatment.

contractility, vaso- and venoconstriction. However, though position D might be reached with upward shifts of both curves, the failing ventricle may be unable to respond to the increased sympathetic drive or to cope with the relative increase in afterload and thus position Z rather than D may eventuate. Steepening the ventricular function curve by the administration of digoxin may help here. Digoxin and other cardiac glycosides appear to act by inhibiting the sarcolemmal Na^+–K^+–ATPase. As a consequence the Na^+ concentration gradient across the membrane falls, there is a decrease in Na^+–Ca^{2+} counter-transport (pp. 12 & 13) and more Ca^{2+} is taken up into the sarcoplasmic reticulum for release during subsequent action potentials.

A further problem during cardiac failure lies in the fact that reduced renal perfusion increases renin release and hence circulating angiotensin II and aldosterone. The consequent fluid retention and oedema shifts the systemic vascular function curve upwards still further. This retained fluid volume can be reduced by the administration of diuretics and the capacitance of the vascular system can be increased by the administration of vaso- and venodilatating drugs. If this is successful, the systemic vascular function curve will move downwards, the ventricular function curve will move upwards to position E and finally the original position A with a normal mean systemic filling pressure will be achieved.

12 · Respiration

12.1 Introduction
Structure of the lung

12.2 Lung mechanics
Lung volumes
Pressure–volume relationships
Pleural space
Lung surfactant
Maximal respiratory forces
Air flow
Work of breathing

12.3 Pulmonary ventilation and gas transfer
Respiratory gas mixtures and gas
    exchange
Conditions of measurement
Dead space and alveolar ventilation
Gas pressures
Physiological dead space
Alveolar ventilation and gas equation
Ventilation and perfusion
Diffusion

12.4 Blood gas transport
Oxygen carriage
Carbon dioxide carriage
Acid–base balance

12.5 Regulation of respiration
Respiratory centres
Respiratory chemoreceptors
Respiratory mechanoreceptors

12.6 Respiration under abnormal conditions
Decrease in barometric pressure
Acute hypoxia
Acclimatization
Increased ambient pressure

12.1 Introduction

The function of respiration is to ensure that the needs of the tissues for oxygen and for the removal of carbon dioxide—the metabolic demands of the body—are met. Thus it may be said that the lungs exist to ventilate the blood, and their structure is such as to enable blood and gas to come into relation with each other so that O_2 is transferred from the gas mixture in the lungs to the blood that flows through them and CO_2 is transferred in the opposite direction. It is true that the lung tissues have other functions: for example, the pulmonary capillary bed may act as a filter preventing particles greater than a certain size from reaching the systemic circulation, airborne particles may be cleared by mucociliary action or coughing or phagocytosis and various metabolic operations may take place such as conversion of angiotensin I to angiotensin II and synthesis and removal of prostaglandins. However, only the respiratory function of the lungs is considered in this book.

Gas transfer is only one of the fundamental respiratory processes. Gas is moved in and out of the lungs as a consequence of the action of the muscles of respiration—*pulmonary ventilation*—while at the same time the heart pumps blood through the lung capillaries—*perfusion*. If gas exchange is to be adequate and efficient, ventilation and perfusion, or rather the *ventilation/perfusion ratio*, must be reasonably uniform throughout the lungs. Analysis of blood that has

465

come into equilibrium with any mixture of O_2 and CO_2 shows that the concentration of these gases in the blood is much higher than would be expected on purely physical grounds. *Carriage of O_2 and CO_2 in the blood* between the lungs and the tissues of the body depends in fact on chemical processes. These involve, very importantly, the haemoglobin in the red cells of the blood.

Finally, while in the normal healthy person at sea level ventilation and such measures of lung function as the pressures of O_2 and CO_2 in the arterial blood remain remarkably constant at rest, the metabolic demands may change drastically, as in severe exercise, or the conditions under which O_2 is supplied may alter, as at high altitude. Under such circumstances, the volumes of gas and blood coming into contact in the lungs for purposes of exchange alter appropriately as a result of reflex action so that we may talk about the *reflex regulation of ventilation* (and also, of course, of the circulation).

The basic respiratory processes may thus be listed in the order in which we shall deal with them, as follows.

1 Lung mechanics.
2 Pulmonary ventilation and gas transfer.
3 Blood gas transport: (i) oxygen carriage
                                      (ii) carbon dioxide carriage.
4 Regulation of respiration.
5 Respiration under abnormal conditions.

For many years, respiratory physiologists and physicians have used an agreed set of symbols and abbreviations. They are set out in Table 12.1.

Structure of the lung

Gas exchange takes place across the *alveolo–capillary membrane* which is on average 1 μm thick and which has a very large total surface area. The passages leading to the *alveoli* serve for the conduction of gas and also, in the case of the inspired gas, for its humidification and warming to 37°C.

On inspiration, gas passes through the *upper respiratory tract* (nose, pharynx, larynx) into the *trachea*, which divides into the two main *bronchi*, one to each lung. These repeatedly subdivide within the lung, becoming smaller and changing in structure as they do so, until after some twenty-three 'generations' the alveoli are reached. The trachea, and the larger and smaller bronchi (down to the eleventh generation) have, in common, cartilaginous rings and bands of smooth muscle in their walls and a columnar ciliated epithelium with many mucus-secreting cells. The twelfth to sixteenth divisions comprise the *bronchioles* in which cartilage is lacking, smooth muscle is predominant and the lining epithelium is cuboidal. The remaining bifurcations

Table 12.1. Glossary of symbols.

Gas exchange*		
I. General variables	V	Gas volume in general. Pressure, temperature and percentage saturation with water vapour must be stated
	$\dot{V}$	Gas volume per unit time
	P	Gas pressure in general
	F	Fractional concentration in dry gas phase
	$\dot{Q}$	Volume flow of blood
	C	Concentration in blood phase
	f	Respiratory frequency—breaths per unit time
	R	Respiratory exchange ratio in general (volume CO_2/volume O_2)
	D	Diffusing capacity in general (volume per unit time per unit pressure difference)
II. Symbols for the gas phase	I	Inspired gas
	E	Expired gas
	A	Alveolar gas
	T	Tidal gas
	D	Dead space gas
	B	Barometric
III. Symbols for the blood phase	b	Blood in general
	a	Arterial
	v	Venous
	c	Capillary
IV. Special symbols and abbreviations	$\bar{x}$	Dash above any symbol indicates a mean value
	$\dot{x}$	Dot above any symbol indicates a time derivative
	s	Subscript to denote the steady state
	STPD	Standard temperature, pressure, dry (0°C, 760 mmHg)
	BTPS	Body temperature, pressure, saturated with water
	ATPD	Ambient temperature, pressure, dry
	ATPS	Ambient temperature, pressure, saturated with water
V. Examples	F_{A,O_2}	= Fraction of O_2 in the alveolar air
	$\dot{V}_{O_2}$	= Volume of oxygen consumed per minute
	P_{c,O_2}	= Partial pressure of oxygen in the capillary blood
Respiratory mechanics†	l	Total lung
	w	Chest wall
	rs	Total respiratory system, i.e. l+w
	ao	Airway opening
	alv	Terminal airspaces in lungs
	pl	Pleural surface
	bs	Body surface
	mus	Respiratory muscles
	P	Pressure in cmH_2O relative to atmospheric unless otherwise specified
	C	Compliance, the slope of a static volume–pressure curve at a point expressed in $l\,cm^{-1}\,H_2O$ or $ml\,cm^{-1}\,H_2O$

* From Otis, A.B. (1964) Quantitative relationships in steady-state gas exchange. In *Handbook of Physiology*, Section 3, *Respiration*, eds. Fenn, W.O. & Rahn, H. Vol. 1, pp. 681–698. American Physiological Society, Washington.
† After Mead, J. & Milic-Emili, J. (1964) Theory and methodology in respiratory mechanics with glossary of symbols. In *Handbook of Physiology*, Section 3, *Respiration*, eds. Fenn, W.O. & Rahn, H., Vol 1, pp. 363–376. American Physiological Society, Washington.

(*respiratory bronchioles, alveolar ducts, alveolar sacs*) give rise to increasing numbers of alveoli and thus increasingly subserve gas exchange, until finally the alveolar sacs end blindly in alveoli (Fig. 12.1).

The alveolo–capillary membrane may be further subdivided, in terms of the layers through which a gas molecule must pass, into a fluid lining layer, the alveolar epithelium, an interstitial space and the capillary endothelium (Fig. 12.2). The alveolar epithelial cells are of two kinds, type I, comprising the great majority, and type II. Type I are thin flat cells. Type II are cuboidal and are believed to be the source of lung surfactant (p. 475).

12.2 Lung mechanics

Gas can flow only along a pressure gradient, developed in the case of the lungs between the alveoli and the airway opening. The amounts of gas that can be moved in and out of the lungs in unit time depend on the force which the muscles of respiration can exert and on the forces which have to be overcome. The latter in turn depend on the stiffness or otherwise of the lungs and chest wall and on the ease with which air can flow through the respiratory passages.

The elevation of the ribs by muscular action causes an increase in both the anteroposterior and side-to-side diameters of the thorax. This is because of the 'bucket handle' geometry of the ribs which slope (a) downwards and forwards and (b) downwards and laterally. Contraction of the *diaphragm* has the double action of depressing the floor of the thorax and raising the ribcage at its point of origin, thus enlarging the thorax and causing an inspiration. Its motor nerve is the *phrenic*, arising from the cervical roots 3 to 5 of the spinal cord. The action of the *intercostal muscles* is primarily to support the chest wall in the intercostal spaces. The contraction of the essentially vertical fibres of the external intercostals favours inspiration and contraction of the more horizontal fibres of the internal intercostals favours expiration. The contribution of these muscles is, however, pronounced only in postural activity, for example, during lateral bending of the trunk. Contraction of such muscles as the *sternomastoid* raises the ribcage in strenuous inspiration. Contraction of *abdominal* muscles can drive the diaphragm upwards to cause expiration. During quiet breathing, however, expiration is caused by passive *elastic recoil* of elements stretched during the previous inspiration. The duration of expiration is prolonged because of some inspiratory muscle action continuing during expiration. Such an expiratory 'brake' effect also arises as a result of constriction of the glottis by the laryngeal muscles.

Lung volumes

The normal subject at rest breathes in and out a volume of gas, the *tidal volume* (V_T), that is very much less than that which can be moved in and out of the

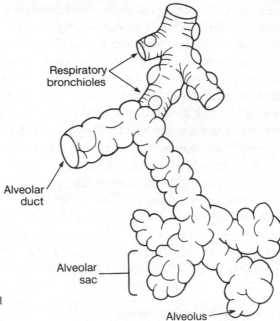

Fig. 12.1. Structure of the terminal airways of the lung.

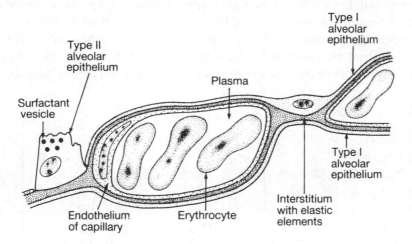

Fig. 12.2. Structure of the shared wall of two alveoli.

lungs with maximal effort. That is to say, there are normally *inspiratory* and *expiratory reserve volumes* (IRV, ERV) of considerable size. The volume of gas that can be exhaled following a maximal inhalation is the *vital capacity* (VC). One cannot, however, expel by voluntary effort all the gas in the lungs; there remains, after a maximal exhalation, a *residual volume* (RV) which can be got rid of only if the lungs are made to collapse completely as they do if the chest wall is opened widely. The volume of gas in the lungs following a maximal inspiration, the *total lung capacity* (TLC), is the sum of the VC and the RV. The volume of gas in the lungs at the resting position of the chest after a quiet expiration is called the *functional residual capacity* (FRC). The *inspiratory capacity* (IC) is the difference between TLC and FRC. These relationships are best seen from the accompanying figure (Fig. 12.3). These measurements depend on the age, sex and build of the subject, and are essentially static in nature. To say that the VC, for example, has a particular value does not tell us whether the subject requires a long or short time to inspire or expire that amount of gas. More information can be obtained by asking him to inspire to the maximum and then to expire as rapidly and completely as possible. This volume is termed the *forced vital capacity* (FVC) and the amount of gas which can be exhaled thus in 1 second, the *forced expired volume in 1 s* (FEV_1), as a percentage of the FVC is a useful measure of lung function. In healthy subjects, the ratio FEV_1/FVC is of the order of 80%.

Pressure–volume relationships

In order to see, quantitatively, just how the lungs and chest wall function as a system and thus just what forces are involved in moving a given volume of gas in and out of the lungs, it is easiest, perhaps, to imagine a simple model comprising a balloon in a bottle (Fig. 12.4). The bottle has had its bottom removed and replaced by a flexible diaphragm. The balloon can be connected by a two-way stopper either to the atmosphere or to a manometer A. A second

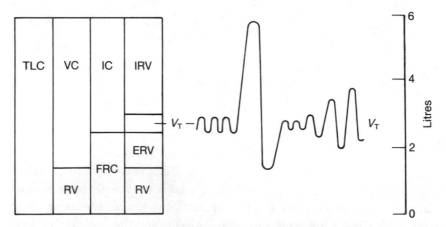

Fig. 12.3. Subdivisions of total lung capacity. IRV, inspiratory reserve volume; ERV, expiratory reserve volume; VC, vital capacity; RV, residual volume; TLC, total lung capacity; FRC, functional residual capacity; IC, inspiratory capacity.

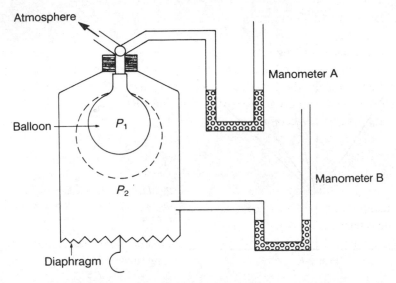

Fig. 12.4. Simple apparatus used to demonstrate pressure–volume relationships. $P_{inside} - P_{outside}$, $(P_1 - P_2)$, is the transmural pressure gradient.

manometer B is connected to the space between the balloon and the bottle wall. Initially, the whole system can be regarded as being at atmospheric pressure.

If now the diaphragm is pulled down, the balloon will expand to an extent that depends on its *distensibility*. If the diaphragm is held in the new position and the balloon remains connected to the atmosphere, the pressure within it is obviously atmospheric while that registered by manometer B will be less than atmospheric. A transmural pressure gradient has developed across the balloon wall. If next the diaphragm is increasingly pulled down, the volume of air in the balloon will increase as the transmural pressure gradient increases. Such measurements permit the drawing of a *pressure–volume curve* which will display the distensibility of the balloon (Fig. 12.5a). If now the balloon is stretched to the same series of volumes but the diaphragm punctured (and repaired again!) at each volume, then the balloon will collapse to its original volume as the transmural pressure gradient is abolished. If, however, just prior to making the puncture at each volume the tap is turned to manometer A, the balloon cannot collapse and the manometer registers a pressure that is now greater than atmospheric by exactly the amount that the pressure registered in manometer B was less than atmospheric at each volume in the first series of readings. The transmural gradient has remained unchanged; it is simply measured in a different way (Fig. 12.5b). In the first instance, P_1 is always zero and each P_2 has a minus value. In the second, each P_1 is positive and P_2 is always zero. Simple algebra makes it clear that either method gives the same

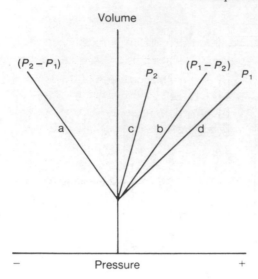

Fig. 12.5. Pressure–volume relationships for the balloon (a & b), diaphragm (c) and balloon and diaphragm as a whole (d). See text for further explanation.

set of answers (i.e. b in Fig. 12.5). However since in real life (see below) it is the equivalent of P_2 that is always measured, it is often convenient to think in terms of $P_2 - P_1$ (i.e. curve a). If, finally, the balloon is expanded to the same series of different volumes but now at each volume the balloon is quickly connected to manometer A and the diaphragm simply allowed to spring back without any question of puncture, manometer B will register a series of pressures greater than atmospheric. These represent the distensibility of the diaphragm and permit the drawing of its pressure–volume curve (Fig. 12.5c). Manometer A will register a series of even greater pressures, each of which will be the sum of the pressure now registered on B and the transmural pressure, b, previously registered on A at that volume. Now the measurements on A permit the drawing of a pressure–volume curve for the system as a whole (Fig. 12.5d). The transmural pressure across the balloon wall at each volume remains as it was in the first two series of measurements and is given by the difference between the pressures registered by A and B. In other words at any volume the horizontal distance from the y-axis of d minus that of c gives, in absolute terms and disregarding sign, that of either b or a.

Let us now turn from model to physiological reality. In Fig. 12.6, the volume of gas (litres) in the lungs, up to TLC, is plotted against pressure (cmH$_2$O). The four curves, P_{pl}, P_1, P_w and P_{rs} clearly correspond to a, b, c, d in Fig. 12.5 obtained from the series of measurements on the model. Naturally, P_1, the *transmural pressure gradient across the lungs*, is not measured by opening the chest wall. It is obtained by measuring the *intrapleural pressure* (P_{pl}), or in practice the oesophageal pressure, with the chest wall held voluntarily at the appropriate position by muscle action and the airway open so that, there being

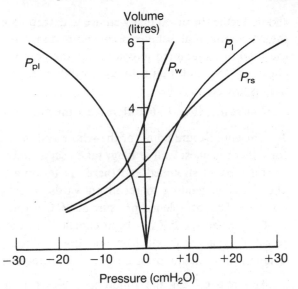

Fig. 12.6. Static pressure–volume curves for the lung (P_{pl} and P_l), for the chest wall (P_w) and for the total respiratory system (P_{rs}).

no air flow, *intra-alveolar pressure* (P_{alv}) is always atmospheric, as was P_1 in Fig. 12.5a. The pressure difference, $P_{pl} - P_{alv}$, (i.e. P_{pl}), thus measured is analogous to $P_2 - P_1$ in Fig. 12.5a. In the same way, just as b in Fig. 12.5 measured $P_1 - P_2$ when P_2 was at atmospheric pressure throughout, so, in Fig. 12.6, P_l has the same value as P_{pl}, but with the sign changed. It measures the tendency of the lungs to recoil, as they would do if the chest were opened and P_{pl} became atmospheric. It is thus clear that, in the intact subject, P_{pl} indicates the force necessary to keep the lungs expanded to any given volume while P_{alv} is at atmospheric pressure.

The *pressure across the chest wall*, P_w, at any lung volume can also be measured from the oesophagus if the respiratory muscles are briefly relaxed and the airways are occluded. Since body surface pressure is atmospheric the various numerical values represent P_{pl} under these particular conditions and correspond to Fig. 12.5c. In the absence of muscle action and with the glottis closed, the alveolar pressure now expresses the combined tendencies of the lungs and chest wall to recoil. The pressure in the alveoli under these conditions (the *relaxation pressure, P_{rs}*) is then the pressure gradient across the system as a whole.

It should be noted that the conditions under which P_{pl} and P_w are measured are mutually exclusive. The first can be measured only if at each point except the equilibrium point (FRC) there is muscle action; the second can be measured only if at each point there is no muscle action and, again except at FRC, the lungs are prevented from altering their volume by brief occlusion of the airway. The slope of each curve measures the *compliance*, that is (the change in volume per unit change in pressure) of the structure in question. It should further be noted that were we to extrapolate in Fig. 12.6 and plot $-P_{rs}$, then the horizontal distance of any point on it from the y-axis would represent the total force required to counteract the combined tendencies of lungs and chest wall to recoil; that is to say it would measure the force being exerted by the respiratory muscles.

The reality (Fig. 12.6) differs from the model (Fig. 12.5) as follows.

1 The equilibrium point for the whole system, FRC, is not the point of rest for either the chest wall or lungs taken separately. At FRC, the lungs are still to some extent stretched and therefore tending to recoil inwards, while the chest wall is tending to recoil outwards, the two forces being equal and opposite. The intrapleural pressure at FRC is thus less than atmospheric.
2 Only when the RV has been expelled are the lungs no longer tending to recoil. This can result only from opening the chest with the airway unoccluded, when the intrapleural pressure becomes atmospheric.

The reasons for the differences between model and reality may be understood if we consider the compliance of lungs removed from the body and filled to different volumes with saline rather than gas. The compliance is much greater under these circumstances, and so most of the 'stiffness' of the lungs *in vivo* is due to the presence of an air–liquid interface, in other words to the force of *surface tension*.

Pleural space

Why is this tendency of the lungs to retract, with the consequent generation of a pressure less than atmospheric, not counteracted by the accumulation of gas or liquid or both in the pleural cavity?

Absorption of liquid from the pleural cavity depends on the balance of osmotic and hydrostatic forces across the capillary walls (Section 1.8) and pleural membranes and on the relative vascularities of parietal and visceral pleurae. Hydrostatic pressure in the systemic capillaries is such that one might suppose filtration to exceed reabsorption through the parietal pleura in the face of a subatmospheric intrapleural pressure. In the pulmonary capillaries, however, the hydrostatic pressure is very low and is substantially exceeded by the plasma colloid osmotic pressure and in addition the visceral pleura is the more vascular. All in all, the balance is in favour of reabsorption of any fluid from the intrapleural space, so much so that total reabsorption is prevented, and a thin lubricating layer of fluid left, only by an increase in the tension and decrease in the permeability of the visceral pleura as the last moiety of fluid is tending to be reabsorbed.

Gas is absorbed from the pleural cavity because, basically, as a result of the shapes of the respective O_2 and CO_2 dissociation curves, the sum of the partial pressures of the gases in capillary blood near the venous end of the capillary is always less than that of any gas introduced into the pleural cavity and thus the pressure gradients always favour absorption (Table 12.3, p. 484). This similarly ensures the absorption of gas from a segment of lung should a bronchiole or bronchus be blocked. The rate of absorption depends on the composition of any such gas pocket. Pure O_2 is absorbed very rapidly. The low solubility and biological inertness of N_2 makes the absorption of air a much slower process.

Since, then, neither gas nor liquid can accumulate in the pleural cavity, but also since the gas-containing lungs tend to recoil, the relatively flexible chest wall is the only structure that can, so to speak, 'move in' to replace the retracting lung. This it does until in the equilibrium position the two pressure gradients, $P_{pl} - P_{bs}$ and $P_{alv} - P_{pl}$ (i.e. P_l), are equal and opposite (of the order of ± 5 cmH$_2$O).

Thus the forces that determine the behaviour of the lungs and chest wall as a respiratory pump owe their existence primarily to the development of an air–liquid interface in the course of the first few breaths at birth, and to the shape of the O_2 dissociation curve (see Fig. 12.13), which is an expression of the peculiar relation that exists between O_2 and Hb.

Lung surfactant

In fact, the force of surface tension is much less than might be expected from the behaviour of a simple air–liquid surface film. If such a film in each tiny alveolus were to obey Laplace's law ($P = 2T/r$, where P is the transmural pressure, T is the surface tension and r is the alveolar radius), we would expect:

1 the lungs to be much stiffer than they actually are;
2 that smaller alveoli would transfer their gas to larger (since it may be assumed that not all alveoli are exactly the same size);
3 that, on expiration, as alveoli anyhow become smaller, this tendency would be accelerated in the direction of total collapse of the lung, necessitating a corresponding inspiratory effort; and
4 that fluid would tend to accumulate in the alveoli, because the additional force ($\simeq 20$ cmH$_2$O across the barrier) would now more than balance the excess of osmotic over hydrostatic pressure which would otherwise keep the lungs dry.

That all these things do not normally happen is due to the presence of an alveolar lining layer of *lung surfactant*, a lipoid substance produced by the type II alveolar epithelial cells. This substance:

1 reduces the absolute value of the surface tension to approximately 4 cmH$_2$O; and

2 behaves anomalously in that its effect becomes greater as the lung surface is reduced in expiration, in other words as the surface energy is concentrated, and conversely on inspiration.

Surfactant thus increases not only the compliance but also the stability of the lungs. It develops late in intra-uterine life and premature infants, lacking it, must exert a large effort with each breath.

Maximal respiratory forces

The pressures shown in the pressure–volume diagram (Fig. 12.6) are by no means the greatest that can be exerted by the muscles of respiration. If we breathe maximally in or out against a closed glottis (Müller's and Valsalva's manoeuvres, respectively), then much greater pressures, both positive and negative, are generated in the alveoli and the pleural cavity, but the transmural pressure, P_l, remains unchanged at any lung volume. Large positive intrapleural pressures impede the venous return to the heart (p. 436).

Air flow

So far we have discussed the effort involved in overcoming purely elastic forces, but under dynamic as opposed to static conditions we must allow for the extra effort necessary to propel gas through tubes.

Resistance to gas flow is not easy to quantify. Flow can be laminar or turbulent or both at the same time in different parts of the respiratory tract. Indeed, the anatomy of the nasal bones is designed to promote turbulent flow. Turbulent flow is also likely to occur if volume per unit time ($\dot{V}$) is high, where the tubes are branched or angled, and especially if there is any undue increase in the thickness of the mucous layer in the tubes. For these reasons, clinical descriptions of resistance to air flow are usually qualitative rather than quantitative.

Airways resistance is critically dependent on the internal diameters of the bronchi and bronchioles. Both the tension in their smooth muscle walls and the thickness of the mucous membrane can be affected by circulating catecholamines or autonomic nerves. Sympathetic activity causes *bronchodilatation* which will naturally increase the anatomical dead space. Conversely, parasympathetic activity causes *bronchoconstriction* and decreases the anatomical dead space. The accompanying increase in airways resistance is of little consequence at rest. However, in conditions such as *asthma*, bronchoconstriction, often mediated by histamine acting on the smooth muscle, may cause very high airways resistance. Here the efficacy of inhaled or injected sympathomimetics, or other bronchodilators, may be life-saving.

In normal quiet breathing, however, air flows into the lungs in inspiration along a pressure gradient, $P_{alv} - P_{ao}$, of approximately -2 cmH$_2$O. During expiration, this gradient is reversed. This is shown in Fig. 12.7. As the lungs fill, in this example by 0·5 l, the intrapleural pressure (P_{pl}) changes as shown by the line ACB; on expiration, the path is shown by BDA. Intra-alveolar pressures will be, correspondingly, to the left or right of the *y*-axis on which they lie when there is no air flow.

Under these circumstances, activity on the part of the expiratory muscles is not required to expel the air. Sufficient elastic energy has been stored up in the lungs during inspiration to supply what is required for expiration. In fact, activity on the part of expiratory muscles is likely to increase air flow only to a certain extent before it is balanced, or indeed more than compensated for, by reduction in airway calibre by compression.

The work of breathing

A constant pulmonary minute volume may be maintained by breathing deeply and relatively slowly or by a breathing pattern that is shallower and relatively rapid. These two modes of breathing do not affect in the same way the amount of respiratory work that is done. The work that is done against elastic resistance is independent of time and increases with increased inflation of the lungs

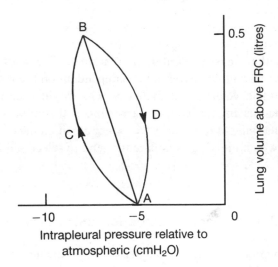

Fig. 12.7. Dynamic pressure/volume loop in which lung volume above FRC (litres) is plotted against intrapleural pressure relative to atmospheric pressure (cmH$_2$O). ACB represents inspiration and BDA expiration. The line (AB) joining the points of no flow corresponds to P_{pl} in Fig. 12.6. The horizontal distance between any point on the curve and that line represents the force required to move air through the resistance of the respiratory passages over and above that required to alter the lung volume.

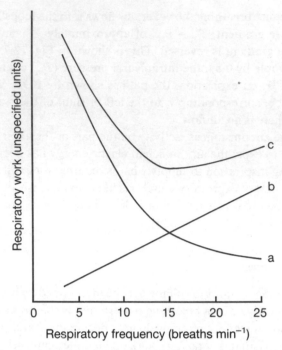

Fig. 12.8. Work done against (a) elastic resistance, (b) air-flow resistance and (c) the sum of the two, at different breathing frequencies. For any given minute volume, $\dot{V}_E$, work will be minimal at higher frequencies the stiffer the lungs and at lower frequencies the greater the resistance to air flow.

particularly if the lungs are stiffer than normal. The work done against resistance to air flow on the other hand is dependent on the rate at which the lungs change their volume and increases with rapid shallow respiration, especially if there is any tendency to narrowing of the airways. There is thus for any particular lungs a respiratory frequency at which the total work will be minimal, and which individuals apparently tend to select automatically (Fig. 12.8).

12.3 Pulmonary ventilation and gas transfer

The respiratory process is determined in essence by the metabolic demands of the tissue mass of the body. Since the end of the eighteenth century it has been accepted that this implies the consumption of O_2 and the production of CO_2 and H_2O in proportions that are fixed by the chemical constitution of the substances that are being combusted, and that N_2 is chemically inert and plays a purely passive part in respiration.

Respiratory gas mixtures and gas exchange

On inspiration, a volume of air from the surrounding atmosphere is taken into the respiratory tract. A portion of it remains in the upper part of the tract, where no gas exchange with the blood takes place (*dead space*), and part of it mixes with the gas that remains in the lungs at the end of the previous expiration. This gas mixture that obtains at any moment in the depths of the lung is called the *alveolar gas*. On expiration, a volume of gas equal to that inspired leaves the respiratory tract. Analysis (Table 12.2) of what goes in and what comes out, the *inspired* and *expired gas*, shows that while the former has a fixed composition from breath to breath (and indeed in the long term unless the atmosphere has in some way been vitiated), the latter may vary if the pattern of breathing changes.

In Table 12.2, which also shows the fractional composition of alveolar gas (F_A), we see that in this example the O_2 fraction has fallen from 0·21 to 0·14 while the CO_2 fraction has increased by a similar, but not precisely the same, amount; moreover, the N_2 fraction (which includes the rare gases) has also changed by a very small amount.

Such an analysis, if we also know the volumes of gas inspired and expired per unit time ($\dot{V}_I$, $\dot{V}_E$), enables us to calculate how much O_2 and CO_2 have been exchanged. Thus for any gas, X, the amount exchanged per unit time is given by the expression:

$$\dot{V}_X = \dot{V}_I \cdot F_{I,X} - \dot{V}_E \cdot F_{E,X}.$$

The calculation is easiest for CO_2 production, $\dot{V}_{CO_2}$, since there is effectively none in inspired air and so:

$$\dot{V}_{CO_2} = -\dot{V}_E \cdot F_{E,CO_2}.$$

Although often neglected, the minus sign shows that, precisely speaking, CO_2 is being produced, O_2 being consumed, the two thus moving in opposite directions. In the case of O_2 consumption, $\dot{V}_{O_2}$,

$$\dot{V}_{O_2} = \dot{V}_I \cdot F_{I,O_2} - \dot{V}_E \cdot F_{E,O_2}.$$

Measurement of $\dot{V}_E$ is easy, but that of $\dot{V}_I$ leads to a difficulty. The ratio of CO_2 produced to O_2 consumed, $\dot{V}_{CO_2}/\dot{V}_{O_2}$, is called the *respiratory exchange ratio* (R). In the steady state, it is common to use the term *respiratory quotient* (RQ)

Table 12.2. Composition of inspired, expired and alveolar gas.

	Fraction of inspired gas (F_I)	Fraction of expired gas (F_E)	Fraction of alveolar gas (F_A)
O_2	0·21 (0·2093)	0·16	0·140
CO_2	0·00 (0·0003)	0·04	0·055
N_2	0·79 (0·7904)	0·80	0·805

which depends purely on the chemical nature of the combusted foodstuffs. In theory, if pure carbohydrate were being burnt, the value of R would be 1.0, while if pure fat were the fuel the value would be about 0.7. Since normally the tissues burn a mixture of carbohydrate, fat and protein, a value of $R = 0.82$ is often assumed when it is not measured directly. In other words, the volume of O_2 taken up into the blood in the lungs is generally rather more than the amount of CO_2 excreted. This means that any volume of expired air has usually been manufactured, so to speak, from a somewhat larger volume of inspired air. Under any conditions, the exact difference is given if we compare the N_2 fractions in inspired and expired air. Since N_2 is inert,

$$\dot{V}_I \cdot F_{I, N_2} - \dot{V}_E \cdot F_{E, N_2} = 0 \qquad \therefore \dot{V}_I = \dot{V}_E \cdot F_{E, N_2} / F_{I, N_2}.$$

The difference is small, but important theoretically and for exact work.

Conditions of measurement

Gas in the lungs is at body temperature and pressure and is saturated with water vapour at that temperature (BTPS). Expired gas is measured, however, at room (or ambient) temperature which as a rule is lower than body temperature and so some of the H_2O vapour condenses. The gas is now saturated at ambient temperature and pressure (ATPS). Since we commonly wish to know the volume a mass of gas occupied when it was in the lung, the measured volume must be corrected appropriately.

On the other hand, it is usually necessary to know the volumes of CO_2 produced and O_2 consumed ($\dot{V}_{CO_2}$, $\dot{V}_{O_2}$) as they would be under standard conditions, that is, dry and under a pressure of $760 \, mmHg$ at $0°C$ (STPD) since only then can volumes and moles of gas be related. Normally, any given values for $\dot{V}_{O_2}$ and $\dot{V}_{CO_2}$ may be taken to be expressed thus. The necessary conversions are made by means of the Boyle–Charles formula,

$$\frac{P_1 V_1}{T_1} = \frac{P_2 V_2}{T_2}.$$

It should be noted that the pressure refers to dry gas and when water vapour is present its pressure must be subtracted. Temperature is measured in K (kelvin).

Dead space and alveolar ventilation

Expired gas, collected as is usual in a large bag, has a uniform composition. However, a single expirate V_E, analysed rapidly and continuously as it is expired, has a changing composition. This is shown for CO_2 in Fig. 12.9. To

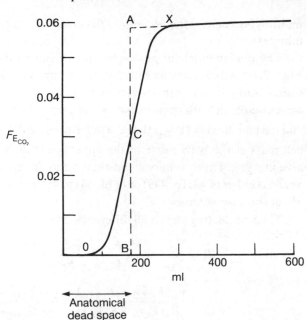

Fig. 12.9. Method for measuring anatomical dead space. See text for further explanation.

begin with, there is no CO_2 in the expired gas; a little later the CO_2 fraction rises rapidly and, finally, in the normal case at least, reaches more or less a plateau. This is because the first portion of the expirate, coming from the upper part of the respiratory tract where no exchange is taking place between gas and blood, is essentially atmospheric air. This portion is followed by gas which, increasingly, has come from the depth of the lungs, that is, *alveolar gas* (V_A). The final part of the expirate may be regarded as pure alveolar gas (Table 12.2). Expired gas is thus a mixture of inspired and alveolar gas and its composition depends in normal healthy subjects simply on the proportion of the one to the other; that is, on the volume of gas which has occupied the non-exchanging part of the respiratory tract (dead space) relative to that which has come into equilibrium with the blood. A perpendicular line ACB drawn so as to make the areas AXC and COB equal (Fig. 12.9) provides one means of measuring the *dead space* (V_D). In the normal adult, this *anatomical dead space* has a value of roughly 150–200 ml and comprises about 0·25 to 0·35 of the tidal volume (V_T). Any increase in V_D indicating a dead space greater than that determined by the structure of the respiratory tract implies in turn *wasted ventilation*.

The volumes that comprise a breath can be written

$$V_E (\text{i.e. } V_T) = V_A + V_D$$

and each volume can be converted into volume/unit time, i.e. *pulmonary ventilation* ($\dot{V}_E$), *alveolar ventilation* ($\dot{V}_A$) and *dead space ventilation* ($\dot{V}_D$) by multiplying each by the *respiratory frequency* (f) in breaths/unit time (usually minutes).

The fundamentals of gas exchange are summarized diagrammatically in Fig. 12.10: Mixed venous blood flowing continuously from the right heart comes into contact with an alveolar gas mixture that is renewed from the surrounding air with each inspiration. In unit time, volumes of oxygen ($\dot{V}_{O_2}$) and carbon dioxide ($\dot{V}_{CO_2}$) that are determined by the metabolism of the total cell mass of the body move in the appropriate direction between blood and alveolar gas. These volumes and their ratio R can then be calculated from analysis either of gas (p. 479) or of blood (p. 490), since we are of course talking about the *same* volumes.

Thus, neglecting any small difference between $\dot{V}_I$ and either $\dot{V}_E$ or $\dot{V}_A$:

$$R = \frac{\dot{V}_{CO_2}}{\dot{V}_{O_2}} = \frac{\dot{V}_E(F_{E,CO_2} - F_{I,CO_2})}{\dot{V}_E(F_{I,O_2} - F_{E,O_2})}$$

$$= \frac{\dot{V}_A(F_{A,CO_2} - F_{I,CO_2})}{\dot{V}_A(F_{I,O_2} - F_{A,O_2})} = \frac{\dot{Q}(C_{\bar{v},CO_2} - C_{a,CO_2})}{\dot{Q}(C_{a,O_2} - C_{\bar{v},O_2})}$$

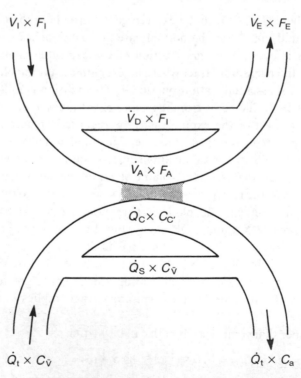

Fig. 12.10. Diagrammatic representation of the flow of gas and blood through the lungs to illustrate the derivation of the dead space and venous admixture equations (see text).

The diagram also shows the basis on which *dead space* and its circulatory analogue, *shunt* or *venous admixture*, may be calculated. In the first case, for either O_2 or CO_2, the volume expired per unit time is clearly the sum of that derived from the dead space gas (in which the concentrations are that of inspired air) and that from alveolar gas. The calculation is easier for CO_2 since there is normally none in inspired air. Thus,

$$\dot{V}_E \times F_{E,CO_2} = (\dot{V}_D \times F_{I,CO_2}) + (\dot{V}_A \times F_{A,CO_2})$$

and since,

$$\dot{V}_A = \dot{V}_E - \dot{V}_D \quad \text{and} \quad F_{I,CO_2} = 0$$

$$\therefore \ \dot{V}_D / \dot{V}_E = \frac{F_{A,CO_2} - F_{E,CO_2}}{F_{A,CO_2}} \quad \textit{(Bohr equation)}.$$

For a single breath we divide by f and write V_D / V_E.

In the second case, the volume of gas transported in the total volume of arterial blood flowing per unit time, i.e. the cardiac output, $(\dot{Q}t)$ comprises that from the shunted blood $\dot{Q}s$ (which has the same gas concentrations as mixed venous blood) and that from blood that has passed through the lung capillaries $(\dot{Q}c)$. It is usual to measure O_2 rather than CO_2 in the blood and to assume complete equilibrium between alveolar gas and end-pulmonary capillary blood $(C_{c'})$. Thus,

$$\dot{Q}t \times C_{a,O_2} = (\dot{Q}c \times C_{c',O_2}) + (\dot{Q}s \times C_{\bar{v},O_2})$$

$$\therefore \ \frac{\dot{Q}s}{\dot{Q}t} = \frac{C_{c',O_2} - C_{a,O_2}}{C_{c',O_2} - C_{\bar{v},O_2}}$$

The analogy with the Bohr equation is obvious.

Gas pressures

Gases diffuse down pressure gradients. By Dalton's law, for any gas X in the mixture M, the partial pressure $P_{M,X} = F_{M,X} (P_B - P_{H_2O})$ mmHg. If body temperature is taken as 37°C, P_{H_2O} has a value of 47 mmHg and so the pressures of the alveolar gases, calculated from Table 12.2, will be as shown in the top row of Table 12.3 and they may be compared with the pressures obtaining in similarly representative samples of end-pulmonary capillary blood $(P_{c'})$, systemic arterial blood (P_a) and mixed venous blood $(P_{\bar{v}})$ respectively.

Three points may be noted from Table 12.3. First, it is usual to regard alveolar gas and end-pulmonary capillary blood as being in equilibrium. Second, although it is usual to assume no difference in P_{CO_2} between alveolar

Table 12.3. Partial pressure of gases in the alveoli and blood.

	Partial pressure (mmHg)			
	O_2	CO_2	N_2	H_2O
Alveolar gas (P_A)	100	39	574	47
End-pulmonary capillary blood ($P_{c'}$)	100	39	574	—
Systemic arterial blood (P_a)	94	39	574	—
Mixed venous blood ($P_{\bar{v}}$)	40	46	574	—

gas and systemic arterial blood, there is a small but distinct difference in the case of oxygen. This difference arises in part because some of the venous blood from the myocardium and that from the bronchial tree drains back into the systemic circulation and is not oxygenated in the lungs (a *physiological* shunt, normally 1–2% of the cardiac output), and in part because ventilation and perfusion are not in the same proportion throughout the lungs. Third, the total gas pressure in mixed venous blood is a good deal less than that of the corresponding alveolar gases in arterial blood. This is crucial for the maintenance of a subatmospheric intrapleural pressure and therefore for the working of the respiratory pump (p. 475).

Physiological dead space

We must now reconsider the concept of dead space. In Table 12.2 and Fig. 12.9, it is, strictly speaking, *end-tidal* rather than alveolar air that has been analysed. In healthy people, the two may be taken as identical, but this is not so in diseased lungs where there may be unequal rates of alveolar emptying or spaces in the lung which are ventilated but not in effect perfused. In the latter case, there may be said to be an *alveolar dead space* lying in parallel, so to speak, with the alveolar gas proper. This is in addition to the anatomical dead space, the two together making up the *physiological dead space*. The latter may be calculated by transforming the Bohr equation into the form:

$$V_D = \frac{P_{a, CO_2} - P_{E, CO_2}}{P_{a, CO_2}} \cdot V_T$$

taking advantage of the equilibrium between alveolar gas and arterial blood in the case of CO_2. The alveolar dead space is then calculated by subtracting an assumed value for the anatomical dead space.

Alveolar ventilation and gas equation

The pulmonary minute volume $\dot{V}_E$ is equal to $f \times V_T$ and an increase in $\dot{V}_E$ can obviously be brought about by an increase in f or in V_T or in both. But the alveolar ventilation $\dot{V}_A = f(V_T - V_D)$ is equally obviously not unaffected by the breathing pattern and will depend on whether a given increase in $\dot{V}_E$ has been brought about mainly by an increase in V_T or in f. For example, if at rest $f = 12$, $V_T = 600$ ml and $V_D = 200$ ml, then $\dot{V}_E = 7.21$ min^{-1}. A 25% increase in $\dot{V}_E$ can be brought about by increasing f to 15 or V_T to 750 ml, but the increase in $\dot{V}_A$ is not the same in both cases and it is $\dot{V}_A$ that matters as far as P_{A,CO_2} and P_{A,O_2} are concerned. This becomes clear when we consider how these two gas pressures are calculated.

From p. 479, $\dot{V}_E \times F_{E,CO_2} = \dot{V}_{CO_2}$. Since there is no CO_2 in inspired air,

$$\dot{V}_{CO_2} = \dot{V}_A \times F_{A,CO_2} \quad \therefore F_{A,CO_2} = \frac{\dot{V}_{CO_2}}{\dot{V}_A} \quad \therefore P_{A,CO_2} = \frac{\dot{V}_{CO_2}}{\dot{V}_A} \times (P_B - P_{H_2O}).$$

Since $\dot{V}_{CO_2}$ is usually measured in ml min^{-1}, STPD, and $\dot{V}_A$ in 1 min^{-1}, BTPS,

$$\therefore P_{A,CO_2} = \frac{\dot{V}_{CO_2} \cdot (P_B - 47)}{\dot{V}_A} \cdot \frac{310 \times 760 \times 10^{-3}}{273 \times (P_B - 47)} = \frac{0.863 \times \dot{V}_{CO_2} \text{ (STPD)}}{\dot{V}_A \text{ (BTPS)}}.$$

Thus, for any steady rate of production of CO_2, P_{A,CO_2} (and by extension P_{a,CO_2}) is inversely proportional to $\dot{V}_A$ and is independent of P_B.

The case of O_2 is less simple because the gas is present in both inspired and expired air but one can see that for any given $\dot{V}_{O_2}$, P_{A,O_2} must approach P_{I,O_2} as $\dot{V}_A$ increases; that is,

$$P_{I,O_2} - P_{A,O_2} \simeq 0.863 \times \frac{\dot{V}_{O_2} \text{ (STPD)}}{\dot{V}_A \text{ (BTPS)}}$$

but $\dot{V}_{O_2} = \dfrac{\dot{V}_{CO_2}}{R} \quad \therefore P_{A,O_2} \simeq P_{I,O_2} - \dfrac{P_{A,CO_2}}{R}$ (see preceding paragraph).

However, it must be remembered that $\dot{V}_A$ calculated from $\dot{V}_E$ is not the same as $\dot{V}_A$ calculated from $\dot{V}_I$ unless $R = 1.0$. When the correction is made, one form of the so-called *alveolar gas equation* is:

$$P_{A,O_2} = P_{I,O_2} - \frac{P_{A,CO_2}}{R} + \left[F_{I,O_2} \times \frac{P_{A,CO_2}}{R} \times (1 - R) \right].$$

It can thus be seen that the alveolar gas mixture (P_{A,CO_2}, P_{A,O_2}) is not just a random matter, but is determined by (i) P_{I,O_2}, (ii) the value of R and (iii) the ratio $\dot{V}_{CO_2}/\dot{V}_A$ (or $\dot{V}_{O_2}/\dot{V}_A$). These alveolar gas pressures are important because they determine the gas pressures in arterial blood.

These relationships between (i) P_{A,CO_2}, $\dot{V}_{CO_2}$ and $\dot{V}_A$ and (ii) P_{A,O_2}, $\dot{V}_{O_2}$ and $\dot{V}_A$ are shown graphically in Fig. 12.11. For any given $\dot{V}_{CO_2}$, P_{A,CO_2} rises

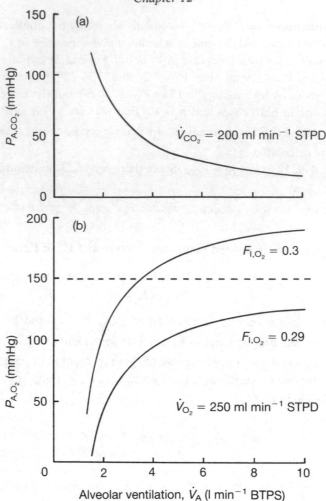

Fig. 12.11. (a) Alveolar P_{CO_2} and (b) alveolar P_{O_2} (mmHg) at different alveolar ventilation volumes, $\dot{V}_A$ (l min^{-1}). The curves have been calculated for $\dot{V}_{CO_2} = 200$ ml min^{-1} and $\dot{V}_{O_2} = 250$ ml min^{-1} ($R = 0.8$). In (b), the lower curve is based on $F_{I,O_2} = 0.209$ and the upper on $F_{I,O_2} = 0.3$.

steeply at the lower levels of alveolar ventilation; similarly, for a corresponding $\dot{V}_{O_2}$, P_{A,O_2} will fall, and quickly become insufficient to saturate the haemoglobin (see Section 12.4). If, however, F_{I,O_2} is stepped up, say to 0.3 from the usual 0.209, P_{A,O_2} may be high enough to saturate the blood at quite low levels of $\dot{V}_A$. Of course, the production of CO_2 is not affected and P_{A,CO_2} will become dangerously high. Oxygen must thus be administered with great care to any patient whose sensitivity to CO_2 may be diminished.

Ventilation and perfusion

No matter what the alveolar ventilation may be or how much blood is sent by the right ventricle to the lungs, unless the two phases, blood and gas, come into effective contact, the arrangement is useless for the business of respiration. We need only think of the extreme example of a block in one main bronchus and a block in that branch of the pulmonary artery supplying the other lung. It is necessary that ventilation, $\dot{V}_A$, and perfusion, $\dot{Q}$, be reasonably matched in each exchanging unit throughout the lungs.

In normal man at rest, a value of $\dot{V}_A$ of about $4.5\,l\,min^{-1}$ is matched with a cardiac output of about $5\,l\,min^{-1}$ giving a $\dot{V}_A/\dot{Q}$ of approximately 0·9. However, this ratio does vary quite substantially in different parts even of the normal lung when the subject is in the upright position; indeed it is not to be expected that all the 300×10^6 exchanging units should behave in exactly the same way. Such variation in ratio as occurs even in normal people is sufficient to account for at least half of the admittedly rather small difference between P_{A,O_2} and $P_{a,O_2}(\Delta P_{A-a,O_2})$.

The reasons for the variations are as follows. The apex of the lung in the erect position is less well perfused than is the base because of the low pressures that obtain in the pulmonary circulation. This fact, combined with the collapsible nature of the pulmonary vessels, means that at times pressures outside may be greater than those inside the apical vessels and perfusion may be at best intermittent.

Ventilation also falls off from base to apex of the lung because, strictly speaking, the lung has weight and contrary to what is usually assumed for simplicity's sake, intrapleural pressure is more subatmospheric at the apex than at the base. Consequently, the apex of the lung is working, so to say, on a 'stiffer' part of the compliance curve (Fig. 12.6) and so for a given expanding force the base fills more easily than does the apex. These discrepancies between base and apex are more marked in the case of perfusion than in that of ventilation so that the ratio $\dot{V}_A/\dot{Q}$ is appreciably higher at the apex than at the base (Fig. 12.12).

This spread of $\dot{V}_A/\dot{Q}$ affects gas exchange in the lung, basically because of the shape of the O_2 dissociation curve (see Fig. 12.13). Units that are well perfused but ill ventilated contribute to the systemic circulation blood that is not fully arterialized. Because the dissociation curve at oxygen pressures greater than approximately 100 mmHg is virtually flat, blood from these units is not compensated for by blood from units where ventilation is more than adequate to arterialize the blood. Moreover, in absolute terms, the contribution from the lung bases is greater than that from the apices. And so a spread of $\dot{V}_A/\dot{Q}$ throughout the lung leads to a measurable $\Delta P_{A-a,O_2}$. In many lung diseases, such inequalities are greatly magnified.

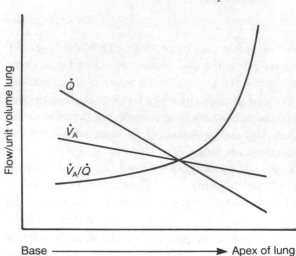

Fig. 12.12. Alveolar ventilation ($\dot{V}_A$) and perfusion ($\dot{Q}$) and the ventilation/perfusion ratio ($\dot{V}_A/\dot{Q}$) from base to apex of the lung. (From West, J.B. (1977) *Ventilation/Blood Flow and Gas Exchange*, 3rd edn, p. 30. Blackwell Scientific Publications, Oxford.)

The same argument of course holds in theory for CO_2, but because the CO_2 dissociation curve is, in the physiological range, virtually a straight line with a slope much steeper than that of the O_2 dissociation curve, and because also of the Haldane effect, hyperventilation can reduce P_{a,CO_2} to compensate or more than compensate for any increase in blood P_{CO_2} leaving units that are relatively poorly ventilated. Moreover, any tendency for P_{a,CO_2} to rise leads reflexly to an increase in ventilation.

There is some automatic compensation for the $\dot{V}_A/\dot{Q}$ mismatching, in that in overventilated zones there will be a tendency for the blood vessels to dilate in response to the high P_{O_2} and low P_{CO_2}. The bronchioles will tend to constrict in the same circumstances. The reverse applies to hypoxic, hypercapnic zones. This compensation can only partially counter mismatching, and is a disadvantage when the whole lung is hypoxic and hypercapnic: then the whole pulmonary vasculature will constrict!

Diffusion

The amount of a gas that will diffuse in unit time across a barrier such as that separating alveolar gas from blood (strictly, from the interior of the red blood cell) depends on the area and thickness of the barrier and on the pressure difference across it. By Fick's law, $\dot{V}_x = A/T \cdot D(P_1 - P_2)$, where A is the alveolar surface area and T is the thickness of the air-to-blood barrier. The diffusion constant D is for any gas proportional to the absorption coefficient α (ml gas/ml solution/760 mmHg gas pressure), the value of which depends on the temperature, and is inversely proportional to the M_r. At 37°C, with plasma as solvent, α has a value for O_2 of 0·023 and for CO_2 of 0·51. Thus the relative diffusibilities of CO_2 and O_2 are given by:

$$\frac{\alpha_1}{\alpha_2} \cdot \sqrt{\frac{32}{44}} \simeq 20 \cdot$$

The high diffusibility of CO_2 is balanced by the rather small pressure gradient between mixed venous blood and alveolar gas, if we bear in mind that similar amounts of O_2 and CO_2 have to be exchanged in the same time.

Because of difficulties in measurement, it is usual to lump A/T and D together as the *diffusion capacity* (D_L) of the lung for, say, O_2. Thus,

$$D_{L,O_2} = \frac{\dot{V}_{O_2}}{P_{A,O_2} - P_{\bar{c},O_2}} \cdot$$

Since $P_{\bar{c},O_2}$, the average O_2 pressure in the lung capillaries, cannot really be measured, the diffusion capacity for carbon monoxide is measured instead, because $P_{\bar{c},CO}$, can be taken as zero on account of the very high affinity between CO and haemoglobin. Thus, $D_{L,CO} = \dot{V}_{CO}/P_{A,CO}$. To obtain D_{L,O_2} we could multiply by 1·23 but in practice clinicians simply compare the measured with the predicted $D_{L,CO}$. This account of diffusion and its calculation in clinical practice takes no account of the fact that part of the 'resistance' to the passage of O_2 (or CO) from gas to final combination with Hb lies in the rate of reaction itself which, though rapid, is not infinitely so. What we measure as D_L has thus a chemical as well as a physical component.

12.4 Blood gas transport

The amount of O_2 or CO_2 taken up or excreted from the alveolar gas must equal that taken up from or added to the blood by the tissues in the same time. In other words, for O_2,

$$\dot{V}_{O_2} = \dot{V}_I \cdot F_{I,O_2} - \dot{V}_E \cdot F_{E,O_2} = \dot{Q}(C_{a,O_2} - C_{\bar{v},O_2})$$

where C_{a,O_2} and $C_{\bar{v},O_2}$ are the concentrations of oxygen in the arterial and mixed venous blood. Since the same applies to CO_2, the respiratory exchange ratio (R) can be obtained from analysis of either gas or blood.

In the normal adult male at rest, $\dot{V}_{O_2}$ is roughly 300 l min^{-1} and $\dot{V}_{CO_2}$ correspondingly less, depending on the respiratory quotient (RQ). If O_2 were carried in the blood only in physical solution, its solubility is such that C_{a,O_2} would be $100/760 \times 0·023 \times 10^3 \simeq 3$ ml l^{-1}, so that even if $C_{\bar{v},O_2}$ were zero, $\dot{Q}$ would have to be impossibly high to provide the necessary O_2 to the tissues. In fact, the amount of O_2 normally carried in arterial blood is of the order of 200 ml l^{-1}, as a result of its combination with haemoglobin.

Oxygen carriage

Because O_2 combines with the haemoglobin in the red blood cells, the amount of O_2 carried in any volume of blood is much greater than one would expect on physical grounds. The O_2 combines with the iron–porphyrin complex or haem part of the haemoglobin molecule to form oxyhaemoglobin. The association is easily reversible and is termed an *oxygenation*. It is not an oxidation; oxidized haemoglobin is useless for the biological carriage of O_2. From the relative molecular mass of haemoglobin (Hb) and the fraction of iron in it, one would expect 1 g of Hb to combine at a maximum with 1·39 ml O_2. Actual measurement shows that 1·34 ml is a more realistic value. Haemoglobin combined with this amount of O_2 is said to be fully saturated with O_2 and thus the O_2 *capacity* of blood depends on its Hb concentration. The O_2 concentration in any blood sample will depend on the extent to which the Hb is saturated, i.e. on the *percentage saturation, S*, and will also include the small amount of O_2 that is physically dissolved. Both depend in turn on the P_{O_2} with which the blood has come into equilibrium. Thus

$$\text{Percentage saturation }(S) = \frac{\text{total } O_2 \text{ concentration} - O_2 \text{ in physical solution}}{O_2 \text{ capacity}}$$

The relation between O_2 bound to haemoglobin and P_{O_2} is given by the *oxyhaemoglobin dissociation curve* (Fig. 12.13). Such a curve is constructed by equilibrating a series of blood samples with gas mixtures over a range of

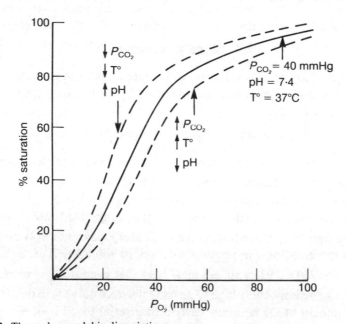

Fig. 12.13. The oxyhaemoglobin dissociation curve.

oxygen partial pressures, the temperature, pH and P_{CO_2} being the same for each. The bound O_2 (i.e. the total O_2 concentration $-O_2$ in physical solution) per unit volume of blood, or the percentage saturation, is then plotted against P_{O_2}. In the former case, the position of the curve will obviously depend upon the Hb concentration. If we take $150 \, \text{g} \, 1^{-1}$ as a reasonably normal Hb concentration, then the bound O_2 concentration at full (100%) saturation, i.e. the oxygen capacity, will be approximately $200 \, \text{ml} \, 1^{-1}$. Blood from an anaemic patient will be fully saturated at the same P_{O_2} as is required to saturate fully normal blood, but the O_2 concentration will be correspondingly less (see Fig. 12.20d).

The shape of the O_2 dissociation curve depends on a number of factors. The Hb molecule has four iron-containing haem groups and its combination with O_2 takes place in four stages. The stages are not, however, independent of each other; in particular, the last of these reactions proceeds at a high velocity which counteracts the tendency of the whole process to slow down as the number of oxygen-binding sites on the Hb molecule diminishes and so oxygenation proceeds more or less uniformly until it is virtually complete (p. 330).

The combination of O_2 and Hb depends, too, on the precise shape of the haemoglobin molecule. This is influenced by such factors as temperature, pH and P_{CO_2}, so that the iron-containing haem groups with which the O_2 combines may become more or less accessible. This is the basis of the *Bohr effect*, the term applied to a shift of the dissociation curve following a change in any of the above factors. Thus a decrease in temperature or in P_{CO_2} or an increase in pH will increase the affinity of Hb for O_2 and thus will shift the curve to the left. The opposite changes will move the curve to the right. Such shifts are greater in the middle part of the curve than at either end. The shape of the Hb molecule and therefore the position of the dissociation curve is also greatly influenced by the concentration in the red cells of the substance 2,3-diphosphoglycerate (2,3-DPG), which is formed in the course of red-cell metabolism as a deviation from what in other cells is the main glycolytic path. An increase in 2,3-DPG concentration decreases the affinity of Hb for O_2 and moves the curve to the right (p. 330).

The rather peculiar shape of the dissociation curve has a number of important biological implications.

1 Above $P_{O_2} \simeq 100 \, \text{mmHg}$, the curve is almost flat. At that P_{O_2}, which is more or less what we would expect to obtain in normal alveolar gas, the Hb is approximately 97% saturated and a further increase in P_{O_2} will make little difference to the O_2 concentration in the blood. The last little moiety of Hb is very hard to saturate and requires a P_{O_2} of approximately 300 mmHg; beyond that the O_2 concentration in the blood will increase at a rate of approximately 3 ml 1^{-1} per 100 mmHg P_{O_2}, purely as a result of increase in the physically dissolved O_2.

2   There is little tendency for the Hb to desaturate until P_{O_2} has fallen to around 60 mmHg. In biological terms, there is thus a considerable reserve, so that, despite a fall in P_{A,O_2}, the saturation of Hb in arterial blood, within limits, remains high.

3   Below $P_{O_2} \simeq 60$ mmHg, desaturation is rapid so that O_2 is readily given off to the tissues with little further fall in P_{O_2}. Thus, when we compare with normally oxygenated blood, blood that has been exposed in the lungs to a P_{A,O_2} substantially lower than normal, the difference in $P_{\bar{v},O_2}$ is surprisingly small.

It is important to realize that whereas P_{a,O_2}, and therefore C_{a,O_2} (or S_{a,O_2}), is a function of P_{A,O_2} modified by the ease or otherwise with which O_2 can diffuse from alveolar gas to blood and also by any undue spread of $\dot{V}_A/\dot{Q}$ (p. 487), $P_{\bar{v},O_2}$, on the other hand, depends on the *arteriovenous oxygen difference* ($C_{a,O_2} - C_{\bar{v},O_2}$) which in turn is a function of $\dot{V}_{O_2}$ and of the cardiac output, $\dot{Q}$. This is implicit in the Fick equation for calculating cardiac output (p. 410).

It is useful to be able to describe the O_2 affinities for shifts in the dissociation curve and for different haemoglobins. This is done by finding the P_{O_2} *at which the Hb is half-saturated* (P_{50}). Normal adult Hb under the usual conditions of P_{CO_2}, temperature, etc., has a P_{50} of $\simeq 25$ mmHg. Haemoglobin with a greater O_2 affinity will have a dissociation curve with a lower P_{50} that lies to the left of the normal curve, and vice versa. Fetal Hb has a curve that lies to the left of that for maternal Hb, i.e. it has a greater affinity for O_2.

If we now consider the uptake of O_2 in the lungs and its transfer to the tissues, we see that these processes are aided by the reciprocal movements of CO_2 as a result of the Bohr effect. The decanting of CO_2 from the mixed venous blood in the lungs assists the uptake of O_2 from the alveolar gas; conversely, in the tissues, as CO_2 is loaded on to the blood, O_2 is more easily released from the Hb.

The P_{O_2} within the individual cells of the tissue mass is hard to measure but is assumed to be very low—at the mitochondria, the actual point of utilization of the transferred O_2 molecules—perhaps 1 mmHg. Thus $P_{\bar{v},O_2}$ provides a rough estimate of the pressure gradient that serves to drive O_2 from blood to tissues.

Haemoglobin can combine with gases other than O_2, for example CO_2 (see below). Very importantly, it can combine with carbon monoxide, CO, to form *carboxyhaemoglobin*. This combination, with the ferrous iron, is similar to that of Hb and O_2 except that the affinity of CO and Hb is very much greater, of the order of 250 times, than that of Hb and O_2. Thus a P_{CO} of only 0·1 mmHg will serve to half-saturate the Hb so that it takes only a small fraction of CO in the inspired air to reduce drastically the ability of the blood to carry O_2. Moreover, as far as the remaining Hb is concerned, the oxyhaemoglobin dissociation curve is shifted to the left so that a subject suffering from CO poisoning is much worse off than an anaemic patient with a comparable reduction in the O_2 capacity of his or her blood.

Carbon dioxide carriage

Just as O_2 must be picked up in the lungs and unloaded in the tissues so must CO_2 be transported in the reverse direction, and, just as in the case of O_2, so with CO_2 there is a discrepancy between what we might expect to find on purely physical grounds and what we do find when blood is analysed.

In arterial blood, we might expect in *physical solution* $39/760 \times 0·51 \times 10^3 = 26$ ml l^{-1}, and a little more in mixed venous blood, whereas we commonly do find something of the order of 500 ml l^{-1}. In many ways, CO_2 carriage is more complicated than O_2 carriage, but it is helpful once more to start by constructing a dissociation curve. The graph of the CO_2 *dissociation curve* includes for purposes of comparison an O_2 dissociation curve drawn to the same scale (Fig. 12.14).

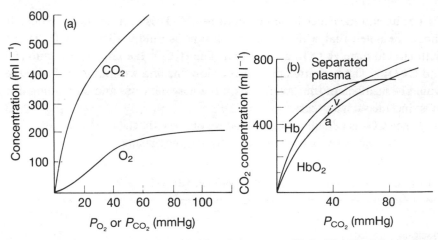

Fig. 12.14. (a) Comparison of O_2 and CO_2 dissociation curves, (b) the CO_2–blood dissociation curves for separated plasma and for whole blood in which haemoglobin is deoxygenated (Hb) or fully oxygenated (HbO$_2$). The line a–v represents the uptake of CO_2 as HbO$_2$ is gradually deoxygenated to 70%.

The following points should be noted.

1 Physically dissolved CO_2 is a sizeable proportion of the total CO_2 (Table 12.4) in contrast with O_2.

2 There is no obvious upper limit to the amount of CO_2 that can be taken up and therefore there is no sense in talking of 'percentage saturation'.

3 If red cells and plasma are separated from each other, about one-third of the total CO_2 is found to be in the red cells and about two-thirds in plasma, whereas nearly all the O_2 is in the red cells in combination with Hb.

4 If a dissociation curve is constructed for plasma, which has been separated from its red cells, the curve is noticeably different from that for whole blood or that for plasma samples separated anaerobically from the red cells after equilibrium has been reached: it is much flatter on the right and does not come down to the origin of the graph on the left. Indeed there is a substantial proportion of the total CO_2 which can be obtained only from such separated plasma following the addition of a strong acid. Thus, despite the fact that most of the CO_2 in whole blood is in fact in the plasma, the presence of the red cells during the equilibration process is clearly essential. Whole blood can, equally clearly (Fig. 12.14), take up a given increment of CO_2 with a smaller rise in P_{CO_2} than can plasma separated from the red cells.

5 Just as in the case of the combination of O_2 and Hb, the P_{CO_2} had to be defined, so with CO_2 carriage P_{O_2} and thus S has to be stated. As the graph shows, whole blood in which the Hb is completely desaturated can carry more CO_2 at a given P_{CO_2} than can fully oxygenated Hb. This is known as the *Haldane effect*.

Let us now consider in more detail how CO_2 is carried in the blood. It should be noted that while blood, or plasma, is mildly alkaline, CO_2 reacts with H_2O to form H_2CO_3, a weak acid. The H_2CO_3 then dissociates into H^+ and HCO_3^-. The hydration process is a slow one and without a catalyst there would be insufficient time available for the necessary on- and off-loading if it must include such a process.

When CO_2 is added to whole blood we observe that:

1 HCO_3^- increases in both the red cells and the plasma;

Table 12.4. Proportions in which CO_2 is transported in the blood.

	% Arterial (a)	% Venous (v̄)	% (v̄ − a)*
Physically dissolved	5·5	5·8	8·0
HCO_3^-	89·6	87·0	62·0
$HbCO_2$	4·9	7·2	30·0

* Percentage of newly produced CO_2 in each form.

2 pH decreases in both the red cells and the plasma;

3 Cl^- increases in the red cells and decreases in the plasma; and

4 the H_2O content of the red cells increases.

These observations can be explained if we note in detail what happens (Fig. 12.15). At a molecular level, when CO_2 is added, the reaction $CO_2 + H_2O \rightleftharpoons H_2CO_3 \rightleftharpoons H^+ + HCO_3^-$ takes place. In the plasma, this is slow, but it does take place to some extent whereupon the H^+ is buffered by the plasma proteins which have a net negative charge within the relevant pH range; thus,

$$H^+ + HCO_3^- + Na^+ + Prot^{n-} \rightleftharpoons Na^+ + HCO_3^- + HProt^{(n-1)-}.$$

CO_2 also readily diffuses into the red cells in which K^+ is the predominant cation. These contain an enzyme, *carbonic anhydrase*, which catalyses the hydration process, which thus takes place very much more quickly. The H^+ is now buffered largely by the globin moiety of the Hb; thus,

$$O_2Hb^{n-} + K^+ + H_2O + CO_2 \rightleftharpoons O_2 + HHb^{(n-1)-} + HCO_3^- + K^+.$$

The red-cell membrane is very much more permeable to anions than to cations. The continuing formation of HCO_3^- establishes an electrochemical gradient down which HCO_3^- moves from the red cells to plasma. Since the K^+ cannot move in association, electrical balance can be maintained only if anions move in, and so Cl^-, the only obvious candidate, moves from plasma to cells (Fig. 12.15). This is known as the *chloride shift* or Hamburger phenomenon. However, although this serves to maintain electrical equilibrium, the total red-cell osmolar content has been increased by the gain of Cl^- anions at the expense of polyvalent Hb^-, and so H_2O moves from plasma into the red cells to restore osmotic balance (Fig. 12.15). The red cells thus increase in volume and the haematocrit increases.

Thus CO_2, or rather H_2CO_3, is buffered by the formation of a weak acid, HHb, so that the H^+ load results in a much smaller change in pH than would otherwise be the case. Furthermore, HHb is a weaker acid than is $HHbO_2$ and so the concomitant loss of O_2 from combination with Hb aids the buffering of CO_2. Nevertheless, venous blood is a little more acid (pH 7·37) than arterial blood (pH 7·40).

Because of its concentration in the red cells and their content of carbonic anhydrase, Hb is much the most important of the substances responsible for the minimal change in pH that occurs with the uptake of CO_2. Its importance is enhanced by the fact that it also combines directly with CO_2 to form a *carbamino-complex*, thus:

$$CO_2 + Hb - NH_2 \rightleftharpoons Hb - NHCOOH.$$

This reaction is inherently rapid and does not depend on enzyme activity; it

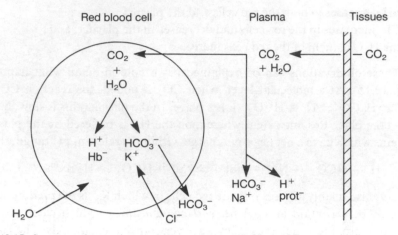

Fig. 12.15. Sequence of events involved in the carriage of CO_2 as HCO_3^-.

also takes place more readily with desaturated than oxygenated Hb. Although this direct combination with Hb comprises a smaller fraction of the total CO_2 in whole blood than was originally thought, it accounts for a substantial proportion of the Haldane effect when the CO_2 concentration of the blood flowing through the tissues is increased at the same time as the O_2 concentration is decreased (Table 12.4).

So far, CO_2 carriage has been described in terms of its uptake from the tissues into the blood. In the lungs, the various processes are reversed as the CO_2 is given off. The Haldane and Bohr effects taken together form a very beautiful example of biological organization. Both depend essentially on the spatial configuration of the Hb molecule and the forces which hold it together.

Figure 12.14 shows that the usual increment of CO_2 can be taken up with a much smaller increase in P_{CO_2} than would ensue were there no concomitant desaturation of the Hb. Nevertheless, it should not be thought that this partial desaturation of Hb is absolutely essential for CO_2 carriage. If we consider the circumstances, unusual but not unheard of, in which a patient may be exposed to very high pressures of pure O_2, for example, in conjunction with radiotherapy, then it can easily be calculated that at a P_{O_2} of about 3 atm the usual resting demands for O_2 can be satisfied by the O_2 in physical solution. In this case, CO_2 carriage has to take place without the assistance of desaturation of Hb. Figure 12.14 demonstrates that the price to be paid is a rise in P_{CO_2}, as arterial becomes venous blood, of about three times the usual rise. The P_{CO_2} matters because of its critical involvement in the acid–base balance of the body.

Acid–base balance

From the law of mass action and the behaviour of buffer solutions, one can derive the expression:

$$pH = pK' + \log \frac{[HCO_3^-]}{S \times P_{CO_2}}.$$

In this equation HCO_3^- is expressed in $mmol\,l^{-1}$ and S in this case is a solubility factor converting P_{CO_2} to $mmol\,l^{-1}$. It has a value of 0·03 at 37°C.

This is known as the Henderson–Hasselbalch equation. The equation makes it clear that pH depends on a ratio rather than on the absolute amount of any particular substance. If we bear in mind the relation of CO_2 to H_2CO_3, we see that the terms of the ratio comprise a 'buffer pair', quantitatively the most important mechanism for the buffering of all acids stronger than H_2CO_3 and bases stronger than HCO_3^-.

It may seem strange from the point of view of the chemist that the most important buffer system in the body should have a pK' (6·1) so far removed from that pH at which it is desirable the body fluids be stabilized. The explanation lies in the fact that both HCO_3^- and CO_2 are under physiological control by the kidneys and the lungs, respectively, so that a primary change in the one is compensated for by a secondary change in the other in the appropriate direction (Section 13.3).

It should be appreciated that the carriage and buffering of CO_2 *in vivo* is a good deal more complicated than the preceding account might suggest, and is by no means yet fully understood. For one thing, the total amount of CO_2 in the body is very large, over 100 litres, much of which is dissolved in fat or stored in the bones. These tissues are relatively poorly perfused with blood and their CO_2 levels can change only slowly so that a new steady state is reached only after a considerable lapse of time. Second, buffering does occur, although again slowly, in cells other than red cells, whereas in the laboratory when dissociation curves are being constructed all the bicarbonate formed is obviously confined to red cells and plasma. Finally, within the blood itself where time for equilibration is not unlimited, changes in HCO_3^-, Cl^-, etc., are the results, really, of two equilibration processes taking place at very different rates, one in the red cells and one in the plasma.

12.5 Regulation of respiration

Breathing occurs rhythmically. The rhythmicity is generated within areas of the brainstem referred to as the *respiratory centres*. Respiration is reflexly regulated by specific *chemoreceptor* and *mechanoreceptor* inputs to the respiratory centres. This results in a level of $\dot{V}_E$ that is appropriate to the O_2 requirements of the moment and to the maintenance of arterial P_{O_2}, P_{CO_2} and acid–base homeostasis. Superimposed on this are short-duration reflexes initiated by, for example, irritants as in coughing or sneezing. The effects of some stimuli, e.g. temperature, are mediated by *higher brain centres* which then influence the respiratory centres. Respiration is also under appreciable *voluntary control* from the cerebral cortex as in breath-holding, taking large breaths or speaking. The cortex can for a short time dominate respiratory centre reflexes but eventually the reflex control takes precedence.

Respiratory centres

In the early 1920s, Lumsden performed brainstem transections and concluded that the respiratory centres were composed of a *pneumotaxic centre* in the upper pons, an *apneustic centre* in the lower pons and an expiratory centre and a gasping or inspiratory centre in the medulla oblongata. The latter two centres were postulated because after removal of the pontine influence breathing was dominated either by expiratory spasms or by a pattern of irregular gasping. More refined neurophysiological techniques have since demonstrated that the medulla when separated from the pons does maintain a relatively normal rhythmic respiration. Rather than an inspiratory and expiratory centre, this medullary centre is composed of two bilateral aggregations of respiratory neurones (Fig. 12.16) known as the *dorsal respiratory group* and the *ventral respiratory group*. Breathing is abolished only by a transection between the medullary centre and the spinal cord.

The *dorsal respiratory group* (DRG) is close to the nucleus of the tractus solitarius from which it receives and integrates afferent information from respiratory mechano- and chemoreceptors. It is composed entirely of inspiratory cells which are upper motor neurones projecting via the ventrolateral column of the cord to the lower motor neurones of the contralateral phrenic nerve which innervates the diaphragm. Respiratory rhythmicity is thought to originate in the DRG. One suggestion is that the inspiratory cells have pacemaker properties and spontaneously discharge in a phasic manner.

The *ventral respiratory group* (VRG) is composed of both inspiratory and expiratory upper motor neurones. When the cells of the DRG are firing in inspiration, they excite VRG inspiratory cells and inhibit VRG expiratory cells. The VRG is located rostrally in the nucleus ambiguus and caudally in the nucleus retroambigualis (Fig. 12.16). The rostral group innervates the ipsilateral accessory muscles of respiration and the caudal group projects via the ventrolateral column of the cord to the contralateral expiratory intercostal

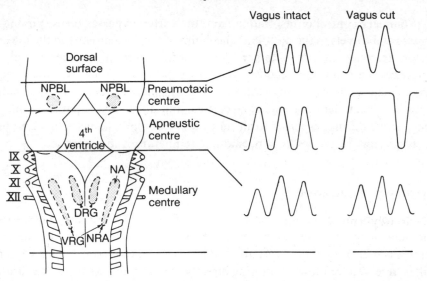

Fig. 12.16. Location of the respiratory centres in the brain and the effects of transections of the medulla and pons on the pattern of breathing in intact and vagotomized animals. NPBL, nucleus parabrachialis; DRG, dorsal respiratory group; VRG, ventral respiratory group composed of nucleus ambiguus (NA) and nucleus retroambigualis (NRA).

and abdominal lower motor neurones and ipsilaterally and contralaterally to the lower motor neurones of the inspiratory intercostal muscles.

The expiratory lower motor neurones in the spinal cord are inhibited during inspiration and vice versa. This prevents reflex contraction of antagonist muscles when agonist muscles actively contract. This *reciprocal inhibition* originates in the DRG and VRG and is not mediated via muscle-spindle reflexes as in other skeletal muscle systems. However, muscle spindles are present in intercostal muscles and modulate lower motor neurone discharge to the same muscle throughout contraction. Thus the strength of contraction can be adjusted to achieve a certain V_T whatever the airway resistance.

The activity of the DRG is modulated by the two centres in the pons. The *pneumotaxic centre* (PNC) is found in the nucleus parabrachialis and contains both inspiratory and expiratory cells. Its function is to tune finely the f/V_T pattern by influencing the switch-over from inspiration to expiration. Without the PNC, breathing is slower and deeper because inspiration proceeds for longer (Fig. 12.16). If removal of the PNC is combined with a bilateral vagotomy, breathing shows a pattern of prolonged inspirations interspersed with brief expirations (Fig. 12.16). This is known as apneustic breathing and led to the postulation of the *apneustic centre* (APC). No specific neuronal population has been identified yet in the APC but it is thought that it is the location of the inspiratory cut-off switch to which project both axons from the PNC and vagal afferents carrying information about lung volume.

Voluntary control of respiration from the cortex bypasses the respiratory centres and travels in the dorsolateral column of the cord directly to the lower motor neurones. An involuntary tonic control from higher centres also influences the respiratory centres. Inhibitory pathways from the cortex and hypothalamus and excitatory pathways from the diencephalon and hypothalamus have been demonstrated. The cortical and diencephalic pathways may also operate during responses to altitude and the hypothalamus is involved in the respiratory responses to pain, emotion, temperature and exercise.

Respiratory chemoreceptors

The receptors for respiratory control are classified usually into chemoreceptors and non-chemical or mechanoreceptors. However, as some of the mechanoreceptors can also respond to certain chemical stimuli, the term chemoreceptor will be reserved for those monitoring directly or indirectly blood gas pressures. Chemoreceptors maintain homeostasis of arterial P_{O_2}, P_{CO_2} and pH and assist in ensuring that $\dot{V}_E$ is appropriate to the level of metabolism. They are stimulated by a rise in P_{a,CO_2} and arterial $[H^+]$ and a fall in P_{a,O_2} and reflexly cause an increase in $\dot{V}_E$. Since $\dot{V}_E$ is more sensitive to increases in P_{a,CO_2} and arterial $[H^+]$ than to decreases in P_{a,O_2}, the controlling link between $\dot{V}_E$ and metabolism is the CO_2 produced rather than the O_2 consumed. The precision of chemical control is such that in healthy individuals the alveolar and arterial P_{CO_2} remain remarkably constant over a wide range of metabolic rates. On the other hand, if the F_{I,CO_2} is raised or the F_{I,O_2} is lowered the reflex increase in $\dot{V}_E$ is such that the initial increase in P_{a,CO_2} (*hypercapnia*) and arterial $[H^+]$ or decrease in P_{a,O_2} (*hypoxia*) is minimized and a new steady state is maintained.

The steady-state responses in hypercapnia show that the $\dot{V}_E$ is linearly related to P_{a,CO_2} (Fig. 12.17a). However, at lower P_{a,CO_2} values, $\dot{V}_E$ is relatively insensitive to P_{a,CO_2} and at higher P_{a,CO_2} values sensitivity is reduced due to CO_2 narcosis of the brain and respiratory centres. By comparison, $\dot{V}_E$ is inversely related to P_{a,O_2} with a curve that steepens progressively as the hypoxia becomes more severe (Fig. 12.17b). At P_{a,O_2} values below 30 mmHg, the hypoxia severely depresses the brain and decreases $\dot{V}_E$. Note that, since an increase in $\dot{V}_E$ *per se* will lower P_{a,CO_2} and elevate P_{a,O_2}, the F_{I,CO_2} and F_{I,O_2} have to be manipulated experimentally to maintain a normal P_{a,CO_2} of 40 mmHg during the hypoxic tests and a normal P_{a,O_2} of 100 mmHg during the hypercapnic tests. When the P_{a,O_2} is held constant at say 50 mmHg, the $\dot{V}_E$ response line to P_{a,CO_2} not only shifts upwards but also steepens (Fig. 12.17a). This increased $\dot{V}_E$ sensitivity is referred to as a multiplicative or

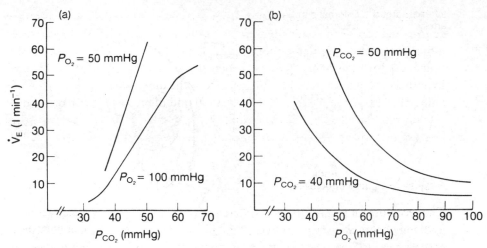

Fig. 12.17. (a) Ventilation ($\dot{V}_E$) in response to changes in arterial P_{CO_2} during established normoxia ($P_{O_2} = 100$ mmHg) or hypoxia ($P_{O_2} = 50$ mmHg). (b) Ventilation ($\dot{V}_E$) in response to changes in arterial P_{O_2} during established normocapnia ($P_{CO_2} = 40$ mmHg) or hypercapnia ($P_{CO_2} = 50$ mmHg).

potentiating interaction between hypoxia and hypercapnia and apparently is well developed only in man. This interaction can also be demonstrated by holding P_{a,CO_2} at say 50 mmHg whilst the sensitivity to P_{a,O_2} is tested (Fig. 12.17b).

There are two groups of chemoreceptors. The first group, the arterial or *peripheral chemoreceptors*, are located in the *carotid* and *aortic bodies* and are stimulated by arterial hypoxia, hypercapnia and acidity. After removal of the peripheral chemoreceptors, all $\dot{V}_E$ responses to acute hypoxia, approximately 20% of the $\dot{V}_E$ response to hypercapnia and nearly all the $\dot{V}_E$ response to acute metabolic acidosis are lost.

The *carotid bodies* are small nodules of tissue found bilaterally at the bifurcation of the internal and external carotid arteries (Fig. 12.18). The vast capillary network of each carotid body is supplied by a branch from the external carotid artery and the venous blood drains into the jugular vein. The blood flow of 20 ml g^{-1} min^{-1} is so great that the $\dot{V}_{O_2}$ of the carotid body causes only a tiny arterial–venous O_2 difference. Thus capillary gas tensions are very close to arterial values and the carotid bodies are said therefore to monitor arterial P_{O_2}, P_{CO_2} and pH. The change in the arterial stimulus reaches equilibrium in the carotid body within seconds because of the massive blood flow. Changes in the blood flow to the carotid body alter its apparent sensitivity to arterial gases. For instance, a reduction in blood flow, caused by local sympathetic vasoconstriction or by a large drop in blood pressure, increases

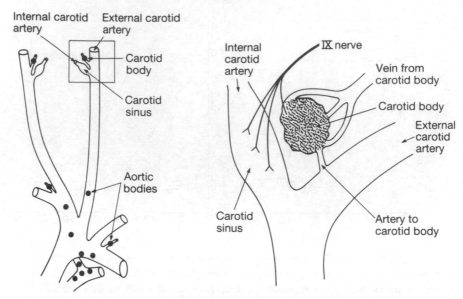

Fig. 12.18. Location of aortic and carotid bodies on the aortic arch and carotid arteries respectively. Enlargement shows the vasculature and innervation of the carotid body.

the P_{CO_2} and decreases the P_{O_2} in the capillaries and stimulates the carotid body without any change in arterial gas tensions. There may also be a parasympathetic vasodilatory control of blood flow.

The capillaries of the carotid body are surrounded by *glomus cells* and sensory nerve endings of cranial nerve IX (*glossopharyngeal nerve*). The latter are believed to be the chemosensitive structures because the neuroma that eventually forms on the stump of the IX nerve after the carotid body is removed shows some chemosensitivity. The glomus cells are high in dopamine and are thought to be inhibitory interneurones modulating the responses of the sensory nerve endings.

The *aortic bodies* are scattered over the aortic arch and the main arteries of that area (Fig. 12.18) and are anatomically very similar to the carotid bodies. The sensory nerve is cranial nerve X (*vagus nerve*). Their reflex effect on respiration is weak or non-existent during hypoxia, hypercapnia or acid conditions. However, as they appear to have a low blood flow, they may have important reflex effects on $\dot{V}_E$ during anaemia, carbon monoxide poisoning and hypotension.

The second group of chemoreceptors comprises the intracranial or *central chemoreceptors* located superficially at a depth of about 500 μm in discrete areas on the *ventrolateral surface* of the *medulla oblongata* (Fig. 12.19). Within these areas, no precise structure has yet been identified anatomically but it is presumed that nerve endings here monitor the brain interstitial fluid (ISF).

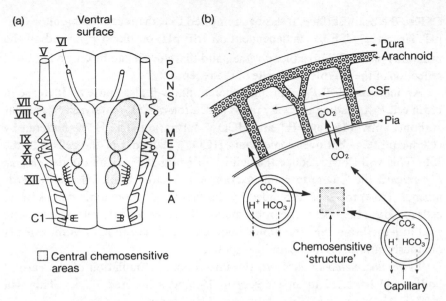

Fig. 12.19. (a) Location of the central chemosensitive areas on the ventral surface of the medulla oblongata. (b) The position of the chemosensitive structure in relation to the capillaries and the cerebrospinal fluid (CSF).

The central chemoreceptors are stimulated by arterial hypercapnia and are responsible for about 80% of the $\dot{V}_E$ response to hypercapnia but are only slightly stimulated by acute increases in arterial acidity and are insensitive to hypoxia. The presence of a blood–brain barrier, which greatly restricts movements of ions (but not of O_2 and CO_2), affects the extent to which arterial H^+ can influence the brain ISF. Since H^+ and HCO_3^- ions do not readily cross this barrier, most of the arterial $[H^+]$ changes in acute metabolic acidosis and alkalosis are not reflected in the brain ISF. During chronic metabolic acidosis or alkalosis, there is a slow leak of HCO_3^- down its concentration gradient between blood and ISF but equilibrium is never reached.

An increase in blood P_{CO_2}, on the other hand, immediately increases P_{CO_2} in the brain ISF although, since the relative blood flow to the brain is about forty times less than that to the peripheral chemoreceptors, it takes 5–10 min to reach equilibrium. This ISF CO_2 is then hydrated to H_2CO_3 which dissociates into the HCO_3^- and H^+ and it is believed that H^+ rather than P_{CO_2} is the actual stimulus to the central chemoreceptors.

The pH of brain ISF is influenced mainly by the composition of the arterial blood but also by the rate of cerebral blood flow and the metabolism of the brain cells. It is also influenced to some extent by the slowly flowing cerebrospinal fluid (CSF) on the surface of the medulla (Fig. 12.19). Since the location of the central chemoreceptors was first identified by applying acidic

CSF to the brain surface, it has been implied that the receptors monitor CSF pH. However, CSF pH is dependent on ISF pH, on blood P_{CO_2} and on the proximity of the CSF to a blood vessel, and therefore it can be only a modified reflection of the ISF pH at the site of the receptor.

An increased ISF P_{CO_2} during hypercapnia will also cause an increase in brain cell P_{CO_2} leading to a very rapid hydration of CO_2 (catalysed by cellular carbonic anhydrase) and H^+ and HCO_3^- formation. The H^+ is buffered by cellular proteins but over some hours HCO_3^- leaks out of the cells into the ISF. This will slowly reduce the acidity of brain ISF pH and under *chronic hypercapnia* the long-term compensation is to restore brain ISF pH towards normal. This process is also aided by the fact that hypercapnia causes brain cells to reduce their basal anaerobic metabolism and hence to release less lactic acid into the brain ISF. In chronic hypercapnia, restoration of brain ISF pH obviously minimizes the acute $\dot{V}_E$ response.

In *chronic metabolic acidosis*, the initial acute stimulation of the carotid bodies will have led to an increase in $\dot{V}_E$ and a lowered P_{a,CO_2}. This will decrease brain ISF H^+ and decrease stimulation of the central chemoreceptors. HCO_3^- will move slowly from ISF to the brain cells and, to a small extent, to plasma whilst the hypocapnia will increase lactic acid production from the brain cells. All this will restore brain ISF pH towards normal. The converse happens in *chronic metabolic alkalosis*.

Acute hypoxia in the absence of the carotid bodies can lead to a delayed increase in $\dot{V}_E$ if the reduced O_2 level does not depress the CNS too much. It does so by increasing brain cell lactic acid production and making ISF pH acidic. Normally, acute hypoxia will stimulate the carotid bodies and hence increase $\dot{V}_E$ leading to hypocapnia and an alkaline ISF pH. Acclimatization to *chronic hypoxia* involves restoration of ISF pH by pathways described for chronic metabolic acidosis and aided by both hypocapnia and hypoxia stimulating brain cell lactic acid production. This leads to a progressive increase in $\dot{V}_E$.

Respiratory mechanoreceptors

Mechanoreceptors having reflex effects on respiration are found in the nose, epipharynx, larynx and trachea of the upper airways, and in the lower airways and alveoli there are *lung stretch receptors*, *lung irritant receptors* and juxtacapillary or *J receptors*. All of these mechanoreceptors provide protective reflexes affecting both $\dot{V}_E$ and bronchomotor tone but some are involved also in determining the f/V_T pattern of breathing. *Limb proprioceptors* appear to cause some of the $\dot{V}_E$ increase occurring in exercise.

Nerve endings of the olfactory and trigeminal nerves in the nasal mucosa are excited by chemical and mechanical irritants and reflexly cause a *sneeze* and broncholaryngeal constriction. A sneeze is a number of superimposed inspirations followed by a strong and rapid expiration and then a short pause in the expiratory position.

Glossopharyngeal nerve endings in the epithelium of the epipharynx are excited by mechanical irritants such as a blockage and this results in the *sniff* or aspiration reflex and bronchodilatation. The sniff is a powerful inspiration pulling the blockage into the pharynx.

Vagal nerve endings in the epithelium of the larynx and trachea are excited by mechanical and chemical irritants and reflexly cause a *cough* and broncholaryngeal constriction. A cough is a large slow inspiration followed by rapid and powerful expirations. Mild stimulation of these receptors causes slow, deep breathing.

Lung stretch receptors (LSR) are vagal nerve endings in the smooth muscle of the trachea and lower airways that are stimulated by the increase in airway pressure accompanying lung inflation. The nervous discharge shows little adaptation and is proportional to the degree of inflation. A sustained inflation followed by occlusion of the airways reflexly causes cessation of further inspiratory efforts before the next inspiration is attempted. The duration of the pause is proportional to the degree of inflation. This response is called the *Hering–Breuer reflex* and demonstrates that the LSR shorten the inspiratory duration, restrict the depth of V_T and lengthen the expiratory duration by inhibiting via the APC the inspiratory cells of the DRG. LSR stimulation also causes bronchodilatation. These reflex effects also occur during normal breathing as the LSR nervous discharge increases and decreases in phase with inspiration and expiration. In the resting V_T range of man but not of other animals, the LSR reflex is weak but at the higher V_T of exercise or of breathing driven by hypoxia or hypercapnia, the LSR reflex in man becomes apparent. The inhibitory effects of LSR on V_T may be reduced in hypercapnia because increased airway P_{CO_2} has been shown to depress LSR activity.

Lung irritant receptors (LIR) are vagal nerve endings in the epithelia of the trachea and lower airways that are stimulated by mechanical or chemical irritants. The reflex initiated is an increase in f and V_T together with broncholaryngeal constriction. Histamine released in *asthma* probably stimulates the LIR. These receptors can also be stimulated by large deflations and large inflations but the response rapidly adapts. The significance of this sensitivity to volume is not clear but it may be part of the known role of LIR in triggering the augmented breaths or *sighs* that punctuate normal breathing. Evidence suggests that the resting discharge of LIR shortens the expiratory duration and that this is stronger than the lengthening effect of the LSR. Thus a vagotomy removing both LSR and LIR inputs causes an increase in V_T and a lengthening of both inspiratory and expiratory duration resulting in a decrease in f.

J receptors are vagal nerve endings thought to be in the interstitium between alveolar epithelium and capillary endothelium. They are stimulated by an increase in interstitial fluid pressure such as occurs in pulmonary congestion and oedema and by large inflations but the response rapidly adapts. The reflex initiated is broncholaryngeal constriction and apnoea followed by rapid, shallow breathing. It is not known what role, if any, the J receptors play in the regulation of normal respiration.

12.6 Respiration under abnormal conditions

In all of the foregoing sections, we have presupposed healthy subjects breathing air of normal composition at or near sea-level, i.e. at a pressure of approximately 1 atmosphere ($=760$ mmHg). Obviously, these conditions need not always hold. Human beings may live temporarily or permanently at altitudes many thousands of feet above sea-level; conversely, they may be exposed in pressure chambers to pressures considerably greater than 1 atmosphere or they may dive for longer or shorter periods under water with the same result. The atmosphere may be polluted, for example, with carbon monoxide (CO) in coal gas poisoning. Finally, the basic respiratory process, the supply of O_2 and excretion of CO_2, may be disrupted by disease.

Individuals with P_{A,O_2} lower than the usual 100 mmHg (give or take 10 mmHg) are said to be *hypoxic* and they will also show *arterial hypoxaemia* even if there is complete equilibration between blood and gas. The term *hypoxia* can be a source of some confusion inasmuch as it is also used to denote an insufficient supply of molecular O_2 to the tissues, irrespective of the cause. In this latter sense of the term, hypoxia may occur in the absence of arterial hypoxaemia, for example, in circulatory failure or in anaemia. The graphs in Fig. 12.20 show the arterial and venous points under normal conditions and in the three main types of tissue hypoxia, assuming an average rate of O_2 consumption and ignoring the Bohr effect. $P_{\bar{v},O_2}$ and, by extrapolation, $P_{\bar{c},O_2}$, which indicates the driving pressure for the transfer of O_2 from blood to tissues, have moved down and to the left in each of b, c and d.

A patient who is hypoxic will exhibit signs and symptoms which are a mixture of:

1 the direct effects of a poor O_2 supply to the tissues;
2 the attempts of the body to compensate for this; and
3 possibly, primary and secondary effects of the pathological process which are unrelated to the hypoxia.

Hypoxia in its purest form has long been studied in healthy individuals acutely and chronically exposed to low P_{O_2} at altitude.

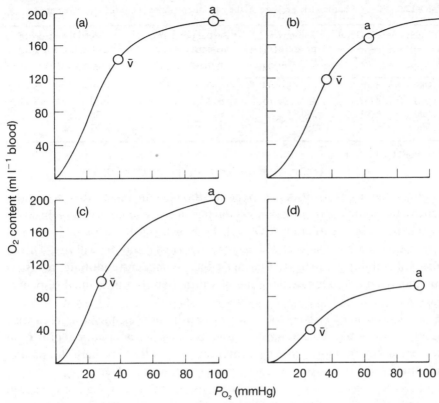

Fig. 12.20. The oxyhaemoglobin dissociation curve and arterial (a) and venous (v) points in normal conditions (a), at altitude (b), in circulatory failure (c), and during anaemia (d). Note that in (d) the anaemic arterial blood is fully saturated although it contains much less oxygen than normal arterial blood.

Decrease in barometric pressure

As we ascend, the barometric pressure falls as does therefore the pressure of O_2 in the inspired air (Table 12.5). As the P_{I,O_2} (moist tracheal inspired air) falls, so inevitably does P_{A,O_2}, although the body reacts in such a way as to minimize this fall. The right-hand column in Table 12.5 has been calculated, assuming a doubling of $\dot{V}_A$ and an RQ of 0·8 (p. 479). In fact, at the higher altitudes, $\dot{V}_A$ has obviously to be more than doubled.

Acute hypoxia

Acute hypoxia most obviously affects the following.

1 *Respiratory function:* there is hyperventilation and respiratory alkalosis. The exact effect is hard to predict because of the interaction of a number of

Table 12.5. Effect of barometric pressure on the partial pressure of inspired and alveolar O_2.

Height above sea level		Barometric pressure (P_B) (mmHg)	Partial pressure of inspired O_2(P_{I, O_2})* (mmHg)	Partial pressure of alveolar O_2(P_{A, O_2}) (mmHg)
(ft)	(m)			
8500	2591	550	103	80
14 000	4267	450	84	59
21 000	6401	350	63	38
28 000	8534	250	42	17

* In moist tracheal air.

different factors. For example, hypoxia directly depresses the respiratory centres but stimulates ventilation via the peripheral chemoreceptors. In so far as this increases ventilation, CO_2 will be blown off in excess of its rate of production. Thus the subject will become *hypocapnic* and pH will rise. This in turn will inhibit ventilation. The net effect at moderate altitude at rest is commonly an approximate doubling of ventilation, but individual responses vary.

2 *Cardiovascular function:* there is tachycardia and an increase in cardiac output, while arterial blood pressure usually remains about normal or is even decreased. Once more, in any individual, the effect is hard to predict quantitatively because of interactive processes (Section 11.4).

3 *Cerebral function:* this is acutely sensitive to the severity and to the speed of onset of hypoxia. Severe and sudden hypoxia will result in rapid loss of consciousness. Moderate hypoxia of relatively gradual onset may cause only such changes as lack of judgement and psychomotor disturbance.

4 *The blood itself:* the degree of deoxygenation of the blood in the small vessels affects skin and mucosal colour. In hypoxia, the darkening of the blood may be detectable as a change in colour, termed *cyanosis*.

These manifestations, combined with dizziness, headache, vomiting, nausea, dyspnoea (difficulty in breathing), incoordination, and so on, constitute the picture of *acute mountain sickness*.

Acclimatization

Acclimatization means the sum of those processes by which the body tries to compensate for the conditions which would otherwise result in a substantial degree of impairment of O_2 supply to the cells. It is what the patient is trying to achieve while burdened with pathology often of long standing. It can be studied 'pure' in those who live for long periods at high altitude.

1 Hyperventilation continues, indeed tends to increase for the next 5–10 days after acute exposure (p. 504), despite the systemic alkalosis, and it is no longer abolished by breathing O_2. $D_{L,CO}$ is increased. Cyanosis may be absent at rest but recur on exercise.

2 Cardiac output and heart rate tend to come back to normal at rest, but the maximum workload is decreased.

3 The kidneys excrete HCO_3^- with eventual correction of the systemic alkalosis; this takes 2–3 weeks.

4 Changes develop in the pulmonary vascular bed; increased muscularity of arterioles and hypoxic vasoconstriction lead to pulmonary hypertension and right ventricular hypertrophy.

5 Over 4–6 weeks there develops an increase in blood volume and bone marrow hyperplasia with increased red-cell count (haematocrit 0·60 or more; Hb 200 g l^{-1} or more). The increased blood viscosity further increases cardiac work.

6 An increase in 2,3-DPG production shifts the O_2 dissociation curve to the right so that O_2 is given up to the tissues a little more easily than would otherwise be the case.

7 There are increases in capillarity and other changes at tissue level, but these are much less well understood.

Over days, weeks and months, most people acclimatize and remain so while at altitude, provided the height is not too great. No human communities live permanently above about 5500 m (18 000 ft). Above that there is a slow deterioration in all aspects of performance.

Some individuals, apparently acclimatized, gradually or even suddenly fail in their responses. Their ventilation falls, with dyspnoea and cyanosis. There may be clubbing of the fingers. Haematocrit and pulmonary arterial blood pressure increase still further, leading to right heart failure. This is the picture of *chronic mountain sickness*. In sudden failure, acute pulmonary oedema may be dominant. In either case, the only course of action is to remove the subject to a lower altitude.

As we might expect, many of the above features may occur in patients who have been hypoxic for a long time.

Increased ambient pressure

Human beings are subjected to pressures greater than 1 atmosphere (760 mmHg) by:

1 submersion in water; for every 10 m (=33 ft) the pressure increases by 1 atm; and
2 exposure to compressed gas in a closed chamber.

In neither case is the pressure as such likely to have any effect on the solid tissues or liquids of the body. It is the gas-filled cavities of the body that are affected—the air-filled sinuses, the middle ear, the gastrointestinal tract and the lungs.

In addition, submersion poses problems arising from:

1 distortion of 'information' reaching special sense organs, i.e. disturbances of vision, hearing and proprioception; and
2 the need to breathe under water.

Discussion is confined here to the specifically respiratory aspects of exposure to increased pressures (*hyperbarism*).

A dive may be: (1) short enough to be accomplished on such air as can be inhaled at the surface—'single-breath' diving; (2) carried out at little depth by maintaining connection with the atmosphere through a tube–'snorkel' diving; or (3) sufficiently long to demand a maintained supply of air, in which case the latter must be supplied at ambient pressure—'conventional' and 'free' (or scuba) diving.

1 This is, basically, breath-holding with the additional complicating factor of submersion. Suppose one takes a deep breath and goes down to 10 m. The P_{A,O_2} and P_{A,CO_2} will approximately double—not quite, because both gases will pass along their gradients into the blood. After a brief period (20–30 s), the combined effect of rising P_{CO_2}, falling P_{O_2} and decreasing lung volume is sufficient to force the diver to the surface. The danger lies in the temptation to hyperventilate before submerging in an attempt to prolong the dive. The initial hypocapnia may permit a much longer stay at depth by which time P_{O_2} has reached a low level. On ascent, this is very rapidly lowered further to levels at which unconsciousness is likely to supervene.
2 In the case of snorkel diving, the problem is essentially one of lung mechanics. The subject is breathing to and from the atmosphere. Alveolar pressure is therefore, on average, atmospheric pressure. The pressure on the chest wall is greater than atmospheric by an amount that depends on the depth. This must be counteracted by the action of the respiratory muscles if the chest is not to collapse. The maximum subatmospheric intrapleural pressure that can be generated is, however, of the order of 100 mmHg ($\simeq 1\cdot2$ m). When we consider that maximum muscular effort can in general be expected only for short periods, and further when we take into account the increased dead space and some increase in resistance to air flow provided by the tube, we see that the 'snorkel' diver is, over any appreciable period of time, confined to a depth of about half a metre under the surface.
3 Where the diver is required to function at a depth for relatively long

periods, he must be provided with a respirable gas mixture at ambient pressure. Technical difficulties apart, the problem is now one of the effects of breathing N_2 and O_2 at pressures greater than atmospheric, and is thus similar to breathing in a compression chamber.

Nitrogen breathed at a pressure greater than about 4 atm ($\simeq 30$ m below the surface) results in narcosis; the effect is very similar to that of N_2O at atmospheric pressure. This difficulty cannot be circumvented by substituting O_2, because it also is toxic under pressure. The breathing of pure O_2 even at atmospheric pressure for long periods (days) leads to pathological changes in lung structure, while O_2 at pressures greater than about 3 atm for much shorter periods affects predominantly the CNS to cause convulsions.

It is worth noting that exposure to N_2 pressures on the safe side of those causing narcosis puts the diver at risk not at depth but as he ascends. The N_2 although inert dissolves physically in the body fluids and especially in the fatty tissues in proportion to the pressure. If this is decreased too quickly, the N_2 comes out of solution with the formation of tiny bubbles causing *decompression sickness*. If these bubbles form emboli in the cerebral, myocardial or pulmonary circulation, or in the substance of the CNS, severe effects may follow such as coughing and dyspnoea ('the chokes'), paralysis, and, where bubbles have formed in the joints, severe pain ('the bends'). Treatment is by recompression in a pressure chamber followed by slower decompression.

13 · Kidney, Water and Electrolytes

13.1 The kidney
Structure
Renal blood flow
  Variations in and control of renal
    blood flow
Glomerular function
  Measurement of glomerular filtration
    rate
  Glomerular filtration rate in man
  Variations in glomerular filtration
    rate
Clearance
Tubular function
  Tubular reabsorption
  Tubular secretion
Renal handling of water and sodium
  Filtration
  Proximal tubules
  Loops of Henle
  Distal tubules
  Collecting system
Generation and maintenance of the
  medullary hyperosmotic gradient
Diuresis

13.2 Body water and electrolytes
Water
  Distribution of water in the body
  Water balance
  Regulation of total body water
  Disturbances of total body water
    content
Sodium
  Distribution of sodium in the body
  Sodium balance
  Regulation of body sodium
  Disturbances of body sodium content
Potassium
  Distribution of potassium in the body
  Potassium balance
  Regulation of body potassium
  Disturbances of body potassium
    content

**13.3 Acid–base metabolism and regulation of
    pH**
Mechanisms of control
Why plasma pH is 7·4
Production, initial buffering and final
  disposal of acids
Final disposal of hydrogen ions and
  regeneration of body buffers
  The driving force for secretion of
    hydrogen ions
  Fate of hydrogen ions secreted into
    the tubular fluid
  Measurement of urinary titratable
    acid and ammonia
Regeneration of buffer stores
Regulation of hydrogen ion secretion
  and urinary acidity and alkalinity
Compensation for metabolic
  disturbances
Renal compensation for respiratory
  disturbances of pH
Some clinical aspects
  Significance of plasma bicarbonate
  Effect of depletion of potassium
  Effect of inhibitors of carbonic
    anhydrase

13.4 Function in diseased kidneys
Disturbances of glomerular function
Disturbances of tubular function
  Handling of sodium
  Handling of potassium
  Handling of hydrogen ions

13.1 The kidney

The kidneys are essential for life; they eliminate in the urine unwanted water and solutes and so regulate both the volume and the composition of the body fluids. The urine contains (a) waste products including inactivated hormones,

(b) foreign substances and their derivatives, and (c) surplus water and normal soluble constituents of the body.

The kidneys also produce a variety of humoral agents—erythropoietin, active metabolites of vitamin D, renin and prostaglandins.

Structure

Each human kidney has about 1 million *nephrons* arranged in parallel. Nephrons are similar but not identical; the chief variants (Fig. 13.1) are:

1 superficial, *cortical nephrons.* These have short *loops of Henle* reaching only into the outer medullary zone and their *efferent glomerular arterioles* supply

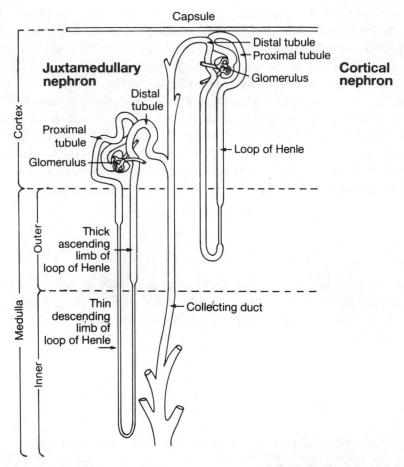

Fig. 13.1. An outline of the anatomical arrangement and organization of the nephrons within the kidney.

their *peritubular capillaries*. About 80% of the nephrons in the human kidney belong to this group;

2 deeper, *juxtamedullary nephrons*. These have long loops of Henle, which plunge deep into the inner medulla. Their efferent glomerular arterioles supply peritubular capillaries and also venous capillary loops (*vasa recta*) which course among the long loops of Henle and the *collecting tubules* deep in the medulla.

Note that all renal tubules receive only postglomerular blood, and that the vasae rectae carry the sole blood supply to the inner medulla.

Renal blood flow

The rate of the renal circulation through both kidneys can be measured by applying the Fick principle to any substance which is transferred from plasma to urine but is not produced, destroyed or stored within the kidney. Let concentrations in arterial plasma, renal venous plasma and urine be P_A, P_{RV} and U, respectively, and let urine flow per min be V. Then the amount entering the kidney per min is the *renal arterial plasma flow* (RPF) $\times P_A$ while the amount leaving is the sum of the amounts leaving in the urine (UV) and in the renal vein (neglecting renal lymph flow which is very slow). It turns out that the volume of the urine per min (about 0·5 to 1 ml min^{-1}) is so much smaller than that of the plasma entering the kidneys each minute (about 650 ml min^{-1}), that the renal venous plasma flow can be taken as equal to the renal arterial plasma flow, RPF. Therefore, the amount of the substance leaving the kidney in the veins equals RPF $\times P_{RV}$. Thus,

$$\text{RPF} \times P_A = UV + (\text{RPF} \times P_{RV})$$

or

$$\text{RPF} = \frac{UV}{(P_A - P_{RV})}.$$

Catheterization studies have shown that *para-aminohippurate* (PAH) is nearly completely removed from the plasma of blood passing through the kidney when its arterial concentration is sufficiently low that the transport mechanism in the proximal tubule is not saturated. Its extraction, E_{PAH}, which equals $(P_A - P_{RV})/P_A$, is 0·9. Thus 90% is excreted and 10% remains in renal venous plasma. We can, therefore, write:

$$\text{RPF} = \frac{UV}{(P_A - P_{RV})} = \frac{P_A}{(P_A - P_{RV})} \times \frac{UV}{P_A} = \frac{1}{E} \times \frac{UV}{P_A}.$$

Hence three practical possibilities for the determination of RPF are:

1 measure U, V, P_A and P_{RV};

2 measure U, V, P_A for PAH and assume $E = 0·9$; or

3 measure U, V, and P_A (for which mixed venous blood suffices), assume $E = 1.0$ and use the formula $ERPF = UV/P$, where ERPF stands for *effective renal plasma flow* (and is, of course, the clearance of PAH).

This assumes that all PAH that reaches tubular epithelium is secreted and that the other 10% of renal plasma flow bypasses the nephrons. This is unlikely. Alternatively, diodone can be used instead of PAH. *Diodone* (introduced into radiology because its high content of iodine makes it radio-opaque) labelled with ^{131}I is easier to estimate than PAH and is also vigorously secreted, but its extraction (0.73) is less complete.

Typical results for ERPF in an adult man are 630 ml min^{-1} estimated from UV/P for PAH. Assuming $E = 0.9$, then $RPF = 630/0.9 = 700$ ml min^{-1}. If packed cell volume is 0.44, *total renal blood flow* (RBF) is $RPF/(1 - PCV) = 700(1 - 0.44) = 700/0.56 = 1250$ ml min^{-1}. Thus, the kidneys, which form 0.5% body weight but account for nearly 10% of the total oxygen consumption of the body, receive 20% of the resting cardiac output. Renal arteriovenous difference for oxygen is low. The large blood supply reflects the need to supply fluid for filtration, not to supply oxygen for renal metabolism.

Variations in and control of renal blood flow

Experiments measuring transit times of dyes, wash-out of gases like ^{85}Kr and lodgement of microspheres in the renal vessels in experimental animals suggest that the cortex receives 94%, the outer medulla 5% and the inner medulla 1% of the total RBF.

Between 80 and 180 mmHg arterial blood pressure as such has little effect on RBF (Fig. 13.2). This is because of *autoregulation* which is believed to be mainly myogenic, the arterioles contracting when transmural pressure increases. Severe reductions of arterial blood pressure as in shock will depress RBF. Sympathetic vasoconstrictor activity is low at rest (warm, recumbent, relaxed) but increases with changes of posture, cold, pain, emotion and exercise and so reduces RBF. The reductions associated with the erect posture and exercise are often exaggerated in patients with cardiac failure. Pathological processes that destroy nephrons chronically reduce RBF.

The renal regulation of the volume and composition of the body fluids involves three processes:

1 filtration at the glomerulus;
2 tubular reabsorption; and
3 tubular secretion.

For any substance, the algebraic sum of the amounts filtered, reabsorbed and secreted by the tubules is equal to the amount excreted in the urine (Fig. 13.3).

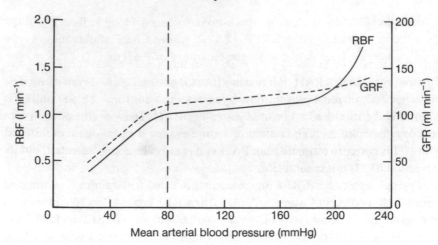

Fig. 13.2. Relationships between mean arterial blood pressure, renal blood flow (RBF) and glomerular filtration rate (GFR). Note the relative stability of both RBF and GFR within the physiological range of arterial blood pressure.

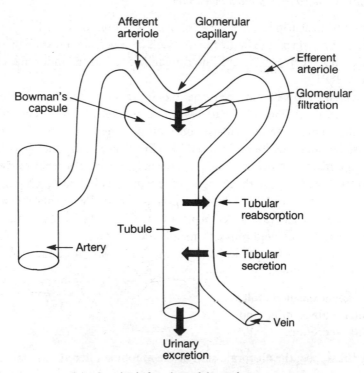

Fig. 13.3. A summary of the three basic functions of the nephron.

Glomerular function

In capillary tufts, blood is exposed at about 40% of mean aortic pressure to a filtering membrane of over 1 m^2 (more than half the external surface area of the body) which separates the plasma from *Bowman's space* (Fig. 13.4). The capillary endothelium is *fenestrated*; its basal lamina (0·2–0·3 μm thick) is composed of loose fibrillar glycoproteins with fixed negative charges. Slit pores between the foot processes (pedicels) of investing podocytes (specialized epithelial cells) provide a diffusion path between plasma and Bowman's space which does not cross cell membranes or cytoplasm. Solutes up to 10 000 M_r pass through the filter freely. For these solutes, which include the ions and metabolites of the extracellular fluid, the concentrations in the filtrate in Bowman's space equal those in the plasma (slightly modified for ions by the Gibbs–Donnan distribution). With larger molecules, diffusion is increasingly restricted and ceases around 70 000 M_r (for albumin, an elongated molecule with negative charge) to 100 000 M_r (for uncharged molecules). For the same head of hydrostatic pressure, an ultrafiltrate passes through this barrier about 100 times faster than through capillaries elsewhere. Thus glomerular capillaries behave as if they possessed pores of the same size as in muscle capillaries but occupying 10% instead of 0·1% of their surface. Glomerular capillaries are also unlike others in that they form a capillary bed in the course of an arteriole. (Contrast this with the situation in other capillaries where arteriolar and venular ends show the Starling equilibrium between ultrafiltration and osmosis, Section 1.8.) In Wistar–Munich rats, which are unusual in that their glomerular capillaries are accessible to micropuncture, hydrostatic pressures around

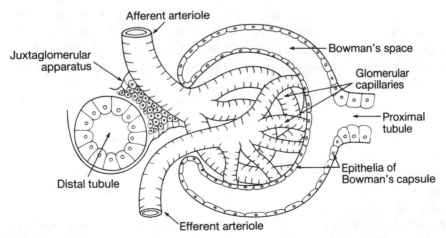

Fig. 13.4. The organization of the glomerulus. Juxtaglomerular apparatus includes juxtaglomerular cells and macula densa.

45 mmHg have been measured, with little drop from the afferent to the efferent end.

The glomerular filtration rate (GFR) should be, in terms of Section 1.7, equal to

$$J_v = L_p A[(P_{gc} - P_t) - \pi_{gc}],$$

where P_{gc} and P_t are the hydrostatic pressures in glomerular capillaries (45 mmHg) and in Bowman's space (10 mmHg), and π_{gc} is the colloid osmotic pressure in glomerular capillaries (25 mmHg at their afferent arteriolar ends).

Substituting the measured values of hydrostatic pressure,

$$GFR = L_p A[45 - 10) - \pi_{gc}],$$

which indicates that filtration should continue until π_{gc} rises to about 35 mmHg. Such increases have been measured in some strains, so that glomerular filtration may not continue along the whole length of their glomerular capillaries. In other species, including man, so large an increase in colloid osmotic pressure cannot be expected because relatively less ultrafiltrate is formed from the plasma.

Regardless of whether filtration proceeds to equilibrium, it is generally accepted that the glomeruli continually produce an ultrafiltrate (less pedantically by common consent called 'filtrate'), and that this is the first step in the production of the urine. Note that the energy required for filtration is supplied by the heart, not by the kidney.

Measurement of glomerular filtration rate

For a substance which has the same concentration in glomerular filtrate and plasma *and is neither removed from nor added to the urine by the tubular epithelium*, the amount filtered per minute, *the filtered load*, must equal the amount excreted per minute, i.e. $P \times GFR = UV$ or $GFR = UV/P$, where P and U are concentrations in plasma and urine *in the same units*, and V is the volume of urine produced per minute.

In dogs, inulin (polyfructosan, M_r 5000), ferrocyanide, creatinine, mannitol and thiosulphate all have the same value of UV/P when excreted simultaneously. Excretion at the same rate for identical concentrations in plasma argues against metabolic handling by cells and favours a common physical process such as filtration through glomerular capillaries and simple conduction along tubules. Other substances which are excreted more rapidly (because tubule cells add them to the urine) or more slowly (because tubule cells remove them) have larger or small values of UV/P which approach that for inulin if tubular cells are poisoned or chilled, or if active transport is saturated by overloading at high plasma concentrations. *Inulin is regarded as the most reliable test* substance for the measurement of GFR, but accurate measurement needs continuous infusion to establish a steady state with constant plasma concentration and rate of excretion. Simpler methods based on single injections, usually of radioactively labelled substances, may serve for routine use. Note that, since the amount of inulin excreted is directly proportional to plasma concentration (Fig. 13.5a), the plot of UV/P against plasma concentration gives a line parallel to the x axis, i.e. UV/P for a substance which is filtered and neither reabsorbed nor secreted is independent of its concentration in the plasma (Fig. 13.5b).

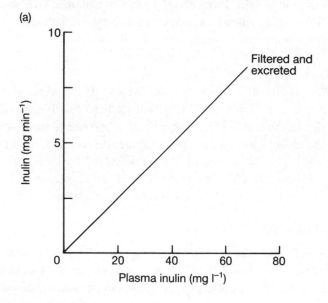

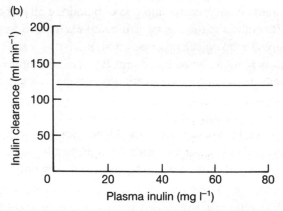

Fig. 13.5. (a) The relationship between plasma inulin concentration and the amounts of inulin filtered and excreted per unit time. (b) The relationship between plasma inulin concentration and the clearance (UV/P) of inulin.

Glomerular filtration rate in man

In adults, GFR is about 125 ml min^{-1} or 180 l/day, of the order of fifty times the volume of plasma in the body. This value represents the sum of the contributions of the individual nephrons from both kidneys. About 650 ml of plasma flows through the kidneys each min (renal plasma flow, RPF). Of this, about one-fifth is filtered; the remaining four-fifths pass into the peritubular capillaries. That is to say, the ratio GFR/RPF, called the *filtration fraction*, is about 0·2. Contrast this with the ratio of ultrafiltration to flow through typical systemic capillaries, which is about 0·005. GFR is better related to body surface area than to weight. From about 2 years of age until after middle-age, when the GFR declines slowly, the average value is 120 ml min^{-1} for each 1·73 m^2 (average body surface area). With 1 million nephrons in each kidney, the average filtration rate per nephron (single nephron glomerular filtration rate) comes to 60 nl/min or 90 μl/day.

Creatinine, produced in the body from the creatine of skeletal muscle, is often used in man to estimate GFR. It is produced at a relatively constant rate throughout the day so that a 24 h urine collection and one plasma sample may be used for the calculation. In fact, renal tubules do secrete a little creatinine but, fortuitously, plasma creatinine is overestimated by the usual chemical method because plasma contains a non-filtered chromogen which reacts as creatinine. Thus the ratio *UV/P* for creatinine approximates that for inulin.

Variations in glomerular filtration rate

The constancy of GFR was probably overemphasized by early measurements with inulin. Like renal blood flow, GFR shows evidence of autoregulation when arterial blood pressure alters (p. 516), although if arterial blood pressure falls below about 60 mmHg, as in shock, it ceases, leading to *anuria*. The erect posture, emotion, pain, cold, exercise and loss of blood are all associated with reductions in GFR, especially the erect posture and exercise in some patients with cardiac failure. Pathological processes which destroy nephrons reduce GFR, and GFR may increase when blood and extracellular fluid volumes are expanded, especially with saline, which dilutes plasma protein and lowers colloid osmotic pressure.

Note that the whole volume of plasma is filtered many times daily. This effectively removes waste products from the blood, but in so doing removes water and all solutes of low M_r at the same time. Hence a major task of the tubules must be recovery of water and solutes needed by the body.

Clearance

The expression *UV/P* is called *renal plasma clearance* or *clearance* for short. It has the dimensions of volume/unit time. The volume calculated from the clearance expression is the smallest volume of plasma that could yield the amount excreted per unit time if the kidneys cleared the substance completely from that volume as it passed through them. The concept of clearance is useful for comparing the renal handling of different substances. The clearance of inulin, as we have seen, estimates GFR, and that of PAH renal plasma flow. If a substance that is filtered has a clearance less than that of inulin, then there must be a net reabsorption of that substance within the renal tubules. If its clearance is greater than that of inulin then there must be a net secretion by the tubular cells into the tubular fluid. This comparison can be made formally by calculating the *clearance ratio*, i.e. clearance of X (C_x)/clearance of inulin (C_{In}), which compares the amounts of a substance in the urine and in the glomerular filtrate from which that urine was formed.

$$\frac{C_x}{C_{In}} = \frac{U \cdot V}{P \cdot GFR} = \frac{\text{Amount of X excreted}}{\text{Amount of X filtered}} \text{ in a period of time.}$$

Hence, if C_x/C_{In} is less than 1·0, there is less X in the urine than was filtered, i.e. X is *reabsorbed*; while if C_x/C_{In} is greater than 1·0, there is more X in the urine than was filtered, i.e. X is *secreted* by the tubules as well as being filtered.

Tubular function

Tubules may reabsorb or secrete. Both processes may be either active or passive. Before discussing some examples of each process, the main factors affecting tubular function need to be summarized. As discussed in Section 1.9, the rate of epithelial transport is determined by surface area × flux, since flux is defined as amount moved/unit time × unit area.

The available surface area is enormous and offers a very favourable ratio of surface area to volume (each tubule handles only about $1/(2 \times 10^6)$ of the total volume filtered per min). In addition, especially in the proximal tubules, there is a very extensive brush border on the apical surface. In contrast to the situation in the gut, there is no motility to control contact time. Normally, the rate of filtration is such that the load presented to the tubules does not exceed their capacity to deal with it. However, if more solute than normal is filtered, or if tubular reabsorption is depressed, the increased flow rate downstream may overload the reabsorptive mechanisms for Na^+ in the distal nephron and result in an increased excretion of sodium chloride and water (*osmotic diuresis*).

In the case of passive absorption, the flux of solutes depends upon the properties of the plasma membranes, of the epithelial cells and of the tight junctions. The proximal tubule is lined by a typical *'leaky' epithelium* (Section 1.9). Glucose, amino acids, phosphate and Cl^- are co-transported across the apical plasma membrane coupled to Na^+ transport and driven by the energy inherent in the electrochemical gradient for Na^+. They pass across the basolateral membrane by either simple or facilitated diffusion. Isosmotic absorption occurs in this segment as expected for 'leaky' epithelia. The epithelia lining more distal portions of the nephron show various degrees of 'tightness' as will be discussed later. The gradient for passive absorption reflects, in part, the rate of removal of reabsorbed solutes by the peritubular capillaries. This is rapid and effective in the cortex but the arrangement of the vasa recta in long loops results in accumulation of reabsorbed solutes within the medulla (see later).

Tubular reabsorption

1 *Urea*. Depending upon the circumstances, the clearance ratio for urea may vary from 0·3 to 0·7, i.e. 30–70% of the filtered load is excreted. Reabsorption of urea is always passive, from a higher concentration in the urine to a lower concentration in plasma, and is greatest when urine volume per minute is small, i.e. when urine is most concentrated and the diffusion gradient steepest. At urine flow rates greater than about 2 ml min^{-1}, the clearance of urea is relatively independent of flow rate and is about two-thirds of the inulin clearance (Fig. 13.6). Note that plasma urea concentration does not necessarily provide a good indication of renal function for it is dependent upon the rate of production of urea (reflecting protein intake and catabolism) as well as on renal excretion (Fig. 13.7).

2 *Glucose*. The clearance ratio of glucose normally is zero. Reabsorption is 'uphill' from tubular fluid, where the glucose concentration reaches zero along the proximal tubule, to 5 mmol l^{-1} in the plasma. Reabsorption involves co-transport with Na$^+$ at the luminal membrane, is inhibited by phlorhizin and

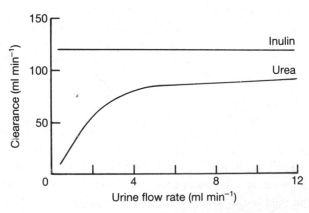

Fig. 13.6. Relationships between urine flow rate and clearance of inulin and urea.

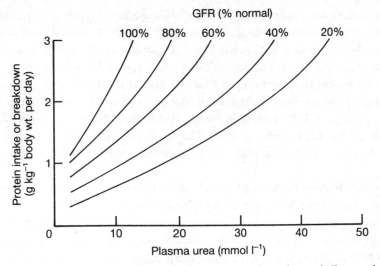

Fig. 13.7. Effects of alterations in renal glomerular function and protein metabolism on plasma urea concentration.

shows saturation effects typical of carrier-mediated transport. The characteristics of glucose handling are illustrated in Fig. 13.8.

If the filtered load is increased by raising the plasma glucose concentration, the transport mechanism in some tubules becomes saturated and glucose begins to appear in the urine. When the transport mechanism in all the tubules is saturated, the maximal reabsorptive capacity of the tubules (*tubular maximum*, Tm) has been reached. In man, when GFR is normal at about 125 ml min^{-1}, the 'threshold' plasma concentration required for glucose to

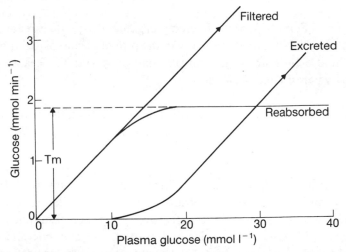

Fig. 13.8. Relationships between plasma glucose concentration and the amount of glucose filtered, reabsorbed and excreted by the kidneys per unit time. (Tm, tubular maximum.)

appear in the urine (the *plasma threshold*) is some 10–12 mmol l^{-1}. Hence the transport mechanism begins to be saturated when the filtered load reaches $10 \times 0.125 = 1.25$ mmol min^{-1}. At plasma glucose concentrations of about 15 mmol l^{-1}, the Tm is reached. This is thus about 1·9 mmol min^{-1} (15×0.125). Note that the plasma threshold reflects the limit of the rate of tubular reabsorption, it is not a fixed plasma concentration. Because filtered load equals $P_{glucose} \times$ GFR, threshold concentration is inversely proportional to GFR. With GFR reduced from 125 to 60 ml min^{-1}, it would take 20 not 10 mmol l^{-1} of plasma glucose to reach the threshold load of 1·25 mmol min^{-1}.

The commonest cause of *glycosuria* is diabetes mellitus, in which plasma glucose concentration is abnormally high. *Renal glycosuria* occurs as an uncommon anomaly when tubular reabsorptive capacity is subnormal. Here plasma threshold will be low so that glucose appears in the urine though its concentration in the blood is not abnormally high.

3 *Calcium*. About 50% (or 1·25 mmol l^{-1}) of Ca^{2+} is bound to plasma proteins and cannot be filtered. The remainder, mostly ionized, is filtered. Of the filtered load of about 200 mmol per day, less than 5 mmol is usually excreted. Some 60% is reabsorbed proximally, most of the remainder in the ascending limbs of the loop of Henle, distal tubules and collecting ducts. The mechanisms involved are not understood, though there may be some relationship between Na^+ and Ca^{2+} reabsorption. *Parathyroid hormone* stimulates reabsorption in the distal tubule though it inhibits reabsorption proximally; and, since it raises plasma ionized Ca^{2+}, its net effect is often to promote urinary Ca^{2+} excretion.

4 *Phosphate*. Some 95% of the filtered inorganic phosphate is reabsorbed by co-transport with Na^+, predominantly in the proximal convoluted tubule with the threshold a little above normal plasma concentration. Reabsorption is inhibited by parathyroid hormone.

5 *Uric acid*. Filtered uric acid is mostly reabsorbed from the proximal tubule but there is also evidence of secretion from plasma to tubular fluid in this same segment.

6 *Amino acids*. About 98% of the filtered load is reabsorbed from the proximal convoluted tubules by co-transport with Na^+. There is evidence of competition within, but not between, five groups: (i) neutral amino acids, (ii) imino acids, (iii) basic amino acids and cystine, (iv) glutamic and aspartic acids and (v) glycine. *Aminoacidurias*, in which large amounts of particular amino acids are lost in the urine, result from deficiencies in specific enzymes, often genetically determined and frequently associated with abnormal metabolism of the same amino acids.

7 *Peptides and small molecular proteins.* Most of the proteins and peptides which cross the glomerular filter are taken up and degraded by proximal tubular cells; only their constituent amino acids are restored to the plasma. Hence the kidney effectively eliminates these substances without excreting them in the urine. The continuous removal of freely filtered peptide hormones at rates proportional to their concentrations in the plasma ensures that their concentrations are largely determined by the rates at which they are released into the circulation.

Tubular secretion

1 *Foreign substances.* Para-aminohippurate (PAH), diodone and some penicillins have clearance ratios of about 5, phenol red (partly bound to plasma protein and incompletely filtered) of 3–4. These substances are secreted by the proximal tubule almost as rapidly as the plasma presents them to the epithelial cells. Secretion is active, steeply 'uphill', and shows saturation, competition between substrates, and inhibition, all typical of carrier-mediated transport. The renal handling of PAH is summarized in Figs 13.9 and 13.10.

2 *Physiological substances.* Metabolic end-products, for example, methyl-nicotinamide, uric acid, urobilin, aromatic sulphates, steroid glucuronides and

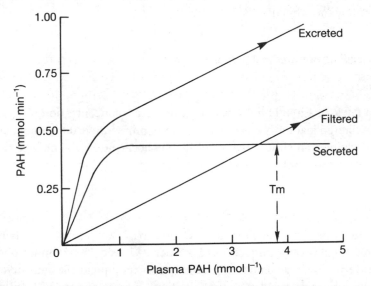

Fig. 13.9. Relationships between plasma para-aminohippurate (PAH) concentration and the amounts of PAH filtered, secreted and excreted by the kidneys per unit time.

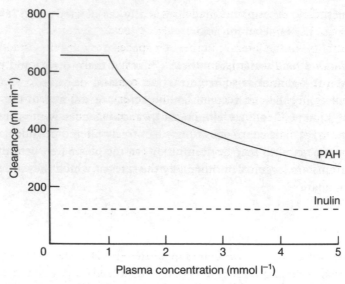

Fig. 13.10. The clearance of PAH as a function of plasma concentration (the clearance of inulin is also shown for comparison).

choline, and autacoids like adrenaline, noradrenaline, acetylcholine and histamine are added to the urine by tubular secretion. Most of the urinary K^+ is secreted by cells in the distal nephron, especially in the connecting tubules and cortical portions of collecting ducts. H^+ and NH_3 are important urinary constituents (Section 13.3) which are not transferred from the plasma but produced by the tubular cells.

Renal handling of water and sodium

Filtration

With a GFR of 120 ml min^{-1} and a plasma Na$^+$ concentration of 150 mmol l^{-1}, the filtered load of Na$^+$ is some 18 mmol min^{-1} or 26 000 mmol per 24 h. Clearly, changes in either GFR or plasma concentration will change the load presented to the tubules.

Proximal tubules

About two-thirds of the filtered H_2O and Na$^+$ are reabsorbed from the proximal tubules each minute, i.e. 120 l per 24 h and 17 500 mmol per 24 h, respectively. The Na$^+$ enters the cells across the apical plasma membrane down its electrochemical potential gradient in association with either co-transported solutes or counter-transported H^+ ions, and is extruded across the basolateral membrane against its electrochemical gradient by Na^+–K^+-ATPase (Fig. 13.11). The Na^+ accumulated locally (with Cl^- to maintain electroneutrality) in the lateral intercellular spaces provides the osmotic force for absorbing salt and water isosmotically. The hydraulic conductivity of this segment of the nephron is so large that an osmotic imbalance as little as 1–2 mosmol kg^{-1} H_2O can account for the observed rate of reabsorption. It seems likely that H_2O follows both cellular and paracellular routes in moving from lumen to interstitium. Reabsorption of salt and H_2O from later portions of the proximal tubules may be assisted by the increased Cl^- concentration resulting from reabsorption of other solutes (glucose, amino acids, HCO_3^-)

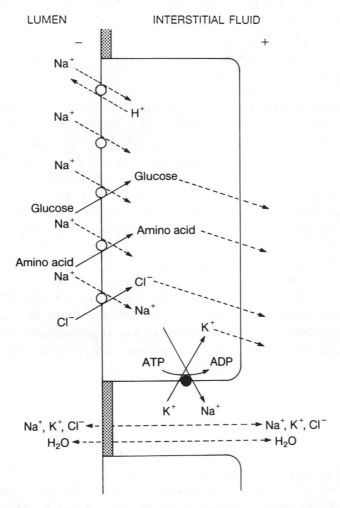

Fig. 13.11. A summary of the major transport mechanisms for solutes and water across renal proximal tubules.

upstream. The increased Cl^- concentration in the lumen favours paracellular passive diffusion of Cl^-, along with Na^+ to maintain electroneutrality, and may contribute to salt absorption in this segment.

   The amounts of salt and H_2O reabsorbed from the proximal tubule are also influenced by the balance of hydrostatic and osmotic forces in the peritubular capillaries. An increased colloid osmotic or a decreased hydrostatic pressure in peritubular capillaries favours uptake of interstitial fluid into capillaries and assists reabsorption. Conversely, decreased colloid osmotic pressure or increased hydrostatic pressure hinders reabsorption. For example, if GFR increases with constant RPF, the filtration fraction will increase and the increased colloid osmotic pressure in the peritubular capillaries will assist in reabsorbing the increased volume of filtrate. Proximal tubular fluid reabsorption matches GFR closely over a wide range of GFR and such adjustments of capillary forces help to account for this '*glomerulotubular balance*'.

Loops of Henle

A volume of about 60 l containing 9000 mmol Na^+ leaves the proximal tubules and enters the *descending limbs* of the loops of Henle each day as an isosmotic solution. Those loops from the juxtamedullary nephrons which run deep into the medulla (some 20% in the human kidney) pass through a region in which the interstitial fluid osmolality increases progressively from isosmotic (285 mosmol kg^{-1} H_2O) at the corticomedullary junction to some 1200–1400 mosmol kg^{-1} H_2O at the tip of the renal papilla in man. In antidiuresis, NaCl and urea contribute in similar proportions to this gradient. The mechanism by which it is generated and sustained is discussed later. The epithelium lining this segment is quite permeable to water, but less so to Na^+ and Cl^-, and impermeable to urea. Therefore, H_2O moves passively and progressively from the descending limb to the interstitium as fluid flows along the tubules, and some Na^+ and Cl^- enter the tubule. Each day there is a net removal of about 10 l of H_2O from this tubular segment, much of it from the short descending limbs of cortical nephrons. There is probably no active transepithelial transport of solutes in the descending limbs.

   In the medullary descending limbs, fluid has equilibrated with the adjacent interstitial fluid and at the tips of the loops now has an osmolality of some 1200 mosmol kg^{-1} H_2O, having lost water and also gained some Na^+ and Cl^-.

   The *thin ascending limb* of the loop differs significantly from the thin descending limb in its permeability, being impermeable to H_2O, highly permeable to Na^+ and Cl^- and moderately permeable to urea. Consequently, NaCl diffuses from the tubules to the interstitium as the fluid flows back towards the cortex, and some urea enters the tubules down its concentration gradient. By the time that the thick ascending segment is reached, much of the Na^+ and Cl^- gained in the descending limbs as the fluid flowed deeper into the medulla has been lost again passively to the interstitium, and the fluid has become somewhat hypo-osmotic compared with the adjacent interstitial fluid.

Unlike the thin ascending limb, the *thick ascending limb*, and its continuation in the cortex as the first part of the distal tubule, avidly reabsorb NaCl from the tubular fluid by an energy-dependent process. The unusual finding that, in this segment of the nephron, the interior of the lumen is positive with respect to the interstitium, led to the proposal that Cl^- rather than Na^+ was transported actively from the lumen. It is now realized that a combination of co-transport of Cl^- at the apical membrane with passive diffusion at the basolateral membrane can account for this electrical phenomenon. The positive charge of the luminal relative to the peritubular fluid helps to promote the reabsorption of cations such as K^+, Ca^{2+} and Mg^{2+} from this site. A possible model of transport in this segment is shown in Fig. 13.12. The impermeability to H_2O means that the removal of some 6500 mmol of NaCl per day (25% of the filtered load) dilutes the luminal fluid which becomes hypo-osmotic. Since, in total, 60 l of fluid with 9000 mmol of Na^+ entered all of the loops of Henle and 50 l with 2500 mmol returned to the distal

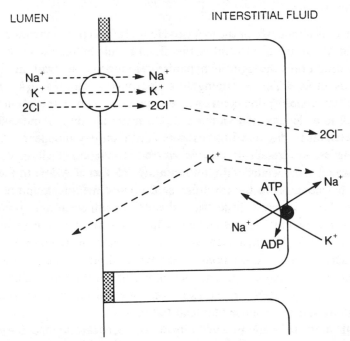

Fig. 13.12. A summary of the major transport pathways for ions across the cells in the thick ascending limb of the loop of Henle.

tubules, Na^+ concentration is now about 50 mmol l^{-1} and osmolality somewhat over 100 mosmol kg^{-1} H_2O.

Distal tubules

It is now appreciated that the distal tubule is not a homogeneous segment. The first part is lined by an epithelium of the same type as the adjacent ascending limb of the loop of Henle and the two together comprise the *diluting segment* of the nephron. The last part is analogous in its properties to the collecting tubule which it joins. NaCl is reabsorbed throughout the length of the distal tubule, some 1200 mmol of Na^+ (5% of the filtered load) being removed from the tubular fluid each day. The permeability of the tubule to H_2O reflects its heterogeneity. While the early part is impermeable to H_2O, the permeability of the last part is determined by the level of circulating *antidiuretic hormone* (ADH). With normal hydration some 20–30 l of water per day are probably absorbed by the distal tubule, leaving 30–20 l containing some 800 mmol of Na^+ to pass through the collecting ducts. The distal tubule is impermeable to urea. Therefore urea entering the thin ascending limb of the loop of Henle is retained within the tubule in this portion of the nephron.

Collecting system

Though only about 20% of the loops of Henle run deep into the medulla, all of the fluid remaining at the end of the distal tubule enters the collecting duct system which runs through the hyperosmotic medulla to drain into the renal pelvis and ureters. The cells lining the collecting ducts form a '*tight*' *epithelium*. Active Na^+ reabsorption continues here and Na^+ concentration in the urine may fall to as low as 10–15 mmol l^{-1}. On a normal diet, about 600 mmol is reabsorbed each day and this reabsorption is largely independent of water handling in this segment. *Water diuresis* (the production of a large volume of a dilute urine) is associated with low circulating levels of ADH. In the absence of this hormone (as in the rare disease diabetes insipidus), epithelial water permeability is very low, little water absorption will occur and some patients have been known to excrete as much as 30–40 l of urine per day (a flow rate of about 25 ml min^{-1}) with a urine osmolality lower than 50 mosmol kg^{-1} H_2O. ADH acts via adenylate cyclase and cyclic AMP to increase the water permeability of the apical plasma membrane of the epithelial cells, opening a cellular, not a paracellular, pathway for water to move down the osmotic gradient from tubular lumen to interstitial fluid.

ADH increases water permeability of both cortical and medullary portions of the collecting ducts, as well as of the most distal portion of the distal tubules. With maximal plasma concentrations of ADH (*maximal antidiuresis*), as little as 0·4–0·5 l of urine is excreted per day (about 0·3 ml min^{-1}) with a urine osmolality about 1200 mosmol kg^{-1} H$_2$O. ADH also increases the permeability of the medullary (but not cortical) collecting ducts to urea and promotes its reabsorption. Therefore, in antidiuresis, the urea that passes from the medulla into the thin ascending limb of the loop of Henle then diffuses back into the medullary interstitium, and its recycling between medullary interstitium and tubular fluid helps to preserve medullary hyperosmolality. During water diuresis, the loss in the urine of much of the urea passing into the ascending limb contributes substantially to the reduction in the medullary osmotic gradient under these conditions. Figure 13.13 summarizes the handling of Na$^+$ and H$_2$O by the different tubular segments.

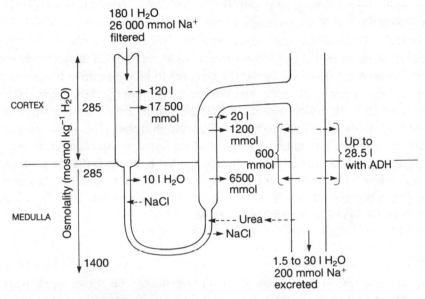

Fig. 13.13. A summary of the contribution of the various segments to the daily handling of Na$^+$ and water by the human kidney. For simplicity, representative values are provided, rather than a range of normal values.

Generation and maintenance of the medullary hyperosmotic gradient

The generation of the gradient of medullary hyperosmolality depends primarily upon the energy-dependent reabsorption of 25% of the filtered load of NaCl in a water-impermeable region of the tubule—the thick ascending limb of the loop of Henle. The solute so reabsorbed would be washed away from the medulla were it not for the fact that the capillaries within the medulla are arranged in loops—the vasa recta—the descending and ascending limbs of which are close to each other and to the adjacent loops of Henle. Reabsorbed solute diffusing into an ascending capillary loop will tend to increase the concentration of NaCl in the capillary at that point. As it is carried towards the cortex, it flows past plasma in the adjacent descending capillary loop in which the NaCl concentration is lower. At every level there will be a tendency therefore for NaCl to diffuse from the ascending to the descending capillary loop and thus to be retained within the renal medulla rather than lost to the cortex. Similarly, at every level, plasma in the descending limb coming from a more dilute region of the medulla will be slightly less hyperosmotic than plasma in the ascending limb which is emerging from the deeper, more hyperosmotic medulla. Thus, in contrast to solute, H_2O will tend at every level to pass from the descending to the ascending limb down its osmotic gradient and, thereby, be shunted away from the deeper medulla. Urea, passing from the medullary collecting ducts into the interstitium during antidiuresis, will also be trapped within the medulla by diffusion from ascending to descending vasa recta. These diffusional movements between the two limbs of the vasa recta are often described as *counter-current exchange*. Note that the arrangement of the vasa recta allows the medullary gradient to be maintained but, were it not for the sources of water-free solute (from energy-dependent NaCl reabsorption in the thick ascending limb and from urea diffusion from the collecting ducts into the medulla during antidiuresis), the large osmotic gradient could not be created. The effect of the latter process is often referred to as *counter-current multiplication*. In effect, what happens is that some of the solute filtered at the glomerulus with H_2O as an isosmotic solution has been trapped in the medulla while its associated H_2O is either lost from the distal tubule in the cortex or excreted in the urine.

Diuresis

An increased rate of production of urine (diuresis) can be of two types: water diuresis and solute or osmotic diuresis. *Water diuresis* results when water is ingested or administered in excess of the body's requirements. ADH secretion is suppressed (p. 536), the collecting ducts become relatively impermeable to water and the excess water is lost without solute. A typical water diuresis is illustrated in Fig. 13.14. There is an inverse relationship between urine osmolality and urine flow rate; flow rate × osmolality (the amount of solute excreted per min) is relatively constant and independent of flow rate. Thus the kidney can adjust its excretion of water without markedly affecting its handling of solutes.

Osmotic diuresis results when more solute is presented to the tubules than they reabsorb, for example, if a non-reabsorbable solute, e.g. mannitol, is filtered, or if the concentration of glucose in the plasma in diabetes mellitus rises so that the filtered load exceeds the tubular maximum, or if tubular function is inhibited (e.g. by drugs which block reabsorption of NaCl in one

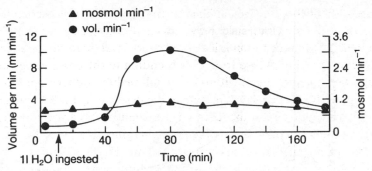

Fig. 13.14. An example of the changes in the rates of excretion of water (vol. min⁻¹) and of total solutes (mosmol min⁻¹) in a water diuresis in man.

or more nephron segments). A typical osmotic diuresis is illustrated in Fig. 13.15. In contrast to water diuresis, urinary flow rate depends upon urinary solute content. Moreover, the greater the rate of solute excretion, the lower is the maximal attainable urinary concentration even with maximal concentrations of ADH (Fig. 13.16), for the larger volume of water reabsorbed from the collecting ducts dilutes the medullary interstitial fluid. Faster flow through the ascending limbs of loops of Henle, especially combined with a decreased concentration of Na^+ when glucose or mannitol is present, may also decrease reabsorption so that less Na^+ is deposited to maintain the medullary osmotic gradient. This gradient cannot be demonstrated in kidneys removed during osmotic diuresis.

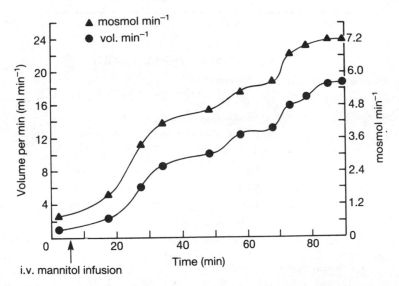

Fig. 13.15. An example of the dependence of the rate of water excretion on solute excretion in an osmotic diuresis induced by the intravenous (i.v.) infusion of mannitol.

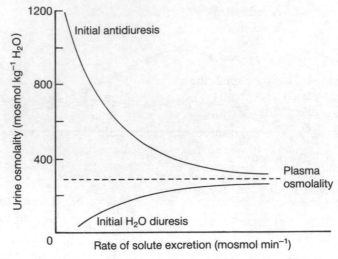

Fig. 13.16. The relationship between the rate of solute excretion and urine osmolality. As solute load decreases, urine osmolality approaches that of plasma. For further explanation see text.

The roles of the different segments of the nephron are summarized in Table 13.1. Micturition is dealt with in Chapter 6, p. 199. Adjustments of salt and water handling to meet the changing needs of the body and the role of the kidney in potassium homeostasis are included in the following sections.

Table 13.1. Summary of the functions of the different segments of mammalian nephrons.

Segment	Functions
Glomerulus	Ultrafiltration: 120 ml min^{-1}, about 20% of RPF
Proximal tubule	Reabsorption: (i) two-thirds filtered Na$^+$, Cl$^-$, and H$_2$O; (ii) filtered HCO$_3^-$; (iii) filtered glucose, K$^+$, phosphate, amino acids and uric acid; and (iv) about half filtered urea. Secretion: organic acids and bases, H$^+$, NH$_4^+$
Loop of Henle	Reabsorption: Na$^+$, K$^+$, Ca^{2+}, Mg^{2+}, Cl$^-$ (ascending limb)
Distal tubule	Reabsorption: Na$^+$, Cl$^-$, H$_2$O Secretion: H$^+$, NH$_4^+$, K$^+$
Collecting tubule	Reabsorption: Na$^+$, Cl$^-$, H$_2$O, (H$_2$O permeability controlled by ADH) Secretion: H$^+$, K$^+$, NH$_4^+$

13.2 Body water and electrolytes

Water

Distribution of water in the body

As outlined in Chapter 1, p. 2, the average 70 kg man has a total body water of about 42 l of which 55% (23 l) is in the cells and 45% (19 l) is extracellular. This extracellular water is further subdivided into plasma (3 l), interstitial fluid and lymph (15 l) and transcellular fluids (CSF, ocular, pleural, peritoneal and synovial fluids) (1 l). The volumes of these compartments can be estimated from the volumes of distribution of substances thought to equilibrate in different compartments. For example, total body water has been estimated from the volume of distribution of urea and isotopes of water (deuterium oxide, tritiated water). Similarly, inulin, sucrose, mannitol and isotopes of sodium and chloride have been used to estimate extracellular water, and isotopically labelled albumin to estimate plasma water. Interstitial water cannot be measured directly but is (ignoring the transcellular volume) the difference between extracellular volume and plasma volume.

The best guide to changes in body water is alteration in body weight, because water constitutes about 60% of body weight. In estimating body water from body weight, however, it must be remembered that fat contains very little water, so that the greater the body fat as a proportion of body weight the lower the body water content. Thus women, in general, have a lower body water content than men of the same body weight.

Water balance

Normally, total body water remains constant. Therefore, over a 24 h period, intake and loss of water must balance exactly. Both intake and loss are controlled, through the thirst–ADH mechanism (discussed below).

Water is taken in drinks as well as up to a litre per 24 h in 'solid food'. It is also formed from the oxidation of metabolites (about 300 ml per 24 h).

Water is lost through the skin (0·5 l) and lungs (0·5 l) by evaporation and, depending upon the need to increase heat loss, in sweat. Faeces contain about 0·1 l of water per 24 h. Urinary water loss is variable. About 600 mosmol of solutes, including end-products of metabolism, must be excreted each day in the urine. The maximal achievable urinary osmolality of about 1200 mosmol kg^{-1} H_2O therefore demands a minimal urinary volume of 0·4–0·5 l per day. Normally, water intake is such that about 1·5 l of urine is excreted each day.

Regulation of total body water

If water loss exceeds gain, with a reduction in total body water content, the osmolality of the body fluids increases. This excites *thirst*, so that water may be ingested, and releases ADH so that water is retained by the kidneys. Conversely, an excess intake of water expands and dilutes the body fluids; the decrease in osmolality eliminates thirst and inhibits the release of ADH, so that water diuresis rids the body of the excess. These mechanisms can hold the osmolalities of plasma and other body fluids constant to within $\pm 1\%$ of 285 mosmol kg^{-1} H$_2$O.

Both thirst and the release of ADH appear to be triggered by the same mechanisms. Of prime importance, normally, is the osmolality of the plasma perfusing the hypothalamus. Neurones in the supraoptic nucleus synthesize ADH which is stored with neurophysin in the nerve endings in the neurohypophysis. The 'blood–brain barrier' is deficient in this region, and these neurones, or nearby osmoreceptors which relay to them, respond to a local increase in plasma sodium concentration. (If plasma osmolality is increased by solutes which penetrate plasma membranes, e.g. urea or ethyl alcohol, thirst and ADH release are not initiated. This suggests that changes in the volumes of the osmoreceptor cells link the changes in plasma osmolality to the response.) Other mechanisms which may stimulate thirst and ADH release include:

1   increased NaCl concentration in the CSF in the third ventricle;
2   increased angiotensin II in the brain (produced locally or delivered via the systemic circulation);
3   decreased arterial blood pressure, signalled via the carotid and aortic baroreceptors; and
4   decreased central venous pressure, signalled via the low pressure receptors in atria and great veins.

The latter two mechanisms, especially **4**, are important in conditions where circulating blood volume is deficient (pp. 442 and 461). Water may even be retained in excess of solute in response to the depleted circulating volume and the control of extracellular osmolality may be sacrificed. Figure 13.17 illustrates the relationship between plasma osmolality, circulating ADH and changes in circulating blood volume.

Disturbances of total body water content

Hypernatraemia. The effectiveness of the thirst–ADH mechanism ensures that a primary loss of water (primary dehydration) only occurs when fluid intake is not possible (e.g. in the absence of a source of water, in the very young, the aged, the confused, the unconscious or in patients with severe vomiting who cannot drink and retain water). It is most likely to be seen when loss of water is excessive, e.g. with hyperventilation (particularly at high altitude where humidity is low), high environmental temperature, fever, or abnormal urinary

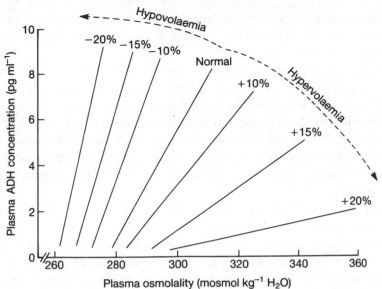

Fig. 13.17. The effect of blood volume and of plasma osmolality on antidiuretic hormone (ADH) secretion. Note the relative contributions of changes in plasma osmolality and changes in circulating blood volume to ADH secretion. (Adapted from Robertson, S.L., Shelton, R.L. & Athar, S. (1976) *Kidney Int.* **10**, 25–37.)

loss resulting either from a failure to produce or secrete ADH (neurogenic diabetes insipidus) or from a decreased responsiveness of the collecting ducts to ADH (nephrogenic diabetes insipidus).

The loss of water increases the osmolality in all fluid compartments and the concentration of Na^+ in the plasma (hypernatraemia). The loss is shared between the cells and the extracellular compartment, so that cell volume is decreased but extracellular volume is better preserved than when there is a loss of extracellular Na^+. Nevertheless, the contraction of extracellular volume stimulates aldosterone secretion promoting renal retention of sodium.

Apart from thirst which may dominate consciousness, symptoms are vague, e.g. weakness and lethargy. The urine is scanty and concentrated except in diabetes insipidus. Plasma osmolality and sodium concentration are increased. It is important that the lost water be replaced by mouth or, if necessary, intravenously (as 5% glucose). In neurogenic diabetes insipidus, ADH is also administered.

Hyponatraemia. This occurs with consumption or administration of inappropriately large volumes of water when the renal excretion of water is impaired, e.g. in anuria, with inappropriate secretion of ADH as in trauma, and with neoplasms that produce ADH. Hyponatraemia may also follow excessive loss of extracellular fluid associated with loss of blood or of sodium when the thirst–

ADH mechanism is stimulated through the low-pressure central venous stretch receptors, and the maintenance of normal osmolality is sacrificed while circulating volume is better maintained.

As the retention of water lowers extracellular osmolality and plasma sodium concentration, water moves from the extracellular fluid into the cells, which swell until the osmolalities of the fluid compartments are equalized.

Depending on the degree of water retention, weakness and muscle cramps may develop, followed by confusion ('water intoxication'), convulsions, coma and death. The plasma hypo-osmolality and low plasma sodium concentration can be reversed by restricting water intake and treating the underlying cause of the condition.

Note that, unlike those of other mammalian organs, the brain cells seem to be able to regulate their volume in anisosmotic media by adjusting their solute content. In hypernatraemia, they shrink initially but gradually increase their solute content over the next few hours and take up water to restore their volume towards normal. Conversely, in hyponatraemia, they first swell but then lose solute and with it water, again partly restoring their volume. This ability to adjust volume has obvious value for an organ enclosed in a rigid container (the skull). However, it means that hyper- and hyponatraemia ought never to be reversed too rapidly. For example, in hypernatraemia, the cells contain more solute than normal and a sudden restoration of normal plasma osmolality will result in pronounced swelling of these cells, raised intracranial pressure and symptoms and signs like those of hyponatraemia.

Sodium

Distribution of sodium in the body

Sodium is the major cation of the extracellular fluid. A typical 70 kg man, with an extracellular Na^+ concentration of 150 mmol l^{-1} and an extracellular volume of 19 l, would have 2850 mmol Na^+ in the extracellular compartment. With an average concentration of 15 mmol l^{-1}, 23 l of cellular water would contain 345 mmol of Na^+. The bones contain a further 2500 mmol of Na^+, less than half of which is readily exchangeable with extracellular Na^+ or other cations; none of this contributes to body fluid osmolality. Total exchangeable Na^+, measured by isotope dilution, is about 50 mmol kg^{-1} body weight.

Sodium balance

On Western diets, 50–300 mmol Na^+ are consumed daily. Almost all of this is absorbed from the gut (together with the much larger amount secreted into the gut, p. 602) and the faeces normally contain only 5–10 mmol daily. The only additional route for Na^+ loss, other than the renal, is from the skin in sweat. This loss is extremely variable and depends solely upon the need to maintain a relatively constant central body temperature. Each litre of sweat contains 30–50 mmol of Na^+ so that the loss of a few litres of sweat can cause a significant Na^+ loss from the extracellular compartment. Renal loss of Na^+ is adjusted to maintain Na^+ balance and can range from a few mmol up to about 500 mmol per day.

Regulation of body sodium

Most of the 'mobile' Na^+ that can vary in amount from day to day is in solution in the extracellular fluid. Hence day-to-day variations in the amount of body Na^+ represent variations in extracellular fluid volume. For every 150 mmol in Na^+ content, there is a corresponding change of 1 litre in extracellular fluid volume. Indeed *the volume of the extracellular fluid is regulated by regulating Na^+ content.*

Unlike the regulation of total body water through the thirst–ADH mechanism, which is comparatively well understood, the mechanisms involved in the regulation of Na^+ and thereby extracellular volume remain controversial. A brief simplified outline is all that can be provided here.

Regulation requires the monitoring of some function of the variable to be controlled, central coordination, and the appropriate control of an effector organ, in this case the kidney.

Sodium concentration is already fixed through the thirst–ADH mechanism, and neither total extracellular Na^+ nor volume is monitored directly. Rather, the volume of a sub-compartment—*central venous volume*—is monitored through the low-pressure stretch receptors in the central veins and atria (p. 442). But how can the volume of this sub-compartment adequately reflect total extracellular volume? Most of the ECF is interstitial fluid which forms a weak gel with mucopolysaccharides, largely hyaluronic acid, and does not move under gravitational forces as a free fluid, though it offers no hindrance to diffusion. One hypothesis proposes that this interstitial gel is unsaturated, because pressures as low as 6 mmHg below atmospheric have been measured within it. 'Imbibition' pressure reflects the tendency for the gel to take up and thereby immobilize fluid, and while the gel remains unsaturated the vascular system and extravascular interstitial spaces appear to have similar compliances. Consequently, fluid gained or lost from the extracellular compartment will be shared between plasma and interstitial fluid, and monitoring the volume of one will monitor the volume of both. However, if sufficient fluid accumulates to saturate the gel (3–4 l), additional fluid in the interstitial compartment is free to move under gravity to the most dependent parts of the body (resulting in oedema), the total extracellular volume can no longer be monitored by changes in central venous pressure, and regulation of volume breaks down.

The carotid sinus and aortic arch baroreceptors, which monitor systemic arterial blood pressure, probably contribute significantly to the regulation of extracellular fluid volume only when intravascular volume is severely depleted.

Information from the atrial low-pressure stretch receptors is processed in the brain but, unlike the well-established role of the hypothalamic osmoreceptors in the thirst–ADH response, there is no established localization within the CNS for the regulation of extracellular fluid volume as such.

The efferent limb of the control mechanism involves regulation of the renal handling of Na^+. It has become customary to describe this as involving three factors: the filtered load, aldosterone which acts in the distal tubules, and a group of so-called 'third factors' which are believed to act in the proximal tubule. Overall, Na^+ excretion = filtered load − tubular reabsorption.

Filtered load. When Na^+ concentration in the plasma remains constant, filtered load varies only with GFR. Therefore factors like blood loss or loss of extracellular volume, exercise and adoption of the upright posture, which alter GFR through cardiovascular reflexes mediated by sympathetic nerves to the kidneys and by circulating catecholamines, change filtered load.

Aldosterone. Secretion of this hormone from the zona glomerulosa of the adrenal cortex is stimulated principally by circulating angiotensin II (ANG II). Of lesser importance are decreased Na^+ and increased K^+ concentrations in the blood perfusing the adrenal gland. ANG II is formed through the following sequence (Fig. 13.18). Renin, a specific proteolytic enzyme produced in the juxtaglomerular apparatus in the kidney, splits a decapeptide angiotensin I (ANG I) off from angiotensinogen, an α_2-globulin synthesized in the liver. ANG I is converted in turn to ANG II (an octapeptide) by *converting enzyme* found chiefly, but not entirely, in the lungs. The rate-limiting step in this sequence is normally the release of renin from the juxtaglomerular apparatus. Three major stimuli induce this release:

1 decreased arterial blood pressure which is thought to act directly on the afferent arterioles;
2 increased sympathetic activity (including circulating catecholamines) acting through β-adrenoceptors; and
3 a decrease in the rate of delivery of fluid past the macula densa in the distal tubule.

As well as promoting the secretion of aldosterone, ANG II is a potent vasoconstrictor and can increase total peripheral resistance. An increase in mean arterial pressure, however, inhibits further release of renin. ANG II also

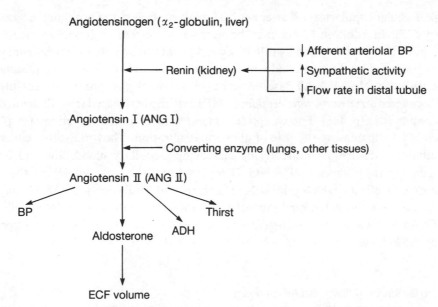

Fig. 13.18. The formation, and major physiological roles, of angiotensin II. BP, blood pressure.

promotes thirst and stimulates the release of ADH, by acting directly on circumventricular organs which are outside the blood–brain barrier. Thus it not only stimulates Na^+ retention through aldosterone, but also the acquisition and retention of water, and so plays a major role in the regulation of extracellular volume. It is inactivated by angiotensinases in the plasma.

Aldosterone promotes the reabsorption of Na^+ by a variety of 'tight' epithelia, principally the distal portion of the nephron, but also the epithelium of the colon and the rectum, and the ducts of salivary and of sweat glands. In all of these, it promotes Na^+ absorption and K^+ (and, in some, H^+) secretion. It appears to initiate the synthesis of protein which promotes Na^+ entry to the cells by facilitated diffusion across the apical plasma membrane. This increases intracellular Na^+ concentration and stimulates active Na^+ transport from the cells across the basolateral membrane via the Na^+–K^+-ATPase. The increased rate of entry also depolarizes the apical plasma membrane and increases the electrochemical potential gradient favouring diffusion of cations (K^+ and H^+) from the cells to the lumen. The effects of aldosterone on Na^+ reabsorption however take about 1 h to become apparent. Therefore this hormone is important for long-term rather than minute-to-minute adjustments of Na^+ excretion. It is inactivated in the liver.

'Third factors'. In some experimental situations, the renal handling of Na^+ cannot be accounted for by changes in filtered load or by aldosterone. There is some evidence that changes in the balance of the Starling forces in the

peritubular capillaries will alter net Na$^+$ reabsorption in the proximal tubules (p. 527). In addition, there may be hormones, *natriuretic hormones*, which inhibit Na$^+$ reabsorption by tubular cells. One example is *endoxin*, which may be produced by the brain and which appears to be an inhibitor of the plasma membrane Na$^+$-K$^+$-ATPase. Another example is the group of recently discovered *atrial natriuretic peptides* (ANP), which are released from distended cardiac atria (p. 443). These peptides appear to increase the excretion of Na$^+$ chiefly by increasing the rate of glomerular filtration. They may also reduce tubular reabsorption indirectly by increasing medullary blood flow and by reducing the secretion of ANG II and aldosterone. In contrast to these hormonal effects, which inhibit Na$^+$ reabsorption, reabsorption of Na$^+$ may be increased through direct sympathetic stimulation of proximal tubular cells. Figure 13.19 summarizes current knowledge of the control of Na$^+$ excretion by the kidney.

Disturbances of body sodium content

These result in changes in extracellular volume.

Oedema. Oedema is a clinically detectable excess of extravascular ECF. It may be either localized or generalized.

Localized oedema need not imply an increased amount of sodium in the body if the affected region is small. It results from local disturbance of the Starling equilibrium (p. 432) by: (a) increased capillary hydrostatic pressure (e.g. with venous stasis, thrombosis or obstruction); (b) increased pericapillary colloid osmotic pressure (e.g. with lymphatic obstruction or inflammatory exudation); or (c) increased capillary permeability, leading to increased loss of protein to the interstitial spaces (e.g. inflammatory and allergic conditions).

Generalized oedema occurs when the disturbance of the Starling equilibrium is widespread; it is associated with increased body weight, the movement of free fluid under gravity, tissue swelling and pitting on pressure. There is typically at least 4–5 l of excess fluid (enough to saturate the interstitial gel and overflow) and hence there must be an excessive amount of sodium in the body—150 mmol for each litre of oedematous fluid or kg of excess weight. This disturbance cannot occur by the mere shifting of fluid from the plasma, for there is less than 4 l of plasma, and so a patient whose whole plasma volume was shifted into the tissue spaces would not be oedematous—and would also not be alive!

Two important causes of widespread loss of fluid from blood vessels are:

1 lack of plasma protein, especially albumin, so that the colloid osmotic pressure is everywhere low. This may occur in gross undernutrition, in diseases

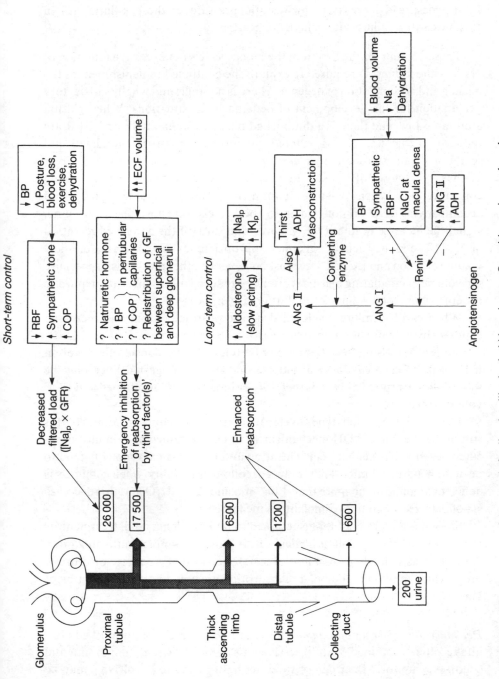

Fig. 13.19. A summary of the control of sodium handling by the kidneys in man. Quantities are in mmol per day.

of the liver which impair production of albumin, and when albumin is lost in the urine faster than the liver can replace it (nephrotic syndrome); and

2 a widespread increase in hydrostatic pressure in the capillaries, as in chronic cardiac failure with venous congestion.

These account for the loss from the blood vessels but excessive retention of Na^+ by the kidneys is required to explain the continued replenishment of the plasma and accumulation of oedema. There is no single and wholly satisfactory explanation for the development of oedema in all situations. When plasma albumin is low, and there is a diminished plasma volume, RBF and GFR are reduced so that less Na^+ is filtered; secretion of renin is stimulated and secretion of aldosterone is increased, promoting Na^+ reabsorption. In liver disease, breakdown of aldosterone may be slower, increasing the circulating concentration. When plasma albumin is depleted, retained NaCl and accompanying H_2O do not remain in the vessels to restore plasma volume, signal the increase in total ECF volume, and turn off the enhanced retention of Na^+. The kidney acts in a manner appropriate for the correction of a low volume of blood and ECF. There must however be other mechanisms, because the volume of circulating plasma is often not reduced in patients with nephrotic oedema whose total volume of ECF is greatly increased.

When cardiac failure results in decreased cardiac output and increased capillary hydrostatic pressure, reflex vasoconstriction reduces RBF and GFR, so that less Na^+ is filtered. Renin may be released and aldosterone secretion increased, while breakdown of aldosterone in the congested liver may be slowed. Reabsorption of Na^+ is therefore enhanced, Na^+ is retained and ECF volume increases.

Retention of Na^+ first tends to elevate plasma Na^+ concentration, but this stimulates the thirst–ADH mechanism so water is retained and, in uncomplicated oedema, plasma Na^+ concentration and total plasma osmolality tend to be within normal limits. (The reduced colloid osmolality when albumin is deficient amounts to no more than 1 or 2 mosmol kg^{-1} H_2O and this decrease is not detected when total osmolality is measured.)

Since it is dietary Na^+ that is retained to make oedema fluid, treatment by restriction of dietary salt is logical. It is however unpleasant, and often ineffective as it tends to mobilize aldosterone, so that a variety of diuretic drugs which inhibit renal tubular reabsorption are used to combat retention of Na^+.

Depletion and deficiency of sodium. This is caused by absent or diminished Na^+ intake often combined with excessive loss. Under extreme environmental conditions, up to 15 l per day of *sweat* with 30–50 mmol l^{-1} of Na^+ may be lost. Normally, about 8 l of fluid containing up to 1 mol of Na^+ are secreted and reabsorbed per day in the *gastrointestinal tract*. This Na^+ can be lost through vomiting, aspiration of the gut contents, by diarrhoea or through fistulae. In cholera and other secretory diarrhoeas, the rate of intestinal secretion may be increased enormously. Normally losses through non-renal routes are countered by the virtual disappearance of Na^+ from the urine. But the *urine* can become a vehicle of loss in Addison's disease (absence of aldosterone), during osmotic diuresis, and in diabetes mellitus when Na^+ is lost not only as a consequence of the osmotic diuresis but also with the conjugate bases of keto-acids.

So long as osmoregulation continues, loss of Na^+ leads to loss of water at the rate of 1 litre per 150 mmol Na^+. This loss of water is shared between the plasma and the extravascular ECF. The cell water is unaffected or may even increase if Na^+ concentration is depressed. Hence, for a given loss of water, the consequences are much more serious for the circulation than are those of primary loss of water, since the ECF and plasma bear the brunt of the water deficiency, as shown by haemoconcentration, with increases in red-cell count, plasma protein concentration and viscosity, and evidence of impaired renal function with retention of urea and other waste products as a consequence of the depressed GFR. Moreover, thirst is less prominent than in pure water loss with its increased osmolality, and the cardinal manifestation is peripheral circulatory failure. There is also the danger of a vicious cycle developing if tissue perfusion falls enough for cells deprived of oxygen to swell, taking up H_2O and Na^+, and further reducing ECF volume, circulating blood volume, and perfusion of the tissues. As the loss of ECF continues, the thirst–ADH mechanism is activated as a consequence of decreased venous volume and, eventually, of decreased arterial blood pressure. This may result in retention of H_2O without Na^+ and consequently in decreased body fluid osmolality and hyponatraemia. Thus it is possible to see hyponatraemia associated with a decreased volume of total body water as well as with pure water retention with overexpansion of total body water. Note that water alone is relatively ineffective in restoring circulating blood volume. In a 70 kg man with about 23 l of cell water and 19 l of extracellular water, less than half of any additional water would be retained within the extracellular compartment, and only one-sixth of this (about 75 ml for each additional litre) would remain in the plasma in the steady state. In contrast to hypernatraemia, it is Na^+ not water that needs to be administered either orally (with glucose to promote intestinal absorption) or as an isosmotic intravenous solution.

Potassium

Distribution of potassium in the body

Potassium is the chief intracellular cation of man and of the animals and plants he eats. The body of a 70 kg man contains about 3600 mmol K$^+$ of which 80 mmol is in the skeleton, 80 mmol is in the ECF and the rest, about 95%, is in the cells, some 3000 mmol of it in skeletal muscle. Total exchangeable K$^+$ is about 50 mmol kg^{-1} of body weight in men and about 40 mmol kg^{-1} in women whose muscles, the largest component of the cell mass, form a lower proportion of body weight. The intracellular concentration (100–150 mmol l^{-1}) is not critical but a two- or threefold increase or decrease in the extracellular concentration (4–5 mmol l^{-1}) can paralyse muscles or stop the heart. Rapid loss of 5% of cellular potassium into the extracellular fluid would be lethal. Extracellular potassium concentrations below 2·5 or above 7 mmol l^{-1} produce weakness of limb and trunk muscles. Below 2 and above 9 mmol l^{-1}, flaccid paralysis occurs and the respiratory musculature is affected and the heart may cease to beat. These changes in cardiac and skeletal muscle activity are consequences of altered K$^+$ gradients with a normal cellular K$^+$. Low plasma potassium concentrations (*hypokalaemia*) increase the diffusion gradient, hyperpolarizing the plasma membrane so that the nerve and muscle cells become less easily depolarized and therefore less excitable. High plasma potassium concentrations (*hyperkalaemia*) decrease the diffusion gradient, depolarizing the plasma membrane. Initially, as threshold is approached, cells may become more excitable but, once the membrane potential has fallen below threshold, action potentials can no longer be generated and paralysis results. Hence acute alterations in plasma potassium, which leave cellular potassium concentration unaltered, are less well tolerated than are chronic alterations in which cellular potassium concentrations change in the same direction and the ratio of cell to plasma [K$^+$] remains relatively constant.

Potassium balance

Ordinary diets supply about 100 mmol. The faeces remove about 10 mmol and the urine 90 mmol per day.

Regulation of body potassium

Though the losses of K$^+$, principally in the urine, maintain body potassium constant despite variations in intake, the details of this regulation are poorly understood. Following the intake of K$^+$, about a half is lost in the urine over the next 6 h or so; the remainder largely disappears from the extracellular fluid into cells and is excreted subsequently. Insulin promotes cellular uptake of K$^+$ and a raised plasma K$^+$ concentration stimulates release of insulin. Insulin may therefore be used in an emergency to lower plasma K$^+$ concentration. Other hormones including adrenaline, aldosterone and glucocorticoids may also stimulate cellular K$^+$ uptake, though their physiological importance is unclear. The acid–base balance also influences exchange of K^+ between cells and extracellular fluid. During acidosis, H^+ ions enter cells in exchange for K^+, during alkalosis, K^+ enters cells in exchange for H^+ ions. These effects are more marked with metabolic than with respiratory disturbances.

The major excretory route for K^+ is through the kidneys. Of the 800 mmol or so of K^+ filtered per day, some 80% is reabsorbed in the proximal tubule by mechanisms which are not yet fully understood. Some of this reabsorption may be a passive consequence of the absorption of NaCl and other solutes together with water, which concentrates luminal K^+ and creates a favourable gradient for diffusion through the paracellular pathway. Some may involve uptake into the cells across the apical membrane. A further 10% of the filtered load is reabsorbed in the thick ascending limb, being co-transported across the apical membrane with Na^+ and Cl^- as discussed earlier (Fig. 13.12).

About 10% of the filtered load reaches the late distal tubule and collecting ducts. Here, in addition to reabsorption, K^+ ions are secreted. Of the two cell types lining the nephrons in this region, it seems that the *principal cells* reabsorb Na^+ and secrete K^+, whereas the *intercalated cells* secrete H^+ ions and reabsorb K^+. Factors which promote K^+ secretion by the principal cells include (1) increased dietary intake of K^+, (2) aldosterone, (3) increased rate of Na^+ delivery and (4) increased flow rate. Factors promoting K^+ reabsorption by the intercalated cells include (1) decreased cellular H^+ activity and (2) a low K^+ diet. The relationship between K^+ secretion and H^+ secretion within this tubular segment is of great importance. Increased availability of luminal Na^+ here promotes secretion of both ions, the relative contributions of each reflecting the acid–base and K^+ status of the individual, but we lack a detailed understanding of the mechanisms involved.

The urinary excretion of K^+, therefore, reflects the relative activities of the reabsorptive and secretory processes and the clearance ratio for K^+ can range between 0·1 (reabsorption predominant) and 3·5 (secretion predominant). In K^+ balance, with dietary intake of 100 mmol per day, the clearance ratio ordinarily approximates to 0·15.

Potassium is also secreted by epithelial cells lining the colon. Again the mechanism is unclear but uptake of Na^+ from the lumen appears to stimulate secretion of K^+, and aldosterone promotes both processes.

Whereas the control of water balance and, to a lesser extent, of Na^+ balance are fairly well understood, there is no comparable framework for discussing the regulation of K^+ balance. Are there receptors sensitive to a function of body K^+? Is there central integration of information about body K^+? How is the function of the effector organs (kidney and colon) regulated so that K^+ excretion matches K^+ intake? These questions are still unanswered. Most discussions of K^+ balance imply that the K^+ concentration in cells of the distal nephron and colon mirrors that of cells elsewhere so that factors affecting cellular K$^+$ generally affect these cells as well. Then if losses of K$^+$ from principal cells to the luminal fluid and from colonic epithelial cells to the gut lumen are adjusted to maintain the concentration of K$^+$ in those cells constant, body K$^+$ content will perforce be regulated. Whether this simple view is adequate remains to be established.

Disturbances of body potassium content

Hyperkalaemia. Potassium is normally excreted by the kidneys so effectively that body K$^+$ content remains constant, any increase in intake being matched by urinary excretion. Plasma K$^+$ concentration *does not* necessarily parallel body K$^+$ content. Clinically hyperkalaemia is usually caused by excessive loss of cellular K$^+$ combined with diminished renal functions as, for example, in:

1 shock where the fall in mean arterial pressure results in hypoperfusion of tissue and a loss of K$^+$ from hypoxic cells, as well as decreased renal perfusion;
2 crush injuries where damaged cells leak K$^+$ into the plasma and myoglobinuria disrupts renal function; and
3 terminal renal failure where increased tissue catabolism releases K$^+$ from cells and depressed renal function prevents its excretion.

In all of these conditions, plasma K$^+$ rises because K$^+$ is lost from cells faster than the kidneys can excrete it. Total body potassium is actually falling, though plasma K$^+$ is elevated.

The symptoms and signs of hyperkalaemia *per se* are vague; cardiac arrhythmias and cardiac arrest may develop before excitability of nerve and skeletal muscle has been clinically affected. Typical, progressive changes in the ECG (Fig. 13.20) give the earliest indications of the severity of hyperkalaemia. In an emergency plasma K$^+$ concentration may be decreased abruptly by insulin which drives K$^+$ (and glucose) into cells. Bicarbonate also lowers plasma K$^+$ by stimulating cellular uptake of K$^+$ in exchange for H$^+$ ions.

Potassium depletion and hypokalaemia. These must be distinguished from each other. Chronic depletion of body potassium may be associated with hyperkalaemia, for example, in a metabolic acidosis where cell K$^+$ is exchanged for the extracellular H$^+$ ions. Conversely, body potassium may be greater than normal in the presence of hypokalaemia, for example, in a metabolic alkalosis. Normally, plasma K$^+$ concentration remains relatively constant, excess or deficit in body potassium being accommodated by the cells. Thus measurements of plasma K$^+$ concentration *do not* allow conclusions to be drawn about body potassium content. Estimates of the latter have been obtained by using isotopes or by extracting potassium from muscle biopsy specimens. A recent finding that the K^+ content of red blood cells mirrors that of other cells offers a simpler way to estimate total body potassium.

Chronic K^+ depletion is associated with degenerative changes in myocardium and skeletal muscle as well as with functional changes (muscular weakness, paralysis). Vacuolar changes in renal epithelial cells accompany an impaired urinary concentrating ability and a polyuria which is insensitive to exogenous ADH.

Since the kidneys normally conserve K^+, loss of body potassium usually reflects a combination of diminished intake with increased loss through the gut or kidneys. Vomiting, diarrhoea, aspiration of gastrointestinal contents, and fistulae may all lead to hypokalaemia. Aldosterone and a variety of diuretics which facilitate K^+ secretion by the epithelial cells through additional distal delivery of Na^+ promote loss of K^+ in the urine. As with hyperkalaemia, typical ECG changes may give early warning of hypokalaemia (Fig. 13.20).

Potassium and acid–base balance. Cells exchange K^+ and H^+ ions with plasma. In metabolic acidosis, plasma K^+ concentration increases even though body potassium may become depleted. In metabolic alkalosis plasma K^+ concentration may decrease. But though cells gain K^+ initially, chronic alkalosis may lead to loss of body potassium because of increased K^+ secretion by renal principal cells (perhaps reflecting their higher K^+ concentration) and of increased delivery of Na^+ to this segment of the tubule. This encourages exchange of tubular Na^+ for cell K^+ with retention of K^+ in the lumen to preserve electroneutrality.

Conversely, chronic K^+ depletion can lead to an alkalosis, when decreased K^+ secretion by depleted principal cells results in a greater portion of Na^+ delivered to the distal tubule being reabsorbed in exchange for secreted H^+ ion. The corresponding transfer of cell HCO_3^- to the plasma may explain the paradoxical association of an acid urine with an alkaline plasma.

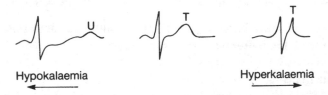

Hypokalaemia   Hyperkalaemia

Fig. 13.20. Characteristic changes in ECG produced by alterations in plasma K^+ concentration. Hypokalaemia results in a reduction in amplitude or reversal of the T wave and the appearance of a U wave; hyperkalaemia results in a reduced P wave, a bizarre QRS complex and a tall narrow T wave.

13.3 Acid–base metabolism and regulation of pH

Cells, especially those of the nervous system, are sensitive to the pH of their surroundings. Too great an alkalinity leads to headache, lassitude, tetany and convulsions. Too great an acidity makes respiration deep and sighing, and causes drowsiness progressing to coma. Plasma pH below 7 or above 7·8 is usually lethal, so that $[H^+]$ must remain between 16 and 100 nmol l^{-1}.

The blood in man is normally kept blandly alkaline at pH $7·4 \pm 0·04$, $[H^+] = 36$–44, i.e. 40 ± 4 nmol l^{-1}. Most cells are believed to be less alkaline than plasma, with pH nearer 7·0.

Note how little free H^+ there is in the body. Forty litres of body water at a pH of 7·0 would contain 4000 nmol or 0·004 mmol of H^+; holding this constant within 10% means a tolerance of 0·0004 mmol; and each day the body produces by metabolism 13 000 mmol of potentially acid-forming carbon dioxide and 50–80 mmol of stronger, non-volatile acids—H_2SO_4 and H_3PO_4. A little arithmetic shows that the precision of control is far better than that of a watch keeping time to 1 s per day.

Mechanisms of control

Stability is achieved by buffers and exchanges of ions with bones and cells which minimize changes of pH pending disposal of acids or alkalis by lungs or kidneys. The important buffer bases that take up H^+ ions in the body are the conjugate bases of three weak acids—carbonic acid, proteins and phosphoric acid. The reactions may be written:

$$\text{Bicarbonate: } H^+ + HCO_3^- \rightleftharpoons H_2CO_3 \rightleftharpoons H_2O + CO_2$$
$$\text{Phosphate: } H^+ + HPO_4^{2-} \rightleftharpoons H_2PO_4^-$$
$$\text{Proteinate: } H^+ + Prot^{n-} \rightleftharpoons HProt^{(n-1)-}$$

or, in general: $H^+ + Buf^- \rightleftharpoons HBuf$. Note that this is reversible. If pH tends to fall, buffer base takes up H^+, while, if pH tends to rise, buffer acid supplies H^+, and so pH is stabilized and changes are 'buffered'. Note also that just as much Buf^- is used up as H^+ is neutralized. For example, when sulphuric acid is added to the plasma:

$$SO_4^{2-} + 2H^+ + 2Buf^- + 2Na^+ \rightarrow 2Na^+ + 2HBuf + SO_4^{2-}.$$

Only the very weak base SO_4^{2-} and the weak buffer acid HBuf remain; pH is controlled, but buffer base has disappeared.

Why plasma pH is 7.4

The Henderson–Hasselbalch equation,

$$pH = pK' + \log_{10}\frac{\text{[conjugate base]}}{\text{[conjugate acid]}}$$

may be applied to any of the plasma's buffer pairs, most conveniently to the predominant one, bicarbonate/carbonic acid. pK' for the sequence $CO_2 + H_2O \rightarrow H_2CO_3 \rightarrow H^+ + HCO_3^-$ is 6·1, and the solubility of CO_2 in plasma is 0·03 mmol mmHg.
Hence,

$$pH_{plasma} = 6 \cdot 1 + \log_{10} \frac{[HCO_3^-]}{0.03\, P_{CO_2}}.$$

With $[HCO_3^-]$ maintained at about 24 mmol l^{-1} by the kidney and P_{CO_2} regulated close to 40 mmHg by the respiratory system (Section 12.5),

$$pH = 6 \cdot 1 + \log_{10} \frac{24}{0 \cdot 03 \times 40} = 7.4.$$

The Henderson–Hasselbalch equation indicates the important variables: pH varies with $[HCO_3^-]$ and inversely with P_{CO_2}. Hence a change in pH that results from an alteration in *either* $[HCO_3^-]$ or P_{CO_2} can be avoided or corrected by changing the *other* variable to preserve or restore the buffer ratio. In a solution in which bicarbonate is the only buffer base, doubling or halving $[HCO_3^-]$ or P_{CO_2} shifts pH by 0·3 unit. In blood, which has other buffers besides bicarbonate, doubling or halving P_{CO_2} produces a smaller final change in pH of 0·2 unit because other buffers release or take up H^+ and $[HCO_3^-]$ changes in the same direction as P_{CO_2}.

The Henderson–Hasselbalch equation also summarizes the basis for physiological regulation of pH: the kidneys control the numerator and the respiratory system is the denominator. The respiratory chemoreceptors are affected by pH as well as by P_{CO_2}. Adding acid to the blood lowers pH as well as $[HCO_3^-]$. The reduction in pH stimulates respiration so that P_{CO_2} is lowered to match the lowered $[HCO_3^-]$, and pH is rapidly returned towards normal.

Finally, the Henderson–Hasselbalch equation offers a rational classification of disturbances of pH. The numerator, $[HCO_3^-]$, is affected by ingestion, production or loss of acid or alkali, giving rise to metabolic (including renal) acidosis or alkalosis. The denominator, $0 \cdot 03\, P_{CO_2}$, is affected by alterations in pulmonary ventilation or in the composition of inspired air, giving rise to respiratory acidosis or alkalosis. Fundamentally, disturbances of acid–base balance fall into one or other of the following categories; clinically, mixed disturbances are sometimes seen.

1 Reduced $[HCO_3^-]$: *metabolic acidosis.* This can arise from : (i) ingestion of HCl or NH_4Cl ($Cl^- + NH_4^- \rightarrow Cl^- + H^+ + NH_3$, converted by the liver to urea); (ii) production of ketone acids in diabetes or of lactic acid in shock; (iii) loss of alkaline secretions such as bile, pancreatic or intestinal secretions; and (iv) failure to generate adequate HCO_3^- in the kidney (*renal tubular acidosis*).

The fall in pH is largely corrected by increased pulmonary ventilation. Hence the chief effect is a reduction in [Buf⁻], especially [HCO₃⁻].

2 Increased [HCO₃⁻]: *metabolic alkalosis*. This can arise from: (i) ingestion of NaHCO₃ or alkaline stomach powders; (ii) loss of acid secretions, notably gastric juice; and (iii) depletion of body potassium. The rise in pH is partly compensated by reduced pulmonary ventilation with retention of CO_2, and the chief effect is increased [HCO₃⁻].

3 Increased P_{CO_2}: *respiratory acidosis*. This can arise from: (i) hypoventilation from, for example, respiratory obstruction or paralysis, drugs or toxins; and (ii) high inspired P_{CO_2}. Retention of CO_2 increases P_{CO_2} substantially. Since total anions must still equal total cations, [HCO₃⁻] can increase no more than other buffer bases are reduced with falling pH; hence initially the main effect is a reduction in plasma pH, later to be corrected by slow replacement of Cl⁻ by HCO₃⁻ from the kidneys if the retention of CO_2 persists.

4 Decreased P_{CO_2}: *respiratory alkalosis*. This results from hyperventilation which may be: (i) voluntary (experimental); (ii) nervous, with tension and anxiety (hyperventilation syndrome); or (iii) hypoxic, at high altitudes, from peripheral chemoreceptor drive (carotid body). P_{CO_2} is substantially reduced, but initially alteration in [HCO₃⁻] is minimal, so that pH is increased with no change in [Buf⁻]. Later, if hyperventilation persists, the kidneys excrete HCO₃⁻ in an alkaline urine and slowly restore pH towards normal by reducing [HCO₃⁻].

Remember that initially, before renal compensation can be effective (which takes days), *metabolic disturbances affect [HCO₃⁻] far more than pH; respiratory disturbances affect pH far more than [HCO₃⁻].*

Production, initial buffering and final disposal of acids

1 *Carbonic acid.* CO_2 is produced at the rate of about 13 000 mmol per day (10 mmol min⁻¹) and represents potential acid that would need 13 000 mmol per day of NaOH to neutralize it to NaHCO₃. In fact, it is excreted by the lungs, and very effectively buffered in transit from tissues to lungs in the blood, mainly by haemoglobin (Section 12.4). Removal of O_2, as CO_2 is being taken up, replaces O_2Hb^- by Hb^- which is a much stronger base. About 0.7 mmol of H^+ for each mmol of O_2 given up (about 25 ml at body temperature) is required to prevent the blood becoming even more alkaline than normal. Hence only about 0·1 mmol of the 0·8 mmol of CO_2 produced from 1 mmol of O_2 (with *RQ* of 0·8) yields H^+ ions which need to be handled by buffers other than Hb, and venous blood is only about 0·03 pH more acid than arterial. In summary:

$$CO_2 + H_2O \xrightarrow{\text{(carbonic anhydrase in rbc)}} H_2CO_3 \rightarrow H^+ + HCO_3^-$$

$$H^+ + O_2Hb^- \longrightarrow HH_b + O_2.$$

Summing these two,

$$O_2Hb^- + H_2O + CO_2 \longrightarrow O_2 + HH_b + HCO_3^-.$$

Because red-cell HCO_3^- exchanges readily with plasma Cl^-, CO_2 taken up into the blood is mostly carried as HCO_3^- in the plasma. The process is reversed in the pulmonary capillaries.

2 *Non-volatile acids.* Each day metabolism yields a total of about 50–80 mmol of H_2SO_4 (from oxidation of S in sulphur-containing amino acids) and H_3PO_4 (derived from phosphoproteins and phospholipids). Organic acids produced in metabolism are normally oxidized to CO_2 and H_2O. In shock, and in exercise severe enough to be partly anaeorobic, oxidation may fail to keep up with production, and lactic and pyruvic acids may also accumulate. In diabetes, as much as 750 mmol of acetoacetic and β-hydroxybutyric acids may be produced daily. The H^+ ions from all these non-volatile acids must be buffered in the body.

Mechanisms which minimize the fall in pH caused by H^+ ions pending final disposal are as follows.

1 *Dilution in total body water.* Acid produced in cells diffuses into ECF and into other cells; each tenfold dilution raises pH by 1 unit.
2 *Buffering in blood.* It would take 28 mmol of H^+ to bring 1 litre of whole blood from pH 7·4 to pH 7·0. The H^+ ions would be shared between the buffers of the blood as follows: HCO_3^-, 18 mmol; inorganic phosphate, 0·3 mmol; plasma protein, 1·7 mmol; and haemoglobin, 8 mmol. Five litres of blood could take up 140 mmol H^+ (twice the normal daily production) without a lethal decrease in pH; but note that $[HCO_3^-]$ would be reduced from 24 to 6 mmol per litre.
3 *Buffering in extravascular ECF.* This is mainly achieved by HCO_3^-, for there is no haemoglobin and much less protein than in the blood. The 14 l of ECF containing 24 mmol of HCO_3^- per litre could take up a further 200 mmol of H^+ before pH was reduced to 7·0.

Processes **1**, **2** and **3** are rapid, being completed in minutes, and total ECF including plasma could take about 350 mmol, or some five times the day's production without disaster. This is ample reserve for normal purposes.
4 *Buffering in cells.* Intracellular buffers, mainly proteinate and organic phosphate, can back up extracellular buffers by coping with H^+ ions that enter

cells which exchange them for K^+ or Na^+. Their capacity is probably similar to that of extracellular buffers but the process is much slower (hours) because of limited permeability of cell membranes to H^+.

5 *Carbonate in bone.* This can contribute some HCO_3^- to the plasma and ECF to replace some of that which is used up in 2 and 3; again this probably takes hours.

6 *Exchange of ions with bone mineral.* Extracellular H^+ ions can exchange with cations from bone and cells, e.g. $H^+ + M^+$ in bone $\rightarrow H^+$ in bone $+ M^+$ where M^+ may represent Na^+ or K^+ or $\frac{1}{2}Ca^{2+}$ or $\frac{1}{2}Mg^{2+}$. These are slow processes, taking hours to days. Prolonged metabolic acidosis depletes bones of Ca^{2+} and cells of K^+.

These six processes all add up to *whole-body buffering*, of which the effectiveness was well demonstrated in 1953 by R.F. Pitts who intravenously titrated a dog with molar HCl. The blood pH decreased from 7·44 to 7·14 after addition of 150 ml of acid (i.e. 150 mmol H^+). By contrast, the addition of the same amount of HCl to 11 kg of water (the estimated water content of the dog) brought the pH to 1·84. Since pH 7·44 corresponds to a $[H^+]$ of 36 nmol l^{-1} and pH 7·14 to a $[H^+]$ of 72 nmol l^{-1}, 36×11 or about 400 nmol of the added H^+ remained free in the dog, i.e. of 150 mmol or 150 000 000 nmol of the H^+ added to the dog, 400 remained free and 149 999 600 was buffered in the body. Of this, about a half was buffered extracellularly and a half by cells and bone.

Final disposal of hydrogen ions and regeneration of body buffers

It falls to the kidneys to rectify the changes produced by additional acid or alkali in the body. Because of respiratory compensation, pH may be not far from normal but the buffer stores, indicated by the concentration of bicarbonate in the plasma, may be either depleted (by acids) or augmented (by excess alkali) and these need to be restored to their normal levels.

In terms of the body's principal buffer base, bicarbonate, what happens when, for example, sulphuric acid is added, may be summarized thus:

$$H_2SO_4 \longrightarrow SO_4^{2-} + 2H^+$$

$$2H^+ + 2HCO_3^- \longrightarrow 2H_2CO_3 \longrightarrow 2H_2O + 2CO_2 \text{ (expired)}$$

After this, two H^+ ions have become hydrogen atoms in body water. Two CO_2 have been breathed out. The pH has fallen a little, but in the plasma two HCO_3^- have been replaced by one SO_4^{2-}. The kidneys easily dispose of sulphate and the conjugate bases of other unwanted acids that are replacing HCO_3^- in the plasma. These pass into the glomerular filtrate and need only to be left unreabsorbed by the tubules to be excreted in the urine. The H^+ ions no longer have a free existence; they are present as parts of buffer bases or of body water, and are not directly available for excretion. Eventually, because all the buffers are in equilibrium, as the kidneys manufacture HCO_3^- which they add to the plasma, H^+ ions temporarily buffered by other buffer bases will finally be transferred to HCO_3^- and converted to H_2O. But for every H_2O molecule so formed, a molecule of HCO_3^- has been destroyed. The role of the kidney is to manufacture HCO_3^- to replace the HCO_3^- used up. The kidney cells make this HCO_3^-, which they restore to the plasma, with H^+ which they secrete into the lumen. The H^+ ions excreted in the urine are simply a byproduct in the manufacture of HCO_3^-. Though the mechanism may be more complex, a simple view is that in the renal tubular cells of the proximal and distal tubules and collecting ducts, carbonic acid formed from CO_2 and H_2O (catalysed by carbonic anhydrase) ionizes to provide H^+ and HCO_3^- ions in equal numbers. The HCO_3^- passes across the basolateral borders of the cells along with Na^+ ions that are being reabsorbed, and the H^+ is secreted into the tubular fluid. Because equal numbers of H^+ and HCO_3^- ions are formed in this way, by the time that the HCO_3^- used up in coping with H^+ ions in the body has been replaced, an equivalent amount of H^+ formed in the kidney will have been excreted in the urine.

The driving force for secretion of hydrogen ions

It is probable that the movement of H^+ from cell to lumen involves a counter-transport of Na^+ from lumen to cell, at least in the proximal and early distal tubules. The urine can be acidified to pH 4·4, which, when the plasma remains at pH 7·4, implies a ratio $[H^+]_U/[H^+]_P$ of 1000:1. Such ratios are achieved only in the 'tight' distal portions of the nephron. Here the intercalated cells secrete H^+ actively, utilizing a H^+-ATPase which differs from that found in gastric oxyntic cells (p. 586) in that it does require K^+ for its activity.

Fate of hydrogen ions secreted into the tubular fluid

Figure 13.21 illustrates three major consequences of the secretion of H^+ ions into the lumen.

1 *'Reabsorption' of filtered bicarbonate* by replacement with newly formed HCO_3^- from tubular cells; this consumes about 4500 mmol H^+ per day. Within the lumen, carbonic anhydrase attached to the brush borders catalyses the breakdown of H_2CO_3 and allows CO_2 to recirculate. Normally, this process is largely complete by the end of the proximal tubule and simply restores to the body HCO_3^- that would otherwise have been lost in the glomerular filtrate.

2 *Production of titratable acid.* Buffers in the glomerular filtrate react with

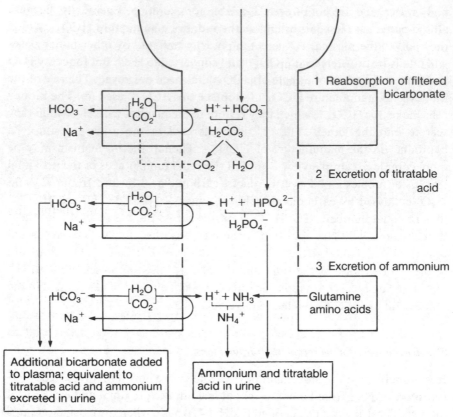

Fig. 13.21. A summary of the three processes by which renal tubular epithelial cells are enabled to synthesize sufficient HCO_3^- (and to secrete H^+ ions) to maintain plasma HCO_3^- constant. (From Robinson, J.R. (1972) *Fundamentals of Acid–Base Regulation*, 4th edn, p. 76. Blackwell Scientific Publications, Oxford.)

secreted H^+ ions. Normally, the predominant filtered buffer is phosphate, but, in diabetes mellitus and in shock, weak organic acids may make a major contribution, and as much as 250 mmol per day of titratable acid may be excreted compared with the normal 20–30 mmol per day. This process consumes less H^+ than **1** but the H^+ is secreted against an increasing gradient as urine is acidified, and an equivalent amount of HCO_3^- is added to plasma over and above the amount rescued from the filtrate.

3 *Production of ammonium.* The traditional mechanism depicted in Fig. 13.21 needs modifying since the recent realization that it is ammonium rather than ammonia that is formed in the tubular cells. Ammonium, derived mostly from glutamine (to a lesser extent from amino acids) in the tubular cells, is secreted into the lumen, where it is trapped in acid tubular fluid and concentrated by recirculation within the medullary counter-current system.

Two ammonium ions are formed from each glutamine deaminated. But two HCO_3^- ions are formed and added to the plasma when the oxoglutarate left over from the deaminated glutamine is oxidized to glucose. The normal range is 30–50 mmol per day, but up to 500 mmol per day of NH_4^+ can be secreted and equivalent HCO_3^- added to the plasma by this mechanism.

Note that **1, 2** and **3** are sequential processes with some overlap and are not confined to anatomically separate segments of nephrons. H^+ ions consumed in **1** are not excreted in the urine. H^+ ions consumed in **2** and **3** are excreted as urinary titratable acid and NH_4^+; process **3** allows excretion of more H^+ in an acid urine without further lowering its pH. The kidney in effect manufactures its own urinary buffer, NH_4^+, and therefore more HCO_3^- can be added to the plasma and more total acid can be excreted before the minimum urinary pH (4·4 to 4·6) is reached.

The amounts of HCO_3^- ions contributed by the three basic processes change under different conditions as shown by typical round figures in Table 13.2. Note that less H^+ is actually secreted during metabolic acidosis than under normal or alkalotic conditions. But during acidosis, since tubular fluid pH falls to a lower level, more H^+ has to be secreted against a substantial gradient. Also note that, in every situation, most of the H^+ secreted is used up in the 'reabsorption' (by replacement) of filtered HCO_3^-.

Measurement of urinary titratable acid and ammonia

Because urine contains buffers its pH cannot provide an estimate of total H^+ excretion in the urine. Urinary titratable acid is measured by adding OH^- ions to a sample of urine until its pH has been restored to the pH of the plasma from which it was formed. Since the pK of the NH_3/NH_4^+ pair is 9·3, only about 1/100 of the NH_4^+ has been titrated at the normal plasma pH of 7.4.

Table 13.2. Secretion versus excretion of H^+.

	Metabolic alkalosis (mmol/day)	Normal (mmol/day)	Metabolic acidosis (mmol/day)
$[HCO_3^-]_{plasma}$	33	25	17
HCO_3^- filtered	6000	4500	3000
HCO_3^- reabsorbed	5600	4500	3000
HCO_3^- excreted	400	0	0
Total H^+ secreted	5600	4580	3400
Tit. acid excreted	0	30	100
NH_4^+ excreted	0	50	300
Total H^+ excreted*	0	80	400

* Total H^+ excreted = additional bicarbonate synthesized by kidney and added to plasma.

The titratable acid does not therefore include NH_4^+, and separate chemical methods are used to measure total urinary ammonia.

The importance of the measurement of titratable acid and total ammonia lies in the fact that their sum is equal to the total acid excreted in the urine, and this is equal to the total amount of additional HCO_3^- added to the plasma after replacing all the filtered HCO_3^- destroyed in the renal tubules. Under steady-state conditions, this additional HCO_3^- is equal to the HCO_3^- consumed in the body in buffering H^+ from non-volatile acids. That is, over a 24 h period, titratable acid + total ammonia = production of non-volatile acid.

Regeneration of buffer stores

Bicarbonate added to the plasma by the tubules first replaces plasma HCO_3^- that had been lost temporarily in the glomerular filtrate. This conserves existing stores of HCO_3^- but does not add to them. This 'extra' HCO_3^-, corresponding to H^+ ions excreted as titratable acid and ammonium:

1   replaces HCO_3^- that had been used in buffering H^+ ions in the body;
2   replaces bone carbonate that had supplied bicarbonate to the plasma and other ECF;
3   regenerates other (non-bicarbonate) buffer bases that had been used up:
$HCO_3^- + HBuf \rightarrow Buf^- + H_2CO_3 \rightarrow H_2O + CO_2$; and
4   reverses ion exchanges with bone and cells:
$HCO_3^- + M^+ + H^+$ in bone or cell $\rightarrow M^+$ in bone or cell $+ H_2CO_3$.

Regulation of hydrogen ion secretion and urinary acidity and alkalinity

Most filtered HCO_3^- is 'reabsorbed' from the proximal tubules and not more than about 5% ordinarily reaches the distal tubules, but process 1 must give place to processes 2 and 3 (see Fig. 13.21) when filtered HCO_3^- has all been destroyed within the lumen. Whether the final urine is to be acid or alkaline depends upon the relation between the rate of filtration of HCO_3^- and the rate of secretion of H^+ ions (Table 13.2).

The filtered load of HCO_3^- equals $[HCO_3^-]_{plasma} \times GFR$ and is proportional to the plasma $[HCO_3^-]$ so long as GFR remains constant. The rate of secretion of H^+ ions is roughly proportional to P_{CO_2} in the blood supplying the kidneys. Increased P_{CO_2} probably enhances H^+ secretion by making more CO_2 available as a source of carbonic acid and also by lowering pH in the cells so that the gradient from cell to lumen becomes more favourable. Factors which increase Na^+ reabsorption in the distal tubules and collecting ducts (e.g. aldosterone, increased Na^+ delivery) also promote secretion of H^+.

Compensation for metabolic disturbances

Alterations in plasma $[HCO_3^-]$ provide automatic adjustments to minimize metabolic disturbances. In *metabolic acidosis*, when plasma $[HCO_3^-]$ is reduced, less HCO_3^- is filtered, less of the H^+ secreted is consumed in 'reabsorption' of filtered HCO_3^-, more H^+ is excreted as titratable acid and NH_4^+ in an acid urine, and corresponding amounts of additional HCO_3^- are added to the plasma restoring $[HCO_3^-]$ towards normal.

In *metabolic alkalosis*, when $[HCO_3^-]$ is increased, the filtered load of HCO_3^- requires more H^+ for its 'reabsorption', and less H^+ goes to form titratable acid and ammonium. When filtered HCO_3^- exceeds the total rate of H^+ secretion, process 1 in Fig. 13.21 cannot be completed, unreabsorbed HCO_3^- is excreted in an alkaline urine, reducing the plasma $[HCO_3^-]$ towards normal.

Alterations in P_{CO_2} caused by ventilatory responses to changed plasma pH retard but do not prevent these automatic adjustments. Thus, in *metabolic acidosis*, stimulation of ventilation lowers P_{CO_2} and reduces the rate of secretion of H^+. In *metabolic alkalosis*, retention of CO_2 by reduced ventilation increases P_{CO_2} and the rate of H^+ secretion (see illustrative values in Table 13.2). The body's store of HCO_3^- is normalized more slowly than it would be if the ventilatory response did not occur. In the meantime, the ventilatory responses protect against considerable alterations in pH, as the nervous system is intolerant of pH change but relatively unaffected by alterations in the amount of buffer in the stores.

The excretion of NH_4^+ ions takes up to a week to reach its maximum rate during sustained acidosis. The delay cannot be satisfactorily explained by slow adaptive increases in activity of the enzymes involved in NH_3 formation, but it might reflect a slow increase in the intensity of H^+ secretion mediated by aldosterone; for Na^+ is inevitably lost to maintain electroneutrality in urines containing excessive amounts of conjugate bases of acids such as H_2SO_4 or HCl (experimentally) or ketone acids (in diabetic keto-acidosis) and there is a threat of Na^+ depletion until sufficient NH_4^+ is available to be used instead.

Renal compensation for respiratory disturbances of pH

The dependence of H^+ secretion upon P_{CO_2} provides slow renal compensation for respiratory disturbances of pH. In *respiratory acidosis*, the increased P_{CO_2} increases H^+ secretion. While the filtered load of HCO_3^- is not increased appreciably, more H^+ is excreted as titratable acid and NH_4^+, and the corresponding HCO_3^- added to the plasma increases plasma $[HCO_3^-]$ and slowly corrects the buffer ratio.

In *respiratory alkalosis*, the low P_{CO_2} slows H^+ secretion so that the initially undiminished filtered load of HCO_3^- is not completely 'reabsorbed'. Bicarbonate is excreted in an alkaline urine and lowers plasma $[HCO_3^-]$ so that the excessive alkalinity of the plasma is slowly corrected.

Some clinical aspects

Significance of plasma bicarbonate

After renal compensation for respiratory disturbances note that *respiratory acidosis* ends up with a *high* plasma $[HCO_3^-]$ whereas *respiratory alkalosis* ends up with a *low* plasma $[HCO_3^-]$. Consequently, an *increased* plasma $[HCO_3^-]$ may indicate either *metabolic alkalosis* or *respiratory acidosis* (chronic and renally compensated). A *decreased* plasma $[HCO_3^-]$ may indicate either *metabolic acidosis* or *respiratory alkalosis* (chronic and renally compensated). Laboratory differentiation between these alternatives requires the measurement of plasma pH or P_{CO_2} as well as HCO_3^-. The conditions can usually be distinguished clinically from observation of the patient and the history of the disturbance.

Effect of depletion of potassium

Potassium-depleted tubular cells are more acid than normal and, especially when stimulated by aldosterone, secrete more H^+; hence the urine tends to be acid and the plasma alkaline with $[HCO_3^-]$ high and $[Cl^-]$ low.

Effect of inhibitors of carbonic anyhydrase

Inhibition of carbonic anyhydrase slows the secretion of H^+ so that HCO_3^- is excreted in an alkaline urine and plasma $[HCO_3^-]$ is lowered. This explains the otherwise mysterious acidosis seen in the early days of treatment with massive doses of sulphonamide drugs, which inhibit carbonic anhydrase. Later the potent inhibitor acetazolamide (Diamox) was tried as an alternative to mercurial diuretics because Na^+ excreted with the HCO_3^- reduced the body's excessive load of Na^+. The diuretic effect ceases however when $[HCO_3^-]$ has been reduced to the level at which the filtered load is not greater than the tubules can 'reabsorb' using H^+ produced through the uncatalysed reaction of CO_2 with water. Bicarbonate then ceases to be excreted and a new steady state is established.

13.4 Function in diseased kidneys

Damaged kidneys may fail to maintain the normal composition and volume of the body fluids because they lose or excessively retain natural constituents of the body or else because they fail to remove waste products.

Disturbances of glomerular function

The glomeruli may leak protein or may fail to produce their normal daily quota of 180 l of protein-free filtrate for the following reasons.

1 *Protein loss from kidney*. In 'nephrotic syndromes', an abnormal glomerular basement membrane allows plasma proteins, especially albumin, to escape; the urine may contain from 5 to as much as 30 g of protein per day. When synthesis in the liver fails to keep up with renal loss, there is less circulating albumin and its concentration in the plasma falls, so that the volume and the colloid osmotic pressure of the plasma tend to be reduced. There is a body-wide disturbance of the Starling equilibrium (Section 1.8), with gross generalized oedema, caused and sustained by excessive retention of Na^+ by the kidneys.

2 *Failure to produce enough glomerular filtrate*. In chronic renal failure, destruction of nephrons gradually reduces GFR. About three-quarters of a person's nephrons can be destroyed before renal function is obviously impaired. The remaining nephrons hypertrophy and adapt so that total function is surprisingly well maintained. Note that for substances excreted primarily by the glomeruli, excretion depends upon filtered load, not upon GFR as such. Consequently, for creatine or urea, the rate of excretion will be maintained if plasma concentration increases in proportion to dwindling GFR. The normal amount of creatinine can be eliminated in one-quarter as much glomerular filtrate if each volume of filtrate contains four times as much creatinine, i.e. if plasma creatinine concentration is four times normal. Similarly, the rising plasma urea concentration with advancing renal failure is not merely a sign of falling GFR; it is an effective compensating mechanism that allows the excretion of urea to be maintained.

In a steady state, the rate of excretion of urea must equal the rate of production. If GFR suddenly falls to half, excretion will at first lag behind production and plasma urea concentration will increase until it reaches twice normal, when excretion will catch up with production. The body's urea pool will then have twice as much in it, but retention of urea will no longer be increasing. If GFR falls to 12 ml min^{-1} (10% of normal), plasma urea will have to rise to 40 mmol instead of 4 mmol and the body will contain 1700 mmol instead of a normal 170 mmol of urea, but the rates of production and of excretion will again be equal and normal at about 500 mmol or 30 g per day.

Disturbances of tubular function

In chronic destructive diseases, renal tissue gradually disappears, but the amount of solute to be excreted every day is not diminished. The hypertrophied and adapted nephrons that are left have to do all the work that was formerly done by a full complement of normal nephrons. If 10% of the nephrons remain, each one of these must on average handle ten times as much solute as a nephron in a normal kidney, like a nephron in fact in a normal pair of kidneys excreting 5000 (instead of 500) mmol of urea each day. Under such conditions of permanent osmotic diuresis, the urine cannot be concentrated. During water deprivation, normal kidneys can make urine concentrated to four times the osmolality of the plasma—but at rates less than about $0 \cdot 6$ ml min^{-1}. The attainable concentration falls as flow rate increases during osmotic diuresis, and at 15 ml min^{-1} the urine can be little more than isosmolal. With only 10% of the nephrons remaining, the urine cannot be more concentrated than the plasma if the rate of production exceeds $1 \cdot 5$ ml min^{-1}. This is about 2000 ml per day, the volume of isosmolal urine required to contain the daily output of urinary solutes, which explains the fixed osmolality and specific gravity of 1010 (isosthenuria) characteristic of renal failure.

Inability to concentrate and to reduce the volume of the urine also accounts for the *nocturia* which may be an early sign of renal damage. Note that a simple test of the power of the kidneys to concentrate the urine (such as the urinary osmolality or specific gravity after 12 h overnight without fluid) is both the simplest and the most sensitive test for detecting early impairment of renal function, especially in diseases like pyelonephritis which attack the medulla. Note too that the increased urinary volume that results from the kidneys' loss of concentrating power promotes the excretion of urea; for the clearance of urea increases with rising flow rate and becomes maximal above about 2 ml min^{-1} (Fig. 13.6).

Handling of sodium

In contrast to urea, retention of Na$^+$ with increasing concentration in the plasma would disturb the balance between cells and ECF. Retention is avoided by reabsorbing a smaller fraction of filtered Na$^+$ and since normally more than 99% of filtered Na$^+$ is reabsorbed, there is a large margin of safety. If GFR fell to 12 ml min^{-1} (one-tenth of normal), Na$^+$ balance could still be maintained by reabsorbing $97 \cdot 5\%$ of the filtered load of 2500 mmol per day. Even with GFR reduced to 3 ml min^{-1} (one-fortieth of normal), the filtered load would be 600 mmol per day and balance could be maintained by reabsorption of five-sixths of the filtered Na$^+$. Possible causes of reduced fractional reabsorption include:

1   osmotic diuresis, since back-flux increases as Na^+ concentration becomes lower in the tubular fluid;
2   reduced availability of NH_4^+, which means that Na^+ must accompany conjugate bases or urinary acids; and
3   possibly the action of a natriuretic substance.

Sometimes, and especially when tubular functions have been damaged more than glomerular, too much Na^+ may be lost, with depletion of the volume of ECF which further depresses GFR and aggravates retention of nitrogenous substances.

Handling of potassium

Since K^+ is secreted by the tubules, its conservation presents no problem to the failing kidney. Excretion of K^+ may however be impaired if a grossly reduced filtered load of Na^+, due to a very low GFR, delivers too little fluid to the K^+-secreting site to sustain the rate of secretion. Increased catabolism of tissues in uraemia releases K^+ from broken-down cells, and failure to excrete this K^+ may lead to a dangerous elevation of plasma K^+ concentration requiring treatment by dialysis.

Handling of hydrogen ions

Renal acidosis is a metabolic acidosis arising from impaired capacity of the renal tubular cells to secrete H^+ ions and add HCO_3^- to the plasma, and is accordingly often characterized as *renal tubular acidosis*. Possible causes include:

1   failure to secrete H^+ into the urine and add HCO_3^- to the plasma at a sufficient rate;
2   failure to bring urine to a sufficiently low pH, so that less H^+ can be excreted as titratable acid and ammonium;
3   failure to supply sufficient NH_4^+ to carry H^+ ions into the urine; and
4   shortage of filtered buffer to make into titratable acid.

Only the last of these depends primarily upon GFR. The rest depend upon metabolic failure to transport H^+ ions against gradients or to provide NH_4^+. The biggest single factor is usually the inadequate supply of NH_4^+ from the dwindling of the mass of metabolizing tubular tissue as nephrons are destroyed.

In summary, the patient with failing kidneys has an undiminished need for the services of his kidneys, and so must overload a dwindling population of functioning nephrons. Nitrogen end-products accumulate (*azotaemia*) because there are too few glomeruli, and GFR falls. Lack of tubular tissue leads to failure to secrete sufficient K^+, H^+ and NH_4^+ and sometimes to retain Na^+. The overloaded remaining nephrons, possibly with a disorganized medulla with its counter-current arrangements wrecked, cannot concentrate the urine; and so the patient must live with an increased turnover of water or risk dehydration. Two well-organized kidneys have been reduced to a motley collection of nephrons.

14 · Digestive System

14.1 Introduction
Structure of the digestive tract
Regulation of the digestive system

14.2 Motility of the digestive tract
Mouth and oesophagus
  Chewing
  Swallowing
Stomach
  Gastric motility
  Control of gastric motility
  Vomiting
Small intestine
  Motility of the small intestine
  Control of small intestinal motility
Large intestine and rectum
  Colonic motility
  Defaecation
  Control of colonic and rectal motility

14.3 Secretions of the digestive system
Salivary secretions
  Functions of saliva
  Formation of saliva
  Control of salivary secretions
Gastric secretions
  Functions of gastric secretions
  Formation of gastric secretions
  Control of gastric secretions

Pancreatic exocrine secretions
  Functions of pancreatic secretions
  Formation of pancreatic secretions
  Control of pancreatic secretions
Biliary secretions
  Functions of biliary secretions
  Formation of biliary secretions
  Control of biliary secretions
Intestinal secretions

14.4 Gastrointestinal hormones
Gastrin
Cholecystokinin
Secretin
Gastric inhibitory polypeptide
Vasoactive intestinal peptide
Other intestinal peptides

14.5 Absorption
Absorption in the stomach
Absorption in the intestine
  Sodium
  Potassium
  Chloride and bicarbonate
  Calcium
  Water
  Sugars
  Proteins
  Lipids
  Vitamins

14.1 Introduction

All the water, nutrients, vitamins, minerals and electrolytes required for growth and maintenance are absorbed from the digestive tract and transported to other tissues by the bloodstream. However, only a small portion of the ingested solid material—small lipid- and water-soluble molecules—may be absorbed unchanged. The complex macromolecules that form the bulk of the diet must first be reduced to simpler, metabolizable forms by *digestion*. This is accomplished in the lumen of the digestive tract by exposing the ingested material to specific sequences of mechanical and chemical events.

The *digestive tract* is a tube extending from the mouth to the anus and includes the pharynx, oesophagus, stomach, small intestine and large intestine (Fig. 14.1). Food entering the tract is mixed and propelled along the tract as a result of skeletal and smooth muscle activity. As ingested material progresses

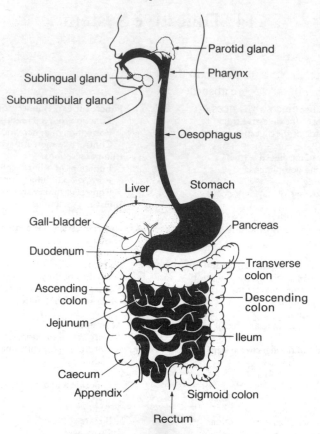

Chapter 14

Fig. 14.1. Anatomy of the human digestive tract.

through the tract, the complex molecular structures may be broken down by hydrolytic enzymes which act in either acidic or neutral environments. The digestive juices are secreted from glandular tissue that forms part of the intestinal wall or from glands that lie outside the intestinal tract (the salivary glands, pancreas and liver). The secretions of the latter enter the digestive tract through ducts.

The *digestive process* starts in the mouth where the food may be ground and mixed with saliva. Here the addition of a salivary amylase to the food begins the breakdown of starches. After passage through the pharynx and oesophagus, this material enters the stomach where it is acted on by acid and gastric proteases. However, digestion in the stomach is limited and one of the main functions of the stomach is to act as a store and hopper that slowly releases partially digested material (chyme) into the small intestine. The final phases of digestion occur in the small intestine and colon. In the small intestine, the chyme is broken down by the actions of bile, pancreatic enzymes (e.g. protease, lipase, amylase) and enzymes released from desquamated epithelial cells. Most of the products of digestion are absorbed by the small intestine—some by specific transport mechanisms—and the residue passes into the large intestine (colon). There is little digestion in the colon but salt, water and some vitamins (produced by bacteria) are absorbed. The waste products, undigested material and a few secreted substances are expelled as faeces.

Structure of the digestive tract

From the mouth to the anus, the digestive canal is lined by a mucous membrane (*mucosa*). In the mouth, oesophagus and anus, where there can be considerable abrasion, the epithelial layer is of a stratified squamous variety some ten to fifteen cells thick. In regions devoted to absorption and secretion, a single-layered columnar epithelium is found (Fig. 14.2). Absorption is enhanced in these regions by folding and by the formation of villi, which are finger-like projections of the mucosa.

The *muscles* surrounding the digestive tract may be striated or smooth muscles. The striated muscles are found in the upper regions of the tract (the mouth, pharynx and oesophagus) and at the anus. In between these regions, the tract is surrounded by an external layer of smooth muscle, the *muscularis externa*, composed of an outer layer with its fibres oriented in a longitudinal direction and an inner layer containing fibres with a circular orientation (Fig. 14.2). This part of the tract also contains a thinner, smooth muscle layer, the *muscularis mucosae*, at the base of the mucous membrane (Fig. 14.2a). The thicker outer coat is involved in the propulsion and mixing of food, while the muscularis mucosae may control the folding and shape of the mucosa.

The *blood vessels, lymphatics* and *nerves* to the intestinal tract travel in the mesenteries. The blood vessels entering the tract branch to supply the external muscle layer, the submucosa and the mucosa. The final vascular supply within these regions may vary from a simple rectilinear network in the muscle layers to a basket-like arrangement in the villi of the small intestine (Fig. 14.2b).

Intrinsic nerves in the tract between the lower oesophagus and the rectum comprise the *enteric nervous system*, the number of neurones (10^8) in which approaches that found in the spinal cord. These neurones are grouped in the *myenteric (Auerbach's) plexus* which lies between the outer longitudinal and circular muscle layer, and in the *submucosal (Meissner's) plexus* which lies between the circular muscles and the muscularis mucosae (Fig. 14.2a). These intrinsic networks are composed of excitatory and inhibitory motor nerves to the smooth muscle fibres and excitatory innervation to secretory cells and hormone-releasing cells. The input to these plexuses comes from the autonomic nervous system and from intrinsic chemo- and mechanoreceptors which may have their sensory terminals in the epithelial sheet (Fig. 14.3).

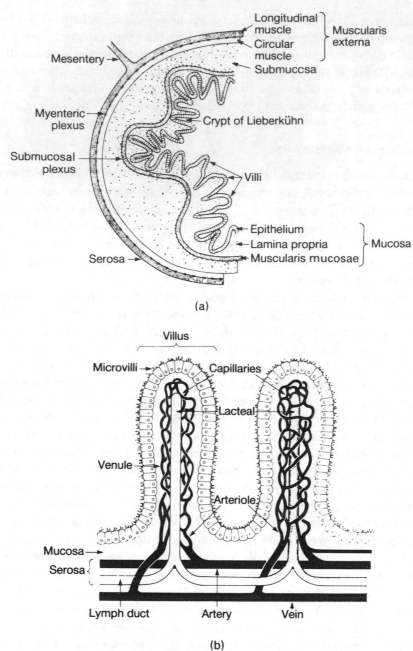

Fig. 14.2. (a) Cross-section through the small intestine. (b) Blood and lymphatic vessels supplying the villi.

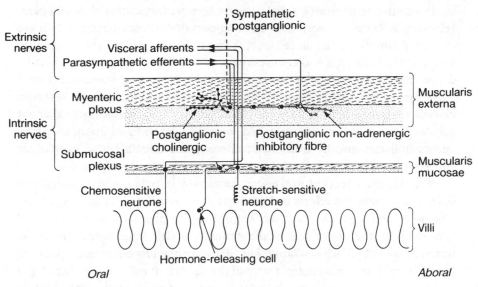

Fig. 14.3. Intrinsic and extrinsic nerves of the gastrointestinal tract.

Regulation of the digestive system

In general, the motor and secretory activity of the gastrointestinal tract is regulated and coordinated by nerves and hormones. Nervous control can be subdivided into intrinsic and extrinsic components. *Intrinsic nervous control* involves local reflexes confined to the wall of the gut itself and effective over rather short distances (cm) along the tract. Receptors responsive to a variety of stimuli (e.g. stretch, pH, osmolality, products of digestion) activate afferent fibres whose cell bodies are located in the intramural plexuses. These neurones synapse within the plexuses with efferent neurones supplying smooth muscle, secretory cells or hormone-producing cells in the locality. A variety of substances have been suggested as neurotransmitters or neuromodulators in the enteric nervous system. As well as acetylcholine and noradrenaline, these include 5-hydroxytryptamine (5HT, serotonin) and a variety of peptides with a putative endocrine function such as somatostatin, vasoactive intestinal polypeptide (VIP), substance P and cholecystokinin (CCK).

Extrinsic nervous control is mediated via receptors and afferent fibres, the cell bodies of which are found within the intestinal plexuses or the dorsal roots. Some of the information conveyed may reach areas within the brain, and the efferent outflow conveyed by *sympathetic* or *parasympathetic* nerves from the central nervous system may reach synapses within the intramural plexuses or secretory organs (salivary glands, liver, pancreas) directly. These extrinsic pathways and reflexes through prevertebral ganglia allow regulation and coordination of gastrointestinal activity over greater distances.

In addition to nervous control, there is *hormonal regulation* of motility and secretion. A variety of hormone-secreting cells of neuroectodermal origin are located within the epithelial cell layer. Some of these cells are exposed directly to the luminal contents and are presumed to be influenced by them; others lie deeper in the epithelial cell layer and would seem to be under less direct control by nerves or by other hormones. Released hormones (e.g. gastrin, secretin, cholecystokinin) diffuse into the wall of the gut and may act locally to alter gut activity. Their local actions are called *paracrine* effects. In addition, they may enter capillaries and be transported through the circulation to exert their effects on more distant parts of the alimentary system or on other organs.

Whereas the effects of nerves can be rapidly initiated or terminated and sharply localized, the effects of hormones are slower in onset and offset and tend to be more diffuse. Nervous control in the digestive system is found exclusively in the upper part of the system (mouth, oesophagus) and at its termination (anus) where rapid but short-lasting responses are required. Nervous and hormonal control play equally important roles in the control and regulation of gastric activity. In the intestine, motility is more dependent on neural mechanisms but, in the regulation of secretions of liver and pancreas, hormonal control is the more important.

14.2 Motility of the digestive tract

Both longitudinal movement and mixing of the luminal contents are essential if efficient digestion and absorption are to be maintained. In the upper regions of the digestive tract, where the movement of food is rapid, motility is controlled directly by the central nervous system. Where the movement is concerned mainly with mixing and slow propulsion (during digestion and absorption), the motility is regulated by mechanisms inherent in the muscle (*myogenic mechanisms*), by *neural mechanisms* and by *hormones*.

The general properties of the skeletal and smooth muscle that are responsible for the motility have been discussed in Chapter 3. The motility of the lower tract is largely under the influence of the myenteric plexus which in turn is influenced by the extrinsic nerves (parasympathetic and sympathetic) and the submucosal plexus (Fig. 14.3). Most *parasympathetic fibres* are *excitatory* and the *sympathetic fibres* are *inhibitory*. The tonic inhibitory influence of the sympathetic nerves, which largely terminate in the myenteric plexus (or on blood vessels), appears to be mediated by presynaptic suppression of transmission at cholinergic synapses in ganglia in the enteric system. These nerves were once believed to be the only fibres involved in the inhibitory control of motility but more recently stimulation of preganglionic 'parasympathetic' fibres (and stretch receptors from the epithelium) has been found to have an inhibitory effect on smooth muscle. This effect is mediated by postganglionic *non-adrenergic non-cholinergic inhibitory fibres* which release an unidentified transmitter, possibly ATP or more likely VIP (p. 600).

Mouth and oesophagus

Chewing

When the food enters the mouth, chewing movements (mastication) reduce the size of the particles and mix them with saliva; both of these actions contribute to the taste, odour and swallowing of food. In man, the mechanical effects are accomplished by the cutting action of the incisor teeth and the crushing movements of the molar teeth.

The chewing movements are under voluntary control and involve the coordinated action of the lower mandible, cheeks and tongue. The lower jaw is usually held closed as a result of reflex (muscle spindle) activity but a conscious decision or the presence of food in the mouth may momentarily inhibit this reflex and initiate jaw opening. The alternate activation and inhibition of the nerves to the opening and closing muscles results in chewing. The appropriate rhythm, force and degree of occlusion are provided reflexly and as the reflex is unilateral the chewing force is exerted largely on the side containing the bolus of food.

Swallowing

Swallowing involves the movement of food from the mouth to the oesophagus (Fig. 14.4), movement down the oesophagus and movement from the oesophagus to the stomach. Initially, the tongue forces a bolus of food to the rear of the mouth. The stimulation of receptors in the posterior wall and soft palate then results in the activation of the *swallowing reflex*. This activity is

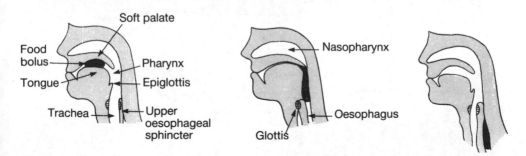

Fig. 14.4. Movement of a bolus of food from the mouth to the oesophagus.

coordinated within a *swallowing centre* in the medulla oblongata and the reflex once initiated cannot be stopped. It involves the elevation of the *soft palate* to close the nasopharynx, the raising of the *larynx* and approximation of the vocal cords to close the *glottis*, the inhibition of breathing, and the relaxation of the *upper oesophageal sphincter*. Lesions in the region of the swallowing centre interfere with this coordinating activity and may be fatal. Once the cam-like action of the tongue has passed the food into the pharynx, the pharyngeal skeletal muscles contract and propel the bolus of food into the oesophagus. The pressure generated, which can be up to 100 mmHg above atmospheric, normally provides the major force propelling the food into the oesophagus. The pressure changes during swallowing are illustrated in Fig. 14.5.

In the oesophagus, a *peristaltic wave* (a ring of contraction preceded by a region of relaxation) propels viscous material toward the stomach. Fluid material travelling under the influence of the pressure generated by swallowing and by gravity may arrive at the stomach before the contractile wave. The peristaltic wave travels at approximately 5 cm s^{-1} and takes approximately 8 s to travel the length of the oesophagus. The pressures generated by peristalsis in the oesophagus range from 30 to 120 mmHg.

The movement of the food bolus from the oesophagus to the stomach is

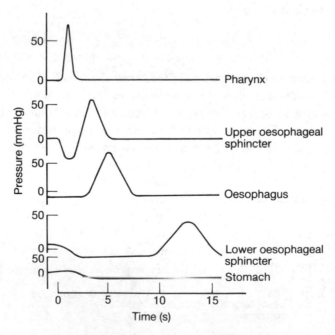

Fig. 14.5. Regional pressure changes during a normal swallowing movement. (Adapted from Davenport, H.W. (1975) *A Digest of Digestion*, p. 8. Year Book Medical Publishers, Chicago.)

accomplished after relaxation of the *lower oesophageal sphincter*. This sphincter is the terminal 4 cm of the oesophagus and is anatomically not much different from adjacent areas. It is normally closed (resting pressure 10–15 mmHg) to prevent reflux of the stomach contents which would otherwise occur because the pressure in the stomach (5–10 mmHg) is higher than that in the thoracic oesophagus (-5 mmHg, p. 475). If the acid contents of the stomach enter the oesophagus, the sensation of 'heartburn' may be experienced. Pressure recordings from this region show that the sphincter relaxes some 1–2 s after swallowing is initiated, remains relaxed for 8–9 s and then contracts. Between swallowing movements, the tone of this sphincter is maintained by slight contraction of the smooth muscle layer. As some tone is maintained in the denervated tissue, intrinsic activity appears to be involved in maintaining this contraction.

The *primary peristaltic wave*, which is initiated by swallowing, normally clears the oesophagus of food but if particles remain in the oesophagus *secondary peristaltic waves* may begin. These waves arise as a result of distension or irritation of the oesophagus and involve the swallowing centre. The secondary waves, which arise without any awareness, commence at the upper oesophageal sphincter and progress to the stomach.

Stomach

The stomach receives, stores and partially digests food prior to its gradual release into the small intestine. The stomach has three outer smooth muscle layers (the longitudinal, circular and oblique layers) which have both mixing and propulsive roles.

Gastric motility

Between meals, the stomach contains a small volume (about 50 ml) and the smooth muscle contracts only occasionally. As food enters the stomach, the intragastric pressure may rise but the increase is only slight due to the elasticity of the stomach wall and to the relaxation of the smooth muscle cells. This *receptive relaxation* of the smooth muscle results from stimulation of stretch receptors in the oesophagus and proximal stomach and results in activation of postganglionic non-adrenergic inhibitory neurones which arise in the myenteric plexus.

For the first hour after the ingestion of a meal, shallow waves of contraction move over the stomach. These gradually deepen until ring-like peristaltic contractions, starting high on the stomach in the fundus, sweep over the body and antrum toward the pylorus (Fig. 14.6). However, little pressure is generated

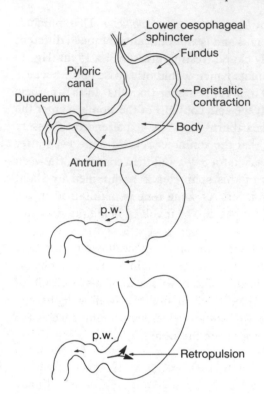

Fig. 14.6. Movement of a peristaltic wave (p.w.) into the pyloric region of the stomach, forcing a small quantity of chyme into the duodenum and the remainder back into the antrum by retropulsion.

within the body of the stomach. As the contractions pass over the antrum and pylorus, they progress more rapidly and become more forceful. During these contractions, the pressure in the antrum may rise to 15–25 mmHg and a small quantity of the gastric contents now referred to as *chyme* enters the small intestine through the *pyloric canal* or sphincter. However, the canal, which is normally relaxed, closes rapidly during each contraction and most of the antral contents are squirted back into the body of the stomach (Fig. 14.6). This is termed *retropulsion* and has a strong mixing action that helps to break down the food.

Control of gastric motility

The contractile activity of the stomach is regulated by *myogenic, neural and hormonal* mechanisms.

The frequency of the contractions (about three per minute) and their coordination depend on the spread of *'slow waves'* of depolarization (20–40 mV) throughout the smooth muscle layers (Fig. 14.7). This rhythmic depolarization (*basic electrical rhythm*) is myogenic in origin. Slow waves occur most frequently high in the body of the stomach and can be initiated in

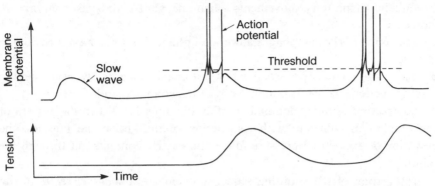

Fig. 14.7. Rhythmic depolarizations (slow waves) of the membrane potential of smooth muscle from the small intestine. On reaching threshold, these initiate action potentials and cause the muscle to contract. In the stomach, the slow waves are usually larger in amplitude and the 'spikes' smaller than those found in the small intestine.

neighbouring regions by passive current spread through gap junctions. Thus the slow waves are conducted around and along the stomach wall. If the slow waves exceed threshold, spike-like potentials are initiated (Fig. 14.7) and a contraction follows. Thus it is the slow waves during normal digestion that determine the *frequency* and *conduction velocity* of the peristaltic contractions. The ionic basis of the slow waves is not understood but it has been attributed to an electrogenic pump and to permeability changes in the plasma membranes of the smooth muscle.

The force of gastric contractions is regulated by neural and hormonal mechanisms. Vagal stimulation of cholinergic motor nerves or gastrin released from the antral mucosa lowers the membrane potential of the smooth muscle cells. Subsequent slow waves then lower the membrane potential further and for a longer time. The resulting potentials in the smooth muscle cells then induce a more powerful contraction (p. 119). In contrast to this, the non-adrenergic inhibitory nerve fibres cause hyperpolarization, and thus inhibit gastric motility (p. 120).

Gastric emptying. The rate at which the stomach empties is determined by its motility. The factors that regulate gastric emptying are the same as those that regulate gastric secretion—they are increased or decreased together. Essentially both are influenced by the physical and chemical nature of the gastric and duodenal contents and will be considered together later (p. 588).

Vomiting

Vomiting results in the rapid expulsion of the stomach contents through the mouth. It involves:

1 forceful inspiratory movements when the glottis and nasopharynx are closed;

2 relaxation of the oesophagus, lower oesophageal sphincter and body of the stomach;

3 contraction of the pyloric region of the stomach and possibly the duodenum forcing duodenal contents into the body of the stomach; and

4 contraction of the abdominal and thoracic muscles. When the descent of the diaphragm coincides with contraction of the abdominal muscles, the elevation of abdominal pressure forces the gastric contents out through the mouth.

Integration of the vomiting sequence occurs in a *vomiting centre* in the medulla oblongata that is anatomically and functionally associated with the centres governing respiration. It is influenced by the adjacent 'chemoreceptor trigger zones' in the area postrema which may be stimulated by a variety of drugs (e.g. morphine and its derivatives) that induce vomiting.

Small intestine

Most of the digestion and absorption of food takes place in the duodenum and jejunum. Thus the motility of this section serves to mix the chyme from the stomach with the digestive juices secreted by the pancreas and liver and also to expose the luminal contents to the intestinal wall across which absorption occurs. In addition, the intestinal contents are slowly propelled toward the large intestine.

Motility of the small intestine

The above functions are accomplished by two motility patterns, namely, segmentation and peristalsis.

1 *Segmentation* is the alternate contraction and relaxation of complete segments of the small intestine (Fig. 14.8). As the chyme from one contracting segment is forced into adjacent relaxed areas, this motility pattern thoroughly mixes the luminal contents. The electrical basis of this contractile pattern appears to be slow waves (Fig. 14.7). The factors that result in segmentation rather than peristalsis are not understood but, since the contractile band does not progress, this pattern has little propulsive action. The contracting segments may be short or long; the latter are more effective in displacing the contents into adjoining regions. The frequency of the segmental contractions decreases down the length of the human small intestine from about 12 contractions per min to 9 contractions per min.

2 Short *peristaltic contractions* travelling 10–15 cm are the main propulsive

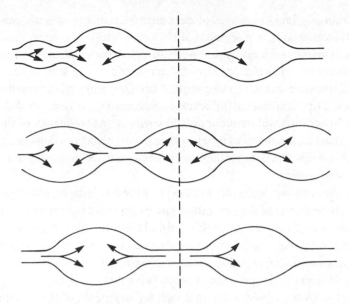

Fig. 14.8. Alternate relaxation and contraction of adjacent segments of the small intestine (segmentation) which cause thorough mixing of the contents.

force in the small intestine. These moving, ring-like contractions are preceded by a more distal region of relaxation as they move towards the colon. They can be elicited by luminal distension. More protracted peristaltic contractions (rushes) that traverse the length of the small intestine are abnormal.

In fasting man (and dog) the antrum and small intestine appear to undergo a characteristic pattern of secretion and of contractility consequent upon electrical activity called the *interdigestive myoelectric complexes*. This interdigestive activity starts with contraction of the gall-bladder and relaxation of the sphincter of Oddi and is followed some 15 min later by increased gastric motility and secretion. This in turn leads to increased duodenal and small intestinal motility and stimulation of pancreatic exocrine secretions. Neither the mechanisms underlying this activity nor its relationship to the patterns of motility and secretion initiated by feeding are understood. Nevertheless, interdigestive activity appears to play an important physiological role in that these patterns of motility effectively 'clean out' the stomach and small intestine.

Control of small intestinal motility

The motility of the small intestine depends on *myogenic, neural* and *hormonal* mechanisms.

As in the stomach, the rhythmic contractile activity of the small intestine results from regular slow waves of depolarization of the smooth muscle cells. Similarly, contraction in a segment of the small intestine is coordinated by the spread of slow wave activity from cell to cell with consequent initiation of action potentials. However, as already mentioned, in contrast to the stomach, the small intestine exhibits a decrease in the frequency of contractions along its length. This decrease is the result of successive portions of the intestine having a lower intrinsic frequency of slow waves, the frequency of slow waves being highest in the duodenum. However, more distal portions are unable to follow this high frequency; they occasionally 'miss' a slow wave and the frequency decreases.

Both the intrinsic and extrinsic nerves play a role in regulating intestinal motility. Stimulation of the postganglionic cholinergic enteric nerves increases the amplitude of contractions while stimulation of the sympathetic or non-adrenergic inhibitory nerves inhibits contractions. Since the motility of the small intestine continues in the absence of extrinsic nerve supply, it has been argued that the neural regulation of motility is mainly intrinsic. This argument is supported by the observation that isolated segments of the small intestine increase their contractile activity in response to distension, acids, the products of digestion and placing of hypertonic solutions in the lumen.

The extrinsic nerves, however, also play a role and their importance is indicated by the presence of the following reflexes.

1   The *intestino-intestinal inhibitory reflex* in which distension of one intestinal segment causes complete intestinal inhibition.
2   The *ano-intestinal inhibitory reflex* in which distension of the anus causes intestinal inhibition.
3   The *gastro-intestinal reflex* in which food entering the stomach causes excitation of intestinal motility.

The role of hormones in the control of intestinal motility is less clear but *cholecystokinin* (CCK) stimulates motility.

For chyme to enter the large intestine, it must pass through the *ileocaecal sphincter* which in man is approximately 4 cm long. This sphincter at rest is closed by a pressure of some 20 mmHg. During normal movements, this pressure is reduced in association with peristaltic contractions in the terminal ileum and the chyme moves into the large intestine.

Large intestine and rectum

The progression of chyme through the large intestine is relatively slow (18–24 h). Little digestion occurs in the colon but salt and water are absorbed.

Colonic motility

The motility of the colon is in some ways similar to that found in the small intestine. The colon is inactive for a large proportion of the time but when contractions occur they may be either of a *segmental* or *peristaltic* nature. The segmental contractions have a slower frequency (3–4 per min) than in the small intestine and the deep but restricted infoldings of the wall form distinct pouches called *haustra*. This movement slowly mixes the luminal contents and improves absorption.

Propulsive activity, *mass movements*, occur infrequently (2–3 times each day) and result in the development of relatively high pressures (80–100 mmHg) that drive part of the colonic contents towards the rectum. These contractions may travel 30 cm or more.

Defaecation

The rectum is normally empty but its distension by the mass movement of faecal material from the colon induces relaxation of the smooth muscle of the *internal anal sphincter* and the urge to defaecate. In paraplegics, this and the contraction of the rectum will lead to automatic defaecation but in normal people contraction of the *external anal sphincter*, which contains skeletal muscle, allows retention of the rectal contents. When conditions permit, voluntary relaxation of the external sphincter allows defaecation to proceed. The expulsion of the faecal mass may be assisted by raising the intra-abdominal pressure as a result of forced expiratory movements against a closed glottis.

Control of colonic and rectal motility

The motility of the colon is largely dependent on *myogenic* and *neural* mechanisms. Slow waves are propagated through the muscle layers but their activity has less directional coordination than that in the stomach and small intestine. Indeed, reverse slow wave activity and contractions can be recorded in the distal colon. The intrinsic nervous plexus plays an important role in the coordination and conduction of contractions as distension can induce coordinated local peristalsis. Loss of the plexus (as in Hirschsprung's and in Chagas' disease) results in pronounced colonic distension. The motility of the colon is also influenced by the extrinsic nerve supply as illustrated by the *gastro-colic* and *duodeno-colic* reflexes which stimulate motility after material has entered the stomach or duodenum, respectively. Mass movements may follow such reflex activity and result in the urge to defaecate.

Defaecation is normally a voluntary act involving higher centres of the brain and also the medulla and spinal cord. Rectal distension causes the rectal muscle to contract and the internal anal sphincter to relax. These actions involve intrinsic and extrinsic cholinergic pathways and, in the case of relaxation, also non-adrenergic inhibitory fibres. Afferent impulses from the rectum pass to the sacral cord and higher centres. Depending on the situation, the higher centres will either augment or inhibit the sacral centre. If defaecation proceeds, the parasympathetic discharge along the pelvic nerves to the colon and rectum is augmented and the somatic motor activity in the pudendal nerve to the external anal sphincter ceases. This defaecation reflex is facilitated by the tactile stimulation which accompanies passage of the faeces through the anus.

14.3 Secretions of the digestive system

There are five major secretory tissues in the digestive system—the *salivary glands*, the *stomach*, the *pancreas*, the *hepatic–biliary system* and the *intestine* (Fig. 14.9). The first four of these have been studied extensively but rather less is known about the secretions of the intestine. Representative values for the compositions of the various secretions are given in Table 14.1.

Salivary secretions

In man, three pairs of salivary glands are found—the *parotid*, *submandibular* (also called the submaxillary) and *sublingual*. There are also numerous small mucus-secreting glands throughout the mouth and pharynx. In the salivary

Table 14.1. Volumes and representative values for osmolality, pH and ionic concentrations of plasma, gastrointestinal secretions and of intestinal fluids.

	Volume (1 per day)	Osmolality (mosmol kg^{-1}H$_2$O)	pH	Na$^+$ (mmol l^{-1})	K$^+$ (mmol l^{-1})	Cl$^-$ (mmol l^{-1})	HCO$_3^-$ (mmol l^{-1})
Plasma	(3·0 l total)	285	7·4	150	5	110	24
Saliva	1·5	100	7·5	40	15	25	30
Gastric juice	3·0	285	1·0	50	10	140	0
Pancreatic juice	1·0	285	7·8	140	8	85	60
Bile	1·0	285	7·3	140	5	80	20
Jejunum	8·0	285	4·0–7·4	140	5	140	8
Ileum	2·5	285	7·4	140	5	120	30
Proximal colon	1·0	285	7·4	125	15	90	60

Note: 1 Volumes given are average values in man for 24-h samples of secretion or, for jejunum, ileum and colon the net amount of fluid entering the segment per day (intake + secretion); 2 Values depend upon flow rates and therefore can vary quite widely; 3 In the jejunum, pH rises from proximal to distal regions.

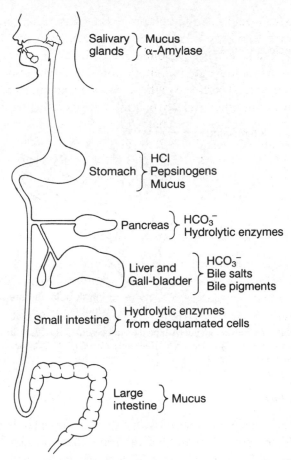

Fig. 14.9. Secretions of the digestive system. Note that secretions also contain electrolytes and water.

glands, acini containing secretory cells drain into ducts which coalesce and finally open into the oral cavity (Fig. 14.10).

About 1·5 l of saliva is secreted per day, of which one-quarter comes from the submandibular glands and two-thirds from the parotids. Secretion occurs at a continuous basal rate of 0·3 ml min^{-1} but when stimulated by acidic foods, such as lemon juice, it can reach a maximum flow of 4–5 ml min^{-1}.

Saliva contains a *mucous secretion*, mainly from the sublingual and submandibular glands, and a *serous secretion* of water and ions mainly from the parotid and submandibular glands. It also contains an enzyme, *α-amylase* (ptyalin), secreted by the parotid gland. Other substances found in the saliva are lysozyme, immunoglobulin (IgA) and blood group antigens (in 'secretors', p. 352).

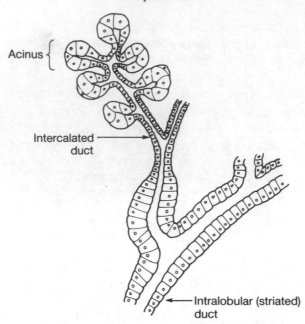

Acinus {

Intercalated
duct

Intralobular (striated)
duct

Fig. 14.10. Structure of a salivary gland. (Adapted from Leeson, C.R. (1967) Structure of salivary glands. In *Handbook of Physiology*, Section 6, *Alimentary Canal*, ed. C.F. Code, vol. 2, pp. 463–495. The Williams & Wilkins Co., Baltimore.)

Functions of saliva

The water, salts and protein secreted by the salivary glands serve several functions. Water and mucin form a *lubricant* that moistens foods, helps swallowing and aids in speech. In addition, the water facilitates *taste* by partially dissolving ingested material. Saliva helps in the maintenance of oral epithelium and teeth by preventing epithelial dehydration and inhibiting the proliferation of bacteria. The bactericidal effect of the enzyme lysozyme may contribute to the latter effect. Dental caries are also inhibited by salivary HCO_3^- which neutralizes residual acid (ingested or produced by oral bacteria).

The digestion of carbohydrates is initiated in the mouth by the secretion of salivary α-amylase. But the rapid passage of food through the mouth and oesophagus means that little digestion of carbohydrate occurs before it reaches the stomach. The effectiveness of the amylase depends on its dispersion through the food bolus and the rapidity of its inactivation by gastric acid. Estimates of its effectiveness range from 5 to 50%.

Formation of saliva

The serous secretion entering the mouth is hypo-osmotic when compared with plasma and has less Na^+ and Cl^- but more K^+ and HCO_3^- than plasma

(Table 14.1). In the acini, Cl⁻ secretion (p. 28) results in the formation of an isosmotic, relatively protein-free secretion with an ionic composition similar to that of plasma. As this primary secretion passes through the ducts, Na^+, Cl^- and HCO_3^- are reabsorbed into the blood and K^+ is secreted. Since the duct epithelium has a low hydraulic conductivity, the reabsorbed ions are not accompanied by water and so a hypo-osmotic secretion is produced. The actual composition of the serous secretion varies with the flow rate; the higher the rate, the closer the ionic composition is to that of the primary secretion (Fig. 14.11). Note that Na^+–K^+ exchange in the ducts is stimulated by aldosterone, a hormone involved in electrolyte balance.

Control of salivary secretions

Increases in salivary flow may result from stimulation of receptors in the mouth, pharynx and oesophagus. Food in the mouth stimulates taste receptors and irritates the oral mucosa while chewing activates a variety of mechano-receptors, such as pressure receptors adjacent to teeth and the muscle spindles of masticatory muscles.

These increases in salivation are reflex actions controlled *solely* by nervous activity. Afferent sensory fibres carry information to *salivatory centres* in the pons and medulla of the brain while efferent autonomic fibres supply the glands. The salivatory centres also receive impulses from higher centres of

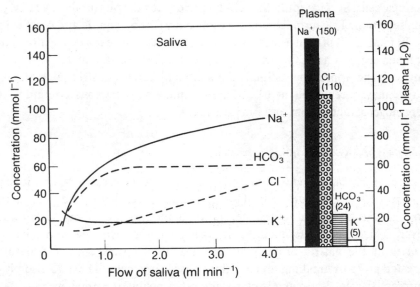

Fig. 14.11. Composition of saliva at different rates of flow. For comparison the ionic composition of plasma is shown on the right. (After Thayson, J.H., Thorn, N.A. & Schwartz, I.L. (1954) *Am. J. Physiol.* **178**, 155–159.)

the brain and so the sight, smell and thought of food may also cause salivation. In contrast to their actions in other viscera, the parasympathetic and sympathetic fibres have a similar, but not identical, effect in salivation. *Parasympathetic* stimulation, which is of major importance in man, causes a prolonged copious secretion accompanied by vasodilatation. The vasodilatation may be an indirect effect due to the release of an enzyme, kallikrein, which converts a plasma protein to the potent vasodilator, bradykinin. Alternatively, vasodilatation could involve VIP released from cholinergic nerves. Acetylcholine released by parasympathetic fibres appears to initiate the secretion of electrolytes and water through a Ca^{2+}-dependent process. *Sympathetic* stimulation produces a small quantity of thick mucous saliva which is accompanied by vasoconstriction. It also increases enzyme secretion by stimulating β-adrenoceptors and elevating cyclic AMP in cells.

Gastric secretions

The fundus and antrum are the major secretory regions of the stomach. Tubular glands in the mucosa of these regions secrete the gastric juice containing pepsinogens, acid and mucus. In the fundic mucosa, *chief cells* secreting *pepsinogens* tend to lie deep within the gland and *oxyntic (parietal) cells* secreting *acid* lie closer to the gland opening (Fig. 14.12). The glands secrete into numerous gastric pits (100 per mm^2) which drain into the stomach. The pits, the glands and the remaining mucosa contain large number of *mucus-secreting* cells. Surface epithelial cells lying between the pits produce a $HCO_3{}^-$-rich secretion. The antral glands are essentially free of oxyntic and chief cells.

In total, 2–3 l of an isosmotic gastric juice is secreted each day. At rest, the stomach secretes a $HCO_3{}^-$-rich isosmotic fluid at the rate of 15–20 ml h^{-1}. But with the appropriate stimuli the stomach secretes gastric juice at rates which may exceed 150 ml h^{-1}. These stimulated secretions contain acid, pepsinogens, mucus, ions and intrinsic factor.

Functions of gastric secretions

Gastric *hydrochloric acid* is produced by the oxyntic cells as an isosmotic fluid (150 mmol l^{-1} HCl, pH 0·8). This secretion mixes with fluid within the stomach so that the final acid concentration of the gastric contents is somewhat lower. This acid environment is necessary for the *activation* and *optimum activity* of the gastric proteolytic enzymes, the pepsins, and it denatures ingested proteins breaking up connective tissue and cells. The acid also plays a protective role, destroying bacteria and other potential pathogens.

The *pepsins* have maximum proteolytic activities between pH 2 and 3 and are formed from precursor pepsinogens by the action of acid or by the

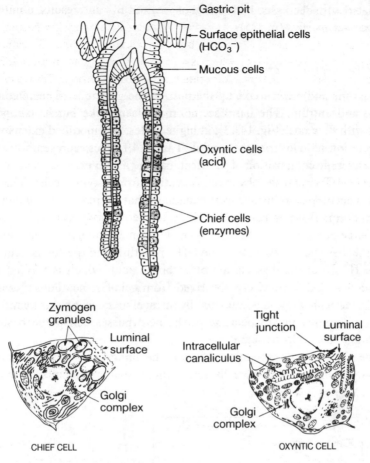

Fig. 14.12. Location and structure of chief and oxyntic cells in a gastric gland. (Adapted from Ito, S. & Winchester, R.J. (1963) *J. Cell Biol.* **16**, 541–577.)

autocatalytic action of pepsins. They initiate the *breakdown of proteins* by cleaving peptide bonds adjacent to aromatic amino acids. This results in the production of polypeptides of widely differing sizes. Note that since peptic digestion is incomplete, significant quantities of protein leave the stomach to be digested in the small intestine.

There is one non-digestive but essential component of gastric juice— *intrinsic factor*. This heat labile glycoprotein (M_r 60 000) is essential for absorption of *vitamin B_{12}* in the ileum. It is secreted by the oxyntic cells and its concentration rises concomitantly to the acid concentration. The amount of intrinsic factor secreted in 24 h is sufficient to bind 50–200 μg of vitamin B_{12} 10–50 times the amount of B_{12} needed for the daily supply of the vitamin. The secretion of intrinsic factor is markedly depressed in gastric atrophy. This is associated with decreased body stores of vitamin B_{12} and results, if untreated, in *pernicious anaemia* (p. 325).

Formation of gastric secretions

Mucus, ions and water make up the bulk of the gastric juice secreted by the fundus and antrum. The basal secretion is plasma-like but its composition varies with flow rate (Fig. 14.13); at higher rates, it is modified extensively by the secretion of acid from the fundus. Table 14.1 gives representative values for the average composition of gastric juice over a 24 h period.

Acid (HCl) is formed by *active* secretion from oxyntic cells. The H^+ is secreted actively into intracellular canaliculi that drain into the lumen of the gastric glands (Fig. 14.12). A H^+-K^+-ATPase is involved in this secretion. The source of the H^+ is debatable; it and OH^- may come from metabolism or from the dissociation of water. The OH^- remaining in the cell is neutralized by the H^+ from the dissociation of carbonic acid, which is formed by the hydration of CO_2. The CO_2 is derived from cellular metabolism and from plasma and its hydration is catalysed by intracellular carbonic anhydrase. The HCO_3^- liberated on dissociation of the acid passes across the basolateral membrane to the bloodstream.

The Cl^- that accompanies H^+ into the lumen may be transported as follows: Cl^- may enter the cells across the basolateral membrane either in

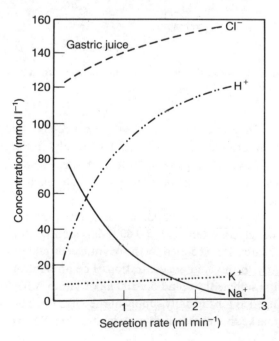

Fig. 14.13. Relationship between ionic composition of gastric juice and its rate of secretion. (After Nordgren, B. (1963) *Acta Physiol. Scand.*, Suppl. **202**, 1–83.)

exchange for HCO_3^- or be co-transported with Na^+ as it moves down its electrochemical gradient. Accumulation of Cl^- within the cell provides a favourable electrochemical gradient for its diffusion across the apical membrane from cell to lumen. Thus Cl^- accompanies H^+ secretion to maintain electroneutrality and water flows down the osmotic gradient so created. The result is in an isosmotic secretion of HCl. Figure 14.14 illustrates one possible model for the formation and secretion of gastric acid.

There are dramatic changes in appearance between resting and stimulated oxyntic cells. The resting cell contains numerous tubulo-vesicular structures, the membranes of which contain H^+-K^+-ATPase units. In the secreting cell, these structures appear to fuse with and increase the area of the apical and canalicular surfaces of the cell. It is suggested that in this way H^+-K^+-ATPase units are incorporated into these surfaces.

The high concentration gradient for H^+ between the lumen of the stomach and the plasma ($10^6 : 1$) appears to be maintained not only by H^+ transport but

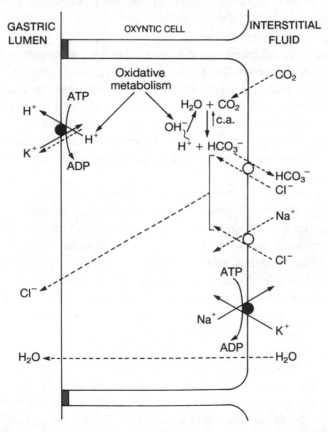

Fig. 14.14. A model of the isosmotic secretion of HCl by the oxyntic cells. c.a., carbonic anhydrase.

also by the general impermeability of the apical membranes and tight junctions of the mucosa. Note that the gastric mucosa is normally resistant to attack by the acid and enzymes secreted by the stomach. The trapping in the mucous layer of HCO_3^- secreted by the surface cells may contribute to the protection of the gastric mucosa, though it is not essential.

Pepsinogens are synthesized, stored in granules and secreted by exocytosis as inactive precursors to avoid digestion of the chief cells.

Control of gastric secretions

In general, stimulation causes an increase in secretion of both acid and enzymes and is accompanied by an increase in motility and hence gastric emptying (p. 575). Thus acetylcholine from postganglionic parasympathetic nerves, the hormone gastrin and also histamine from the mucosa stimulate gastric secretion and motility. These mechanisms, neural and hormonal, appear to interact. The control of gastric secretion is described classically in three phases.

1 *Cephalic phase.* Chewing and the presence of food in the mouth or pharynx stimulate gastric secretion. Receptors in these areas send information to nuclei in the pons. Efferent parasympathetic fibres in the vagus activate secretory cells in the fundus and antrum directly and also activate cells in the pylorus which liberate gastrin. This hormone in turn further stimulates the secretion of gastric juice. The pontine centres of the brain are also innervated by higher centres so that the sight, smell or thought of food may increase secretion.

2 *Gastric phase.* Stretch, products of protein digestion, and possibly changes in osmolality of the gastric contents stimulate gastric secretion. They do so by stimulating receptors in the stomach which generate impulses through either the vagal or intrinsic nerves that synapse in the myenteric or submucous plexuses. These impulses stimulate gastric secretion. In addition, gastrin, released as a result of the same stimuli, increases secretion of acid and enzymes. Gastrin release is inhibited by low intraluminal pH, thus gastric secretion is decreased as free acid accumulates. This inhibition of gastrin release is mediated via somatostatin-producing endocrine cells which are exposed directly to the luminal contents. These cells have extensive branching processes up to $40\,\mu m$ long containing somatostatin granules which are released in the immediate proximity of gastrin-releasing cells when the luminal contents become acid. Histamine released locally in the stomach wall also seems to stimulate acid secretion in this phase. Thus blockers of histamine receptors (H_2 receptors) in the mucosa, such as cimetidine, are used to decrease the production of gastric acid in people with peptic ulcers.

The precise interrelationship between the factors controlling gastric acid secretion and their mechanism of action at the cellular level is complex and remains unclear. It seems that the basolateral membrane of the oxyntic cell has distinct receptors for acetylcholine, gastrin and histamine. Whereas histamine acts via cyclic AMP, the other agents seem to act through Ca^{2+}-mediated mechanisms. At least some of the effects of gastrin are a consequence of its promotion of histamine release from cells within the epithelium. In addition, gastrin and acetylcholine, as well as noradrenaline, may release somatostatin from local endocrine cells which, in turn, inhibits acid secretion.

3 *Intestinal phase.* Chyme entering the intestine is associated normally with a decrease in gastric secretion and motility. This negative feedback thus matches the delivery of chyme to the handling capacity of the upper small intestine and is of great physiological importance. A decreased pH, fat and hyperosmolality in the duodenum cause the suppression of gastric activity and also inhibit gastric emptying. This is largely hormonally mediated and a variety of hormones released from the intestine, including secretin, cholecystokinin (CCK) and gastric inhibitory polypeptide (GIP), have been implicated (Section 14.4). A role attributed in the past to a hypothetical hormone, enterogastrone, may reflect the effects of a combination of these and other substances. Of less overall importance, hormones, possibly gastrin liberated from the duodenal mucosa, may be involved in the continuation of acid secretion, albeit at a reduced rate, during the intestinal phase.

Pancreatic exocrine secretions

The pancreas consists of two portions, exocrine and endocrine, of which only the exocrine will be considered here. The digestive secretions of the pancreas are initially formed in acini which are similar to those in salivary glands. A system of ducts from the acini unites with the bile duct to form a common duct which enters the upper duodenum. The entrance of this common duct into the duodenum is controlled by the *sphincter of Oddi*.

The pancreas secretes about 1 litre of fluid each day. At rest (low flow rates) the secretion is plasma-like; with higher flow rates, an *alkaline fluid* rich in HCO_3^- is secreted. Representative values for its volume and composition are shown in Table 14.1. With the appropriate stimulus the pancreas will also secrete *hydrolytic enzymes* that act on all the major groups of nutrients.

Functions of pancreatic secretions

The HCO_3^--rich secretion of the pancreas and also, to an extent, of the liver contributes to the neutralization of the acid chyme, which enters the duodenum from the stomach. Thus a favourable pH is established for the pancreatic enzymes which have optimal activities in the 6·7–9·0 pH range. A pH profile for the gastrointestinal tract is shown in Fig. 14.15.

The pancreatic enzymes or their precursors are secreted along with the aqueous secretion and include:

1 *protease precursors*—trypsinogen, chymotrypsinogen and procarboxypeptidases—which are secreted by the pancreas and activated in the small intestine. The activation is initiated by the conversion of trypsinogen to trypsin by the intestinal enzyme, *enterokinase*, found on the apical plasma membranes of the intestinal epithelial cells. The trypsin in turn activates trypsinogen autocatalytically and also activates the other precursors;
2 an active *α-amylase* which is similar in action to the salivary α-amylase and splits α-1, 4-glycosidic bonds and hydrolyses starch to mainly maltose;
3 *lipases* which act on triglycerides and phospholipids;
4 a number of other enzymes, e.g. *ribonuclease, elastase* and *collagenase*, that act on specific macromolecules.

The major pancreatic enzymes, their substrates, actions and products are listed in Table 14.2.

Formation of pancreatic secretions

Two groups of cells contribute to the secretion—the acinar cells and the duct cells. The basolateral membranes of the acinar cells contain receptors for both

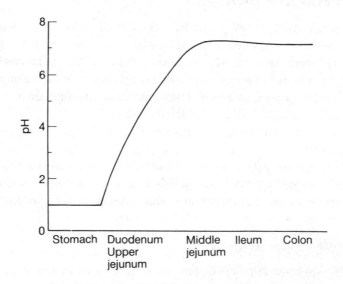

Fig. 14.15. A pH profile of the gastrointestinal tract.

Table 14.2. Pancreatic enzymes: their substrates and products.

Precursor	Enzyme	Substrate	Action	Products
Trypsinogen	Trypsin	Protein and polypeptides	Endopeptidase that cleaves peptide bonds on the C-terminal side of arginine or lysine	Peptides
Chymo-trypsinogen	Chymo-trypsin	,,	Endopeptidase that cleaves peptide bonds on the C-terminal side of tyrosine and phenylalanine	Peptides
Procarboxy-peptidase	Carboxy-peptidase	,,	Exopeptidases that hydrolyse peptide bonds adjacent to C-terminal amino acids	
A	A	,,	Hydrolyses nearly all C-terminal peptide bonds	Amino acids
B	B	,,	Hydrolyses mainly bonds adjacent to C-terminal arginine or lysine	Amino acids
Proelastase	Elastase	Elastin and protein	Hydrolyses peptide bonds adjacent to neutral amino acids	Peptides
	Lipase	Tri-glycerides	Cleaves fatty acids from positions 1 and 3 of triglycerides	Free fatty acids and 2-mono-glycerides
	Phospho-lipase A	Lecithin or cephalin	Cleaves off a fatty acid	Free fatty acid and lysolecithin or lysocephalin
	Deoxyribo-nuclease	DNA	Hydrolyses DNA	Nucleotides
	Ribonucleases	RNA	Hydrolyses RNA	Nucleotides
	α-Amylase	Poly-saccharides	Hydrolyses α-1,4-glycosidic bonds	Maltose and glucose

acetylcholine and for CCK. Their occupancy leads to an increase in cell Ca^{2+} via inositol trisphosphate (p. 256) and this in turn activates Ca^{2+}-sensitive K^+ channels in the basolateral membrane which therefore becomes hyperpolarized. Associated with these changes are exocytotic secretion of enzyme precursors and the production of an isosmotic NaCl secretion, a possible mechanism for which is illustrated in the upper portion of Fig. 14.16. Note that, as in other models of this type, the basolateral membrane may have Na^+–K^+–$2Cl^-$ co-transporters rather than those for Na^+–Cl^-.

The enzymes of the pancreatic secretions are all synthesized in the same cells and released in parallel. However, the ratio is not necessarily constant and examination of the pancreatic secretions of individuals on widely differing diets (e.g. high protein or carbohydrate) shows that these secretions can change to suit the diet.

The cells lining the intra- and interlobular ducts produce an isosmotic HCO_3^--rich secretion which is stimulated via a cyclic AMP-mediated mechanism. One simplified model of possible pathways is illustrated in the lower portion of Fig. 14.16. As the fluid flows through the larger ducts, $HCO_3^- - Cl^-$ exchange occurs to an extent which varies between species and is influenced by flow rate. Due to variations in three processes—isosmotic Cl^-

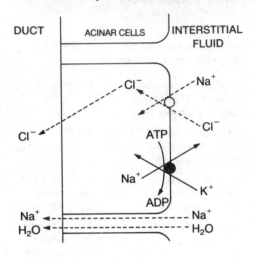

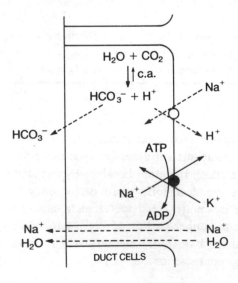

Fig. 14.16. A model of serous secretion by the exocrine pancreas. A relatively small volume of Cl^--rich secretion accompanies the enzyme-rich secretion of the acinar cells. A larger volume of HCO_3^--rich fluid is secreted by the duct cells.

secretion from acinar cells, isosmotic HCO_3^--rich secretion from ductile cells and ion exchange occurring across cells lining the larger ducts—the Cl^- and HCO_3^- composition of the secretion which emerges from the pancreas is not fixed.

Control of pancreatic secretions

As with gastric secretion, a *cephalic phase* may be recognized during which pancreatic secretion is stimulated by vagal activity. The early phase of secretion is supplemented by the release of gastrin from the antrum during the *gastric phase* of digestion. The major control is, however, exerted during the *intestinal phase* when nerves and, more importantly, intestinal hormones determine pancreatic control.

The entry of chyme into the duodenum is generally followed by the secretion of a HCO_3^-- and enzyme-rich fluid from the pancreas. The hormones mostly responsible for this secretion are secretin and CCK. *Secretin* (and a small quantity of CCK) is released from the mucosa of the duodenum and upper jejunum in response to *acid* chyme. It acts on the duct cells and stimulates secretions of the HCO_3^--rich fluid probably by a cyclic AMP-dependent process. The release of CCK is stimulated by the *products of digestion* in the upper small intestine. It stimulates the release of an enzyme-rich secretion from acinar cells by a Ca^{2+}-dependent process.

Biliary secretions

Bile is produced in the *liver* and stored and concentrated in the *gall-bladder*. Bile contains both *secretory* products important in digestion and *excretory* products, such as cholesterol and bile pigments (mainly bilirubin) derived principally from the breakdown of haemoglobin and cytochromes. It also contains certain hormones and drugs which are either metabolized or conjugated within the hepatic cells. However, we will largely ignore the metabolic, detoxifying and filtering actions of the liver and mainly consider those secretions which are involved in digestion. Bile is initially secreted by the *hepatic cells* into the *bile canaliculi*. From there, it passes to the larger ducts which drain each liver lobule and then to the main ducts and common duct. The cystic duct from the gall-bladder joins with the common duct before it enters the duodenum. The entrance to the duodenum is controlled by the *sphincter of Oddi*.

About 1 litre of bile is secreted each day. It is produced continuously but between meals the contraction of the sphincter of Oddi causes it to accumulate in the gall-bladder (volume 20–50 ml). In the gall-bladder, an isosmotic electrolyte solution is reaborbed leaving behind a concentrated solution of bile salts, bile pigments, lecithin and cholesterol. Too much cholesterol or insufficient bile salts or lecithin in bile can result in the precipitation of cholesterol and the formation of *gall-stones*. Bile is ejected by contraction of the gall-bladder and enters the duodenum after relaxation of the sphincter of Oddi. The bile contains two components which are important in digestion, namely, *bile salts* and *bicarbonate*. The remainder is largely bile pigments, other ions and water.

Functions of biliary secretions

The HCO_3^- secreted by the liver aids in the *neutralization* of acid chyme which enters the duodenum from the stomach. The bile salts are important in the *digestion* and *absorption* of *fats* and in the *absorption* of *fat-soluble vitamins*. Their role in fat digestion and absorption can largely be attributed to:

1   their emulsifying action, particularly in the presence of lecithin and monoglycerides; and
2   the ability of conjugated bile acids to lower the optimum pH of pancreatic lipase from approximately 9·0 to 6·7.

Formation of biliary secretions

At rest, the liver secretes a plasma-like fluid rich in HCO_3^- that contains the bile acids and bile pigments (mainly bilirubin). The secretion of bilirubin is discussed on p. 333.

Bile acids (cholic and chenodeoxycholic acids in man) are synthesized in the liver from cholesterol at a rate of 0·5 g per day and are conjugated with taurine or glycine to form bile salts. Conjugation has two advantages: it increases the solubility of the bile acids at low pH and it limits passive absorption in the upper small intestine.

Of particular importance in the formation of bile is the recycling of bile anions that are reabsorbed by secondary active transport from the ileum and carried to the liver in the portal circulation. Approximately 90% of the bile salts secreted into the small intestine are returned to the liver. This circulation of the bile salts is economic in that the pool of bile salts is recycled as much as six to eight times in a 24 h period.

The hepatocytes re-accumulate the bile anions by a Na^+-dependent co-transport mechanism in their basolateral membrane. The reabsorbed and newly synthesized bile anions are then secreted across the canalicular membrane of the hepatocytes. HCO_3^- is also secreted by the hepatocytes. The precise mechanisms involved in these secretory processes are not understood. The tight junctions which hold adjacent hepatocytes together and form the only barrier between the plasma and the canaliculi are relatively leaky.

Therefore, the secretion of the bile salts and HCO_3^- is accompanied by the movement of water and low M_r solutes from the blood to the canaliculi and this primary secretion is isosmotic. The duct cells also secrete a HCO_3^--rich solution by a mechanism which may be similar to that in the pancreatic duct cells.

The bile salts are concentrated in the gall-bladder by the active reabsorption of Na^+. As water follows passively, the bile salts are concentrated about threefold depending on the time between meals. However, because the sodium salts of the bile acids form relatively large aggregates (micelles), the osmolality of gall-bladder bile is not substantially greater than that of plasma.

Control of biliary secretions

Biliary secretion is controlled by autonomic nerves and circulating hormones and by bile salts delivered to the liver by the portal circulation. The most potent stimulators (cholemetics) of bile salt secretion are the *bile salts themselves* which are reabsorbed in the terminal ileum and returned to the liver in the portal circulation.

Nervous control plays only a minor role in regulating biliary secretion. Activity in postganglionic parasympathetic fibres produces a small increase in secretion whereas activity in sympathetic splanchnic nerves causes vasoconstriction and a decrease in secretion.

Hormonal control of biliary secretion is exerted largely by secretin, CCK and gastrin. *Secretin*, released as a result of acid chyme in the duodenum, increases the production of the HCO_3^--rich secretion from the duct cells (as in the pancreas). CCK, released as a result of fats and peptides in the duodenum, causes contraction of the gall-bladder; *gastrin* has a similar action. These hormones (CCK and gastrin) may also stimulate the secretion of bile salts but their importance in this role is in doubt.

Intestinal secretions

As well as mucus-secreting cells which are found throughout the intestinal tract, Brunner's glands in the *duodenum* produce a highly viscid secretion which may serve to trap alkaline (HCO_3^--rich) fluid secreted by the duodenal cells at the surface. This may protect the duodenum from the acid chyme entering from the stomach. The rate of secretion is increased by parasympathetic activity and by secretin.

In the jejunum and ileum, there is also the variable secretion (perhaps 2 l or so) of an isosmotic NaCl solution from the intestinal crypt cells. The cells here accumulate Cl^- from the interstitial fluid which then diffuses from the cells to the lumen. The leaky paracellular pathway allows Na^+ and H_2O to follow readily (Fig. 14.17). This secretion is stimulated by a variety of peptides

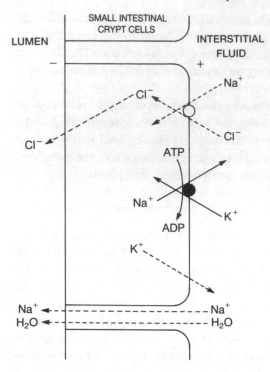

Fig. 14.17. A model of isosmotic Cl⁻ secretion by small intestinal crypt cells. Luminal membrane Cl⁻ permeability seems to be modulated by a cyclic AMP-dependent mechanism. The basolateral membrane co-transporter may be a Na^+–K^+– $2Cl^-$ rather than a Na^+–Cl^- one.

of which VIP, acting through increased cell cyclic AMP, may be the most relevant physiologically, and by cholinergic nerves which increase cytosolic Ca^{2+}. A variety of bacterial endotoxins (cholera toxin, toxins from some strains of *Escherichia coli*) can activate this secretory process by stimulating the G protein and increasing cyclic AMP (p. 257), causing diarrhoea.

Fluid collected from the small intestine, in the absence of gastric, pancreatic and biliary secretions, contains a variety of enzymes. Many of these, e.g. aminopeptidases, amylases and phosphatases, are important in the digestive process and are found in the brush-border region. They are not, however, intestinal secretions but are released from desquamated intestinal cells.

In the colon, mucus and a small volume of alkaline solution rich in HCO_3^- are produced. This may involve an HCO_3^-–Cl^- exchange so that the colonic contents are higher in HCO_3^- than plasma but lower in Cl^-. However, some of the HCO_3^- may be neutralized by organic acids in the lumen. In addition, there is a Na^+–K^+ exchange which results in the colonic contents being rich in K^+. This exchange is enhanced by aldosterone.

14.4 Gastrointestinal hormones

The gastrointestinal hormones are *peptides* produced by enterochromaffin cells in the gastrointestinal mucosa and are involved in the control of gastrointestinal motility and secretion. These cells are related to those endocrine cells (e.g. adrenal medulla) which are of neuroectodermal origin and are sometimes referred to as APUD cells (Amine content, Precursor Uptake, ability to Decarboxylate). A number of these gastrointestinal peptides are also found in central and peripheral neurones where they may function as neurotransmitters or neuromodulators. For example, there is as much CCK in the brain as in the gut and, though its roles in the central nervous system are ill-defined, there is considerable evidence that it may play a part in the control of food intake.

Though a number of hormones have been postulated and several peptides identified in the gastrointestinal mucosa, to date only four—*gastrin, cholecystokinin* (CCK), *secretin* and *gastric inhibitory polypeptide* (GIP)—fulfil the criteria necessary to establish their importance as regulators of gastrointestinal function under physiological conditions, namely:

1 the physiological stimulus involving the postulated hormone (i.e. one arising as a result of the intake of food) applied to one part of the gut must change the activity at another part;
2 this change must persist after all nervous connections between the two parts have been interrupted;
3 a substance isolated from the part of the gut to which the stimulus was applied, must, when purified and injected into the blood, mimic the effects of the physiological stimulus; and
4 this substance must be identified chemically.

There are other peptides of gastrointestinal origin that may be important, e.g. vasoactive intestinal peptide (VIP), motilin, bulbogastrone, enteroglucagon, and somatostatin, but as yet they do not fulfil all of these criteria.

The gastrointestinal peptides can be classified on the basis of their chemical structure into two groups:

1 gastrin and CCK, the five amino acids at the carboxyl end of which are identical; and
2 secretin, GIP, VIP (also glucagon) which have sequences of amino acids in common.

Because of their structural similarity gastrin and CCK may interact at the same receptor site. The result will depend on the concentration of the hormones, on their affinity for the receptor and their efficacy in activating the target cell. Thus the combined effects of gastrin and CCK may be additive or, if one hormone is less effective than the other at a particular site, then its effect may appear to be inhibitory.

The endocrine cells found in the gastrointestinal mucosa are exposed on their luminal surfaces to the contents of the gastrointestinal tract and it is likely that constituents of the chyme can cause the release of their hormones into the bloodstream. In addition, both intrinsic and extrinsic nervous reflexes can cause release of some of these hormones, which not only act through the bloodstream but also may diffuse locally and have so-called *paracrine* actions.

The main actions of the individual gastrointestinal hormones on the motility and secretion of the digestive system are outlined below. In addition, many of the hormones, e.g. GIP, gastrin, CCK and secretin, appear to stimulate the release of insulin. It should also be noted that the gastrointestinal hormones may have an important trophic effect on their target glands, e.g. gastrin stimulates the growth of the gastric mucosa and CCK stimulates the growth of acinar cells in the pancreas.

Gastrin

Two related gastrins, I and II, have been identified. Both are 17-amino acid chains (G17) and differ only at position 12. As well as G17 gastrin, smaller forms (G14) and larger less potent forms (G34) of gastrin (big gastrin) have been isolated from man. The properties of gastrin are determined by the terminal tetrapeptide sequence (–Try–Met–Asp–Phe–NH$_2$). Pentagastrin is a synthetic product containing this tetrapeptide sequence that has all the properties of gastrin but is less potent. It is used clinically to stimulate gastric secretion, as for example in assessing inability to secrete HCl (*achlorhydria*).

The gastrins are produced by cells, called G cells, which are in the mucosa lining the antral region of the stomach and the upper small intestine. Gastrin is released as a result of:

1   stimulation by *products of digestion* (especially peptides), caffeine and alcohol in the stomach, either as a direct effect on the cells or by local intrinsic nerve reflexes;
2   *nervous activity* in extrinsic pathways during the cephalic phase of gastric secretion; or
3   *local distension* of the antral region mediated by local intrinsic nerve reflexes.

Release of gastrin is inhibited by increasing gastric *acidity* and by the intestinal hormones, *secretin* and *gastric inhibitory peptide*. Gastrin may be produced by certain tumours in the gastrointestinal tract and this can result in marked peptic ulceration (Zollinger–Ellison syndrome).

Gastrin is carried in the bloodstream and stimulates *gastric secretion* of HCl, pepsinogen and intrinsic factor. It enhances *gastric motility* and increases the tone of the lower oesophageal sphincter (though the latter may not be observed at physiological concentrations). It produces a small increase in pancreatic secretion and causes contraction of the gall-bladder. It also stimulates the growth of the gastric mucosa.

Cholecystokinin

Cholecystokinin (CCK), also known as pancreozymin, has the same active terminal tretrapeptide sequence as gastrin but a total chain length of 33 amino acids. Therefore, the effects of CCK and of gastrin are qualitatively similar but differ quantitatively and may be competitive. It is produced by cells (I cells) found in the mucosa lining the duodenum and jejunum. It is released into the blood as a result of stimulation by *products of digestion* (especially fats, peptides and amino acids) in the duodenum and jejunum either as a direct effect on the cells or through local intrinsic nerve reflexes.

CCK is carried in the bloodstream and stimulates enzyme-rich secretion from the acinar cells of the pancreas. It causes contraction of the *gall-bladder* and relaxation of the *sphincter of Oddi*. It increases *small intestinal motility* but inhibits *gastric motility*. In some species, e.g. dog, it inhibits gastric secretion but whether it does so in man is debatable. It also stimulates the growth of acinar cells in the pancreas.

Secretin

Secretin was the first hormone to be discovered by Bayliss and Starling in 1902. Secretin contains 27 amino acids; of these 14 occupy the same relative position as in glucagon. Its structure is unrelated to that of gastrin or CCK and it inhibits gastric acid secretion in a non-competitive way. Secretin is produced by cells (S cells) found in the mucosa lining the duodenum and jejunum. It is released into the blood as a result of *increased acidity* in the duodenum and jejunum, either via a direct effect on the cells or indirectly through local nerve reflexes.

It is carried in the bloodstream and stimulates the duct cells in the *pancreas* and *liver* to produce increased volumes of *bicarbonate-rich secretions*. In the stomach, secretin stimulates *pepsinogen* secretion and inhibits *acid* secretion by direct action on oxyntic cells and by inhibition of gastrin release.

Gastric inhibitory peptide

Gastric inhibitory peptide (GIP), also called glucose-dependent-insulin-releasing peptide, is a 43-amino acid polypeptide with some structural similarity to secretin. GIP is produced by cells (K cells) found in the mucosa lining the duodenum and upper intestine. Its release into the blood is stimulated by the presence of *fat* and *glucose* in the upper intestine.

GIP travels in the bloodstream and inhibits *gastric acid secretion* by a direct action on the oxyntic cells and by inhibition of gastrin release. It also inhibits *gastric motility*. It releases *insulin* from β cells of the pancreas and may be of more physiological importance in this respect than other gastrointestinal hormones. GIP may also stimulate secretion from the small bowel.

Vasoactive intestinal peptide

Vasoactive intestinal polypeptide (VIP) contains 28 amino acids and has some structural similarity to secretin. It is produced in the upper small intestine and is released into the blood, probably as a result of increased acidity in the duodenum. VIP is also found in the nervous system where it may be important as a neurotransmitter, both centrally and peripherally.

VIP released from the intestine travels in the bloodstream and stimulates secretion of serous fluid from the pancreas. It inhibits gastric acid secretion and motility. It may also cause vasodilatation in systemic arterioles. When secreted in excess, for example by a non-insulin-secreting islet-cell adenoma of the pancreas, VIP can cause the 'H_2O-diarrhoea' syndrome.

Other intestinal peptides

Other peptides produced by the intestine may also play roles as hormones, for example, *motilin, bulbogastrone, enteroglucagon* and *somatostatin*. Motilin is produced by the mucosa of the duodenum and jejunum and it may increase motility in the stomach. It is probable that the functions ascribed to enterogastrone, i.e. inhibition of gastric motility and secretion when fat is present in the duodenum, are in fact a reflection of the effects of GIP, of inhibitory nerve reflexes, or of the combined action of other hormones. Somatostatin, first identified as a hypothalamic inhibitor of the release of growth hormone from the anterior pituitary (p. 260), inhibits gastric and pancreatic secretion of enzymes and electrolytes and also gastric and intestinal motility. It also plays an important role in inhibiting gastrin release from antral G cells when luminal pH decreases. Its importance may lie more in its paracrine and neuromodulator roles than as a hormone.

As well as the hormones, other chemicals may play a role in regulating and coordinating gastrointestinal function. For example, the *prostaglandins* are synthesized, released and degraded in the gastrointestinal tract and when ingested or injected may influence its function, but their role in the normal physiology of the tract remains debatable.

14.5 Absorption

Absorption may be defined as the net passage of a substance from the lumen of the gut across the epithelium to the interstitial fluid. Two important factors influencing absorption are the available *surface area* and the *flux* of molecules across the epithelium, i.e.

amount absorbed per min = area × flux (amount per min per unit area).

In the gut, the *surface area* available for absorption is greatly expanded by *loops or coils* of the gut, by *mucosal infoldings* (Fig. 14.2a) which increase the surface area of the loops by some threefold, by *villi* (Fig. 14.2b) which further increase the surface area some thirtyfold, and by *microvilli* which form the brush borders of the luminal surface of the cells and increase the available surface area some 600-fold; in the gut of an adult man, the total surface area available for absorption is approximately 2×10^6 cm^2, which is equivalent to a hundred times the external surface area of the body.

The *flux* of molecules across the epithelium depends on their ability to penetrate the epithelial membranes and also on the driving force for their transport, i.e. their electrochemical potential gradient. The rate of *penetration* of the epithelium depends on the properties of the substance to be absorbed and on the nature of the epithelial barrier. The ability of a substance to cross the epithelium will depend on its *lipid solubility*, its *size*, and on the presence of specific *carrier molecules* in the membrane (Section 1.4). Carrier-mediated transport may be 'downhill', i.e. facilitated diffusion, or 'uphill', i.e. active transport, co-transport or counter-transport as discussed in Section 1.5. Furthermore, the leakier the paracellular pathway between the epithelial cells then the more readily solutes of small relative molecular mass, ions and water can cross the epithelium.

The *concentration* of many substances in the lumen will be determined by their rate of *digestion*. Absorption will also depend on the time in contact with the absorptive surfaces and so control of *motility* is essential to allow normal absorption to take place. The concentration of the products of digestion in the interstitial fluid will depend on their rate of *removal* by the blood, which will depend on the blood flow. The blood vessels in the villi are arranged in loops (Fig. 14.2b) so that the distance for diffusion in the interstitial fluid from the basolateral epithelial plasma membrane to the capillaries is minimized. The volume of blood flowing to the gastrointestinal tract represents 20–30% of the cardiac output at rest. The rate of capillary blood flow is a thousand times greater than that of the lymphatic flow. Thus substances such as glucose, amino acids, ions and water, which readily enter the capillaries, are removed from them and pass through the portal circulation to the liver. Only those substances, chylomicrons (small lipid aggregations, p. 608), which do not readily cross the capillary basement membrane are transported predominantly by the *lymphatic* circulation. Such substances enter the venous blood via the thoracic duct and are thus not exclusively transported directly to the liver after absorption. There is some recent evidence which suggests that the hydrostatic pressure in the capillaries in the villi is less than the plasma colloid osmotic pressure. If this is so, it indicates that the driving force for fluid movements in the villi always favours net absorption from the interstitial fluid to the plasma and that ultrafiltration does not normally occur across the capillaries. Therefore, absorption of water and solutes from the gut lumen will be favoured.

Normally, little of the constituents of a meal are absorbed in the mouth. However, some lipid-soluble drugs can be adminstered by placing them either under the tongue or next to the cheek.

Absorption in the stomach

In the stomach, the H^+ concentration is up to one million times greater than that in the plasma while the Na^+ concentration is as low as one-tenth of that in the plasma. This implies the existence of a *gastric mucosal barrier* in which both the apical plasma membranes of the epithelial cells and the 'tight' junctions between these cells are very impermeable to ions. It is not surprising, therefore, that *little* absorption normally occurs in the stomach. However, lipid-soluble substances, e.g. alcohol and organic acids such as acetylsalicylic acid (aspirin) in their non-ionized form, are absorbed to some extent across the stomach wall.

A number of agents can disrupt the gastric mucosal barrier. These include bile salts, short-chain fatty acids, acetylsalicylic acid, ethanol and perhaps corticosteroids. This has important clinical implications, for the entry of H^+ into the interstitial fluid acidifies the fluid and leads to damage to the epithelial cells and to the underlying capillaries resulting in local haemorrhage. Such acute ulceration may progress to chronic ulceration.

Absorption in the intestine

The products of digestion are absorbed predominantly in the small intestine (Fig. 14.18). Absorption of salts and water occurs in the small intestine and also in the large intestine. We shall deal first with the absorption of electrolytes and water and then the absorption of sugars, proteins, lipids and vitamins. (The absorption of iron is discussed on p. 322.)

Sodium

Under normal conditions, up to about 1 mol of Na^+ (equivalent to 58 g of NaCl) is absorbed per day. Only about one-sixth of this is derived from the diet, the remaining five-sixths having been secreted into the gastrointestinal tract. Since total exchangeable body Na^+ is approximately 3 mol, the importance of Na^+ absorption is obvious, as is the rapid depletion of ECF

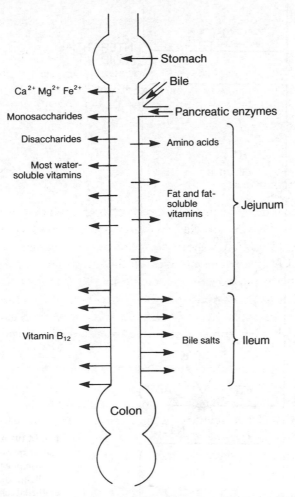

Fig. 14.18. Absorption in the small intestine. (After Booth, C.C. (1968) Effect of location along the small intestine on absorption of nutrients. In *Handbook of Physiology*, Section 6, *Alimentary Canal*, ed. C.F. Code, Vol. 3, pp. 1513–1527. The Williams & Wilkins Co., Baltimore.)

volume if Na^+ is lost from the gut in severe vomiting or diarrhoea. Approximately 5–10 mmol of Na^+ is excreted in the faeces daily.

About 90% of Na^+ absorption occurs in the *small intestine*; the remaining 10% occurs in the *large intestine*. Absorption is thought to occur across mature cells at the tops of the villi. Absorption from the large intestine is modified by *aldosterone*, which stimulates the uptake of Na^+ in exchange for K^+ which is secreted. Na^+ absorption is an *active process* involving passive entry of Na^+ from the lumen to the cell and active transport from the cell to the interstitial fluid (Fig. 14.19). Note that in the small intestine, in particular, the paracellular pathway is 'leaky' and therefore considerable passive movements occur

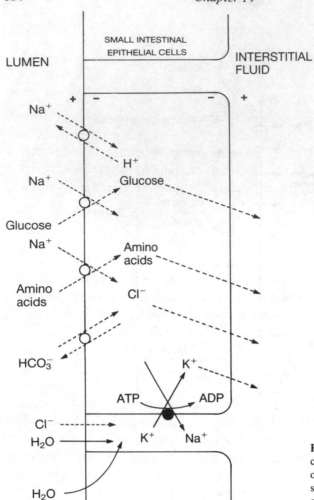

Fig. 14.19. A model of apical co-transport mechanisms and of isosmotic fluid absorption in small intestinal villous epithelial cells.

between the gut lumen and the interstitial fluid. These movements contribute to the rapid equilibration of luminal contents with interstitial fluid. Thus luminal fluid quickly becomes isosmotic with plasma.

Na^+ absorption in the small intestine, but not in the colon, is increased by *glucose* and *amino acids* in the lumen, reflecting co-transport. Na^+ and glucose, and Na^+ and amino acids appear to share common carriers for cellular entry from the lumen. It is believed that under normal circumstances this entry is driven by Na^+ diffusing down its electrochemical potential gradient thereby carrying sugar or amino acid into the cell. This coupling of Na^+ and organic solute movements has important therapeutic implications in the management of cholera and other secretory diarrhoeas. In such cases, the increased net

Na^+, Cl^- and H_2O secretion can be offset by stimulating Na^+ absorption with glucose taken orally.

Potassium

K^+ is *absorbed* throughout the intestinal tract. In the jejunum, the mechanisms involved are not well characterized. In the ileum and colon, there is both active K^+ movement through the cells and passive movement through a paracellular pathway. K^+ may also be *secreted* both actively and passively. In the colon, K^+ secretion is stimulated by Na^+ in the lumen and is increased by *aldosterone*.

Chloride and bicarbonate

Throughout the small and large intestines, the dominant pathway for Cl^- absorption appears to be by counter-transport across the luminal membrane with cell HCO_3^- (or OH^-), which is secreted (Fig. 14.19). At the bases of the crypts of the villi, Cl^- is secreted, not absorbed (Fig. 14.17).

Calcium

Ca^{2+} absorption from the intestine occurs largely in the *duodenum* and is regulated by the hormone, 1,25-dihydroxycholecalciferol (1,25-DHCC), which stimulates Ca^{2+} entry to the cells from the lumen (p. 277). The mechanism involved, however, is not understood. Ca^{2+} entry may be carrier-mediated and there have been claims for the presence of a Ca^+-ATPase in the brush border suggesting that such uptake might be an active process. A Ca^+-ATPase has also been identified in the basolateral membrane and Ca^{2+} extrusion from the cell to the interstitial fluid is probably an active process. A Ca^{2+}-binding protein has been isolated from intestinal epithelia but its relationship to Ca^{2+} transport remains to be clarified. Absorption of Ca^{2+} is impaired at an alkaline pH and by the presence in the gut of fat and of other substances such as oxalates and phytates which form insoluble complexes with Ca^{2+}.

Water

Of the approximately 10 l of water absorbed per day, about 8 l (20% of total body water) has been secreted into the gastrointestinal tract (Fig. 14.20). Approximately 90% of the water absorption occurs in the *small intestine* with only about 1 l absorbed in the colon, leaving 0·1 l to be excreted in the faeces. As much as 2 l of water can be absorbed by the colon per day.

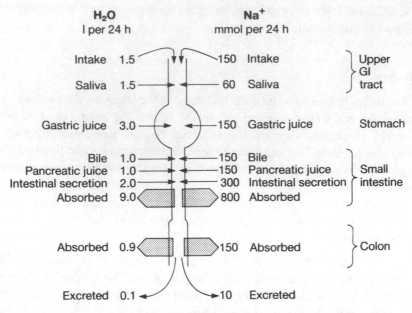

Fig. 14.20. Approximate H_2O and Na^+ movements across the gut over a 24 h period.

There are rapid movements of water across the small intestine in response to osmotic gradients. Ions also move freely through the 'tight' junctions. Thus the luminal contents rapidly attain the osmolality of the interstitial fluid. Throughout the intestine, water absorption from an isosmotic luminal solution is secondary to *solute* absorption and is driven by the resulting osmotic gradient (Fig. 14.19). How solute and water absorption are coupled is obscure, though the coupling probably occurs within the lateral intercellular spaces. Movement of this absorbed water from the interstitial fluid to the capillaries is driven by the plasma colloid osmotic pressure as expected. Changes in the capillary hydrostatic or plasma colloid osmotic pressure may influence net water movement between blood and lumen.

Sugars

These are absorbed almost entirely as monosaccharides, the final conversion from the di- to the monosaccharide involving disaccharidases in the brush border of the epithelial cells themselves. Monosaccharides are largely absorbed in the *duodenum* and *upper jejunum*. The rate of absorption of glucose (and galactose which competes with glucose for the co-transport carrier) is greatly influenced by Na^+ in the lumen. As already discussed, Na^+ and glucose appear to share a common mode of entry from lumen to cell, the diffusion of

Na^+ down its electrochemical gradient providing the driving force for glucose entry into the cell which may be against its concentration gradient. The absorbed monosaccharides seem not to be metabolized by the cells but leave across the basolateral membrane by a carrier-mediated process, which is not dependent upon Na^+, and then enter the capillaries.

Proteins

Before absorption takes place, the proteins are broken down to peptides or amino acids. About 50% of the digested protein comes from ingested food, 25% from proteins in the gut secretions and 25% from desquamated epithelial cells. Gastric and pancreatic enzymes hydrolyse protein to short-chain peptides (up to 6 residues long) which are further hydrolysed at the brush border to free amino acids or to smaller peptides ($M_r < 250$) which enter the intestinal mucosa. Absorption normally occurs in the *duodenum* and *upper jejunum*. Amino acid absorption, like sugar absorption, seems to be increased by luminal Na^+. There seem to be at least four specific mechanisms by which amino acids are absorbed: one for neutral amino acids, one for basic amino acids, one for acidic amino acids and one for the imino acids (proline, sarcosine). Di- and tripeptides may be absorbed from the lumen as such and are hydrolysed within the epithelial cells. In neonates, antibodies and other proteins contained in colostrum may be absorbed in their intact form by pinocytosis.

Lipids

The dietary lipids, up to 150 g per day, are primarily triglycerides, phospholipids, cholesterol and plant sterols. The absorption of lipid is normally an efficient process, and faeces contain less than 5 g fat per day. It occurs in a number of stages.

1 *Digestion and micelle formation.* Large lipid droplets must be converted to small lipid droplets and these stabilized to prevent coalescence. This process of *emulsification* requires a shearing force provided by intestinal motility and the presence of stabilizers, of which the physiologically most important are *bile salts*. Pancreatic *lipase* acting at the oil–water interface of the emulsion converts triglycerides to free fatty acids and 2-monoglycerides. These, together with bile salts, lecithin, cholesterol and fat-soluble vitamins, form *micelles*. Micelles are small aggregations (4–6 nm in diameter) of about 20 fat molecules whose hydrocarbon chains interdigitate within a fluid interior and whose polar groups form a negatively charged spherical shell surrounded by cations in aqueous solution.

2 *Passage from lumen to cell.* It is believed that the micelles diffuse to the apical cellular membranes and then free fatty acids and monoglycerides leave the micelles and diffuse across the apical membrane. The micelles themselves do not cross the membrane. The bile salts are not absorbed to any extent in the duodenum or jejunum. Absorption occurs in the terminal ileum and involves a carrier-mediated active process.

3 *Cellular metabolism and synthesis of chylomicrons.* The lipid crossing the brush borders of the epithelial cells is transferred to the endoplasmic reticulum. Fatty acids may pass as water-soluble coenzyme A derivatives; alternatively, it has been suggested that binding to a soluble protein may facilitate their passage. Within the endoplasmic reticulum, fatty acids are re-esterified to triglycerides. Then chylomicrons are assembled in the region of the Golgi apparatus of the cells. *Chylomicrons* are complexes, about 100 nm in diameter, of triglycerides (87%), cholesterol esters (3%) and fat-soluble vitamins, all of which are enveloped in a hydrophobic coat composed of a specific apolipoprotein B (1%), phospholipid (9%) and free cholesterol.

4 *Passage from cell to interstitium.* The release of the chylomicrons from the cell occurs by poorly understood mechanisms which may involve exocytosis.

5 *Removal by the lymphatic system.* The chylomicrons enter the lymphatic system via the lacteals in the villi. Removal occurs by the lymphatic system rather than by the blood capillaries because the former is permeable to the chylomicrons, perhaps reflecting a lack of tight junctions between the endothelial cells of the terminal lymphatics. The chylomicrons pass through the lymphatic system to the systemic circulation via the thoracic duct. In the blood, the chylomicrons bind apolipoprotein C, derived from high-density lipoproteins (HDL), to their coat. The triglyceride component (87%) of the chylomicrons is taken up by adipose tissue and muscle, whereas the cholesterol component (3% of total) passes on through the circulation to enter liver cells.

In contrast, short- to medium-chain fatty acids (< 12 carbons) are absorbed from the lumen without requiring prior hydrolysis or micelle formation and tend not to be re-esterified within the epithelial cells. They pass into the capillaries where they bind to albumin, and thence into the portal circulation rather than into the lymphatics.

Vitamins

The pathway followed by vitamins during absorption from the gastrointestinal tract is to a large extent dependent on their lipid and water solubilities. The *fat-soluble* vitamins (A, D, E and K) and their precursors are incorporated into micelles from which they diffuse into the epithelial cells. Like the dietary lipids the absorption of these vitamins is dependent on the presence of adequate bile salts.

The absorption of many *water-soluble* vitamins is facilitated by the presence of specific metabolic-dependent carriers. In some cases, these uptake processes show a Na^+ dependence similar to that for the uptake of amino acids and sugars.

The absorption of *vitamin B_{12}* is further specialized in that it requires the formation of a B_{12} intrinsic factor complex. The intrinsic factor, a glycoprotein secreted by the oxyntic cells of the gastric epithelium, is absorbed along with the B_{12} by a pinocytotic mechanism. This mechanism is facilitated by luminal Ca^{2+} or Mg^{2+} at an alkaline pH. Within the epithelial cells the complex is dissociated and B_{12} passes from the cells to the interstitial fluid.

15 · Energy Metabolism and Temperature Regulation

15.1 Energy metabolism
Metabolic heat
Energy storage and utilization
Control of food intake
Obesity

15.2 Temperature regulation
Heat and temperature
Heat transfer
  Conduction
  Convection
  Radiation
  Body temperature
  Core temperature
  Peripheral (shell) temperature

Thermal comfort and thermoneutrality
Thermoreception
Thermoregulatory effectors
  Shivering thermogenesis
  Non-shivering thermogenesis
  Cutaneous heat transfer
  Sweating
Fever
Exercise
Survival and temperature

15.1 Energy metabolism

Ultimately the body acquires all its energy from outside itself, mostly in the form of food, although thermal energy is also exchanged with the environment. Because energy is neither created nor destroyed,

Absorbed food energy = metabolic heat + stored energy
+ external work + energy losses in urine.

The *chemical energy of food* can be determined by combusting it with pure oxygen in a bomb calorimeter. Since the final oxidation products of this process are similar to those of human metabolism, at least for carbohydrates and fats, the heat produced by combustion is a measure of the calorific value of the food. The end-products of protein metabolism in man are urea and ammonia, and not the nitrogen or its oxides which are produced in the bomb calorimeter, but a correction can be made for this.

Metabolic heat

Providing no external work is being performed and body weight is constant, the chemical energy of food is transformed into heat. Even when external work is being done no more than 20% of the total available chemical energy of food is in fact converted to work. It is possible to measure the rate of body heat production (*metabolic rate*) in a whole-body calorimeter. However, such equipment is not readily available and it is usual to estimate metabolic rate indirectly from respiratory measurements of O_2 uptake and CO_2 production and their ratio, the *RQ* (p. 479). For example, the oxidation of 1 mol of glucose requires 6 mol of O_2 and produces 6 mol of CO_2, i.e. the *RQ* is 1. The total energy produced is 2880 kilojoule (kJ) for each mol of glucose utilized. There are thus 480 kJ of energy released for each mol of O_2 involved. For pure carbohydrates, which have an *RQ* of 1, 21·4 kJ of energy are released for each litre of O_2 (measured at STPD) that is consumed. This is called the *energetic equivalent of oxygen* and it varies for different substrates (Table 15.1). Given the actual *RQ* and also the urinary nitrogen excretion, the mix of the substrates being metabolized can be estimated and the appropriate energy equivalent of oxygen can be assigned.

For the purposes of comparing individuals, the *basal metabolic rate* (BMR) is determined while the subject is resting in a comfortable environment, some 12 h after eating. The BMR is calculated as watts (1 W = 1 J s^{-1}) or W m^{-2} of body surface because it correlates better with surface area than with weight and height. The resting metabolic rate of a standard 70-kg man is 100 W (equivalent to 1·433 kcal min^{-1}, i.e. 58 W m^{-2} of surface area, which has been defined as one *met*). Other factors that affect BMR include age, sex, race and disease. It is important that the subject be fasting for at least 12 h because food, particularly protein, increases heat production by its so-called *specific dynamic action* (SDA). Most of this SDA is due to oxidative deamination of amino acids in the liver. Hormones, e.g. adrenaline and thyroxine, also affect the BMR and the determination of BMR was formerly used in the diagnosis of hypothyroidism and hyperthyroidism (p. 273).

The metabolic rate of a person can increase by as much as fifteen times during severe exercise. Indeed one of the major ways of combating a cold environment is by shivering. This is an involuntary contraction of skeletal muscle fibres which does no external work so that all energy appears as heat. There are also non-shivering mechanisms for increasing metabolic rate and these are discussed in Section 15.2.

Energy storage and utilization

For a given individual, weight remains remarkably constant during adult life and it is clear that there is a very precise control mechanism involved. Over a

Table 15.1. Energetic equivalent of oxygen for different substrates

	Substrate		
	Glucose	Palmitic acid	Alanine
Respiratory quotient (*RQ*)	1·0	0·7	0·8
Energetic equivalent of O_2 kJ mol^{-1}	480	425	440
kJ l^{-1} STPD	21·4	19·0	19·6

large population weights vary widely, with a normal distribution. Body composition, however, cannot be determined with the precision that weight can. Thus it is not clear whether or not a constant weight implies a constant composition. The statement that 'within every fat man a thin man is striving to get out' illustrates the concept of a *lean body mass* of muscle, skeleton and essential organs embedded in fat. The lean body mass is related to the amount of regular physical activity whereas the fat represents the stored energy. Fat stores in an average person are equivalent to the amount of energy normally consumed over some 40 days. By comparison, carbohydrate stored as glycogen provides for about a day's energy needs and glucose for about one hour's energy needs. The advantage of storing energy in the form of fat is that fat contains about twice as much energy per gram as protein or carbohydrate. Furthermore, fat cells contain relatively little water and so the volume in which energy is stored is minimized, and, as we are all aware, there is no apparent limit to the amount of energy that can be stored in this form!

Table 15.2 summarizes some aspects of energy storage and utilization. Note that although the energy stored as carbohydrate represents only about 1% of the total, carbohydrate metabolism contributes about half of the basal energy consumed each day. Also, though about 20% of the energy is stored as protein, the central role of proteins in cellular structure and in skeletal muscle activity precludes the use of these stores to any extent until relatively late in starvation. Without mobility, food cannot be sought!

Given the preferred role of glucose as an immediate source of energy, and the fact that glucose is stored in cells as glycogen, it is not surprising that

Table 15.2. Body composition, energy stores and energy consumption of a typical 70 kg man at rest. The energy content of protein and carbohydrate is 17 kJ g^{-1} and of fat 37 kJ g^{-1}

Component	Site of storage	Body weight (%)	Amount (kg)	Total energy (J × 10^6)	Daily consumption (kg)	Daily energy consumption (%)
Protein	Mainly skeletal muscle	12	8·4	143	0·07	17
Fat	Adipose tissue	18	12·6	466	0·06	33
Carbohydrate	As *glycogen* in liver (1/3) and skeletal muscle (2/3), as *glucose* in body fluids	0·5	0·4	7	0·2	50

several hormones play key roles in regulating carbohydrate metabolism. Principal amongst these are insulin, glucagon, adrenaline, growth hormone and cortisol (p. 290). Insulin, released from pancreatic β islet cells when blood glucose rises, acts to *lower* blood glucose; it facilitates glucose entry into most cells (other than brain, gut and kidney) and favours carbohydrate storage by stimulating glycogenesis and inhibiting glycogenolysis. In contrast, the other hormones, whose secretion is favoured by a decrease in blood glucose (amongst other stimuli), tend to *increase* plasma glucose by collectively inhibiting glucose uptake and facilitating glycogenolysis and gluconeogenesis.

In addition to glucose, metabolism of free fatty acids also provides a substantial proportion of daily energy requirements. Free fatty acids are derived either directly from the diet, or are released from the fat stored in adipose tissue. When plasma glucose levels are high, uptake of fatty acids by adipose tissue is enhanced, as it is by insulin, whereas glucocorticoids inhibit it. Conversely, the release of fatty acids from adipose tissue is inhibited by insulin but stimulated by adrenaline, glucagon and growth hormone. Thus the hormones involved in regulating blood sugar also regulate fat metabolism.

Finally, a dynamic balance is also maintained between protein and amino acid metabolism. Amino acids absorbed in the gastrointestinal tract are resynthesized into new proteins to replace those continually being broken down. This process is stimulated by growth hormone and insulin. In addition, some of the amino acids are converted, principally in the liver, to glucose via a variety of intermediates (gluconeogenesis). Once again, glucagon, glucocorticoids and growth hormone favour this process, while insulin opposes it.

There is, therefore, a dynamic balance, finely regulated by a variety of hormones, between anabolism and catabolism. A summary of energy storage and utilization during the absorptive and post-absorptive states is given in Fig. 15.1. When energy is readily available, as after feeding, blood glucose and amino acid levels tend to rise, insulin is secreted and cellular uptake of glucose, glycogenesis, protein synthesis and fat storage are favoured. The secretion of glucagon is suppressed and so too may be that of glucocorticoids, growth hormone and adrenaline. Conversely, when an animal is deprived of food, blood glucose falls, insulin secretion is depressed, and glucose uptake by most cells is inhibited ensuring that what glucose is available is supplied to the cells of the brain. In addition, the hypoglycaemia stimulates secretion of glucagon and perhaps also glucocorticoids, growth hormone and adrenaline. Consequently glycogenolysis and gluconeogenesis are stimulated, increasing the supply of glucose, and fatty acids are released from adipose tissue providing an alternative source of energy to cells deprived of glucose. Furthermore, the sensation of hunger is aroused and food intake thereby encouraged.

ABSORPTIVE STATE

Energy source for most tissues is glucose.

Energy storage occurs in
    liver and muscle as glycogen and in
    adipose tissue as triglycerides obtained from
       — ingested triglycerides
       — triglycerides synthesized from glucose
         in adipose tissue
       — triglycerides synthesized from glucose
         and amino acids in liver and transported
         to adipose tissue.

POST–ABSORPTIVE STATE

Brain is an obligate user of glucose which is provided
in the short term by glycogenolysis

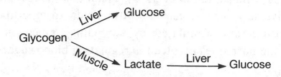

and in the long term by gluconeogenesis

Amino acids ⎫
         ⎬ —Liver→ Glucose
Glycerol ⎭

Other tissues undergo glucose-sparing reactions and
utilize free fatty acids and ketones for their energy needs.

Fig. 15.1. Summary of energy storage and utilization during the absorptive and post-absorptive states.

Control of food intake

It is generally held that energy intake rather than energy expenditure is regulated to maintain a steady state. Though such regulation has been described in terms of hypothalamic 'satiety' and 'hunger' centres, a more complex picture is emerging. Between meals, blood glucose remains remarkably constant, reflecting neural and endocrine matching of hepatic glucose production to peripheral uptake. The neural contribution may involve participation of *glucose-sensitive neurones* in the *nucleus of the tractus solitarius*, the *vagal nuclei* and the *hypothalamus* in the regulation of autonomic outflow to the pancreatic islets and adrenal glands.

    Feeding is initiated by a complex interaction of factors. Experimentally, a fall in blood glucose can be demonstrated several minutes before feeding begins, and this fall may be detected by neurones in the *lateral hypothalamus*.

This area also receives a wealth of olfactory, gustatory and visual inputs and is believed to be responsible for initiating food intake. In contrast, the *ventromedial hypothalamic area* appears to regulate the utilization and restoration of body fat stores. When this area is activated, sympathetic outflow inhibits pancreatic insulin secretion and stimulates lipolysis in adipose tissue. Destruction of this area, therefore, raises plasma insulin concentration and inhibits lipolysis leading to increased food intake, increased body fat, weight gain and obesity. It is unclear what normally controls the activity of this area and therefore determines the balance between lipolysis and lipogenesis. However, it is suggested that the availability of glucose to cells in the area may be critical. With glucose availability diminished, insulin secretion would be inhibited and lipolysis favoured; with sufficient glucose, insulin secretion would be stimulated and lipogenesis would predominate.

Though the lateral hypothalamic area initiates feeding, the amount of food eaten is determined by a variety of factors. These include the palatability of the food and the presence of food in the mouth, stomach and intestine, as well as a range of psychological and learned responses. Eating normally ceases *before* much of the ingested food has been absorbed. Other regions of the CNS also affect food intake (p. 246).

Obesity

People are regarded as being obese if their adipose tissue is more than 30–40% of the total body weight. The causes of obesity are obscure. Genetic factors play a role in certain strains of obese animals. Hormonal disturbances can sometimes cause obesity; for example, it is common in eunuchs, in hypothyroidism and in hyperadrenocorticalism. Most experts, however, consider that psychological factors are the main cause of obesity. Eating habits are a learned phenomenon and cater for a multitude of sensory and emotional needs over and above energy balance. Obesity is a health hazard because it predisposes to cardiovascular disease and diabetes mellitus.

15.2 Temperature regulation

Life is a low-temperature phenomenon existing in a universe in which temperatures range from near absolute zero to greater than 10^6 K. The rate of chemical reaction varies with temperature, increasing about 2·5-fold for each 10°C rise (Q_{10}). However, at temperatures above 45°C, the enzymes that catalyse living reactions lose their structural integrity and are said to be denatured. Unlike lower animals, which do not regulate their body temperature (*poikilotherms*), mammals and birds derive an advantage by maintaining their body temperature constant (*homoiotherms*) and close to denaturing temperatures. Enzyme reactions of the body function optimally within a narrow range. Conscious intelligence requires an even narrower range of approximately 35–42°C. Body temperatures below 35°C are associated with loss of memory and body temperatures greater than 42°C with delirium and hallucinations.

Heat and temperature

The concepts of heat and temperature must be clearly distinguished. *Heat* is a form of energy and is measured in joules (J) or in calories (1 kilocalorie = 4184 joules). The same quantity of heat can be distributed over a larger or smaller amount of material. The former will have a lower temperature than the latter. *Temperature* is a measure of the average kinetic energy per degree of freedom of the constituent molecules. As it is an average, the concept of temperature can be applied only to objects consisting of a large number of molecules.

As the thermal energy in an object increases, the temperature rises. The *specific heat capacity* is the amount of energy required to be added per unit amount of material to raise the temperature by one unit. At 37°C, the addition of $4 \cdot 178 \times 10^3$ joules of energy to 1 kilogram of water will raise its temperature by 1 degree (Kelvin or Celsius), i.e. the specific heat capacity of water at 37° C is $4 \cdot 178$ kJ kg^{-1}. The specific heat capacity of body tissue is about 85% that of water, i.e. $3 \cdot 6$ kJ kg^{-1}.

The velocity with which molecules move is dictated by their thermal energy in relation to any attractive forces between molecules. Adding energy permits the more energetic molecules to escape the attractive forces, leaving behind the less energetic molecules. In this way, energy is carried away from a liquid by evaporation. The balance between the attractive forces between molecules can alter suddenly at certain critical temperatures. For example, the state of water can change from solid to liquid to gas. Each change requires an addition of thermal energy (*latent heat*), but while the change is occurring the temperature remains constant. Thus, to melt ice, $333 \cdot 7$ kJ kg^{-1} of thermal heat (*specific heat of fusion*) is required. The evaporation of water requires $2 \cdot 257$ MJ kg^{-1} (*specific heat of vaporization*).

Heat transfer

Heat is lost from the body through:

1   conduction, convection, radiation and evaporation of water from the skin;
2   evaporative loss and direct warming in the upper respiratory tract; and
3   urination and defaecation.

During the resting state, about 50% of heat is lost by radiation, about 15% by conduction and convection and up to 30% by evaporation from the skin. Heat lost by respiration and in urine and faeces accounts for about 5%.

Conduction

Within a volume of material, the random movement of the molecules results in molecular collisions. These partition the kinetic energy amongst the molecules. In this way, the thermal energy diffuses from areas of higher heat concentration, i.e. higher temperature, to those of lower. This is the process of *heat conduction*. Materials differ in the velocity of heat conduction and they can be compared under standard conditions.

Thermal conductivity is the flux of energy per second through a slab of material 1 m² in area when the temperature is 1°C over a distance of 1 m. The units of thermal conductivity are therefore $J\,s^{-1}\,m^{-2}\,(°C\,m^{-1})^{-1}$ which reduces to $J\,s^{-1}\,m^{-1}\,°C^{-1}$ or $W\,m^{-1}\,°C^{-1}$ since 1 joule s^{-1} equals 1 watt. As shown in Table 15.3, the thermal conductivity of a slab of fat or muscle is about the same as that of cork.

Thermal insulation (i.e. thermal resistance) is the inverse of thermal conductivity. It is the temperature difference across the slab defined above which will drive unit heat flux and has the units $°C\,m\,W^{-1}$. Those interested in clothing insulation find it more convenient to use the unit called the *clo*. The value of 1 clo has been arbitrarily set at $0.155°C\,m^2\,W^{-1}$. This enables the insulation of clothing to be given in clo per cm thickness. Clothing with the insulation of 1 clo (about that of a business suit) will maintain a standard 70 kg man, whose metabolic rate is about 100 W, comfortable indefinitely while sitting at rest in an environment of 21°C, a relative humidity less than 50% and an air movement of 10 cm s^{-1}.

Convection

External transfers. A warm object surrounded by cooler fluid transfers heat to the adjacent fluid by conduction, explained above. In addition, as the fluid

Table 15.3. Rate of heat flow through a slab of material 1 m thick, 1 m² in area per °C temperature difference between faces.

Material	Conductivity ($W\,m^{-1}\,°C^{-1}$)
Felt	0.004
Air (static)	0.025
Cork	0.054
Fat (unperfused)	0.06
Fat (perfused)	0.60
Silver	418.6

warms it expands lowering its density, and an ascending flow, termed *natural convection*, results. Fluid flow resulting from wind or movement of the object is *forced convection*.

However produced, flow over the skin surface will modify heat transfer between the body and the environment. The direction of heat flux depends on whether the temperature of the external fluid is higher or lower than the body. The amount transferred will depend on the specific heat capacity of the fluid, the total volume flow and the degree of stirring which enhances the removal of heat from the surface. The expression for heat exchange by convection is similar to that for conduction but, since no distance is involved when surfaces are in contact, it is measured in $W m^{-2} °C^{-1}$. The value depends on the nature of the fluid and the pattern of flow. In nude subjects, and at a flow velocity of $0·1 m s^{-1}$, heat loss by convection in air at 33°C is about $2·6 W m^{-2} °C^{-1}$ whereas in water at 33°C it is $62 W m^{-2} °C^{-1}$. The specific heat of water is 1000 times greater and its heat conductance 25 times greater than that of air.

Internal transfers. As most of the body has a very low thermal conductivity its core would become very hot if heat were not redistributed by blood flow. During the passage of blood through an actively metabolizing region, heat is conducted to and warms the capillary blood. The amount of heat in an organ, and thus its temperature, is a balance between production and removal. Heat transfer depends on intercapillary distance, thermal conductivity, blood flow per mass of tissue and specific heat capacity of the blood.

The fatty subcutaneous tissue has variable thermal properties. Unperfused it has the thermal conductivity of cork (Table 15.3). However, there are loops of blood vessels perpendicular to the skin and blood flow can vary from between $0·2$ and $2·5 l min^{-1}$ at normal temperatures up to about $8 l min^{-1}$ at elevated body temperatures. Such enhanced blood flow can increase the thermal conductivity of subcutaneous tissue tenfold (Table 15.3). By this means, the effectiveness of the insulating layer between core and environment and also the amount of heat delivered to the skin can be varied.

Radiation

All molecules, unless at absolute zero temperature, are continuously emitting quanta of energy in the form of electromagnetic energy. They are simultaneously absorbing such energy. In this way, energy is transferred at a distance without collision between molecules. The quantity and wavelength of the radiated energy depend on the absolute temperature of the object and also on the ability of the surface to emit radiation, i.e. the emissivity. As the temperature rises the wavelength of the radiated energy falls and, with small temperature differences within the physiological range, the net energy exchange by radiation is directly proportional to the temperature difference.

The quality of the surface determines the fraction of the incident energy which is absorbed and the fraction of the maximum possible radiant energy which is actually emitted. An efficient emitter is also an efficient absorber. The *emissivity* is a measure of this efficiency and is rated as 1 for a surface absorbing all incident radiation (a 'black body') whereas a perfect reflector absorbs no energy and emits no energy, having an emissivity of 0. All real surfaces fall between these. Very dark skin absorbs about 82% of incident solar radiation while very fair skin only 65%. This is modified by any clothing that is being worn.

Not all of the surface area of the naked human body is available to radiate energy. The *effective* surface area varies as body posture changes. The net radiant energy loss is 49 W from a 70 kg person, whose effective surface area is 1·5 m² with a skin temperature of 33°C in an environment at 29°C, and this is about half of the resting heat production (100 W).

Body temperature

There is a range of temperatures from the body core to the hands and feet, and core temperature can also vary (Fig. 15.2). Nevertheless, over a wide range of environmental conditions there is remarkably little variation in core temperature. This constancy implies a very precise control of heat balance and has given rise to the concept of a 'set-point' for temperature regulation. Deviation of core temperature from the set-point produces an 'error signal' that drives heat-loss or heat-retention mechanisms.

Core temperature

There are only very small temperature differences between the main internal organs and it is their temperature that is referred to as deep body temperature or *core temperature*, usually measured in the rectum. Body temperatures obtained from the mouth and axilla are about 0·5°C lower than rectal temperature and show greater variation. There is normally a diurnal variation in core temperature, the temperature being the highest in the evening (37·3°C) and lowest in the early morning (35·8°C). In females, core temperature varies during the menstrual cycle, being about 0·5°C higher in the latter half of the cycle.

Peripheral (shell) temperature

Skin temperature is somewhat less than core temperature and varies much more widely. The extremities have a large surface area relative to their mass.

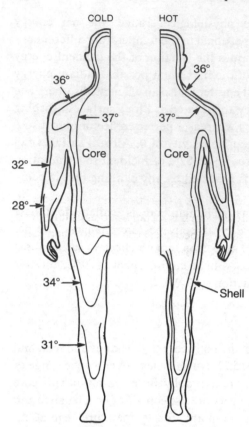

COLD   HOT

36°   36°

36°

← 37°   37° →

Core   Core

32° →

28° →

34° →

31° →

—— Shell

Fig. 15.2. Isotherms in the body in a cold (left) and hot (right) environment. The core, which is surrounded by 37° isotherm, contracts in a cold and expands in a hot environment. (Adapted from Aschoff, J. & Weaver, R. (1958) *Naturwissenschaften* **45**, 477.)

Other than active muscle their only endogenous heat source is the blood flowing from the trunk. The volume of the limb which remains close to the core temperature depends on the size of the heat load (Fig. 15.2). When cold environmental conditions favour heat loss, the core is restricted to the head and trunk, and the shell includes a large proportion of the limb volume. The total body heat content is reduced and if it were evenly distributed core temperatures could fall to unacceptable levels. When total body heat content is high, increased blood flow enlarges the volume of the core.

Thermal comfort and thermoneutrality

Thermal comfort is a conscious experience. It is a mental phenomenon elaborated by the nervous system from a variety of inputs related to temperature. It is an awareness of the thermal state plus an emotional content. Thermoregulatory centres in the hypothalamus monitor central core temperature and receive inputs from peripheral thermal receptors. Thermal comfort provides the motivation for thermoregulatory behaviour. The temperature chosen by a subject as comfortable is the *preferred comfort zone*, which covers a very narrow span. It is that combination of air temperature, air movement, humidity and radiation intensity that leads to a skin temperature of about 33°C. It is this skin temperature that is aimed at by varying clothing or environmental factors. For lightly clothed sedentary adults, the preferred ambient temperature is about 21°C when air movement is less than 30 cm min^{-1} and relative humidity is less than 50%. The elderly and infirm who have reduced thermal responses require a higher minimal temperature.

Below a certain *critical temperature*, the metabolic rate rises linearly as temperature is lowered further (Fig. 15.3). This rise, reflecting the activity of skeletal muscle and other mechanisms (see later), is sufficient to maintain deep body temperature until the total heat loss becomes too great. Then core temperature begins to fall as the *zone of hypothermia* is entered, and regulation is lost. With increasing environmental temperatures above the critical temperature, the metabolic rate remains steady and minimal until at an upper limit the rate rises again. Here the *zone of hyperthermia* is reached as body temperature rises and regulation is lost. The range of environmental temperatures in which the metabolic rate is minimal and steady defines the *thermoneutral zone*.

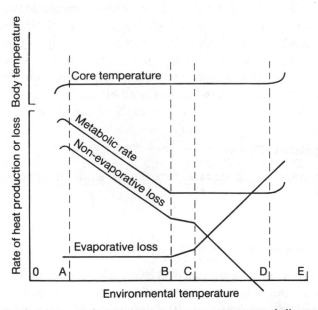

Fig. 15.3. Effects of environmental temperature on core temperature, metabolic rate, and rates of evaporative and non-evaporative heat loss. 0–A, zone of hypothermia; B, critical temperature; B–C, zone of least thermoregulatory effort; B–D, thermoneutral zone; D–E, zone of hyperthermia. (From Mount, L.E. (1974) The concept of thermal neutrality. In *Heat loss from Animals and Man*, eds. Monteith, J.L. & Mount, L.E., pp. 425–439, Butterworth, London.)

Thermoreception

The skin, mouth and pharynx contain thermal receptors (*peripheral thermo-receptors*), the signals of which reach consciousness. There are both warm and cold thermoreceptors whose properties have been described on p. 141. As well as the superficial receptors, there is evidence for *deep body thermoreceptors* located within the oesophagus, stomach and perhaps duodenum and possibly in the intra-abdominal veins. The afferent pathways from thermoreceptors are discussed on p. 141.

Thermoregulatory responses in a variety of animal species can be elicited by local thermal stimulation of various areas in the CNS—the spinal cord, medulla oblongata and midbrain reticular formations and most importantly the *preoptic–anterior hypothalamic* region; from this latter region, maximal responses to both warming and cooling can be obtained. It seems to act as a thermostat as far as core temperature is concerned, causing cooling or heat retention as necessary.

The thermal drives from various parts of the body may interact synergistically or antagonistically. For instance, during heavy exercise in cold weather, skin temperature can be low at the same time as heat production is high and core temperature is above resting levels. Thus information from temperature-sensitive neurones in the hypothalamus is combined with information from other parts of the body. These combinations can alter the threshold temperature of the hypothalamus at which thermoregulatory responses occur. As well as this, the combined information can modify the sensitivity of the hypothalamus to the input and thus the briskness of the reflex response. However, if core temperature moves significantly away from normal, the hypothalamic responses increasingly dominate.

Thermoregulatory effectors

Short-term control is initiated via efferent nervous pathways both somatic and autonomic.

Shivering thermogenesis

This is an involuntary contraction of skeletal muscle fibres which can raise the metabolic rate threefold and is controlled from the hypothalamus. The muscle activity produced commences in the upper half of the body as a gradual increase in contraction of muscle fibres and can extend to the whole body. As shivering becomes more intense, cyclical muscle rigors at the rate of 10–20 Hz appear. Purposeful movements inhibit shivering.

Non-shivering thermogenesis

Under cold stress, there is an increased discharge of the sympathetic nervous system. One consequence of this is lipolysis which in the case of white adipose tissue increases plasma glycerol and free fatty acids. The free fatty acids are then available as energy sources for skeletal muscle and myocardium. In the newborn, there are also deposits of highly vascular brown fat situated between the scapulae and around the intra-abdominal vessels. Brown fat itself metabolizes the free fatty acids generating heat which is distributed to the rest of the body via the blood. In the newborn, this, rather than shivering, is the principal means of increasing heat production and this non-shivering thermogenesis can increase metabolic rate to about two-and-a-half times the resting value. As a healthy baby can produce heat only at about 0.34 kJ kg^{-1} min^{-1}, it requires help to reduce heat loss which immediately after birth, while it is still wet, can reach $0.8 \text{ kJ kg}^{-1} \text{min}^{-1}$.

Although cooling the preoptic region of the hypothalamus in animals has been shown to increase the rate of secretion of thyroid hormone, the effect in adult humans is uncertain. Variations in thyroid function do not seem to play a part in long-term adaptation to cold environments but pathological alterations in thyroid function certainly alter heat production.

Cutaneous heat transfer

For heat to be lost from the body, it must be conveyed by the blood to the skin. If ambient temperature is below the skin temperature, heat can be lost by conduction, convection and radiation. If it is above skin temperature, heat can be lost only by evaporation of sweat. Blood flow to the skin is therefore important not only to provide the heat but also the water for sweat.

The control of blood flow through the skin in response to changes in temperature is discussed on p. 453. The adjustments of the cardiovascular system on the strains imposed by thermal stress are discussed on p. 446. During exercise, cutaneous vasodilatation is confined to the profusely sweating areas of head, trunk and proximal limbs, while blood flow in distal portions of limbs does not alter much.

Cutaneous veins have a specialized function in thermoregulation. Their calibre is under noradrenergic sympathetic control and varies directly with heat load. The pathway for blood returning from the limbs can be either via superficial or deep veins. In cold conditions, it is predominantly in the deep veins and this permits transfer of heat directly from artery to vein thus preventing heat loss from body surfaces. This is a *thermal counter-current* system for heat conservation.

Sweating

This is an active secretory process from eccrine sweat glands which are widely distributed over the surface of the body. They are innervated by sympathetic postganglionic cholinergic fibres and produce a hypo-osmotic secretion. Their activity is blocked by atropine, the administration of which can lead to elevated body temperature.

Sweating can result in a fluid loss of as much as 1 litre per hour. This can rapidly diminish body water and electrolyte content and if these are severely depleted the rate of sweating diminishes and body temperature can rise to fatal levels. For effective heat loss, complete evaporation is needed. Therefore, sweating becomes more obvious but is less effective as environmental humidity increases.

Fever

A variety of substances resulting from infections (pyrogens) can elevate the set-point. The subject, even though body temperature is still normal, feels cold. Peripheral vasoconstriction and shivering then ensures that heat balance is re-established at a higher temperature. This is called fever or *pyrexia*. Aspirin lowers the temperature and, since it also inhibits prostaglandin synthetase, prostaglandins appear to be involved in the mechanism of temperature regulation. The sensation of thermal discomfort associated with fever seems to be present when heat balance has not been achieved and actual temperature and set-point are different. The degree of discomfort is related to the size of the discrepancy.

Excessive temperatures can also arise following damage to the CNS, from excessive heat production related to varieties of muscular dystrophy and from failure of the heat loss mechanism such as in dehydration.

Exercise

During continuous exercise the core temperature varies directly with the work rate, and it can be elevated by as much as 3°C. At a given work rate the increase in core temperature remains constant over a wide range of environmental temperatures. This can be interpreted to mean that the set-point is raised; but it is also possible that the increased core temperature simply provides the error signal needed to drive the mechanisms that dissipate the heat generated during exercise. Note, too, that the higher temperature associated with exercise is not lowered by aspirin.

Survival and temperature

Body temperatures above 40°C cause heat stroke, mental confusion and unconsciousness; temperatures below 34°C give rise to amnesia and also to unconsciousness. Lowering cardiac temperature below about 30°C causes arrhythmia and then cardiac arrest. However, if the circulation is externally supported, the heart can survive for a period and functions well on rewarming. In fact, the whole body can be cooled to 10°C with consequently no circulation and yet with a very high probability of unimpaired survival on rewarming. Cooling of individual organs and tissues is deliberately used to permit survival in the absence of circulation.

16 · Exercise

16.1 Energetics

16.2 Circulatory responses in exercise

16.3 Respiratory responses in exercise

16.4 Applied exercise physiology

Exercise consists of voluntary activation of skeletal muscle. This chapter deals first with the accompanying increased metabolic activity in the muscle, then the changes that occur in the circulatory and respiratory systems to meet the needs for supply of more O_2 and for removal of CO_2, heat and metabolites. Finally some aspects of performance, training and fitness are dealt with in a section on applied aspects of exercise physiology.

16.1 Energetics

The energy for muscular work is derived from the hydrolysis of adenosine triphosphate (ATP). As stores of ATP are limited, exercise lasting for more than a few seconds demands an increased supply from *aerobic* and *anaerobic metabolism* (p. 102). In mild exercises ATP is produced by the aerobic pathway by oxidation of fatty acids and ketones, while in more vigorous exercise glycogen is oxidized. At higher levels of exercise the additional demand for ATP is met by the anaerobic pathway, which produces ATP more rapidly by converting glycogen to lactic acid. The relative contributions from the aerobic and anaerobic pathway depend on the duration of the exercise. Aerobic metabolism contributes very little of the total energy utilized during brief explosive exercise. The contribution increases to about 33% for events such as a 400 m race lasting under a minute, and exceeds 90% in events lasting 10 min or more.

Fatigue is the sensation associated with increasing difficulty or inability to sustain a given level of work output or exercise. Fatigue occurring in the first few seconds of very intense exercise is caused by depletion of ATP and creatine phosphate stores in the active muscles. For slightly less intense exercise, fatigue occurs within a few minutes as the build-up of lactic acid or other metabolites in the muscles inhibits glycolysis or the contractile mechanism. Fatigue also results from depletion of intramuscular glycogen stores after long periods of exercise ('hitting the wall' in the marathon).

The *efficiency* of exercise is a measure of the conversion of chemical energy into external work. In estimating efficiency, the exercise physiologist is usually content to measure the steady-state O_2 consumption of a subject in conditions where the rate of external work can be estimated reasonably accurately. It is assumed either that aerobic metabolism is the sole source of power, or that any products of anaerobic metabolism are oxidized. The efficiency can then be calculated after converting the O_2 cost of exercise ($\dot{V}_{O_2}$ in exercise $- \dot{V}_{O_2}$ at rest) measured e.g., in ml s^{-1}, into units of power. On a mixed diet, 1 ml s^{-1} of oxygen (STPD) yields 20 W of total power (determined using whole body calorimetry), and therefore

$$\% \text{ efficiency} = \frac{\text{rate of external work}}{(\dot{V}_{O_2} \text{ in exercise} - \dot{V}_{O_2} \text{ at rest}) \times 20} \times 100.$$

The external work of exercise may be very difficult to estimate, as for example in running on the level, where work is done to accelerate and decelerate the limbs and to raise and lower the centre of gravity with each step. On the other hand, the external work of cycling is easily determined with a cycle ergometer, and the efficiency of cycling is found to be 20–25%. This means that someone doing 200 W of external work on a cycle ergometer is also generating 800 W of heat, which must be dissipated by increased blood flow to the skin and by the evaporation of sweat.

If exercise is started abruptly and a constant moderate work output is maintained , the O_2 uptake takes several minutes to reach a steady state (Fig. 16.1). The missing O_2 that would have been consumed, if the body's response to the O_2 demand of exercise was immediate, is known as the *oxygen deficit*. For mild to moderate exercise the oxygen deficit is accounted for by:

1   The lower content of ATP and creatine phosphate in the active muscles.
2   The reduction in O_2 content of myoglobin in the active muscles.
3   The reduction in O_2 content of the venous blood leaving the muscles.

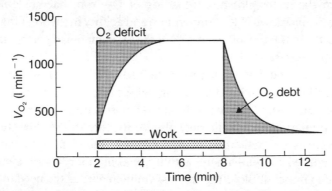

Fig. 16.1. Time-course of O_2 consumption during moderate exercise illustrating the O_2 deficit and debt.

The *oxygen debt* represents repayment of the O_2 deficit (Fig. 16.1), but when the exercise is vigorous or prolonged the debt is greater than the deficit. Following such exercise the ATP and creatine phosphate stores and the O_2 content of myoglobin and venous blood return to normal within a minute or two, but there is a residual elevated O_2 consumption which can last for minutes to hours depending on the intensity and duration of the exercise. This extra O_2 consumption is attributed to the energy needed to dissipate heat and restore intracellular electrolytes to normal concentrations, and to the higher metabolic rate caused by the increase in body temperature and in circulating levels of catecholamines and thyroxine. Regeneration of intramuscular glycogen stores also produces a very small increase in O_2 consumption over a period of a day or two following the exercise. Lactate produced by the anaerobic pathway is metabolized within 30 min after exercise but, contrary to popular belief, metabolism of lactate following anaerobic exercise does not contribute significantly to the O_2 debt.

16.2 Circulatory responses in exercise

Rhythmic exercise produces an increase in blood flow in the active muscles which at maximum can reach 20 times the resting blood flow. The vasodilatation that gives rise to the increase in blood flow may be due, in part, to an increase in activity of the cholinergic sympathetic nerves to blood vessels in muscles, but most of it is caused by local factors such as the decrease in P_{O_2} and pH, the increase in P_{CO_2}, extracellular $[K^+]$ and temperature, and release of other metabolites such as adenosine (p. 429). The exercise-induced vasodilatation results in a large increase in the number of capillaries carrying blood through the muscle; this shortens the mean path length between capillaries and mitochondria in the muscle fibres, allowing a greater extraction of O_2 from the haemoglobin. Off-loading of O_2 from haemoglobin is also assisted by the increased P_{CO_2}, temperature and acidity in the working muscles, resulting in the extraction of about 90% of the O_2 from blood perfusing muscles at maximum exercise.

Changes in blood flow to the rest of the body during exercise are shown in Table 16.1. Blood flow to heart muscle increases to meet the extra O_2 demand of the increased cardiac output; as in skeletal muscle this is under local control although circulating adrenaline may also contribute to the dilatation of the coronary vessels via activation of β_2-receptors. Blood flow to the gut and kidney is reduced by sympathetic activity, making more blood available for the exercising muscles. Blood flow to the skin increases as the need to dissipate heat increases, but at maximum exercise skin flow decreases, giving the subject a pale or ashen pallor just before exhaustion; this indicates that heat dissipation

Table 16.1. Changes in blood flow distribution during exercise.

	Rest		Exercise			
	(ml min^{-1})	(%)	Light (ml min^{-1})	Medium (ml min^{-1})	Heavy (ml min^{-1})	(%)
Cerebral	750	13·0	750	750	750	3·0
Coronary	250	4·5	350	650	1000	4·0
Renal	1100	19·0	900	600	250	1·0
Splanchnic	1400	24·0	1100	600	300	1·2
Skin	500	8·5	1500	1800	600	2·4
Others	600	10·5	400	300	100	0·4
Skeletal muscle	1200	20·5	4500	10 800	22 000	88·0
Cardiac output	5800	100·0	9500	15 500	25 000	100·0

is compromised when cardiac output cannot be increased further to meet the demands of exercise. The constancy of blood flow to the brain reflects the unchanged metabolic rate of the brain whatever the level of exercise, and there appears to be no reserve here that can be drawn upon for the exercising muscles.

The effects of exercise on circulatory and respiratory variables are summarized in Table 16.2 and Fig. 16.2. The increase in O_2 consumption (about fifteenfold at maximum in a young adult) is accommodated by an increase in the arteriovenous O_2 concentration difference (up to threefold) and an increase in cardiac output (up to fivefold). The increase in cardiac output results from venoconstriction, vasodilatation and increased myocardial contractility, as well as from the repeated squeezing of the veins by the active muscles (the 'muscle pump'). The increase in heart rate limits the increase in stroke volume to a factor of 1·5. Most of this increase in stroke volume occurs in light to moderate levels of exercise, and mainly by an increase in end-diastolic volume, but at higher levels of exercise stroke volume is also increased by a decrease in end-systolic volume due to the higher contractility of the heart (see Fig. 11.27, p. 405). Mean blood pressure in healthy young people increases only slightly during exercise, although systolic pressure increases and diastolic pressure decreases, giving a pronounced increase in pulse pressure.

The nature of the primary signal activating the sympathetic nervous system during exercise is still not entirely clear, but is is probably activity in sensory nerves which detect the factors inducing vasodilatation in the active muscles. This means that control of the circulatory system via the sympathetic nervous system is by negative feedback, because the circulatory response (increased cardiac output or blood pressure) acts to try to remove the initiating stimulus (metabolites released in the muscles) by increasing perfusion of the muscles. There may be additional stimulation of the sympathetic system from neural

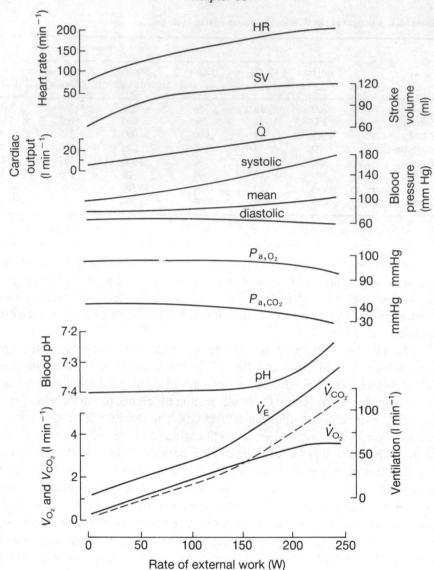

Fig. 16.2. Changes in metabolism, ventilation, arterial pH, arterial gases and cardiovascular variables of a healthy male subject as a function of steady-state work on a cycle ergometer. HR, heart rate; SV, stroke volume; $\dot{Q}$, cardiac output.

activity in the motor cortex or in sensory pathways carrying information about movement, and also from neural activity in the respiratory system. The baroreceptor reflex appears to have a minor role by exerting a 'braking' effect on cardiovascular stimulation at the start of exercise in order to limit increases in blood pressure. Nevertheless, in exercise where perfusion of muscles is

Table 16.2. Cardiorespiratory variables and indices at rest and during exercise.

	Rest	Exercise		
		Light	Medium	Heavy
Oxygen consumption (ml min^{-1})	250	1500	2800	3500
Oxygen consumption of muscles (ml min^{-1})	50	1250	2200	3150
Cardiac output (l min^{-1})	5	13	16	21
a–v̄ Oxygen difference (ml l^{-1})	50	100	127	130
$P_{\bar{v}O_2}$ (mmHg)	40	26	23	20
Heart rate (beats min^{-1})	72	144	160	190
Diastolic filling time (ms)	500	200	170	150
Stroke volume (ml)	70	90	100	110
Mean blood pressure (mmHg)	85	90	95	100
Pulse pressure (mmHg)	40	60	90	120
Systolic/diastolic pressure (mmHg)	110/70	130/70	155/65	180/60

absent or limited (e.g. sustained static contractions or arm work with arms raised above the head), vasodilator factors accumulate in the muscles, stimulate sensory nerves and produce dramatic increases in blood pressure.

The time course of the changes in cardiac output during a bout of moderate exercise is shown in Fig. 16.3. There is a sudden increase in cardiac output at the start of exercise, followed by a gradual exponential rise to the steady state; a sudden decrease in cardiac output and an exponential fall occur when exercise is stopped. The sudden changes at the onset and cessation of exercise are attributed to the effect of the muscle pump on venous return, and also to stimulation of the cardiovascular system from the neural activity associated with movement. The slower changes reflect the time course of vasodilatation

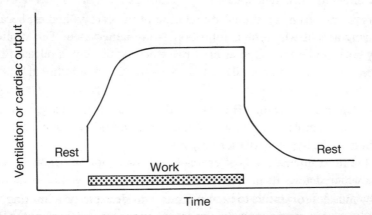

Fig. 16.3. Time-course of abrupt and slow changes in ventilation and cardiac output during exercise.

in active muscles and of stimulation of the cardiovascular system via the sensory nerves which detect the vasodilator factors in the muscles.

16.3 Respiratory responses in exercise

Pulmonary ventilation ($\dot{V}_E$) increases with light to moderate exercise in proportion to the increased demand for the supply of O_2 and the excretion of CO_2 (Fig. 16.2). This involves increases in both the frequency of breathing and tidal volume. The transient response of $\dot{V}_E$ during a bout of exercise (Fig. 16.3) shows rapid and gradual changes similar to those for cardiac output.

The nature of the stimuli that produce the increase in ventilation with exercise is still uncertain. The sudden changes in $\dot{V}_E$ at the beginning and end of exercise appear to result from neuronal input to the respiratory centre from 'movement' receptors in the exercising limbs and possibly also from neural activity in the motor cortex. However, this neural drive appears to play only a minor role in the steady-state response, which has been shown to be coupled very closely to the delivery of CO_2 to the lungs. Detection of CO_2 by a chemoreceptor is therefore implicated, but a venous or pulmonary CO_2 receptor driving ventilation cannot be demonstrated. Arterial chemoreception also seems to be ruled out, because an increase in mean arterial P_{CO_2} would be necessary to provide the stimulus for the increased $\dot{V}_E$, and this is not observed (Fig. 16.2). This has led to the suggestion that the sensitivity of chemoreception by the carotid bodies is altered during exercise; another suggestion is that the increase in venous CO_2 load increases the rate of change of arterial pH associated with the breath-to-breath oscillations in arterial P_{CO_2}, and that this is sensed by the carotid bodies and provides the primary stimulus to ventilation.

At higher levels of exercise there are disproportionate increases in $\dot{V}_E$ and $\dot{V}_{CO_2}$ compared with increases in $\dot{V}_{O_2}$ (Fig. 16.2). This 'hyperventilation' appears to arise from additional stimulation of the carotid bodies, because it is virtually absent when the carotid bodies are denervated. The stimulus is usually taken to be the fall in arterial pH when lactic acid is released by the muscles, and the hyperventilation is viewed as an additional buffering mechanism which lowers arterial P_{CO_2} and thereby prevents arterial blood from becoming too acidic. However, other substances released by active muscles, for example K^+, may also stimulate the carotid chemoreceptors and contribute to the hyperventilation response.

At the most intense levels of exercise, O_2 consumption begins to reach a plateau which defines its maximum value, the $\dot{V}_{O_2max}$ (Fig. 16.2). Utilization of O_2 by muscles (oxidative capacity) is not considered to be a limiting factor for $\dot{V}_{O_2max}$, and in most people the limitation appears to be the supply of O_2 to muscles by cardiac output. However, in some highly trained athletes significant arterial hypoxaemia ($P_{O_2} < 70$ mmHg) develops as $\dot{V}_{O_2max}$ is approached, indicating that for these individuals the limitation may be in the lungs. The hypoxaemia results from a combination of a reduced hyperventilation response and a reduced transit time for blood in the pulmonary capillaries.

16.4 Applied exercise physiology

Physiological indices of performance. Some measurements made by the physiologist in the laboratory under defined conditions can be useful indices of the current performing ability of athletes. However, in sprint or power events the relationship between physiological measures and physical performance is not good. For example, oxygen deficit, oxygen debt and blood lactate concentration in short-term maximal exercise tests might be expected to correlate with performance in sprint or power events, but the relationship turns out to be poor. On the other hand, the performance of endurance athletes correlates well with several physiological measures. One of these is the *maximum oxygen uptake*, $\dot{V}_{O_2max}$, which is usually determined by exercising the subject gradually to maximum on a treadmill or cycle ergometer over a 5–15 min period. A subject with a high $\dot{V}_{O_2max}$ would be expected to perform endurance exercise at a high intensity, and the correlation between endurance performance and $\dot{V}_{O_2max}$ is found to be approximately 0·8. However, this is not a sufficiently high correlation to allow the $\dot{V}_{O_2max}$ to be used to compare the performing ability of one athlete with another. Currently this can be done only with one physiological index, the *anaerobic threshold*. This is essentially the workload above which the concentration of lactic acid in the blood builds up rapidly. The high correlation between the anaerobic threshold and endurance performance (approximately 0·95) arises most probably because the anaerobic threshold represents the highest level of exercise that can be performed without fatigue occurring from the accumulation of acidity in the muscles. It is not clear whether the anaerobic threshold is the threshold for the onset of hypoxia in exercising muscles, because it is possible for muscle cells to produce ATP and lactic acid via anaerobic metabolism when aerobic metabolism is not limited by the supply of O_2.

Training can be for strength or short-term power (anaerobic training) or for endurance (aerobic training). The various changes that occur can be regarded as adaptations to the stress of exercise, but the mechanisms that bring about the changes are poorly understood. Anaerobic training, which consists of many repetitions of brief, intense exercise, results in increases in muscle fibre diameter and activities of enzymes of anaerobic metabolism. Increases in blood pressure during anaerobic training may also produce thickening of the left ventricular wall of the heart. Aerobic training is achieved by exercising at moderate to high intensity for at least 20 min several times per week. The most well-known adaption to such training is an increase in the $\dot{V}_{O_2max}$, which is attributed to an increase in maximum cardiac output and maximum blood flow to the muscles. The increase in cardiac output is itself a consequence of an increased stroke volume with no change in maximum heart rate (see Table 11.2, p. 409). The stroke volume is increased at all work loads, including rest, as a result of an increase in circulating blood volume and in the volume of the ventricles. Cardiac output at rest does not change, so the increased stroke volume entails a reduced resting heart rate. Endurance-trained muscles have significant increases in the concentration of oxidative enzymes and myoglobin and in the density of mitochondria and possibly also of capillaries (p. 112); these changes may raise the anaerobic threshold, but in maximum exercise there is no decrease in the P_{O_2} of the venous blood leaving trained muscles.

Physical fitness for the athlete usually means current ability to perform, and for endurance athletes this can be assessed as the anaerobic threshold or less well as the $\dot{V}_{O_2max}$. Tests which estimate the $\dot{V}_{O_2max}$ directly or indirectly have been used widely to assess fitness for non-athletes, but these are not necessarily appropriate for the average person. Some middle-aged people regard being fit as being the optimum weight for their height, while older people regard it as being physically healthy. However, the majority of people consider fitness to mean their level of physical activity. Unfortunately neither $\dot{V}_{O_2max}$ nor any other physiological parameter correlates well with the level of physical activity, so there is currently no physiological index that can be used as an objective measure of the average person's physical fitness.

Appendix: Units of Measurement

Throughout this text the units are those commonly used by physiologists. Unfortunately, physiologists and their clinical colleagues have been slow to adopt the now accepted SI units familiar to chemists and physicists. To assist readers trained in this system (and those who believe that physiology should use a logical and consistent system), a summary of SI units and conversion for common units used by physiologists follows.

1. Basic SI units

	Unit
Length	metre (m)
Mass	kilogram (kg)
Time	second (s)
Electric current	ampere (A)
Thermodynamic temperature	kelvin (K)
Amount of substance	mole (mol)
Luminous intensity	candela (cd)

2. Derived SI units and relationships to other units in common use

	Unit	Definition	Other units
Area		m^2	
Volume		m^3	1 litre (l) $= 1$ dm^3
Velocity		$m\,s^{-1}$	
Concentration		$mol\,dm^{-3}$	
Force	newton (N)	$m\,kg\,s^{-2}$	1 dyne (dyn) $= 10^{-5}$ N
Pressure	pascal (Pa)	$N\,m^{-2}$ $(m^{-1}\,kg\,s^{-2})$	1 atmosphere (atm) $= 101\cdot325$ kPa $= 760$ mmHg; 1 mmHg $= 1$ torr $= 133\cdot32$ Pa; 1 $cmH_2O = 98\cdot07$ Pa
Energy	joule (J)	$N\,m$ $(m^2\,kg\,s^{-2})$	1 calorie (cal) $= 4\cdot19$ J
Power	watt (W)	$J\,s^{-1}$ $(m^2\,kg\,s^{-3})$	
Dynamic viscosity		$Pa\,s$ $(m^{-1}\,kg\,s^{-1})$	1 poise (P) $= 10^{-1}$ Pa s
Surface tension		$N\,m^{-1}$ $(kg\,s^{-2})$	1 dyne $cm^{-1} = 10^{-5}$N cm^{-1}
Electric charge	coulomb (C)	$A\,s$	
Electric potential difference	volt (V)	$J\,C^{-1}$ $(m^2\,kg\,A^{-1}\,s^{-3})$	
Electric resistance	ohm (Ω)	$V\,A^{-1}$ $(m^2\,kg\,A^{-2}\,s^{-3})$	
Electric conductance	siemens (S)	Ω^{-1} $(A^2\,s^3\,m^{-2}\,kg^{-1})$	
Electric capacitance	farad (F)	$C\,V^{-1}$ $(m^{-2}\,kg^{-1}\,A^2\,s^4)$	
Frequency	hertz (Hz)	s^{-1}	

3. Physical constants

Name	Symbol	Numerical Value
Gravitational constant	G	$6 \cdot 67 \times 10^{-11} \text{N m}^2 \text{ kg}^{-2}$
Universal gas constant	R	$8 \cdot 314 \text{ J K}^{-1} \text{ mol}^{-1}$
		(at $P_o = 101 \cdot 3$ kPa, $T_o = 273 \cdot 15$K)
Molar volume of ideal gas	V_o	$2 \cdot 24 \times 10^{-2} \text{ m}^3 \text{ mol}$
		(at $P_o = 101 \cdot 3$ kPa, $T_o = 273 \cdot 5$K)
Avogadro constant	N_A	$6 \cdot 02 \times 10^{23} \text{ mol}^{-1}$
Faraday constant	F	$9 \cdot 65 \times 10^4 \text{ }^\circ\text{C mol}^{-1}$

Note: R is also sometimes expressed as $8 \cdot 206 \times 10^{-2}$ l atm K^{-1} mol^{-1}.

4. Decimal multiples and submultiples of units

kilo	(k)	for 10^3
hecto	(h)	for 10^2
deca	(da)	for 10^1
deci	(d)	for 10^{-1}
centi	(c)	for 10^{-2}
milli	(m)	for 10^{-3}
micro	(μ)	for 10^{-6}
nano	(n)	for 10^{-9}
pico	(p)	for 10^{-12}
femto	(f)	for 10^{-15}
atto	(a)	for 10^{-18}

Note: 1 Angstrom (Å) = 10^{-10} metre

Index

A band 98
a wave 395, 398, 443, 459
Abdominal muscles 467
Abducent nerve 80, 150
ABO system of blood grouping 350, 351–352
Absolute refractory period
  of cardiac muscle 382–383
  of nerve 43, 45
Absorption
  definition 25
  in gastrointestinal tract 600–609
Absorption coefficient (gas) 488
Absorptive epithelia 26–28
Absorptive state 290, 613–614
Acceleration
  receptors 140
  reflexes evoked by 204–205
Acclimatization 504, 508–509
Accommodation 148
Acetazolamide 560
Acetylcholine 61
  actions on
    autonomic ganglion cells 68
    biliary secretion 595
    blood vessels 430, 454
    gastric secretion 588, 589
    heart 384
    intestinal secretion 596
    intestinal smooth muscle 119
    motor endplate 61
    pancreatic secretion 591, 593
    salivary secretion 584
  blocking of 62, 72
  degradation 63, 72
  receptors 9, 94
    muscarinic 68, 72
    nicotinic 62, 68
    patch-clamp channel recording 31
  release 61
  synthesis 61
Acetylcholinesterase 63
Achlorhydria 598
Acid–base balance 497, 550–560
  and potassium 547, 549, 560
  renal regulation of 551–552
  respiratory regulation of 497, 551–552
Acidosis 550–552, 559
  effect on
    cardiovascular system 400, 444
    cellular K^+ uptake 547–549
    renal tubular cells 555

respiratory system 501, 503, 504, 551, 552, 559, 560
Acquired immune deficiency syndrome (AIDS) 341, 344
Acromegaly 262
Acrosome 297
ACTH see Adrenocorticotrophic hormone
Actin
  in axonal transport 56
  in cardiac muscle 378, 399
  in platelets 357
  in skeletal muscle 97–101, 106
  in smooth muscle 116
Action potential 13, 40, 43–52
  in cardiac muscle 380–382
  delayed rectification 45
  ionic basis 44–46, 93
  in nerve 43
  propagation (conduction) 46–48
  in skeletal muscle 102
  in smooth muscle 117–118
  spontaneous 52
  threshold 43
Active immunity 347
Active transport
  importance of 12–13
  mechanism 11
  primary 9–12
  secondary 9, 12
Acupuncture 145
Acute inflammation 341
Adaptation
  dark 152
  light 153
  of receptors 126
Addison's disease 285, 461, 545
Adenohypophysis 258
Adenosine 3′, 5′-monophosphate (cyclic AMP) 72, 254–256, 530, 584, 592, 596
Adenosine triphosphate (ATP)
  in active transport 11
  in skeletal muscle 100, 102, 109, 626
Adenylate cyclase 255, 530
ADH see Antidiuretic hormone
Adrenal glands 251, 279–287
  cortex 280, 281–287
    function disorders 285
  medulla 75, 279–281
Adrenaline 66, 70–71, 75, 280–281, 290
  action on
    blood vessels 430
    cardiac contraction 399–401, 406

Adrenaline (*continued*)
　action on (*continued*)
　　coronary circulation　450
　　glycogenolysis　74, 281
　　heart　384, 406
　　insulin release　289
　　skeletal muscle circulation　454
　　stroke volume　462
　and basal metabolic rate　611
　and cardiac failure　463
　and exercise　628
　and hypertension　460
　and postural changes　459
Adrenoceptors　70–75
　in blood vessels　430, 431, 434, 453–456
　in cardiac muscle　399
Adrenocorticotrophic hormone
　　(ACTH)　258, 260, 264, 282, 285–287
Adrenogenital syndrome　287
Adsorptive endocytosis　16
Aerobic metabolism
　in exercise　626–628
　exercise training　112, 633–634
　in skeletal muscle　102, 112
Afferent inhibition　134
After-images　154
Afterload
　aortic　402–403
　arteriolar　404
　in cardiac muscle　400, 437–464
　in skeletal muscle　107
Ageing
　effect on arterial distensibility　417, 424
　effect on skeletal muscle　113
Agglutination, red cells　349
AIDS *see* Acquired immune deficiency
　　syndrome
Air flow　476–478
　laminar　476
　turbulent　476
Airways resistance 476–477
Albumin, in generalized oedema　542, 544
Alcoholic cerebellar degeneration　218
Aldosterone　282
　in blood volume control　442, 459, 462
　and cardiac failure　464
　control by renin-angiotensin system　285,
　　442, 540–542
　effects on
　　gastrointestinal tract　596, 603, 605
　　renal distal tubule　537
　　salivary secretion　541, 583
　　sweat gland secretion　541
　and hypertension　460
　and potassium　546, 547, 549
　third factors　541–542
Alkalosis　504, 507, 551–552, 559–560
　effect on cellular K^+ uptake　547, 548–549
Allelomorphic genes　351
Allergic reactions　342

Alloxan　290
Alpha motor neurones　190, 194, 197, 212,
　　220
Alpha rhythm　224
Alpha-amylase　581, 582, 590, 591
Alpha-granules, platelets　357
Alpha-reductase deficiency　294
Altitude　500, 507–509
　and pulmonary hypertension　448, 509
　renal response to　509, 552
　and water loss　536
Alveolar
　dead space　484
　ducts　467
　epithelium　467
　gas　479, 481–484
　pressure　472, 510
　sacs　467
　surface area　488
Alveolar ventilation　480–486
　and gas equation　484–486
　and perfusion ratio　487–488
Alveolar–arterial
　CO_2 gradient　484
　O_2 gradient　483, 484, 487
Alveolo-capillary membrane　466, 467
Alveolus　466
　wall structure　469
Amacrine cells　154–155, 158
Ambient pressure, increased　509
Ambient temperature and pressure (STPD),
　　dry, gas　480
Ambiguus nucleus　498
Amino acids
　and insulin release　288
　intestinal handling　604
　renal handling　524–525
Aminoacidurias　524
Ammonia　557–558
　renal synthesis and secretion of　526
Ammonium, production　556–557
Amnesia　235, 238
　anterograde　238
　post-traumatic　231
　retrograde　239
Amorphic genes　350, 351
Ampullae of semicircular canals　178, 180
Ampullofugal movement　205
Ampullopetal movement　205
Amygdaloid nucleus　237, 244
Amylase　567
Anaemia　334–337
　causes　334–336
　compensatory adjustments in　337
　red-cell indices in　336–337
　types of
　　haemoglobinopathies　336
　　haemolytic　336, 349
　　hypoproliferative　335
　　iron-deficiency　324

macrocytic normochromic 337
microcytic hypochromic 337
pernicious 325, 586
sideroblastic 336
thalassaemia 336
Anaerobic metabolism
in exercise 626–628
in skeletal muscle 102
Anaerobic threshold 663
Anaesthetic, local 46, 51, 129
Anal sphincter 199
external 579
internal 579
Analgesia 144–145
Anatomical dead space, respiratory 481
Androgen-insensitivity syndrome 294
Androgens 282
Androstenedione 282, 302
Aneurysm 414
Angina pectoris 463
Angiotensinogen 540
Angiotensins I and II
action on blood vessels 431, 434
in blood volume control 434, 442, 462
in cardiac failure 464
control of secretion 442, 540–541
effect on aldosterone secretion 285, 540–541
interconversion in lung 465
in thirst-ADH mechanism 536
Anisocytosis 328
Ankylosis of ossicle chain 177
Annulospiral (primary) endings 193
Ano-intestinal inhibitory reflex 578
Anomic aphasia 242
Anterior cardiac vein 337
Anterior pituitary 258–265
hormones 258–260
secretion control 260–261
Anterolateral pathway 129, 131–134
Anti-inflammatory properties of
corticosteroids 284
Antibody
blood-group 350–351
functions 347
production 344–346
response 344–345
Anticoagulants 336
Antidiuresis 265–267, 530–531, 536–537
Antidiuretic hormone (ADH) 265, 267, 285, 431, 530–534
actions on blood vessels 266, 431
on renal tubules 266, 530
in blood volume control 434, 442–443, 462
control of secretion 266–267
in hypernatraemia 536–537
in hyponatraemia 537–538
and thirst and body water regulation 535–538
Antigen-binding site 345

Antigenic determinant 342
Antigens 342
blood-group 349–350, 581
histocompatibility 348
Antiglobulin test 351
Antiplasmins 366–367
Antithrombin III 355, 365
Antitrypsin 365
Antrum 302
Anuria 520, 537
Anykin 328
Aorta 374
flow in 414, 421–423
pressure in 395–397, 402–403, 425–428
pressure pulse 423
properties of walls 417–420
stretch and recoil of 374, 423–424
Aortic
baroreceptors 440–442, 536, 540
body 444, 501–502
semilunar valve 376, 395, 396, 402
incompetence 397
stenosis 397, 451, 461
Aphasia 240–242
anomic 242
conduction 241
motor (Broca's) 241
non-fluent 241
sensory (Wernicke's) 241
Apneustic centre 498–499
Apraxia 208
APUD cells 597
Aqueous humour 146
Arachidonic acid 357, 360
Arachnoid mater 84, 451
Arbor vitae 81
Arrhythmia, cardiac 385–387
Arterial
baroreceptors 440–442
chemoreceptors 444, 501–502
gases 483–488, 490, 493, 507
hypoxaemia 506
mean blood pressure (MAP) 425, 439, 441–447
pH 495
renal plasma flow (RPF) 514
Arteries
blood pressure
determinants 427–428
effect of posture 457–458
effect on renin release 540
effect on thirst-ADH mechanism 536
measurement 425–426
regulation of 440–442
blood volume 420
compliance (distensibility) 417
dimensions of 418
flow pulse 423
pressure pulse 423–424
properties of walls 417–420

Arteries (*continued*)
  systemic circulation   422–428
Arterioles
  blood pressure   442
  critical closing pressure   417
  functions of   374–375, 412, 417, 428–431
  glomerular   513
  properties of walls   417–420
Arteriovenous anastomoses   432, 453
Arteriovenous oxygen difference   492
Articulation   240
Aspirin   602
  and body temperature   624
Asthma   476, 505
Astigmatism   148
Astrocytes   38
Ataxia   218
Atherosclerosis, in diabetes   290
Athetosis   212
Athlete, cardiovascular variables   409, 631
ATP *see* Adenosine triphosphate
ATPS, gas volume   480
Atrial
  'A' receptors   443
  'B' receptors   442, 536, 539
  contraction   393, 395, 405
  depolarization   381, 389
  fibrillation   386
  flutter   386
  natriuretic peptide (ANP) 372, 443, 542
  paroxysmal tachycardia   386
  pressure   395–397
    mean right atrial   435, 437, 458
  repolarization   381, 389
  septal defects   397
Atrioventricular (AV)
  node   379–382, 385–386
  ring   377, 396
  valve   376, 395, 396
    incompetence   397
    stenosis   397
Atrium   375
Atropine   72
Audiometer   178
Auditory
  central pathways   176
  cortex   83
  nerve   174–176
Auerbach's (myenteric) plexus   567
Aura, in epilepsy   232
Auricle   170
Auscultatory gap   426
Autoimmune diseases   348
Autonomic nervous system   3, 64–75
  control of
    arteriolar resistance   430
    bronchomotor tone   476
    gastrointestinal motility   570
    heart rate   383–385
    stroke volume   406

cotransmitters   70
  enteric division   64
  ganglia   64–8
    synaptic transmission   68
  neuroeffector transmission   69
  parasympathetic division   64
  sympathetic division   64–66
Autonomic reflex(es)   67, 199–200
Autoregulation
  of cerebral blood flow   452
  of gastrointestinal blood flow   455
  metabolic   429, 432
  myogenic   417, 429
  of renal blood flow   456, 515
  of smooth muscle   120
Auxotonic contraction   402
Avogadro's number   92
Axon   36
  electronic spread of potential   43
  growth cone   57
  structure   60
Axon hillock   36
Axon–axonal reflex   454
Axonal growth   57
Axonal transport   56–57
Azotaemia   563
Azothioprine   348

B-lymphocytes   317, 342–343
  and antibody production   344–346
Babinski's sign   222
Bainbridge reflex   443
Balance disorders   218
Ballismus   212
Barometric pressure, decrease in   507–508
Baroreceptor(s)   375
  aortic   440
  arterial   440
  carotid   440
  reflexes   440–442, 459, 462, 463, 536
Barr body   293
Basal ganglia   82, 209–211
  functions   211–212
Basal metabolic rate (BMR)   271, 611
Basement membrane   6, 59
Basilar membrane   171–174
Basket cells   214
Basophils   316, 338–339
Bed-wetting   230
Behaviour
  organization of   242–249
  subcortical control   245 246
Bends   511
Berger waves   224
Bernoulli's equation   413–414
Betz cells   220
Bezold–Jarisch reflex   443
Bicarbonate   552–554

in acid–base balance 497, 548, 550–551
in bile 594
in carbon dioxide carriage 494–495, 497
filtered load 555–556
plasma 560
renal handling of 534, 554–556
Biceps muscle 198
Bicuspid valve 376
Bile, canaliculi 593
Bile pigments 594
Bile salts 594–595, 607
Biliary secretions 593–595
formation 594–595
Bilirubin 333, 593, 595
Biliverdin 333
Binocular vision 161–162
Bioassay, hormonal 291
Bipolar cells 154–156
Bladder, urinary
reflexes in spinal shock 199
Blastocyst 310
Bleeding disorders 367–369
Bleeding time 369
Blind spot 154
Blood 314–370
buffering in 553
cells 315
production 316–318
red 318–337
white 338–348
coagulation 360–366
extrinsic system 362–363
factors 361, 364
inhibitors 364–365
intrinsic system 362–363
compatibility tests 353
composition 315
cross-matching 349, 353
group(s) 349–354
ABO system 350
antigens and genes 349–350
Rhesus system 349
tests 349, 353
haematocrit 315, 415, 495
plasma 2, 3, 315
reservoir 420
transfusion 349, 353–354
viscosity 411, 415
Blood flow
capillary circulation 431–432
control by arterioles 428–431
deviations from predicted 414–417
distribution 372
effective viscosity 415
in exercise 628–632
gravitational potential energy 414
hormonal control 430–431
humoral control 429
kinetic energy 413
laminar 411, 414

linear velocity 412, 421
metabolic autoregulation of 429, 432
myogenic autoregulation of 417, 429
nervous control 430
physics 410–414
potential energy 413
pulsatile 414, 421
turbulent 414–415
Blood group, antigens 349, 581
Blood pressure 422
diastolic 425
in exercise 629–631
hydrostatic 456–458
mean arterial (MAP) 425, 439, 441–447
pulse 423
systolic 425
Blood vessels
cross-sectional area 420
elastic properties 417–418
passive stretch 416
passive wall tension 419
pressure-flow relationships 416
pressure–volume relationships 418
relative dimensions 418
resistance 422
transmural pressure 419
types 374–375
Blood volume 316, 420, 432
central 420
control after haemorrhage 461–462
distribution 420, 459
long-term regulation 442, 443
receptors 442
reservoir 420, 448
Blood–brain barrier 84, 451, 503
Blood–CSF barrier 84
Body temperature 563–568
Body water 2, 535–538
Bohr
effect 329, 330, 491
equation 483
Bone 276, 277
Bone-marrow production lines 317
Bony labyrinth 178
Botulinus toxin 61
Bowman's capsule 517
Boyle–Charles formula 480
Bradycardia 384
Bradykinesia 211
Bradykinin 119, 364, 429, 453, 455, 584
Brain 76–85
capillaries 6
EEG 223–225
diagnosis in death 225
Brainstem 78–81
transection, effects 205–207
Breathing, work of 477
Broca's aphasia 241
Bronchi 466
Bronchial circulation 449, 484

Bronchioles 446, 488
Bronchoconstriction 476, 505
Brunner's glands 595
BTPS, gas volume 480
Buffers 497, 550–555
Bulbgastrone 600
Bulbospinal fibres 200
Bulbourethral glands 295
Bundle of His 379–380
Bursa of Fabricius 343
Burst-forming unit (BFU-E) 317

c wave 395, 398, 459
Calcarine fissure 83
Calcitonin 268, 273, 278
Calcitriol 274, 276–277
Calcium 273–279
  balance 275
  in blood coagulation 358–360
  in cardiac muscle 380–385, 399–400
  in exocytosis 15, 61, 252
  extracellular, regulation of 276–278
  in hormone action 256
  intestinal absorption of 277, 605
  intracellular, regulation of 13–15
  metabolism 274–276
    disorders of 278–276
  plasma 278
  renal handling of 524
  in skeletal muscle 100–101
  in smooth muscle 117, 120
Calmodulin 14, 116, 256
Caloric test of vestibular function 187
Capacitance of cell membrane 89
Capacitation 310
Capillary(ies)
  blood (hydrostatic) pressure in 22–23, 422, 423
  circulation through 420, 431–432
  diffusion across 4–5, 432
  exchange across 5–6
  fenestrations 5, 517
  fluid exchange across 21–24, 432–434
  network 375, 431
  properties of wall 418–419
Carbamino-complex 495
Carbon dioxide
  absorption coefficient 488
  alveolar-arterial gradient 484
  blood concentration 493
  blood dissociation curve 493
  carriage 493–497
  fraction of respiratory gases 479
  narcosis 500
  partial pressures in alveoli and blood 484
  production 479
    in exercise 632
  solubility in plasma 497, 551
  total amount in body 497

Carbon monoxide poisoning 493, 502
Carbonic acid 552
Carbonic anhydrase 86, 495, 504, 555, 560
Carboxyhaemoglobin 493
Cardiac *see also* Heart
  arrhythmias 385–387, 549
  conduction system 377–380
  cycle 393–397, 402
  failure 463–464
  function curve 404, 437
  muscle
    electrical characteristics of 377–393
    length–tension curve 398–400
    load–velocity curve 400–401
  output 409–410
  stretch receptors 442–443
  tamponade 406
Cardiovascular centres 444–447
Cardiovascular system 371–464
  basic design and functions 372–375
Carotid
  baroreceptors 440–441
  body 444, 501–502
  chemoreceptors 444, 501
  sinus 440
Carrier-mediated transport 9
Castration 298
Catechol-O-methyltransferase (COMT) 73, 281
Catecholamines 71–73, 280–281
  actions 281
  circulating 75
  control of secretion 281
  synthesis 69–70
Caudate nucleus 209, 210
Cell
  death, neuronal 57
  dimensions 1
  volume regulation 18–21
Cell-mediated immunity 344
Central blood volume 420
Central nervous system (CNS) 34
  design of 76–88
  regeneration 58
Central sulcus 83
Central venous pressure 435, 459
Centrencephalic epilepsy 232
Cerebellar
  cortex 81, 212–214
  climbing fibres 213
  hemisphere 81
  mossy fibres 215
  peduncles 81
Cerebellum 76, 78, 81–82, 212–218
  alcoholic degeneration 218
  functional zones 217
  functions 217–218
  synaptic relationships 214
Cerebral
  aqueduct 84

blood flow 374, 451–453, 458, 460
  cortex 83, 208–209
  hemispheres 76, 83, 208–210, 237
  peduncles 79
Cerebrospinal fluid (CSF) 84–88
  composition 86
  flow 85–86, 451
  functions 87–88
Cerebrum *see* Cerebral hemisphere
Cervix 301
Channels 8–9, 29–31
  gated 8, 41–42
  leak 8
  neurotransmission 41, 55
Chemoreceptors 123–124
  respiratory
    central (intracranial) 444, 502–504
    peripheral (arterial) 444, 501–502
Chemotaxins 341
Chewing 571
Chloride 2
  in gastric acid secretion 586–587
  in intestinal secretion 586, 605
  in pancreatic secretion 592
  renal handling of 527–529, 534
  shift 495
  transport across epithelia 26–29
Chlorolabe 151
Cholecalciferol 276
Cholecystokinin (cck) 578, 591, 593, 595,
    597, 598, 599
Cholera toxin 257, 596
Cholesterol 282, 608
Choline 63, 93
Choline acetyltransferase 61
Cholinoceptors 68, 72
Chordae tendineae 376
Chorea 212
Chorion 310
Choroid of eye 146
Choroid plexuses 85–87, 451
Chromatolysis 58
Chromogranins 69
Ciliary body 146
Ciliary muscle 148
Cingulate gyrus 84, 244
Cingulotractotomy 244
Circadian rhythm 309
Circumventricular organs 84
Circus movement, heart 387
Cisterna magna 84
Claustrum 209
Clearance 520–521
Climbing fibres 216
Clitoris 301
Clo 617
Clone 343
Clonic phase of epilepsy 232
Clonus 221
Clot retraction 359

Clubbing of fingers 509
Coagulation factors 361–366
  inherited deficiencies 369–370
Cobalamin 324
Cocaine 73
Cochlea 171, 172
Cochlear
  duct 171
  fibres, stimulation 172–176
  nucleus 176
Cogwheel rigidity 212
Cold receptors 141
Colliculi
  inferior 79, 176
  superior 79
Colloid osmotic pressure 20, 432, 433
  colon 567
  motility in 579
  secretions of 596
Colon 567, 578–580
Colony-forming units (CFU) 317
Colony-stimulating factors 340
Colour sensation 162–163
  trichromatic theory 162
Colour vision, defects 163
Coma 231
  diabetic 290
  insulin 290
Compensatory pause, heart 386
Compliance
  blood vessel 417
  chest wall 473
  lung 473, 474, 476
  ventricular 406
Complement, functions 347
Complex cells in visual cortex 159
Compound action potential 49–50
Concave lens 149
Concussion 231
Conductance of cell membrane 89
Conduction
  of action potential 42—51
  block 46, 51, 129
  in anterolateral pathway 134
  aphasia 241
  deafness 176–178
  in dorsal column 131
  saltatory 48
  velocity 47, 51, 128
Congenital adrenal hyperplasia 287, 294
Cone-bipolars 155
Cones 150–151
Conn's syndrome 286–287
Consciousness 223–226
  disturbances of 231–234
  variations in level 225–226
Consensual light reflex 149
Contraception, hormonal 308
Contractile process
  cardiac muscle 378

Contractile process (*continued*)
  skeletal muscle   103–109
  smooth muscle   116–117
Convergence
  of eyes   148
  neuronal   52
Convex lens   149
Convulsion   232
Coombs' test   351
Cornea   146, 147
Coronary blood flow   377, 449–451
  atherosclerosis   451, 463
  sinus   377
Corpus callosum   83, 84, 236–237, 246–249
Corpus luteum   303, 304, 307
Corti, organ of   171
Corti cells   171
  stimulation of   172–174
Corticospinal tract   200–201
  direct   201
  indirect   201
Corticosteroid-binding globulin   283
Corticosteroids   73, 280, 282–287
  in sleep   228
Corticosterone   282
Corticotrophin *see* Adrenocorticotrophic
    hormone
Corticotrophin-releasing hormone
    (CRH)   260, 285
Cortisol   282, 285
  and blood glucose   290
  fetal secretion   313
Co-transport   12, 13, 27–29
Cough   505
Counter-current mechanism
  in kidney   532
  in skin   623
Counter-transport   12, 13, 28
Cramping pains   143
Cranial nerve(s)   80–82
Creatine phosphate   102
Creatine phosphokinase   102
Creatinine clearance   520
Cretinism   267, 273
Crista ampullaris   180, 181
Critical closing pressure   417
Critical fusion frequency   154
Crossed extensor reflex   203
Cross-matching   349, 353
Cryoprecipitate   354
Cryptorchidism   296
Cuneocerebellar tract   215
Cupula   180
Curare   62
Cushing reflex   452
Cushing's syndrome   286
Cutaneous
  blood flow   374
  circulation   453–454
  deep venous plexus   453

heat transfer   446, 453, 623
  triple response   454
Cyanolabe   151
Cyanosis   463, 508, 509
Cyclic AMP   72, 254–256, 530, 584, 592, 596
Cyclic GMP   153, 256
Cyclic nucleotide phosphodiesterase   255
Cyclo-oxygenase   360
Cyclosporin   348
Cylindrical lens   149

Dalton's law   483
Dark adaptation   152–153
Dark cells   180
Daydream   223
Dead space   481, 483
  alveolar   484
  anatomical   481
  physiological   484
  ventilation   481
Deafness   176–178
  conduction   176–178
  sensorineural   176, 178
Decerebrate rigidity   206, 207
Decibel   168
Decompression sickness   511
Decorticate rigidity   206
Defaecation   199, 579
Defence reaction centre   446
Defibrillation (heart)   387
Dehydration   433, 536
Dehydroepiandrosterone   282, 283
Deiter's nucleus   186
Delayed rectification   45
Delayed-type hypersensitivity (DTH)   344
Delta waves in sleep EEG   226, 227
Demyelination   51
Dendrites   35
Dendritic cells   343
Denervation, of skeletal muscle   113–114
Dentate nucleus of cerebellum   215
Depolarization of membranes   42
Depressor area, cardiovascular   445, 446
Depth perception   162
Dermatome   129
Descending inhibition   136
Desipramine   73
Detrusor muscle   199
Deuteranomaly   163
Deuteranopia   163
Diabetes insipidus   267, 530
  nephrogenic   537
  neurogenic   537
Diabetes mellitus   287, 290, 532, 545, 551,
    556, 559
Diabetic coma   290
Diacylglycerol   256, 257
Dialysis, renal   563
Diamox   560

Diaphragm 467
Diarrhoea 545, 596, 600, 604
Diastasis 397, 405
Diastole 373, 394–397
Diastolic
  end volume 395, 401
  filling 394
  murmur 397
  pressure (arterial) 425–426
  regurgitation 397
Dichromats 163
Dicrotic
  notch 424
  wave 424
Diencephalon 76, 78, 81, 83
Diffusion 4–5
  capacity 489
  coefficient 4
  concentration gradient 4
  'downhill' transport 5
  facilitated 9, 601
  fluxes 4
  liquid solubility 601
  of gas 289, 488–489
  potential 16–18
  rate 5
Digestion 565–567, 580–596
Digestive tract
  absorption from 600–609
  anatomy 565–567
  intrinsic and extrinsic nerves 569
  motility 570–580
  muscles 567
  pH profile 590
  secretions 580–596
  structure 567–569
Digoxin 464
Dihydrotestosterone 294, 298
Dihydroxycholecalciferol (1,25-DHCC) *see*
    Calcitriol
Dilution principle 315
Diodone 515
  clearance 525
Diphosphoglycerate (2,3-DPG) 330, 491, 509
Diphtheria 51
Direct light reflex 149
Discrimination, immune response 342
Discriminative pathway 130
Disorientation 231
Distensibility
  of blood vessels 417, 427, 434
  of lung 471
  ventricular 406
Diuresis 266, 290, 443, 530–534
Divergence, neuronal 52
Diving
  breathing in 510
  reflex 444
Dizziness 452, 460, 508
DNA, impaired synthesis 326

Doll's head phenomenon 204
Dominance, hemispheral 240, 249
Donnan excess 20, 22, 32
Donnan ratio 31
Dopamine 70, 209, 212, 260, 502
Dorsal, respiratory group (DRG) 498–499
Dorsal column
  nucleus 130
  pathway 129–131
Dorsal horn 129–130
Dorsal root 76
Dorsal spinocerebellar tracts 215–216
Dreaming 229
Duodeno-colic reflex 579
Duodenum
  absorption in 602–609
  hormones 599–600
  secretions 595
Dura mater 84, 451
Dwarfism 261, 263
Dynorphins 145, 264
Dysarthria 240
Dysphasia 241
Dyspnoea 448, 463, 508, 509, 511

Ear, functional anatomy 170–172
Early receptor potential 152
Eating, control by amygdala 246
ECF *see* Extracellular fluid
ECG *see* Electrocardiogram
Echocardiography 410
Ectopic foci 386
EEG *see* Electroencephalogram
Eidetic image 234
Einthoven triangle 388
Ejaculation 300
Elastic coefficient, blood vessels 417
Electro-oculogram 154
Electro-olfactorogram 167
Electrocardiogram (ECG) 387–393
  in hyper- and hypokalaemia 549
Electrochemical potential gradient 10, 25,
  27
Electroconvulsive shock therapy 232
Electrocorticogram 223
Electroencephalogram (EEG) 223–225
Electrogenic pump 18
Electromyography 110
Electroretinogram 152, 154
Electrotonic potentials 42–43, 91
Emboliform nucleus of cerebellum 215
Embolism 354
Embryo, development 310
Emission of semen 300
Emotion 244
  and cardiorespiratory responses 447, 454,
    500
  and motor neurones 202
Encephalitis lethargica 225

End-diastolic volume 395, 401–406, 409
End-systolic volume 396, 401–405, 409
Endocardium 377
Endocrine system 4, 250–292
  basic concepts 250–251
  hormones 251–258
    actions 253–257
    control of secretion 257–258
    synthesis and secretion 252–253
    transport and inactivation 253
Endocytosis 15–16
  adsorptive 16
Endolymph 171, 180–181, 205
Endometrium 304–305
Endoneurium 59, 60
Endorphins 145, 259, 264
Endplate 59
  chemical transmission 61–63
  miniature potential (MEPP) 63–63
  potential (EPP) 63–64
Energetic equivalent of O₂ 611
Energy
  storage 289–290, 611–614
  utilization 289–290, 611–614
Enkephalin 70, 144, 264
Enteric nervous system 64, 66–67
Enteroglucagon 600
Enterohepatic circulation 333–334
Enuresis, nocturnal 230
Eosinophilia 341
Eosinophils 316, 338, 339
Ependyma 85
Epicardium 377
Epididymis 294, 297
Epilepsy 232–234
  EEG in diagnosis 225, 232–233
  focal 232–233
  generalized 232
  grand mal 232
  Jacksonian 233–234
  petit mal 232–233
  psychomotor 233
  surgery for 238
Epimysium 115
Epinephrine *see* Adrenaline
Epineurium 59
Epipharyngeal receptors 444, 505
Epiphyseal closure 262
Epithelia
  absorptive 26–28
  leaky 25, 27
  movement of water and solutes across 24–29
  secretory 28–29
  tight 25–26
EPSP *see* Excitatory postsynaptic potential
Equilibrium potential 17–18, 31, 44–45
Erection 299–300
  in sleep 228, 229
Ergocalciferol 276

Erythroblasts 318
Erythrocytes *see* Red blood cells
Erythrogenin 319
Erythrolabe 151
Erythropoiesis 318–321
  control 319
  ineffective 319
  megaloblastic 326
Erythropoietic factors
  and haemoglobin synthesis 320–321
  renal 319
Erythropoietin 317, 319, 336
Eustachian tube 170
Evaporation
  heat loss by 616–617, 621, 624
  water loss by 535, 624
Evoked potentials 138
Excitatory junction potential (EJP) 120–121
Excitatory postsynaptic potential (EPSP) 53–54
Exercise 626–634
  and blood flow distribution 374
Exocytosis 15–16, 252
Exophthalmia 273
Expiration 468
Expiratory
  duration 505
  muscles 468
  reserve volume (ERV) 469
Extensor plantar response 222
Extensor reflex, crossed 196
External auditory meatus 170
Extracellular fluid (ECF) 2–3
  volume, control of 443, 539–542, 545
Extracellular recording
  biphasic potential 49
  compound potential 50–51
  monophasic potential 49–50
  nerve 48–52
Extrafusal muscle fibres 192
  innervation 193
Extraocular muscles 150
  in sleep 227
Extrasystole 386
Eye
  anatomy 146
  defects in focusing 148–149
  image formation by 146–150
  movements 150–151
    rapid 150, 227–229
    saccadic 150
    smooth pursuit 150
  muscles 148, 150
  receptor mechanisms 150–154
  receptor potentials 153

Fab fragments 345
Facial nerve 165
Facilitated diffusion 9

Fainting 452, 460–461
Fallopian tubes 300–301
Fascicles, nerve 58
Fastigial nucleus of cerebellum 215
Fat
  absorption of 607
  brown 623
  control of synthesis 288, 613
Fatigue 108, 626
Fatty acid
  mobilization 263, 281, 284, 613
  metabolism 102, 613
Fc fragments 345
Female ovarian cycle 293
Female reproductive system 300–309
  control of 307
Fenestrations, endothelial cells 6, 517
Ferritin 321
Fertility and semen 299
Fertilization 310
Fetoplacental unit 312
Fever 624
Fibrillation 386
Fibrin 355, 361, 362
Fibrinogen 315, 357, 358, 361
Fibrinolysis 366–367
Fick principle 410, 514
Fick's law of diffusion 4, 6
Fight or flight response 74, 281
Filtered load 518
Filtration fraction 520
Fitness 634
Fixation reflex 184
Flare 454
Flexor reflexes 196
Flocculus 81
Flowerspray (secondary) endings 193
Fluctuation analysis 29
Fluid mosaic model 6–7
Flux, definition of 4, 25
Folia, cerebellar 81
Foliate papillae 163
Folic acid 321, 325–326
  deficiency 336
Follicle-stimulating hormone (FSH) 258,
    260, 298, 302, 303, 305–309
Follicles, ovarian 301–303
  development control 303, 307
Follicular phase 304
Force–velocity relationship
  in cardiac muscle 401
  in skeletal muscle 106
Forced expiratory volume in 1 s (FEV$_1$) 470
Forced vital capacity (FVC) 470
Forebrain 76
Fornix 237
Fourth ventricle 84
Fovea centralis 147, 156
Frank–Starling law of the heart 403
Frostbite 454

FSH *see* Follicle-stimulating hormone
Functional residual capacity (FRC) 470
Fungiform papillae 163
Fusimotor neurones 194
Fusion frequency of contraction 104

G proteins 256–258
Gait, disorders 218
Gamma aminobutyric acid (GABA) 55, 209
Gamma motor neurones 191, 194–195, 212,
    220
Ganglia 34
  autonomic 64–66
  basal 83, 209–211
  enteric 64, 66–67
  parasympathetic 64
  sympathetic 64–5
  vertebral 65
Ganglion, spiral 171
Ganglion cells 154–155
  receptive field 157–158
Gap junctions 52, 116, 378
Gas
  end-tidal 484
  transfer 465, 478–490
  volume (ATPS, BTPS, STPD) 480
Gastric
  emptying 575
  inhibitory peptide 288, 589, 597–599
  motility 573–575
  mucosal barrier 602
  secretions 584–589
Gastrin 575, 588, 593, 595, 597–598
Gastrointestinal hormones 570, 596–600
  and motility 574–575, 577–578
  and secretion 589, 593, 595
Gated channels in membrane 8, 41–42
Generator potential 123, 125
Genes and blood-groups 349–350
Geniculate nucleus
  lateral 156, 158
  medial 176
Gestation 310
Gibbs–Donnan membrane distribution 22,
    29–33
Giddiness 187
Gigantism 261
Glial cells 35, 38–39
Globose nucleus of cerebellum 215
Globulins 315, 344
Globus pallidus 209, 210
Glomerular
  arterioles 516–518
  capillaries 516–518
  filtration rate (GFR) 518–520
  function 517–520
    disturbances of 561
Glomerulus 513, 515–517, 534
Glottis 467, 572
Glucagon 287, 289, 290

Glucocorticoids 282–285
  actions 284
  control of secretion 285
Gluconeogenesis 263, 284, 289, 613–614
Glucose 287–291, 613–614
  absorption in intestine 604, 606
  blood levels of 289
  co-transport of 12–13, 527
  sparing reactions 290, 613, 614
  tolerance test 291
  tubular reabsorption 522–524
Glucose-sensitive neurones 614
Glycogen 284, 288, 304, 613–614
Glycogenolysis 289, 613–614
Glycosaminoglycans 355
Glycosphorin A 328
Glycosuria 524
Goitre 267, 273
Goldman equation 17
Golgi apparatus 252
Golgi cells 214
Golgi tendon organs 140–141, 192, 194
  innervation 193
  'in series' behaviour 194
Gonadotrophic hormones 258, 303, 305–308
Gonadotrophin-releasing hormone (GnRH) 260, 298, 303
Gonadotrophins *see* Gonadotrophic hormones
Gonads 251, 295, 300
Graafian follicle 302
Grand mal seizure 232
Granule cells 214
Granulocytes 338, 353
  production and life-span 339–340
Granulosa cells 302
Graves' disease 273, 348
Gravitational potential energy 414, 458
Gravitational reflexes 203–204
Gravity
  effect on
    blood pressure 457–458
    blood volume distribution 420, 457, 458
    ventilation/perfusion ratio 448, 478
Grey matter, CNS 76
Growth 262–263
Growth hormone 258, 260–263
  actions 263
  and blood glucose 263, 290, 613
  control of secretion 263
  human synthetic 263
  in sleep 230, 263
Growth hormone release-inhibiting hormone (GHRIH, somatostatin) 260, 263
Growth hormone-releasing hormone (GHRH) 260, 263
GTP-regulatory protein 254–257
Guanosine 3′,5′-monophosphate (cyclic GMP) 153, 256

Guanosine triphosphate (GTP) 254, 256–257
Gustation (taste) 163–165
Gustatory nucleus 165
  pathways 165
  receptors 163–165
Gyrus 76
  cingulate 84
  postcentral 83, 136
  precentral 83, 219
  superior temporal 176

H band 98
H–Y gene 294
H_2O diarrhoea syndrome 600
Habituation 183
Haem–haem interaction 329
Haematocrit 315, 326–327, 415, 495
Haematopoiesis 316–318
Haemoglobin 329–331
  in anaemia 334
  carbon dioxide carriage by 330, 493–496
  and carbon monoxide 492–493
  concentration 326–327
  degradation 333
  and 2,3-diphosphoglycerate 330, 491
  fetal 329, 330, 492
  as H^+ buffer 495
  half saturation (P_{50}) 492
  maternal 492
  molecule 331
  oxygen affinity for 329–330, 490, 492
  oxygen carriage by 466, 490–493
  O_2 dissociation curve 490–492
  synthesis 320
Haemoglobinopathies 336
Haemolysis 349
Haemolytic anaemia 336
Haemolytic disease of newborn 349
Haemophilia A 364
Haemorrhage 461–463
  and ADH release 266, 536–537
  reflex response to 442–443, 459
Haemosiderin 321
Haemostasis 314, 354–370
Haemostatic function, tests of 369–370
Haemostatic plug 355, 358–359
Haemostatic reactions of platelets 357–359
Hagen–Poiseuille equation 411, 414
Hair cells
  cochlear 171–174
  vestibular 178–183
Hair follicle receptor 139, 140
Haldane effect 488, 494, 496
Hamburger phenomenon 495
Hamilton indicator dilution technique 410
Haptens 342
Hashimoto's disease 273

Haustra 579
Hearing 167–188
Heart *see also* Cardiac
  block 385
  ejection
    fraction 396, 397
    velocity 403, 408, 427
  external work 402
  functional anatomy 375–377
  hypertrophy 409
  and law of Laplace 408
  rate 409, 634
    efferent regulation 383–385
    in exercise 409, 629–631, 634
  sounds 395–397
    abnormal 397
  stroke volume 396
    in exercise 409, 629–631, 634
    regulation of 401–408
  suction effect 435
Heartburn 573
Heat
  conservation 623
  definition 616
  emissivity 619
  loss 621
  metabolic 610–611
  production,
    by shivering 611, 622
    non-shivering 623
    thyroid function in 623
  transfer 616–619
    by conduction 617
    by convection 617–618
    cutaneous 623
    by radiation 618–619
Helicotrema 171, 172
Helper T cells 344
Hemiplegia 221
Henderson–Hasselbalch equation 497, 550–551
Heparan sulphate 355
Heparin 366
  neutralizing factor 357
Hepatic blood flow 455
Hepatic cells (hepatocytes) 321, 365, 593
Hepatitis 354
Hering–Breuer reflex 505
Heterometric autoregulation, of heart 403
High-density lipoproteins (HDL) 608
Hindbrain 76
Hippocampus 236–240
Hirschsprung's disease 579
Histamine 119, 341, 430, 476, 505, 588–589
  clearance 526
Histocompatibility antigens 348
HLA system *see* Human lymphocyte antigen system
HMWK *see* Kininogen, high molecular weight

Hodgkin–Katz equation 17–18
Homeometric regulation, of heart 407
Homeostatis 3–4, 372
Homeostatic mechanisms 3–4
Homoiotherms 616
Horizontal cells 154–155
Hormonal contraception 308
Hormones 250–313
  assay of 291–292
  definition of 251
  gastrointestinal 570, 589, 593, 595, 596–600
  inactivation of 253
  paracrine actions 598
  plasma protein binding 253
  receptors for 253
  structure 251
  synthesis and secretion 252
  trophic 258
Human chorionic gonadotrophin (HCG) 310, 311
Human chorionic somatomammotrophin (HCS) 311, 312
Human lymphocyte antigen (HLA) system 348
Hyaluronic acid 539
Hydraulic conductivity 20
  of renal tubules 527
Hydrocephalus 453
Hydrogen ion
  ATPase 15
  buffering 15
    by bicarbonate 497, 550
    by haemoglobin 495, 552
    by phosphate 550
    by plasma proteins 495, 550
    and Henderson–Hasselbalch equation 497, 550–551
  mechanisms 553–554
  as central chemoreceptor stimulus 503
  counter-transport 15
  disposal of 554–558
  effect on cellular potassium uptake 547, 549
  exchange in bone 554
  increase in exercise 632
  intracellular concentration 15
  production 552–553
  secretion
    gastric 586, 602
    in diseased kidneys 563
    renal 527, 534, 555–558
Hydrostatic pressure 20
  across capillary wall 22–23, 432–433
  across glomerular capillaries 518
  in blood vessels 456–458
  gradient 411
  of interstitial fluid 22
  in villi capillaries 601–602
Hydroxyapatite 276

Hydroxyindole-O-methyltransferase
  (HIOMT) 309
Hydroxytryptamine (5HT, serotonin) 202,
  355, 357–358, 430, 569
Hymen 301
Hyperalgesia 143
Hyperbarism 510
Hypercalcaemia 279
Hypercapnia 429, 444, 500
  chronic 504
  effect on lung stretch receptors 505
Hypercomplex cells in visual cortex 160
Hyperkalaemia 354, 546, 548
Hyperkinetic circulation 337
Hypermetropia 148, 149
Hypernatraemia 536–537
Hyperpathia 142
Hyperplasia
  congenital adrenal 287, 294
Hyperpolarization of membranes 42
Hyperprolactinaemia 309
Hypersensitivity, delayed type 344
Hypertension 424, 433, 460
Hyperthermia, zone of 621
Hyperthyroidism 273
Hyperventilation 488, 536, 552
  in exercise 632
  in hypoxia 507, 508
  syndrome 552
Hypocalcaemia 278
Hypocalcaemic tetany 277
Hypocapnia 504, 508
  effect on cerebral blood flow 452
Hypochromic red cells 327
Hypokalaemia 546, 548–549
Hyponatraemia 537–538, 545
Hypophyseal portal system 260–261
Hypophysectomy, effects 259
Hypophysis 258
Hypoproliferative anaemia 335
Hypotension 460–461
Hypothalamic neurohormones 260–261
Hypothalamus 67, 76
  and behaviour 236–237, 244–246
  in cardiovascular control 444, 446, 453,
    454
  in control of endocrine glands 251
  control of food intake 614–615
  defence reaction centre 446
  depressor areas of 446
  and emotion 244
  function of 83
  pressor areas of 446
  in respiratory control 500
  temperature regulating centre 446, 622
Hypothermia, zone of 621
Hypothyroidism 273
Hypotonia 218, 221, 279
Hypoventilation 552
Hypoxia 332, 336, 500

acute 504, 507–508
cardiovascular response to 429, 444
chronic 504
pulmonary vasoconstriction 448
and respiratory alkalosis 552
respiratory response to 500, 506
tissue 506

I band 98
ICF *see* Intracellular fluid
IgA 345
  in saliva 581
  structure 346
IgD 345
  structure 346
IgE 339, 345
  structure 346
IgG 345
  antibodies 350–351
  complement fixation 347
  structure 346
IgM 345
  antibodies 350
  complement fixation 347
  structure 346
Ileocaecal sphincter 578
Image formation 146–150
Imbibition pressure 539
Immune response 284
  abnormalities 348
  suppression by glucocorticoids 284
Immune system 341–348
  effector cells 342–344
Immunity 341–348
  active 347
  cell-mediated 344
  passive 347
Immunoglobulins 314, 344–346
Impedance cardiography 410
Implantation of embryo 310
Impotence 299
Incisura 396, 424
Incoordination 218, 508
Incus 170
Infarcts 354
Inflammation, acute 341
Inhibin 298, 303
Inhibition
  descending 136
  postsynaptic 53–54, 134
  presynaptic 53–55, 134
  proactive 236
  recurrent 197
  Renshaw 197
  retroactive 236
  surround 134–135
Inhibitory junction potential (IJP) 120–121
Inhibitory postsynaptic potential (IPSP) 53–
  54

Inositol trisphosphate 15, 165, 256, 257, 266, 591
Inotropy 399, 406–407
Inspiration 467
Inspiratory
capacity (IC) 470
duration 505
muscles 467
reserve volume (IRV) 469
Inspired gas 479
Insulin 262, 287–289
actions 287–288, 290, 546, 613
coma 290
deficiency 290–291
secretion control 288–289, 546, 598, 599
Insulin-like growth factor (IGF-1) 263
Integration
in neurones 54
retinal 158
Intercalated cells of nephron 547
Intercalated discs of heart 378
Intercostal muscles 467
Interdigestive myoelectric complexes 577
Interdigitating cells 343
Interferon 341
Internal capsule 222
Internal environment 3
Interneurones 134–135, 191
Internodal bands 380
Interpositus nucleus 215
Interstitial cells of Leydig 296, 297
Interstitial fluid 2
gel properties 539
hydrostatic pressure 22
protein in 23
volume 2, 539
Intestinal
absorpiton 600–609
motility 576–580
mucosal infoldings 568, 601
secretions 589, 595–596
surface area 601
villi 568, 601
Intestino-intestinal inhibitory reflex 578
Intra-alveolar pressure 473, 477
Intracellular fluid (ICF) 2
buffers 553
volume 535
Intracellular messenger 14–15, 254–256
Intracranial pressure 452
in hypernatraemia 538
Intrafusal muscle fibres 192–194
Intralaminar nucleus (of thalamus) 211
Intrapleural pressure 406, 435, 472, 477, 487, 510
Intrathoracic pressure 406, 435, 461
Intrinsic factor 324, 585, 609
Inulin, clearance 518–21
Ion channels 8–9, 29–31
gated 8, 41–42

leaky 8
neurotransmission 55
Ionic permeability 18
IPSP *see* Inhibitory postsynaptic potential
Iris 146
Iron 321–324
binding-capacity, serum 322
deficiency 324, 335
exchanges and requirements 323
intake and absorption 322–323
negative balance 323
serum 315
Ischaemia 354
cerebral 446, 452
myocardial 451
neuronal sensitivity to 129
Ischaemic heart disease 385, 463
Islets of Langerhans 287–288
Iso-electric line 389
Isometric contractions
cardiac muscle 398
skeletal muscle 103
Isoprenaline 71
Isosthenuria 562
Isotonic contractions
cardiac muscle 103, 106
skeletal muscle 398
Isovolumetric contraction of heart 395, 401
Isovolumetric relaxation of heart 396, 402
Itch 143

J receptors 444, 504, 506
Jacksonian epilepsy 233–234
Jaundice 334, 336
Jet-lag 285
Jugular venous pulse (JVP) 397–398, 458–459
Junction
gap 52, 116, 378
potentials 120
skeletal neuromuscular 59
smooth neuromuscular 69
tight 6, 25, 84
Juxtacapillary (J) receptors 444, 504, 506
Juxtaglomerular apparatus 517, 540
Juxtamedullary nephrons 513–514

K cells 599
K complexes 226
Kallidin, vasodilating effects 429
Kallikrein 362, 363, 364, 429, 584
Ketones 290, 545, 551
Kidney *see* Renal 512–534
function in diseased 561–564
Kinaesthesia 140–141
Kinetic energy 413
Kininogen 429
high molecular weight (HMWK) 362, 363, 364

Kinins 341, 429
Kinocilium 179, 180
Klinefelter's syndrome 293
Knee jerk 221
Korotkoff sounds 425–426
Korsakoff's psychosis 239, 243

Labia 301
Labyrinth 81
  bony 178
  membraneous 178
Lactation 313
Lactic acid 551, 553
  in exercise 628, 632, 633
  in metabolic acidosis 551, 553
  in metabolic autoregulation 429
  in respiratory drive 504
  in skeletal muscle 102
Laminar flow of blood 411, 414
Laplace's law
  for blood vessels 419–420
  for heart 408–409
  for lung 475
Large intestine *see* Colon
Laryngeal constriction 467, 505, 506
Laryngeal receptors 444, 505
Larynx 572
Latent heat 616
Lateral fissure 83
Lateral geniculate nucleus 156
  responses in 158
Lateral hypothalamus 614
Lateral inhibition 134
Lateral intercellular space 24, 27
Lateral ventricles 84
Lateral vestibulospinal tract 201
Leak channels in membrane 8
Leaky epithelium 25, 521
Learning 234–240
Lecithin 594
Lemniscal system 130
Length–tension relationship
  in cardiac muscle 398, 400, 408
  in skeletal muscle 104–106
Lens 147–149
Lentiform nucleus 209
Leprosy 344
Leu-enkephalin 264
Leukaemia
  chronic granulocytic 340
  chronic lymphocytic 340
Leukocytes 315
  polymorphonuclear 338
Leukocytosis, neutrophil 340
LH *see* Luteinizing hormone
Light adaptation 153
Light reflex 149
Limb proprioceptors 44, 444, 504, 629, 632

Limbic system 167, 236–238, 244, 246
  and cardiovascular control 445, 447
Linear velocity of blood flow 412
Lipase 567, 590, 594, 607
  intestinal absorption 607–608
  regulation of synthesis 288, 615
Lipotrophins 259
Liver
  fluid exchange across capillaries 22, 455
  inactivation of hormones 253, 541, 555
  production of
    plasma proteins 544, 561
    urea 551
  secretions (bile) 595–595
Load–velocity curves
  in cardiac muscle 398, 400–401
  in skeletal muscle 106–108
Lobotomy 243
Local anaesthetics 46, 51
Local potentials 40, 42–43
Locus coeruleus 202, 225
Long-acting thyroid stimulator (LATS) 273
Long-sightedness 149
Long-term memory 235–236
Long-term potentiation (LTP) 239
Loops of Henle 513, 528–531
  descending limbs 528
  thick ascending limb 529, 531
  thin ascending limb 528
Loudness, perception of 168, 175
Lumbar spinal cord 300
Lung (*see also* Pulmonary)
  blood and gas exchange 482
  compliance 473, 474, 476, 487
  diffusion capacity 489, 509
  distensibility 471
  irritant receptors 444, 504, 505
  law of Laplace in 475
  mechanics 467–478
  non-respiratory functions 504
  perfusion 448, 465, 487
    stability 476
  pressure–volume relationships 470–474
  stretch receptors 443, 504, 505
  structure of 466–467
  surfactant 467, 475–476
  terminal airways structure 469
  volumes 468–470
Luteal phase 304
Luteinizing hormone (LH) 258, 260, 298, 303
Luteinizing hormone-releasing hormone (LHRH) 298
Lymph 23–24
  capillaries 438
  drainage 433
  ducts 438
  node 438
  return 439, 459
  vessels 438

Lymph flow 433, 438
Lymphatic system 23–24, 375, 432
  characteristics 438–439
  gastrointestinal 567, 601, 608
  pulmonary 448
Lymphatic valves 438
Lymphocyte circulation 343
Lymphocytes 317, 338, 339, 342–348
Lymphocytosis 340
Lymphoid stem cell 317
Lymphokines 344
Lymphopenia 340
Lysosomes 357
Lysozyme 581

M line 98
Macrocytic normochromic anaemias 337
Macroglobulin 365
Macrophages 338, 341, 343
Macula
  densa 540
  retinal 147
  of utricle and saccule 182–184
Magnesium 3, 12, 100, 609
  renal handling 529
Major basic protein 338
Major histocompatibility complex (MHC)
  348
Male reproductive system 295–300
  control of 298
Malleus 170
Mammary glands 306, 307, 313
Mammillary body 237–238
Mannitol 518, 432, 533
MAP *see* Mean arterial pressure
Mass movements, colonic 579
'Mass' reflex 200
Mast cells 341, 430
Mastication 571
Maximal respiratory forces 476
Mean
  arterial pressure (MAP) 425, 439–440
    cerebral 452
    control of 441–447, 459
    in cardiac failure 463–464
    in exercise 463–464, 629
    in haemorrhage 461–463
    in kidney function 456, 515
  cell haemoglobin (MCH) 327
  cell haemoglobin concentration (MCHC)
    327
  cell volume (MCV) 327
  electrical axis 392
  right atrial pressure 435, 437, 439, 458,
    462
  systemic filling pressure (MSFP) 435,
    437–438
  venous pressure 435
Measurement units (SI) 635–636

Mechanical summation of contraction 104
Mechanically gated channels 42
Mechanoreception 139–140
Mechanoreceptors 123–124, 128, 139–140
  respiratory 497, 504–506
Medial lemniscus 130
Medial vestibulospinal tract 202
Median forebrain bundle 245
Medulla oblongata 78–79
  in cardiovascular control 430, 439, 444
  and respiratory control 498–499
  ventrolateral surface, central
    chemoreceptors on 502–503
Medullary
  collecting ducts, renal 530
  hyperosmotic gradient 531–532
  vomiting centre 576
Medulloblastoma 218
Megakaryoblast 317
Megakaryocyte 317
Megaloblastic erythropoiesis 326
Megaloblasts 326
Meissner's corpuscles 128, 139, 140
Meissner's (submucosal) plexus 567
Melanocyte-stimulating hormone 259, 264
Melatonin 309
Membrane
  basilar 171–174
  capacitance 47, 89
  conductance 18, 89
  'noise' 29
  patch-clamping 8, 29–31
  permeability 6–13, 89
  potentials 41
    local (electrotonic) 42, 90–92
    resting 13, 16–18, 41
  red-cell 328
  Reissner's 171
  resistance 47, 88
  semi-permeable 18
  space constant 47, 90
  structure 6–7
  tectorial 171–173
  time constant 47, 91
  transduction mechanism 123
Membrane transport
  active 11–13
  carrier-mediated 9
  co-transport 12
  counter-transport 12
  endocytosis and exocytosis 15
  facilitated 9
  passive 4
Membraneous labyrinth 178
Memory 234–240
  immune response 342
  long-term 235–236
  neurological basis 236–240
  sensory 234
  short-term 235

Memory (*continued*)
  suppression by ECT 239
Menarche 295
Ménière's disease 187
Meninges 84
Menopause 295
Menstrual cycle 303–305
Menstruation 305, 307
Merkel's discs 128, 139
Meromyosin 98
Met 611
Met-enkephalin 264
Metabolic acidosis 290, 501, 503–504, 548–
    549, 551–552, 557–560, 563
  chronic 504
  compensation 559
  hydrogen ion secretion in 557
Metabolic alkalosis 503–504, 548–549, 552,
    559
Metabolic autoregulation 429, 432, 449,
    452, 454, 555
Metabolic heat 610–611
Metabolic rate 610
  basal 611
Metabolism, in exercise 626–628, 630, 631
Metamyelocyte 339
Metarterioles 431–432
Methaemoglobin 332
Micelles 607–608
Microcytic hypochromic anaemias 337
Microglia 38, 39
Microtubules 56, 357
Micturition reflex 199
Midbrain 76, 78–80, 245
Milk
  immunity provision 347
  let-down 313
Mineralocorticoids 282
  actions 285
  control of secretion 285
Miniature endplate potential (MEPP) 63
Mitral cells 167
Mitral stenosis 448, 461
Mitral valve 376
Modalities of sensation 138–145
Modiolus 171–173
Monoamine oxidase (MAO) 72, 281
Monochromats 163
Monoclonal antibody 348
Monocytes 316, 317, 338, 339, 340, 341
Monocytosis 341
Monosynaptic stretch reflex 195
Monro–Kellie doctrine 452
Morphine and ADH release 266
Morula 310
Mossy fibres 213, 214, 215–216
Motilin 597, 600
Motility of digestive tract 570–580
  gastric 573–575
  large intestine 578–580

oesophageal 571–573
  small intestine 576–578
Motion sickness 188
Motivation 245–246
Motor
  cortex 83, 207, 218–221
    in cardiovascular control 447
    in exercise 630, 632
    in respiratory control 498, 500
  nerves, somatic 300
  neurones 36, 190–192, 194–195, 197, 212,
      217
    supraspinal pathways 200–202
  nuclei 190
    definition 190
    function organization within 191–192
    organization between 192
  pathway, lesions 221–222
  potential 209
  unit 110, 189
    I (MI) 219–220
    II (MII) 220
    III (MIII) 220
Mountain sickness 508
  chronic 509
Movement, control of 189–199, 207–221
Mucociliary action in lungs 465
Mucosa of digestive tract 567
Mucus-secreting cells 581, 584, 591
Mullerian ducts 294
Mullerian-inhibiting factor (MIF) 294
Müller's manoeuvre 476
Multiple myeloma 348
Multiple sclerosis 51
  evoked potentials in 138
Murmurs 397
  continuous 397
  diastolic 397
  systolic 397
Muscarinic receptors
  in autonomic ganglia 68
  in cardiac muscle 384, 400
  in smooth muscle 72
Muscle *see* Skeletal muscle, Cardiac muscle
    and Smooth muscle
Muscle spindles 140–141, 192
  innervation 193–194
  in respiratory muscles 499
  in spinal reflexes 192–194
Muscle tone 205
Muscularis externa 567
Muscularis mucosae 567
Myasthenia gravis 63–64, 348
Mydriasis 228
Myelin 39
  and action potential conduction 48
  loss effects 51
Myeloblast 317, 339
Myelocyte 317, 339
Myeloid stem cell 317

Myenteric (Auerbach's) plexus 567, 570
Myocardial
   contractility 400, 406–408, 437, 444, 459,
      463–464
   ischaemia 451
Myocardium 377
   potassium depletion 549
Myoclonus, nocturnal 231
Myofibrils 96, 378
Myofilaments
   cardiac 378, 399
   skeletal 96–100, 106
   smooth 116
Myogenic
   activity in smooth muscle 118–119, 120
   autoregulation of blood flow 417, 429,
      451, 452, 454, 455, 456, 515
   rhythmicity in cardiac muscle 378, 431,
      438
Myoglobin 109, 627
Myoglobinuria 548
Myopia (short-sightedness) 148, 149
Myosin
   in cardiac muscle 378, 399
   in platelets 357
   in skeletal muscle 97–101, 106
   in smooth muscle 116, 117
Myotatic (stretch) reflexes 195–196
Myxoedema 273

N proteins 256
Na$^+$–K$^+$-ATPase (pump) 11–12, 93, 527,
   541
Narcolepsy 230
Nasal receptors 444, 504
Natriuresis 443
Natriuretic hormones 443, 542
Near point of vision 149
Near response 147
Neck reflexes 204
Negative feedback in hormonal regulation
   257, 261, 305
Neospinothalamic tract 132–133
Neostriatum 209
Nephrogenic diabetes insipidus 537
Nephrons 513–514
   destruction 561
   diluting segment 530
   distal tubule 513, 530, 534, 541, 543
   functions 516, 534
   intercalated cell 547
   loop of Henle 513, 528–531, 534, 543
   principal cell 547
   proximal tubule 513, 526, 534, 543
Nephrotic syndrome 544, 561
Nernst equation 11, 15, 17, 31
Nerve fibres, properties 45
Nerve growth factor (NGF) 57

Nervous activity, extracellular recording 48–
   52
Neurofilaments 36, 56
Neurogenic diabetes insipidus 537
Neuroglia 38–39
Neurohormones 260
Neurohypophysis 258, 536
Neuromodulation 55
Neuromuscular transmission 59–64, 120
Neuronal uptake of catecholamines 72
Neurones 34–37
   cell death in 57
   chemical transmission 52
   chromatolysis 58
   development 57–58
   electrical transmission 52
   enteric 66–67, 567
   initial segment 54
   membrane potential 43
   motor 36, 190–192, 194–195, 197, 212,
      217
   pacemaker 225
   parasympathetic 64, 66
   sensory 40, 127–129
   signal integration 54
   sympathetic 64–66
Neurophysin 265, 536
Neuropil 38
Neurosecretion 52, 265
Neurotoxins
   α-bungarotoxin 62
   botulinus 61
   α-cobratoxin 62
   scorpion venom 46
   tetrodotoxin 46
Neurotransmitters
   acetylcholine 55, 61–64, 67, 70, 72, 209
   actions 29, 53, 54–55, 62, 70–72, 94, 120
   central synaptic 55
   cotransmitters 70
   dopamine 55, 209, 212
   gamma-aminobutyric acid 55, 209
   inactivation 53, 63, 68, 72
   noradrenaline 66, 69–72, 202, 309
   purinergic 67
   release 52, 61, 69, 70
Neurotrophic factors 112
Neutropenia 340
Neutrophilia 340
Neutrophils (polymorphonuclear leukocytes)
   316, 317, 338, 339, 340, 341
Nicotinamide adenine dinucleotide
   phosphate (NADPH) 332
Nicotinic receptors
   in autonomic ganglia 68, 72
   in skeletal muscle 62
Nigrostriatal pathway 209
Nissl substance 35
Nitrogen
   in decompression sickness 511

Nitrogen (*continued*)
  end-products, accumulation 563
  fraction of respiratory gases 479, 480
  narcosis 511
  partial pressures in alveoli and blood 484
Nociceptors 123–124, 128
  polymodal 143
Nocturia 562
Nodes of Ranvier 39, 48, 51
Nodule of cerebellum 81
Non-adrenergic non-cholinergic inhibitory
    fibres 67, 119, 571, 575, 578
Noradrenaline 66, 69–72, 75, 202, 280–281,
    309
  actions on
    blood vessels 430
    heart rate 384
    stroke volume 281, 349, 406
  circulating 75
  inactivation 72
  in stress 74, 281
  synthesis 69–70
  uptake 72
Norepinephrine *see* Noradrenaline
Normochromic red cells 327
Nuclear bag fibres 192
Nuclear chain fibres 193
Nucleus(ei)
  ambiguus 498–499
  amygdaloid 237, 244
  caudate 209, 210
  cochlear 176
  dentate 215
  emboliform 215
  fastigial 215
  globose 215
  interpositus 215
  intralaminar (of thalamus) 211
  lateral geniculate 156, 158
  lentiform 209
  medial geniculate 176
  parabrachialis 499
  raphé 202, 225
  red 201
  reticular 201, 215
  retroambigualis 498
  supraoptic 265, 446, 536
  of tractus solitarius 498, 614
  vagal (dorsal motor) 446
  vestibular 185–187, 215, 216
Null cells 342
Nystagmus 205, 218
  optokinetic 185, 205
  vestibular 185, 205

Obesity 615
Occipital leads in EEG 224
Occipital lobe 83
Ocular dominance 161

  columns 160–161
Oculomotor nerve 80, 81, 140–150, 181
Oedema 433
  generalized 539, 542, 544
  in heart disease 463, 464
  in kidney disease 561
  localized 459, 542
  pulmonary 448, 463, 506, 509
Oesophageal sphincter,
  lower 573
  upper 572
Oesophagus 571–573
  peristaltic wave 572, 573
Oestradiol 302, 305
Oestriol 312
Oestrogen 302, 303, 304, 305, 312
  effects 305–306
  secretion control 306
Oestrone 302
Oestrus 293
Off-centre fields 157
Ohm's law 88
Oil–water partition coefficient 8–9
Olfaction (smell) 165–167
Olfactory epithelium, structure 166
Olfactory pathways 167
Olfactory receptors 166–167
Oligodendrocytes 38–39
Olivary nucleus,
  inferior 216
  superior 176
Olivocochlear bundle 172
On-centre fields 157
Oocytes 301
Oogonia 301
Opiates 144
Opioid peptides 144–145, 263–264
Opioid receptors 264
Opponent cells 162
Opsin 151
Optic
  chiasma 156, 246–247
  disc 154
  nerve 154
  radiation 159
Optokinetic nystagmus 185, 205
Osmolality 19
  extracellular fluid 3, 536
  intracellular fluid 3
  plasma 536–538, 543
  in renal cortex and medulla 528–530
Osmoreceptors 266, 536
Osmosis 18–19
  across capillary walls 22–23
Osmotic
  activities of solutes 19
  diuresis 290, 521, 532, 545, 562, 563
  pressure 19
    and ADH release 266
    colloid 20

effective 19–20
Ossicles 170
Osteoblasts 276
Osteocytes 276
Osteomalacia 276, 279
Osteoporosis 279
  and oestrogen 306
Otitis media 177
Otoconia 182, 183
Otolith 182, 183
Otosclerosis 177
Ouabain 12
Oval window of ear 170, 171, 174
Ovarian follicles 302, 303
Ovarian hormones 305–307
  during ovulatory cycle 308
Ovaries 300–303
Oviducts 300
Ovulation 302–303, 307
  suppression 308
Ovum 301
Oxygen
  absorption coefficient 488–489
  affinity for haemoglobin 329–331, 490–492
  alveolar-arterial P_{O_2} gradient 483–484, 487
  arteriovenous difference 449, 492
    in exercise 629
  capacity of blood 490
  carriage in blood 490–493
  concentration in blood 489, 491
  consumption 479
    in brain 231
    in different organs 449
    in exercise 627, 630, 631
  debt 628
  deficit 627
  diffusion capacity 489
  energetic equivalent of 611
  fraction in respiratory gases 479
  maximum uptake 633
  partial pressure
    alveolar 484, 507, 510
    arterial 484, 507
    barometric effect 508
    inspired gas 485, 507
    in mitochondria 492
    venous 484, 507
  percentage saturation 490
  toxicity 511
Oxygenation 490
Oxyhaemoglobin 330–331
Oxyhaemoglobin dissociation curve 329, 337, 475, 490, 507
  in anaemia 337
  in exercise 628
  factors influencing 337, 491–492
  and ventilation/perfusion ratio 487
Oxynctic cells 584, 585, 587

Oxytocin 120, 265, 267
  and milk let-down 267, 313

P wave 389–393, 395, 549
P_{50} 492
Pacemaker activity
  in cardiac muscle 378, 381, 384, 385, 386
  in smooth muscle 118
Pacemaker neurones 225
Pacinian corpuscle 123, 125–126, 128, 139, 140
Packed cell volume (PCV) 315, 327
Paget's disease 278
Pain 142–145
  cramping 143
  increased sensitivity to 142
  opioid peptides in 144–145, 264
  phantom limb 144
  referred 144
  somatic 142–143
  visceral 143–144
Palaeospinothalamic tract 132–133
Pancreatic exocrine secretions 580, 589–593
Pancreatic islets 251, 287–291
Pancytopenia 340
Papillae of tongue 163
Papillary muscles 376
Para-amminohippurate (PAH)
  clearance 514, 521, 525–526
Parabrachialis nucleus 499
Paracellular pathway 24
Paracrine secretions 251, 598
Paraesthesia 279
Parafollicular cells 273
Paraplegia 221
Parasympathetic nervous system 64, 66
  effects of stimulation 73–75
    on airways 476
    on arterioles 430, 448, 452
    on gastrointestinal motility 569, 575, 588, 593
    on heart rate 384, 441
    on penile erection 300
    on salivation 584
    on stroke volume 354, 360
Parathyroid glands 251, 273–279
Parathyroid hormone 273, 277–278
  actions 277–278
  control of secretion 278
Paraventricular nucleus 265
Parietal lobe 82
Parietal pleura 474
Parkinson's disease 209, 211–212, 218
Paroxysmal atrial tachycardia 386
Paroxysmal ventricular tachycardia 386
Parturition 312–313
Passive force of muscle 104

Passive immunity 347
Passive transport 5
Patch-clamping 8, 29–31, 55
Pedunculopontine nucleus 210, 211
Pepsinogen 584
Pepsins 584, 585
Peptide hormones 251–254
Perforant pathway 239
Pericardial space 377, 406
Pericardium 377
Perikaryon 35
Perilymph 171
Perimysium 115
Perineurium 59
Peripheral nervous system (PNS) 34
  regeneration 58
Pernicious anaemia 325, 586
Permeation, mechanisms 8–9
Permissive hormone action 253
Peroxisomes 357
Perseveration 243
Pertussis toxin 257
Petechiae 368
Peyer's patches 343
pH
  buffering 497, 550
  central chemoreception 503–504
  intracellular 15
  plasma 550
    arterial 495, 552
    venous 495, 552
  regulation of 550–560
Phaeochromocytoma 281
Phagocytes 338
Phagocytosis 16, 338
Phagosomes 338
Phantom limb pain 144
Phenylethanolamine-N-methyltransferase
    75
Phonation 240
Phonocardiography 395
Phospholipids 357, 607
Phosphorus
  in bone 274, 276
  plasma concentration 275
    regulation of 277, 278
  renal handling of 524
Photopia 152
Photopigments 151–152
Photoreceptors 123–124, 150–151, 154–155
Phrenic nerve 467, 498
Physiological dead space 484
Physiological shunt 484
Pia mater 84, 451
Piloerection 74
Pineal gland 309
Pinocytosis 16
Pitch discrimination 168, 175
Pituitary gland 251, 258–267
  anterior 258–265

posterior 258, 265–267
Placenta 251, 311
Placental hormones 311–312
Plantar reflex 222
Plasma 315
  colloid osmotic pressure 21–22, 432
  composition 2–3
  proteins 315, 462
    deficiency of 542, 561
  volume 2, 315, 535
Plasma cells 344
Plasma membrane 4
  passive movement across 6–10
Plasmin 366–367
Plasminogen 366
  activators 366–367
Plate endings 196
Platelet 315, 356–360
  -activating factor 360
  concentrates 354
  count 356, 369
  defects of 367–369
  -derived growth factor 357, 360
  factor 3 357, 359
  haemostatic reactions 353, 357–359
  plug formation 367–368
  ultrastructure 356
Pleura 474
Pleural space 474–475
Plug flow of blood 415
Pluripotent cells 317
Pneumotaxic centre 498, 499
Poikilocytosis 328
Poiseuille equation 411, 412, 414, 417
Polychromasia 328
Polydipsia 267, 290
Polymodal nociceptors 143
Polymorphonuclear leukocytes 338
Polyphosphoinositide system 256
Polyuria 267, 549
Pons 78–80, 444–446, 498–499
Pontine nuclei 216
Position sense 139
Positive feedback mechanism 305
Post-absorptive state 290, 613
Portal vein 373, 455
Postganglionic fibres 64, 67
Postsynaptic inhibition 53, 54, 134
Posture
  and cardiovascular adjustments 420, 433,
      447–448, 456–460
  control of 202–203
  and ventilation/perfusion ratio 448, 487
Potassium 545–550
  and acid–base balance 549
  balance 546
  cellular uptake 288, 546
  diffusion potential 17
  distribution 545
  disturbances of 548–549

in action potential 44–46
intracellular 2
regulation 546–548
Potential
action 40, 43–52
in auditory nerve 174–176
in cardiac muscle 378–385
in sensory fibres 126
in smooth muscle 117–118
diffusion 16–18
early receptor 152
electrochemical 10, 25
endplate (EPP)
equilibrium 17–18
evoked 138
excitatory junction (EJP) 120–121
excitatory postsynaptic (EPSP) 53–54
generator 123, 125
inhibitory junction (IJP) 120–121
inhibitory postsynaptic (IPSP) 53–54
local (electrotonic) 40, 42–43, 91
membrane 41
miniature endplate (MEPP) 63
motor 209
pacemaker 381, 384
readiness 208
receptor 41, 123–124
resting membrane 16–18, 33, 44
reversal 94
synaptic 53
transepithelial 26
Potentiation, long-term 239
Potential energy 413, 458
PQ interval 389
PR interval 389
Precapillary sphincter 431
Preload, heart 400–401, 403, 404, 436
Preprohormones 252
Precentral gyrus 83, 219
Preferred comfort zone 621
Prefrontal cortex 242–243
Prefrontal leucotomy 243
Preganglionic fibres, autonomic 64, 280
Pregnancy 310–313
iron deficiency in 324
Premotor cortex 220
Preoptic region 229–30
Presbyopia 148, 149
Pressor areas 445, 446
Pressure
alveolar 472, 477, 510
aortic 395–397, 402–403, 425–428
arterial 422–428
arteriolar 422
atrial 395–397
barometric 507–508
blood 422, 425–426
capillary 422, 432–434
colloid osmotic 20, 432–433
critical closing 417

diastolic 425–426
hydrostatic 20, 456–458
imbibition 539
interstitial fluid 22
intracranial 452
intrapleural 406, 477, 487, 510
intrathoracic 406, 435, 461
mean systemic filling 435
osmotic 19–20
partial 484
pulmonary blood 395, 447
pulse 425
systolic 425, 427
transmural 408, 419, 458, 471, 472
units of 410, 635
venous 422, 435
ventricular 377, 395–397, 402
Pressure sensation 139
Presynaptic inhibition 55–56
Proactive inhibition 236
Prodynorphin 264
Proenkephalin 264
Proerythroblast 317, 318
Progesterone 303, 304, 305–307, 312
effects 307
Prohormones 259
Proinsulin 287, 289
Prolactin 258, 260, 309, 313
Prolactin release-inhibiting hormone
(PRIH) 260
Promyelocyte 317, 339
Proopiomelanocortin 264
Proprioception 140, 215
Proprioceptors 140
in cardiorespiratory control 444, 504
Prosopagnosia 249
Prostacyclin 355, 360
Prostaglandins (PG) 121, 299, 313
in platelets 359–360
Prostate gland 294, 295
Protanomaly 163
Protanopia 163
Protein
catabolism 284
intestinal handling of 607
loss from kidney 561
Protein 3 328
Protein C 355, 365
Protein kinase A 255
Protein kinase C 256
Protein S 365
Prothrombin 361, 362, 370
Prothrombin time (PT) 370
Pseudobulbar palsy 240
Psychogenic syncope 461
Psychomotor epilepsy 233
Pteroylglutamic acid 325
Puberty 295
Pulmonary *see also* Lung
arterial pressure 395, 447

Pulmonary (*continued*)
  arteriole 416, 448
  artery 373, 419, 447
  capillaries 447–449, 465, 474
  circulation 372, 421, 447–449
  congestion 463, 506
  hypertension 448, 509
  lymphatic system 448
  minute volume (ventilation) 484
  oedema 448, 463, 506, 509
  perfusion 448, 465, 487
  reservoir 448
  resistance 422, 447
  semilunar valve 376, 395–396, 397
  vein 376, 419, 447–449
  ventilation 465, 478–488
  volume 420, 435, 447–448, 463
Pulse pressure 425
  in exercise 629
Pupil, constriction 148
Purinergic nerves 67
Purkinje
  fibres 379–380
  network 379–380
Purkinje cells 213, 215, 217
Purkinje shift 152
Purpura 368
  thrombocytopenic 369
Putamen 209, 210
Pyelonephritis 562
Pyloric canal 574
Pyramidal cells 220
Pyramidal tract 200–201
Pyrexia 624
Pyridoxine 320
Pyrogens 624

QRS complex 389–393, 395, 549
QS interval 389
QT interval 389
Quadriplegia 221

Radiation of heat 618–619
Radioimmunoassay 251, 292–292
Radionuclide imaging 410
Radioreceptor assay 251, 292
Rami communicantes 65
Raphé nuclei 202, 225
Rathke's pouch 258
Reabsorption
  in capillary exchange 22–23, 433–434,
    448
  in kidney 522–525
Reactive hyperaemia 429
Readiness potential 208
Receptive field 127
  disparity 162
  orientation 160

Receptive relaxation 573
Receptor cell 123, 125
Receptor potential 41, 123–124
Receptors
  adrenoceptors 71–75
  auditory 171–174
  cardiovascular 440–443
  cholinergic
    muscarinic 68, 72
    nicotinic 62, 68
  cutaneous 128, 139–140
  hormone 253–254
  in muscle and joints 140–141, 192–194
  opioid 264
  osmotic 266, 536
  pain 142–143
  postsynaptic 52–53, 55
  respiratory 443–444, 500–506
  sensory 122
    adaptation of 126, 152–153
    classification of 124
    coding by 127
    transduction 123–126
  smell (olfactory) 166–167
  taste (gustatory) 163–165
  temperature 141–142, 622
  vestibular 178–183
  visual 150–154
Reciprocal innervation 196
Recruitment
  of motor units 110–111
  of sensory units 127
Rectification 89
Rectum 578–580
  motility control 579–580
Recurrent inhibition 197
Red blood cells 315, 318–337
  abnormal 336
  antigens 349
  characteristics 326–332
  count 326–327
  deformability 415
  indices 327, 336–337
  lifespan and breakdown 332–334
  membrane 328–329
  metabolism 332
  permeability 6, 8
  production 318–321
  volume 315
Red muscle 109
Red nucleus 201
Red reaction 454
Red-cell concentrates 353
Referred pain 144
Reflexes
  autonomic 199–200
  crossed extensor 196, 203
  fixation 184
  flexor 196
  Hering–Breuer 505

mass 200
micturition 199
neck 204
righting 204
spinal 192–199
static 203–204
stretch (myotatic) 195–196, 202
tonic stretch 203
withdrawal 40
Reflexion coefficient 20
Refractory channels 42
Refractory period
of cardiac muscle 382–383
of nerve 43, 45, 129
Reissner's membrane 171
Relative refractory period
of cardiac muscle 382–383
of nerve 43
Relaxation pressure, respiratory 473
Relaxin 313
Relay nuclei, function 134
Release-inhibiting hormones 260
Releasing hormones 260
REM (rapid eye movements) sleep 227–230
and dreaming 229
Renal *see* kidney
arterial plasma flow (RPF) 514–515, 520
blood flow (RBF) 374, 514–516
autoregulation 456, 515
in cardiac failure 463, 464
control of 442, 456, 515–516
in exercise 374, 628, 629, 631
buffer regeneration 554–558
collecting tubules 514, 530–531
distal tubules 513, 530
extraction 514
failure 548, 561–564
glycosuria 524
hydrogen ion secretion 555–560
plasma clearance 520–521
plasma flow, effective (ERPF) 515
plasma threshold 524
protein loss 561
proximal tubules 513, 526–528
regulation of body fluids 515
tubular
acidosis 551, 563
disturbances 562–564
function 521–526
maximum 523
Renal erythropoietic factor 319
Renin–angiotensin system 251, 540–541
in blood volume control 442, 459
in cardiac failure 464
Renshaw cells 197
Renshaw inhibition 197
Residual volume (RV) 469, 474
Resistance (electrical) 89–91
Resistance (vascular) 411–412
pulmonary 422, 447

total peripheral 422
Respiratory
abnormal conditions 506–511
acidosis 551, 552, 559, 560
alkalosis 507, 551, 552, 560
bronchioles 467
centres 446, 497–500
chemoreceptors 444, 497, 500–504
central (intracranial) 502–504
peripheral (arterial) 501–502
exchange ratio (R) 479, 482, 485, 489, 633
frequency 478, 482
gas mixtures 479–480
mechanics 467
mechanoreceptors 497, 504–506
glossary of symbols 468
quotient (RQ) 479, 489, 611
regulation 497–506
responses in exercise 632–633
tract 466
dead space 479
dead space and alveolar ventilation 480–483
gas exchange in 479–489
venous pump 436
voluntary control 498
Resting membrane potential (RMP) 16–18, 33, 44
in cardiac muscle 380
in smooth muscle 117
Reticular
activation theory 226
formation 79, 225, 229
lamina 172, 173
nucleus 201, 215
Reticulocyte count 319
Reticulocytes 318–319
Reticulospinal tract 201
Retina 146, 147
responses in 157–158
structure 154–156
Retinal 151
Retroactive inhibition 236
Retroambigualis nucleus 498
Retrograde amnesia 239
Retropulsion 574
Reversal potential 94
Rheogenic pump 18
Rhesus system of blood grouping 349, 350, 352–353
Rhodopsin 151
Ribonuclease 590, 591
Rickets 276
Righting reflexes 204
Rigidity 211
cogwheel 212
decerebrate 206
decorticate 206
Rigor mortis 99–100
Rod-bipolars 155

Rods 150–151
Rostrospinocerebellar tract 215
Rubrospinal tract 201
Ruffini endings 128, 139, 140

S cells 599
Saccadic eye movements 150
Saccule 178, 179, 182–184
Sacral spinal cord 300
Saliva
    composition 580–581
    formation 582
    functions 582
        control of 541, 580–584
Salivary gland 566, 569, 571
    structure 582
Salivatory centres 583
Salt bridges, haemoglobin 330
Saltatory conduction 48
Sarcolemma 96
Sarcomere 96–97, 106
Sarcoplasmic reticulum 96, 98, 101
Satellite cells 113
Scala
    media 171
    tympani 171
    vestibuli 171
Scarpa's ganglion 185
Schistocytes 328
Schizophrenia 243
Schwann cell 38–39, 60
Sclera 146
Scorpion venom 46
Scotopia 152
Second messenger 255
Secretin 589, 593, 595, 597, 598, 599
Secretors 352
Secretory epithelia 28–29
Segmentation of gut 576–577
Self-tolerance 342
Semen 298–299
Semicircular canals 178–182
    function test 187
Semilunar valve 376, 395, 396, 397, 402
Seminal vesicles 294, 295
Seminiferous tubules 296
Semipermeable membrane 18
Sensation, modality 127, 128, 138–145
Sensitization, to pain 143
Sensorimotor cortex 219
Sensorineural deafness 176, 178
Sensory aphasia 241
Sensory cortex 83
Sensory decussation 130
Sensory memory 234
Sensory nerve fibres, cutaneous 128
Sensory neurones 127–129
    somatic 127
    visceral 128

Sensory receptors 122–127
    adaptation 126
    information coding 127
    specificity 123
    structure 125
    synaptic transmission 125
Sensory unit 127
    recruitment 127
Septal nuclei 236, 237, 245
Septum 237, 246
Serotonin (5-hydroxytryptamine) 202, 355,
        357, 358, 430
    in digestive system regulation 569
Sertoli cells 296
Serum 315
Sex
    chromatin ('Barr body') 293
    chromosomes 293
Sex hormones 262, 282, 295, 297–298, 300,
        305–307
    synthesis 283
Sexual characteristics, secondary 295, 298,
        306
Sexual development 294–295
Shivering 611, 622
Shock
    circulatory 461, 463
        metabolic acidosis in 551, 556
        and potassium loss 548
    spinal 206
Short-sightedness 149
Shunt
    in gas exchange 482–484
    physiological 484
Shunt pathway in epithelia 24
SI units of measurement 635–636
Sickle cells 328
Sideroblastic anaemias 336
Sighing 505, 550
'Simple' cells in visual cortex 159
Sinoatrial (SA) node 378–379, 380–381,
        383–384, 385–386
Sinus arrhythmia 446
Skeletal muscle 95–114
    in aging 113
    atrophy 113
    blood flow 374, 429–431, 449
        in exercise 455, 628, 629
    cellular structure 96–98
    contraction 98–109
        regulation 109–111
    development and maintenance 111–114
    effect of damage 113–114
    effect of training 112, 633–634
    energy balance 102
    excitation–contraction coupling 100–102
    fatigue 109, 626
    force–velocity relationship 106–108
    functions 95
    gradation of tension 110

hypertrophy 112
length–tension relationship 104
motor neurone recruitment 111
motor units 110
oxygen consumption 454
power 107
properties 109
regeneration 113–114
type II (fast) 108–109
type I (slow) 108–109
Skin *see* Cutaneous
Sleep 226–231
apnoea 230
disorders 230–231
EEG recordings in 227
need for 230
neural basis 229–30
rapid eye movement (REM) 227–230
slow wave 228
spindles 226
Sleep-walking 230
Slow wave activity in smooth muscle 118–119, 574–575
Slow wave (SW) sleep 228
Small intestine 576–578
absorption in 602–609
blood and lymphatic vessels 567–569
cross-section 568
fluid absorption in 604
motility 576–577
control 577–578
secretions 581, 595–596
Smell (olfaction) 165–167
Smooth muscle 95, 114–122
action potential 117–118
activation 120–122
contractile activity 116–117
electrically inexcitable 119
excitation–contraction coupling 120
functions 114
gap junctions 116
innervation 69
intermediate 119–120
myogenic activity 118
pacemaker potentials 118
resting membrane potential 117
slow waves 118–119, 574–575
spontaneously active 118–119
structure 115–116
types 118–120
Smooth pursuit eye movements 150
Sneezing 504–505
Sniff reflex 444, 505
Sodium
balance 538–539
conductance
in action potential 44–46
in cardiac potentials 380–381
in endplate potential 63
in synaptic potentials 54

distribution 538
disturbances of 542–545
extracellular 2, 3, 539
filtered load 540
in gastrointestinal secretions 545, 595
inactivation 45
intestinal absorption 578, 602–605
intracellular 2, 3
movement across epithelia 26–28
regulation 539–542
renal handling 526–531, 534, 543
in diseased kidneys 562–563
Sodium chloride, secretion 28
Sodium–potassium-ATPase (pump) 12, 93, 527, 541
Soft palate 572
Somatic nervous system 3, 59–64
Somatic sensory neurones 127
Somatomedin 263
Somatosensory complex 130, 136–138
Somatosensory cortex
primary (SI) 136
secondary (SII) 136
Somatosensory pathways 129–136
Somatostatin 70, 287, 289, 569, 588–589, 597, 600
Somatotopy 130–131
in neospinothalamic tract 134
in somatosensory cortex 137–138
Somatotrophin *see* Growth hormone
Somnambulism 230
Sound 167
amplitude 168–169
direction of source 169
duration 169
frequency 168
intensity 169
as stimulus 168–169
Space constant 90
Spastic tone 221
Spatial discrimination 140
Specific
heat capacity 616
heat of fusion 616
heat of vaporization 616
Specific dynamic action (SDA) 611
Specific membrane capacitance 89
Specific membrane resistance 88, 90
Specificity, immune 342
Spectrally opponent cells 162
Spectrin 328
Speech 240–242
emotional 242
staccato 218
Sperm, capacitation 310
Spermatids 296
Spermatocytes 296
Spermatogenesis 296–297
control of 298, 299
Spermatogonia 296

Spermatozoa (sperm) 296
  structure 297
Sphincter
  anal 119, 579
  of bladder 199
  ileocaecal 578
  of Oddi 577, 589, 593
  oesophageal
    lower 573
    upper 572
  precapillary 431
  pyloric 574
Sphygmomanometer 425–426
Spinal cord 76–78
  laminae 132
  pathways 190–192
    ascending 216
    descending 200
Spinal reflexes 192–199
Spinal shock 206
Spinal transection, effects 205–207, 221
Spinocerebellar tracts 130
Spinoreticular tract 133
Spinothalamic tract 132
Spiral ganglion 171
Splanchnic
  blood flow 374, 459
    autoregulation 455
    control 429–431, 455–456
    in exercise 628–629
    to liver 455
  blood pressure 455
  blood vessels 455
  venous capacity 456
Spleen 316, 343, 455
Squint 58, 150, 162
Stapedius muscle 170
Stapes 170
Starling curve 403–404, 407, 436
Starling equilibrium 22–24, 432, 455, 463,
    517, 541, 561
Starling's law of the heart 403, 439
Static reflexes 203
Stellate cells 214
Stem cells and haematopoiesis 317–318
Stenosis 397, 398
Stercobilin 333
Stercobilinogen 333
Stereocilia 171–172, 174, 178
Sternomastoid muscles 467
Steroid hormones 251, 254
Stomach
  absorption in 602
  functions 566
  motility 573–574
    control 574–575
  secretions 584–588
    control 588–589
STPD, gas volume 480
Strabismus *see* Squint

Stress
  autonomic responses to 74, 281
  corticosteroids in 284
Stretch (myotatic) reflexes 195–196, 202
  pathway 197–198
  phasic (tendon jerks) 196
Stretch receptors
  cardiac 442–443, 459
  lung 443, 505
  in muscle and joints 139, 192–194
Stria terminalis 237
Stria vascularis 171
Striate cortex 156, 161
Striatonigral pathway 209, 210
Stroke, cerebral 221, 452
Stroke volume 373, 396
  and cardiac output 409–410
  control 401–408
  effect of respiration 405, 435
  ejection velocity 403, 404, 407, 427
  in exercise 409, 629–631
  in haemorrhage 462
Subarachnoid space 84, 451
Sublingual glands 580
Submandibular glands 580
Submaxillary glands 580
Submucosal (Meissner's) plexus 567, 570
Substance P 145, 454, 569
Substantia nigra 209, 210, 211
Subthalamic nucleus 209, 211
Suckling 267, 313
Sugars, intestinal handling 606–607
Sulcus 76
  central 83
Sulphonylurea drugs 289
Summation
  of contractions 104
  of local potentials 42, 53–54
Supplementary motor area 220
Suppressor T cells 344
Supraoptic nucleus 265, 446, 536
Supraspinal pathways to motor neurones
    200–202
Surface tension, alveolar 474, 475, 476
Surfactant (lung) 467, 475–476
Surround inhibition 134–5
Swallowing 571–573
  centre 572
  reflex 571
Sweat glands 429, 453
  and sodium 539, 541, 544
Sweating 624
  and water loss 535, 544, 624
Sylvian fissure 176
Sympathetic cholinergic pathway 66, 430,
    446, 447, 453, 454
Sympathetic nervous system 64–66
  chemical transmission 71
  control of
    airways 476

blood flow 430
  digestive secretions 584
  gastrointestinal motility 569
  heart rate 383–384
  renin release 540
  stroke volume 406–408
  venous smooth muscle 434
  effects of stimulation 73–75
Synapses 36–37
  autonomic ganglia 64–65
  central 52–56
  electrical 52
  excitatory 53, 54
  inhibitory 53, 54
  neuromuscular 59–64
Synaptic bouton 52
Synaptic cleft 37
Synaptic delay 53
Synaptic potential 53
  ionic basis 54–55, 94
Synaptic relationships, in cerebellum 214
Synaptic terminals 36–37
  on motor neurone 191
Synaptic transmission 52–56, 68
  reversal potential 54, 94
  spatial summation 54
  temporal summation 53
Synaptic vesicles 36
Syncope 452, 461
Systemic
  arterial circulation, characteristics 422–428
  blood flow, regulation 428–431
  filling pressure, mean 435, 437–438
  vascular function curves 437–438, 462–464
  venous circulation, characteristics 434–438
Systole 373, 394, 395, 396
Systolic
  end volume, ventricular 396
  murmur 397
  pressure 425, 427

T lymphocytes 317, 342–344
  cytotoxic 344
  effector 344
  helper 344
  subsets 344
  suppressor 344
T wave 389–293, 396
Tachycardia 384
  paroxysmal 386, 387
Taste buds 163–165
Taste cells 163–165
Taste (gustation) 163–165, 582
Technetium 273
Tectorial membrane 171–173
Tectospinal tract 201–202

Telencephalon 76
Temperature
  body 615, 619–620, 624
  core 619
  critical 621
  definition 616
  isotherms 620
  peripheral (shell) 619–620
  receptors 123–124, 141–142, 444, 622
  regulating centre 446, 453, 498
  regulation 615–625
  and survival 625
Temporal summation 53
Tendon jerks 196
Tendon organs, Golgi 140, 192, 194
Tensor tympani muscle 170
Testes 294–296
Testicular feminization 294
Testosterone 294, 295, 297–300
  male sexual characteristic development 298
  secretion in sleep 228
Tetanic contraction 104
Tetanus toxin 57
Tetany
  in alkalis 550
  hypocalcaemic 277
  latent 279
  manifest 279
Tetraethylammonium 46
Tetrodotoxin 46
Thalamic nuclei 83, 210, 226
  and memory 236–237
  ventrolateral 218
Thalamus 76, 81, 83
Thalassaemias 336
Thebesian veins 377
Theca externa 302
Theca interna 302
Thermal
  comfort 620–621
  conductivity 617
  counter-current system 623
  insulation 617
Thermogenesis 271, 622–623
Thermoneutral zone 621
Thermoneutrality 620–621
Thermoreception 141–142, 622
Thermoreceptors 123–124, 128
  deep body 622
  peripheral 622
Thermoregulatory effectors 622–624
Theta waves 226
Thiosulphate 518
Third ventrical 83, 84
Thirst 267, 290, 442, 535
  and ADH 462, 535–537, 539–540, 544–545
Thoracic compliance 473
Thoracic duct 438

Thoroughfare channels   432
Thrombin   361–363, 364–365
Thrombin time   370
Thrombocytes   315
Thrombocytopenia   369
Thrombocytopenic purpura   369
Thromboglobulin   357
Thrombomodulin   365
Thromboplastin   363
Thromboplastin time with kaolin, partial
   (PTTK)   370
Thrombopoietin   317
Thrombosis   354
Thrombospondin   357, 359
Thromboxane A$_2$ 121, 355, 358, 359–360
Thromboxane synthetase   360
Thrombus   354
Thymus gland   64, 342
Thyroglobulin   269
Thyroid gland   251, 267–273, 623
   disorders of function   273
Thyroid hormones   268
   actions   271
   secretion control   272
   synthesis and secretion   269–270
   transport and inactivation   270
Thyroid isthmus   267–268
Thyroid-stimulating antibodies   273
Thyroid-stimulating hormone (TSH)   258,
   259, 260, 270, 272
Thyroiditis, autoimmune   273
Thyrotoxicosis   273
Thyrotrophin *see* Thyroid-stimulating
   hormone
Thyrotrophin-releasing hormone (TRH)
   260, 272
Thyroxine (T$_4$)   262, 268–272, 319
   and basal metabolic rate   611
Thyroxine-binding globulin   270
Tickle   139
Tidal volume   467, 481, 485, 499, 505
   in exercise   632
Tight epithelium   25–26, 530
Tight junctions   6, 24, 84
Time constant   91
Tissue factor   355, 361, 362, 363
Tissue transplantation   348
Tonic receptors   140
Tonic stretch reflex   203
Tonicity, definition of   21
Tonotopic organization   176
Tonsils   343
Total fluid energy   414, 458
Total lung capacity (TLC)   470
Total peripheral resistance (TPR)   422, 428,
   439
   control of   439–447, 459
   in hypertension   460
Touch   139, 140
   points   140

Trachea   466
Tracheal receptors   444, 505
Tract of Lissauer   131
Tract
   corticobulbar   220
   corticospinal (pyramidal)   200–201, 220
   cuneocerebellar   215
   dorsal spinocerebellar   215–216
   dorsolateral (of Lissauer)   131
   lateral vestibulospinal   186, 201
   medial vestibulospinal   186, 202
   neospinothalamic   132
   palaeospinothalamic   132
   reticulospinal   201
   rostrospinocerebellar   215
   rubrospinal   201
   spinocerebellar   130
   spinoreticular   133
   spinothalamic   132
   tectospinal   201–202
   ventral spinocerebellar   215–216
Tractus solitarius, nucleus of   498, 614
Trail endings   196
Training (exercise)   112, 633
Transcellular fluids   2–2, 535
Transcobalamin   324
Transcytosis   6, 15–16
Transducin   153
Transduction by receptors   123–126
Transepithelial potential difference   26
Transfer function   124
Transferrin   320, 321
Transition point of heart   393
Transmitters *see* Neurotransmitters
Transmural pressure   408, 419
   blood vessels   419, 458
   heart   408
   lung   471, 472
Transplantation antigens   348
Transport *see* Membrane transport
Transverse (T) tubules   96–98, 101
Tremor   211
Trephine   316
Triad, skeletal muscle   97
Trichromatic theory of colour vision   162
Trichromats   163
Tricuspid valve   376
Trigeminal nerve   80, 504
Triiodothyronine (T$_3$)   268–273
Triple response   454
Tritanomaly   163
Tritanopia   163
Trochlear nerve   80, 150
Trophic hormones   258
Trophoblastic layer   310
Tropomyosin   99, 101
Troponin   14, 99, 101
TSH *see* Thyroid-stimulating hormone
Tuberculosis   344
Tubocurarine   62, 72

Tubular, maximum (Tm) 523
Tubular (renal)
  function 521–526
  functions, disturbances 562–564
  reabsorption 521, 522–525
  secretion 521, 525–526
Tufted cells 167
Turner's syndrome 293
Twitch contraction 104
Tympanum 170, 177

U wave 549
Ultimobranchial bodies 273
Ultrafiltration 19
  across glomerular capillaries 517–518
  in capillary exchange 22–23, 433, 438
Upper motor neurone syndrome 222
Uraemia 563
Urea 535, 551
  plasma in renal failure 561–562
  renal handling 522–523, 528, 531, 532,
    534
Urethra 199, 300
Uric acid, renal handling 524, 534
Urinary
  acid measurement 557–558
  acidity regulation 558–560
  ammonia 556–558
  collecting system 530
  flow rate 514, 522, 532, 535
  osmolality 532–534
  potassium excretion 547
  sodium excretion 543
Urobilin 525
Urobilinogen 333
Urokinase 366
Uterus 301
Utricle 178, 179, 182–184

v wave 396, 397–398, 442, 459
Vagal nuclei 614
Vagal tone 384
Vagina 301
Vagus nerve 66, 384, 440, 442, 502, 505, 506
Vallate papillae 163
Valsalva manoeuvre 476
Valve
  aortic semilunar 376, 395, 396, 402
  atrioventricular 376, 395, 396
  bicuspid 376
  incompetence of 397
  lymphatic 438
  mitral 376
  pulmonary semilunar 376, 395–396, 397
  stenosis of 397, 451, 461
  tricuspid 376
  venous 406, 436, 459

Van't Hoff equation 19, 20
Varicose veins 460
Varicosities, autonomic postganglionic 69
Vas deferens 294, 300
Vasa recta 514, 522, 531–532
Vascular system 410–439
Vasectomy 300
Vasoactive intestinal polypeptide (VIP) 67,
    70, 569, 571, 596, 597, 600
Vasoconstriction 355, 428, 433, 435, 438,
    442, 448, 459, 464
Vasoconstrictor area 445
Vasodilatation 341, 428, 433, 438, 441, 448,
    461
Vasodilator area 445
Vasomotor tone 428, 461
Vasopressin *see* Antidiuretic hormone
Vasovagal syncope 461
Veins
  blood volume 420
  capacitance 417
  compliance (distensibility) 375, 417
  cutaneous, in thermoregulation 453, 623
  dimensions of 418
  pressure in 422, 435
  properties of wall 417–420
  transmural pressure 419, 434, 436
  varicose 460
Vena cava 376, 418, 419
Venoconstriction 434, 438, 459, 461
Venodilatation 434, 438
Venomotor tone 434
Venous
  admixture 482–483
  capacity 434
    alterations in 405, 434
    effects of posture 457–459
  gases 483–484, 489, 493, 507, 627, 631
  pressure, central 435, 459
  pulse 397–398, 434, 459
  pump
    respiratory 436
    skeletal muscle 405, 436, 459, 460
  renal capillary loops (vasa recta) 514
  return 434
    determinants 435–436
    relationship to atrial pressure 436–438
  sinuses in dura mater 451, 458
  valves 406, 436, 459
  volume 420
Ventilation 465
  alveolar 480–483
    and gas equation 484–486
  in exercise 630, 632–633
  pulmonary 478–488
  reflex control of 497–506
  to apex and base of lung 487–488
  wasted 481
Ventilation/perfusion ratio 448, 465, 487–
    488

Ventral horn, function organization   191
Ventral respiratory group (VRG)   498–499
Ventral root   76
Ventral spinocerebellar tracts   215–216
Ventricles
   of brain   76, 84–86
   of heart   375–376
Ventricular
   compliance   406
   contractility, index of   408
   contraction   386, 393–396
      isovolumetric   395, 401
   defibrillation   387
   depolarization   380, 388, 395
   ejection   396, 397
   failure   463–464
   fibrillation   386
   filling   395, 397
   function curve   403, 437, 462, 463, 464
   hypertrophy   393, 397, 406, 509, 634
   output imbalance   404
   paroxysmal tachycardia   386
   pressure   377, 395–397, 402
      effect on end-diastolic volume   406
   receptors   443
   relaxation
      isovolumetric   396, 402
      recoil   405
   repolarization   380, 388, 396
   septum   380, 397
   volume–pressure curves   401–402, 404,
      405
Ventrobasal nuclear complex   130
Ventromedial hypothalamic area   615
Venule   375, 418, 419, 420
Veratrine   443
Vermis   81
   lesions   218
Vertebral ganglia   65
Vertigo   187
Vestibular central pathways   185–187
Vestibular function   178–188
   disorders of   187–188
Vestibular hair cells   178–179
Vestibular nerve   181, 186
Vestibular nuclei   185–187
Vestibular nystagmus   185, 205
Vestibular placing reaction   204
Vestibulospinal tract   201, 202
Vibration sensation   139
Visceral nerves   64
Visceral pain   143–144
Visceral pleura   474
Visceral sensory neurones   128
Vision   146–163
Visual cortex   83
   organization   160–161
   primary   156
   responses in   159–161
Visual pathway   154–161, 246

   neural responses   157–161
Visual placing reaction   204
Vital capacity (VC)   469
   forced (FVC)   470
Vitamins
   fat-soluble   594, 608
   intestinal absorption   608–608
   water-soluble   609
Vitamin A   151, 608
Vitamin B_6 (pyridoxine)   320
Vitamin B_{12} 321, 324–325, 336, 585, 609
Vitamin D   274, 276–277, 513, 608
Vitamin E   608
Vitamin K   364, 375, 369, 608
   -dependent factors   364, 369
Vitreous humour   146
Voltage clamping   93
Voltage-gated channels   41
Vomiting   508, 536, 545, 549, 575–576
von Willebrand factor   355, 357, 364, 369
von Willebrand's disease   369

Warfarin   366
Warm receptors   141
Water
   activity   18
   balance   266, 535
   distribution in body   2–3, 535
   disturbances of total body water
      content   536–538
   diuresis   530, 531, 532, 533, 536
   intestinal handling   578, 605–606
   intoxication   538
   movement across epithelia   24–29
   net flow across membranes   20
   passive exchange   4–10
   regulation of total body   443, 535–536
   renal handling   526–531, 534
   vapour pressure   483
Weal   454
Weber–Fechner law   126
Wernicke's aphasia   241
White blood cells   315, 316, 338–348
   abnormal counts   340–341
   immune system   341–348
   life span   339–340
   normal count   338
   production   339–340
   role in acute inflammation   341
   types of   338–339
White matter   76–78
White reaction   454
Withdrawal reflex   40, 196
Wolffian ducts   294

X chromosome   293

Y chromosome   29

Young–Helmholtz theory of colour
    vision 162

Z line (band, disc) 97, 98
Zollinger–Ellison syndrome 598

Zona fasciculata 282
Zona glomerulosa 282, 540
Zona pellucida 302
Zona reticularis 282
Zonule (lens ligament) 148
Zygote 310